Ecological Studies

Analysis and Synthesis

Edited by

W. D. Billings, Durham (USA) F. Golley, Athens (USA)
O. L. Lange, Würzburg (FRG) J. S. Olson, Oak Ridge (USA)

Volume 36

C.B. Osmond O. Björkman D.J. Anderson

Physiological Processes in Plant Ecology

Toward a Synthesis with Atriplex

With 194 Figures and 76 Tables

Springer-Verlag Berlin Heidelberg New York 1980

Professor C. B. Osmond
Research School of Biological Sciences
The Australian National University
P.O. Box 475, Canberra City, A.C.T. 2601/Australia

Professor O. Björkman
Carnegie Institution of Washington
Department of Plant Biology
Stanford, CA 94304/USA

Professor D. J. Anderson
School of Botany
University of New South Wales
Sydney N.S.W. 2033/Australia

ISBN 3-540-10060-1 Springer-Verlag Berlin Heidelberg New York
ISBN 0-387-10060-1 Springer-Verlag New York Heidelberg Berlin

Library of Congress Cataloging in Publication Data. Osmond, Charles Barry. Physiological processes in plant ecology. (Ecological studies; v. 36) Bibliography: p. Includes index. 1. Atriplex. 2. Plant physiology. 3. Botany—Ecology. I. Björkman, Olle, 1939– joint author. II. Anderson, Derek John, 1935– joint author. III. Title. IV. Series. QK 495.C46084.583'.913.80-13074.

Typesetting, printing, and binding: Brühlsche Universitätsdruckerei, Giessen.
2131/3130—543210

Preface

In the spring of 1969 a small meeting was convened at the CSIRO Riverina Laboratory, Deniliquin, New South Wales, to discuss the biology of the genus *Atriplex*, a group of plants considered by those who attended to be of profound importance both in relation to range management in the region and as a tool in physiological research. The brief report of this meeting (Jones, 1970) now serves as a marker for the subsequent remarkable increase in research on this genus, and served then to interest the editors of the Ecological Studies Series in the present volume.

This was an exciting time in plant physiology, particularly in the areas of ion absorption and photosynthesis, and unknowingly several laboratories were engaged in parallel studies of these processes using the genus *Atriplex*. It was also a time at which it seemed that numerical methods in plant ecology could be used to delineate significant processes in arid shrubland ecosystems. Nevertheless, to presume to illustrate and integrate plant physiology and ecology using examples from a single genus was to presume much. The deficiencies which became increasingly apparent during the preparation of the present book were responsible for much new research described in these pages.

It was a happy coincidence that the authors had been exposed at times to the philosophies of pioneers in several areas of plant ecology; pioneers who themselves had experimented with members of the genus *Atriplex*. We have in mind of course the studies by Turesson in Scandinavia, of the genotypic response to habitat; the work of Hall and Clements at the Carnegie Institution on the phylogenetic approach to plant taxonomy; and the field and laboratory studies of *Atriplex* by Wood and by Beadle in Australia. It was a happy coincidence too, that the authors found themselves in the same laboratory in 1971–72.

Since then we have depended a great deal on the support of our colleagues to build on these foundations and to bring the enterprise to its present form. Preeminent among these helpers have been Malcolm Nobs, whose advice we respect rather more than has been acknowledged in Chapter 2. We are particularly grateful for the stimulating collaboration over several years with Joe Berry and Hal Mooney, to Jim Charley and Ulrich Lüttge for their enthusiasm and encouragement, and to David Williams and many others whose direct contributions are acknowledged in the text. Jelena Emmerick helped with the preparation of Chapter 2 and as the manuscript has taken shape we have been indebted to those prepared to comment emphatically on various chapters. We particularly thank Paul Armond, John Cram, Bob Davis, Graham Farquhar,

Laurel Fox, Hank Greenway, Peter Grubb, Mats Gustafsson, Bernard John, Detlef Kramer, Andrae Läuchli, Ian Noble, Geoff Parr-Smith, Carl Pike, Don Potts, Pertti Uotila, John Waterhouse and Paul Wilson. Dorothy Lee, Jill Hardy, Kate Keller, Frank Nicholson, and Aida Wells have been more than generous with their help in preparing the manuscript. Most of all, we are grateful to Sue, Monika, and Margaret for their understanding and support throughout.

July, 1980   C. B. OSMOND
                                                        O. BJÖRKMAN
                                                        D. J. ANDERSON

Contents

**Chapter 1. Physiological Processes in Plant Ecology:
the Structure for a Synthesis** 1

1.1 The Space-Time Scale 2
1.2 Utility of the Space-Time Scale 7
1.3 The Quest for Adaptive Significance Among Physiological Processes . 9
1.4 Structure for a Synthesis 10

Chapter 2. Systematic and Geographical State of Atriplex 12

2.1 The Habit and Morphology of *Atriplex* 14
    2.1.1 The Fruiting Bracts 14
    2.1.2 Seed Morphology 17
    2.1.3 Epidermal Trichomes 18
    2.1.4 Leaf Morphology and Anatomy 20
2.2 A Species List for *Atriplex* 24
2.3 Geographical Distribution of Species 31
    2.3.1 Diversification of Species in Relation to Aridity 32
    2.3.2 The Tropics as Barriers to Migration 36
    2.3.3 Origins of the Genus 36
2.4 Perspectives from the Systematic and Geographical States 38

Chapter 3. Genecological Differentiation 41

3.1 Intra-Specific Variation and Ecotypic Differentiation 42
3.2 Morphological Aspects of Genecological Differentiation 43
3.3 Genetic Coherence Between Functional and Morphological
    Characteristics 47
3.4 Physiological Aspects of Genecological Differentiation 51
    3.4.1 Light Intensity 52
    3.4.2 Temperature 54
    3.4.3 Drought 58
    3.4.4 Edaphic Factors 59
3.5 Concluding Remarks 65

Chapter 4. Genetic and Evolutionary Relationships in Atriplex 66

4.1 Chromosome Numbers in *Atriplex* 66
4.2 Genetic Structures and Evolutionary Trends in the *Atriplex triangularis*
    Group 75
    4.2.1 Prezygotic Hybridization Barriers 77
    4.2.2 Zygotic Hybridization Barriers 79
    4.2.3 Polymorphism 81
    4.2.4 Evolutionary Trends 83
4.3 Interspecific Relationships and Distribution of C_4 Photosynthesis in
    Atriplex 84
    4.3.1 C_4 Photosynthesis 84
    4.3.2 A Phylogenetic Scheme for *Atriplex* 86
    4.3.3 Inter-Specific Crossing Experiments in *Atriplex* 89
    4.3.4 Inter-Specific Hybridization in Natural Populations 91
    4.3.5 DNA Sequence Comparisons in *Atriplex* 92
    4.3.6 Intraspecific C_3/C_4 Differentiation? 96
4.4 Inheritance of C_4 Photosynthesis in *Atriplex* 97

**Chapter 5. Atriplex Communities: Regional Environments
and Their Ecological Analysis** 111

5.1 Communities of Arid Continental Interiors 111
    5.1.1 Arid Shrub Steppe 112
    5.1.2 Arid Halophytic Communities 118
5.2 Coastal Communities 122
    5.2.1 Sand Dune Communities 123
    5.2.2 Salt Marsh Communities 126
5.3 Weed and Ruderal Communities 128
5.4 The Major *Atriplex* Habitats – a Resumé 131
5.5 Regional Variation in *Atriplex* Shrublands 132
    5.5.1 Ordination of Large-Scale Variation in Plant Communities . . . 132
    5.5.2 Ordination of the Australian Arid Shrublands 135
5.6 Local Variation and Pattern in *Atriplex* Shrublands 138
    5.6.1 Pattern Analysis 138
    5.6.2 Pattern Analysis in Arid *Atriplex* Shrublands 139
5.7 Regularity in Arid Vegetation 144
5.8 Patterns of Biomass and Nutrient Distribution 147
    5.8.1 Biomass Distribution in Arid Shrublands 148
    5.8.2 Nutrient and Salt Distribution 150
    5.8.3 Distribution of Microbiological Activity 152
5.9 Ecological States in Perspective 153

Chapter 6. Germination and Seedling Establishment 155

6.1 Germination Processes 156
    6.1.1 Water Relations of Germinating Seeds 157

6.1.2 Fruit Morphology and the Soil to Seed Water Potential
      Gradient 158
6.1.3 Seed Dimorphism and the Conductance of the Water Transport
      Pathway 161
6.1.4 Temperature Responses of Germination 164
6.1.5 Germination and Light 166
6.1.6 Gas Exchange and Seed Germination 167
6.1.7 Other Factors Regulating Germination 168
6.2 Establishment Processes 169
6.2.1 Water Relations During Establishment 170
6.2.2 Nutrition During Establishment 171
6.2.3 Metabolic and Morphological Changes During Establishment . 173
6.3 Physiological Processes of Germination and Establishment in the
    Context of Ecological Processes 174
6.3.1 Germination Processes and Reproductive Strategy 175
6.3.2 Dispersal Factors in Germination and Establishment 177
6.3.3 Herbivory in Germination and Establishment 179
6.3.4 Evaluation of Germination and Establishment Processes . . . 182
6.3.5 Establishment Processes and Population Dynamics in
      Natural Conditions 184
6.4 Conclusions 190

Chapter 7. Absorption of Ions and Nutrients 191

7.1 Ionic Concentration, Distribution, and Pathways in the
    Soil-Plant System 192
7.1.1 States of Ions and Nutrients in the Soil 192
7.1.2 Pathways and Compartments for Ions in Plants 197
7.1.3 States of Ions and Nutrients in Plants 199
7.2 Ion Absorption Processes in Plants 202
7.2.1 Ion Absorption and Concentration 203
7.2.2 Passive and Active Processes of Ion Absorption 206
7.3 Absorption of Nutrient Anions and Interactions with
    Cell Metabolism 210
7.3.1 Nitrogen Absorption and Incorporation 211
7.3.2 Metabolic Consequences of Nitrogen Incorporation 212
7.3.3 Interactions Between NO_3^- and Cl^- During Absorption and
      Incorporation 214
7.3.4 Efficiency of Nitrogen Utilization 217
7.3.5 Phosphate Absorption 218
7.3.6 Micronutrient Ions and Toxicity 220
7.4 Salinity Responses in *Atriplex* 222
7.4.1 Salt Exclusion, Salt Accumulation, and Salinity Tolerance . . . 223
7.4.2 Salt Absorption by *Atriplex* Seedlings as a Function of
      External Concentration 224
7.4.3 NaCl Uptake and Transport 227

7.4.4 K^+ and Na^+ Selectivity in Halophytes 230
7.4.5 Ion Uptake and Excretion in Leaves 232
7.4.6 Salinity, Metabolism and Growth 236
7.5 Ion and Nutrient Absorption Processes in the Context of
Ecological Processes 242
7.5.1 Mineral Cycling Within the Plant Community 242
7.5.2 Mineral Cycling and Stability of Arid Shrublands 247
7.5.3 Consequences of Plant Ion and Nutrient Absorption Processes
at Higher Trophic Levels 248
7.6 Conclusions 249

Chapter 8. Water Movement and Plant Response to Water Stress 251

8.1 Water Transport 253
8.1.1 The States of Water 253
8.1.2 Water Potentials in the Atmosphere, Plant, and Soil 254
8.1.3 Pathways for Water Movement 260
8.1.4 Dynamics of Water Movement Through the Plant 265
8.1.5 Water Movement in Relation to Energy Balance 267
8.1.6 Components of Water Potential 270
8.2 Response and Adaptation to Water Stress 273
8.2.1 Stomatal Regulation of Water Loss and Water
Use Efficiency 273
8.2.2 Changes in Leaf Inclination and Reflectance 280
8.2.3 Root Extension and Water Uptake 282
8.2.4 Osmotic Adjustment in Response to Low Tissue . . . . . .
Water Potential 284
8.2.5 Conclusions 290

Chapter 9. Photosynthesis 291

9.1 The Photosynthetic Process 292
9.1.1 Photochemistry 292
9.1.2 CO_2 Fixation and Metabolism 295
9.2 Photosynthetic Responses to Light 301
9.2.1 Photosynthetic Light Acclimation in *Atriplex triangularis* . . . 302
9.2.2 Photoinhibition 312
9.2.3 Light Utilization by C_3 and C_4 *Atriplex* spp. 316
9.3 Photosynthetic Responses to Temperature 321
9.3.1 Photosynthetic Characteristics of Cold-Adapted and
Heat-Adapted C_3 and C_4 Plants 323
9.3.2 Environmentally Induced Acclimation to Temperature 328
9.3.3 Leaf Factors Determining Photosynthetic Capacity at
Low Temperatures 333
9.3.4 High Temperature Stability of Photosynthesis 340

9.4 Photosynthesis in Relation to Water Use and Water Stress 347
    9.4.1 Photosynthetic Water Use Efficiency 347
    9.4.2 Photosynthesis in Relation to Water Stress 355
9.5 Photosynthesis in Relation to Salinity and Mineral Nutrition 367
    9.5.1 Response to Salinity 368
    9.5.2 Response to Mineral Nutrition 371
    9.5.3 Photosynthesis and Micronutrients 376
9.6 Conclusions 377

Chapter 10. Productivity and Environment 378

10.1 Photosynthesis and Photosynthate Partitioning in Relation to
     Primary Productivity 378
10.2 Productivity in Relation to Light 380
10.3 Productivity in Relation to Temperature 386
     10.3.1 Performance in Thermally Contrasting Habitats 386
     10.3.2 Response of Growth to Temperature Under Controlled
            Conditions 391
     10.3.3 Effect of Temperature on Photosynthate Partitioning 394
10.4 Productivity in Relation to Mineral Nutrition and Salinity 397
10.5 Productivity in Relation to Water 403
10.6 Productivity in Plant Communities 411
     10.6.1 Annual Patterns of Productivity in Cool-Desert
            Shrub Communities 411
     10.6.2 Long-Term Productivity in Arid Ecosystems 414
10.7 Effect of Future Increase in CO_2 Level on Water Use and
     Productivity 419

Epilog 426

References 429

Taxonomic Index 459

Subject Index 463

Chapter 1. Physiological Processes in Plant Ecology: the Structure for a Synthesis

"Niggle was a painter. He had a number of pictures on hand; most of them were too large and ambitious for his skill. He was the sort of painter who can paint leaves better than trees. He used to spend a long time on a single leaf, trying to catch its shape, and its sheen, and the glistening of dewdrops on its edges. Yet he wanted to paint a whole tree, with all of its leaves in the same style, and all of them different."

(Tolkien, 1964)

The study of physiology – "science of the normal functions and phenomena" – of living plants should be inseparable from the study of their ecology which is concerned with "the habits, modes of life, and relations to their surroundings" (Fowler and Fowler, 1949) of individual plants in communities and ecosystems. Physiological plant ecology serves to bridge physiology and ecology which represent a continuum of biological research method and philosophy, and this bridging discipline has proved a possible and profitable exercise, generating new insights into both physiology and ecology. However, the canvas of physiological ecology is so vast and the scale of the exercise is so immense that there is a real danger of themes being lost and contributors becoming increasingly unable to relate their individual efforts to those of others. Each of us, like Niggle, has a clear concept of how particular leaves should be represented. Yet in striving to paint these within a particular context we too are frequently reduced to tacking our canvasses to the dimly perceived tree. It is this dilemma, the dilemma of context, relevance, and relationship between the different contributions, which poses the greatest challenge to physiological plant ecology.

In an effort to present our own experiences with the physiological ecology of *Atriplex* in a natural format which minimizes these difficulties, we have attempted to develop a central organizational structure which can be used as a basis for synthesis. We believe this structure makes for an easier assessment of the relevance and contribution of the diverse research activities that are embraced by physiological ecology. The essence of this structure is that its nodes and spans serve to delineate exciting, legitimate, and largely equivalent contributions to the vast canvas of the subject. Like Niggle's leaves, some of these appear at the present time to loom larger and brighter, but in the light of another day, may be almost painted out.

The structure we have adopted and the tree that it helps us to compose are both far from complete. Although this structure is based on what seem to us to be logical and intrinsically useful concepts, it is important to emphasize that it is a frame of reference only. It is a convenient abstraction for our present purposes and should not be taken to reflect or describe intrinsic properties of biological systems. Related to this is the notion that the structure is useful only in that it helps to assess the relevance of research activity throughout the continuum of physiological ecology. The measure of its usefulness is the extent to which it may foster imaginative and testable hypotheses.

1.1 The Space–Time Scale

Physiological plant ecology effectively represents a continuum of biological research activities and its practitioners may be discriminated more easily by the problem-scales on which they most frequently operate than by differences in conceptual philosophy or, indeed, by differences in methodology (Anderson, 1971c). Ecologists or physiologists essentially describe the *state* of a biological system and the response of that state to environmental parameters during experimental perturbation. The state description is simply an analysis of the numbers and distribution of units, whether they be molecules or individual organisms, within a larger unit, whether it be a cell or ecosystem. The rearrangements which occur during transitions between states constitute biological *processes*. Thus the state is essentially a space description, while the process involves the description of changing state through time.

Research methodology in physiology or ecology in effect differs only in the scale of space and time at which the investigator is operating. It is frequently argued that because the actual number of boundary interactions must increase with the scale of the experiments, an ecologist is confronted with an immensely more complex system. Although one cannot deny this apparently axiomatic assertion, it is questionable whether it is helpful to accept it as a philosophical, much less pragmatic, basis for research. Rarely would a biochemist simultaneously consider many more parameters than would an ecologist in any particular exercise. More usually the number of significant boundary interactions considered in the two instances is not greatly different. Rather, in biochemistry, as in ecology, the satisfaction derived by the investigator is measured by his success in recognizing and evaluating the significant boundary interactions.

Practically all of the states and processes of plant physiology and ecology are encompassed by the span of approximately 10^{-1} to 10^{14} microns or seconds in space and time. It seems then that a space–time continuum forms a practicable structural basis for presenting analyses and interpretations of physiological and ecological states and processes. The scale of this continuum is vast, extending from seconds to millions of years and from microns to thousands of kilometers. If the time and space scale is based on units of seconds and microns, as shown in Fig. 1.1, then a logical progression of related biological phenomena falls into place. For convenience, the scale is calibrated logarithmically, each power of ten in space or time serving as a useful reference point or node in the continuum. The context of the different

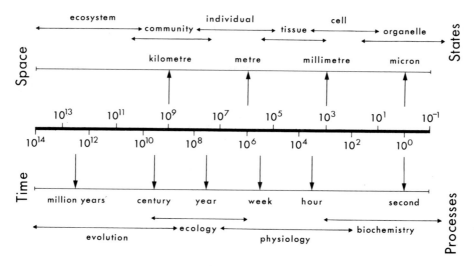

Fig. 1.1. Arbitrary scaling of state descriptions by spatial dimensions and of processes by relaxation times showing the continuum of research activities which comprise physiological ecology

biological states (the organelle, the cell through to the ecosystem) and different biological processes, which have given rise to specific research areas (such as biochemistry, physiology, and ecology) is indicated in Fig. 1.1.

The space–time scale is not unique and is suggested by a range of similar ideas developed in other areas of biology. Kamen (1963) used a time scale of 10^{-18} to 10^{1} s as a structural basis for his biophysical account of primary processes in photosynthesis. Photosynthetic processes with half-times greater than 10^{1} s were relegated by Kamen to "ecology" and this post Benson-Calvin cycle chauvinism was seminal to the space–time scale presented here.

Time scales are also frequently employed in hierarchical formulations in biology (Koestler and Smythies, 1972) but it is the arbitrary matching of the scales in space and time for states and processes which distinguishes our formulation from other familiar levels of organization treatments (Miller, 1957).

It is relatively easy to arrange state descriptions in the orderly sequence shown in Fig. 1.1 according to the dimensions of the system. Process descriptions are more difficult to arrange in sequence but can be loosely characterized on the basis of their response times, or relaxation times. These two measures, which are about equivalent, are the times taken to reach a new state, or to return to the original state, following a perturbation.

Part of the range of state descriptions relevant in physiological ecology is illustrated in Figs. 1.2 and 1.3. The vegetation analysis (Fig. 1.2) provides a state description of *A. vesicaria* vegetation in the vicinity of a sheep-watering trough in South Australia (Barker and Lange, 1970). It shows the number of plants in different age categories in each of 256 quadrats distributed radially through a distance of more than 1 km from the watering point (shown on the small-scale

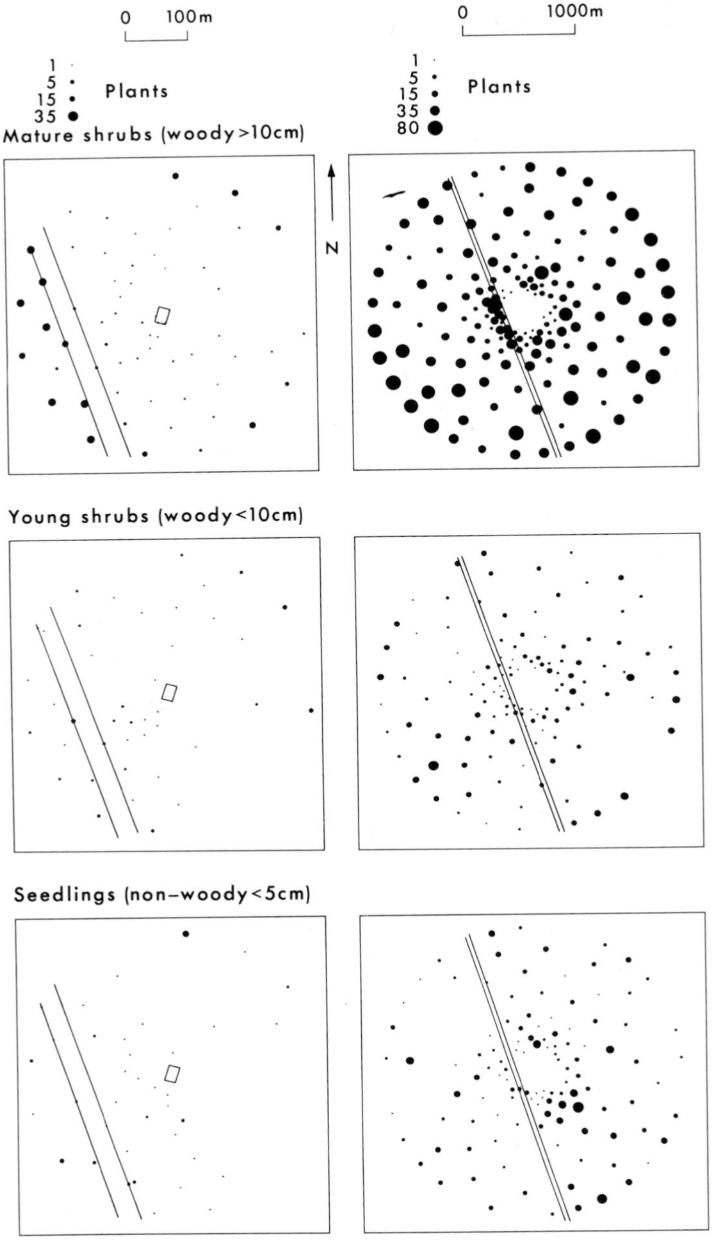

Fig. 1.2. A large-scale state description of the distribution and size category of *Atriplex vesicaria* shrubs in radially arranged quadrats adjacent to a sheep watering trough. The diagrams on the *left* show the location of the trough and shrub distribution up to about 2×10^8 microns radius and the diagrams on the *right* show the shrub distribution in the segment between radius about 2×10^8 and 1.2×10^9 microns from the trough. (Redrawn from Barker and Lange, 1970)

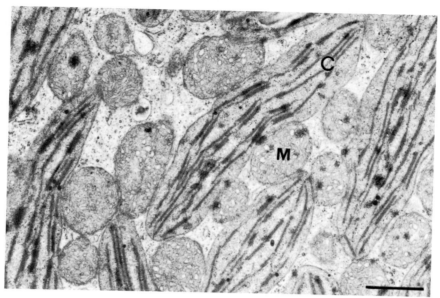

Fig. 1.3. A small-scale state description evident in a transmission electron micrograph (× 15,000) of a bundle-sheath cell in the leaf of *Atriplex spongiosa*. Chloroplasts are marked *C* and mitochondria are marked *M*. The *scale bar* in the lower right corner is one micron in length. (Courtesy D.J. Goodchild and W.J.S. Downton)

diagram). The data are relevant to scales 10^6 to 10^9 microns describing the age structure and population density states considered in Chap. 5. The vegetation gradient away from the watering point indicates the impact of herbivores on these shrubs, and points to processes discussed in Chap. 6.

The electron micrograph (Fig. 1.3) shows photosynthetic cells in the bundle sheath of *Atriplex spongiosa* leaves (Downton, 1971). This state description provides information relevant to the scale of 10^{-1} to 10^2 microns, discussed in Chap. 9. One can distinguish membrane organization within mitochondria and chloroplasts (scale 10^{-1} microns or less), the relative proportions of organelles themselves (scale 1 to 10 microns), and by comparisons with micrographs of other cells in the leaf, recognize that these states are a particular property of bundle-sheath cells (scale 10 to 100 microns).

Processes at scales corresponding to these state descriptions are illustrated in Figs. 1.4 and 1.5. Regeneration of an overgrazed *Atriplex* shrubland in arid regions similar to that documented in Fig. 1.2 has a relaxation time of about 10^8 s. Figure 1.4 shows the increase with time in the number of *A. vesicaria* and *A. stipitata* shrubs measured along a line transect at the Koonamore Vegetation Reserve, South Australia, after this population was isolated from grazing exotic herbivores in 1925 (Hall et al., 1964). It suggests a more vigorous early establishment by *A. stipitata*, a short-lived perennial, as well as the subsequent decline of this species during several years of above annual rainfall which was accompanied by an increase in the *A. vesicaria* population (Chap. 6).

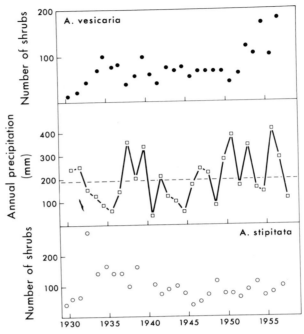

Fig. 1.4. A large-scale process description showing the increase in number of *Atriplex vesicaria* and *Atriplex stipitata* shrubs along line transects 1 and 2 and line transects 3 and 4 in the Koonamore experiment. The relaxation time of these processes is about 10^8 s. (Redrawn from Hall et al., 1964)

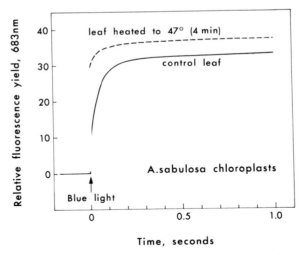

Fig. 1.5. A small-scale process description showing fluorescence kinetics of isolated *Atriplex sabulosa* chloroplasts prepared from control leaves and leaves heated to 47° for 4 min. The relaxation time of this process is approximately 10 s. (Redrawn from Berry et al., 1975)

Relative fluorescence yield of isolated chloroplasts from *Atriplex sabulosa* leaves, similar to those shown in Fig. 1.3, has a relaxation time of 10^{-1}s and is a sensitive indicator of the state of reaction centers associated with photosystem II (Chap. 9). Figure 1.5 shows that the kinetics of this fluorescence process change when *A. sabulosa* leaves are heated to $47°C$, indicating that damage to photosystem II reaction centers is one component of thermal damage in leaves (Berry et al., 1975). It is conceivable, of course, that processes of the sort illustrated in Fig. 1.5 may contribute to the stability and response of plant populations illustrated in Fig. 1.4.

Such a willingness to move along the scales of the continuum is one of the important characteristics of physiological ecologists; a characteristic which is a most important source of new insights in both physiology and ecology. The principal task of this book is to examine critically such possibilities and to assess their context and significance.

1.2 Utility of the Space–Time Scale

The logical association of states and processes which underlies the arbitrary scaling of Fig. 1.1 provides a framework within which speculation as to the relevance of relationships between states and processes at different places in the continuum can be assessed. The space–time scale serves as a filter in the formulation of hypotheses and in designing tests for significant relationships. The principal difficulty arises when extrapolations are made from small-scale states to large-scale processes. In physiological plant ecology this dilemma owes much to the stimulus given by Haberlandt (1884) and Schimper (1893) whose "post-Darwinian interpretations of the adaptive significance of the morphological features of organisms were often based mainly on speculation and infrequently on experimentation" (Mooney, 1974).

With the aid of a hand lens, an aspiring modern-day Haberlandt may establish the presence of the "Kranz" arrangement of chlorophyll-containing cells in a leaf transection and extrapolate with some confidence to the molecular details of the remarkable division of labor which distinguishes the C_4 photosynthetic pathway in the plant from which the leaf was taken. He would now, as in the past, be tempted to assign adaptive significance to these states and processes in the performance or survival of these plants. However, one cannot affirm that a particular state or process is of adaptive value without at the same time specifying the relevant time and space perspectives (Eckardt, 1975).

In the enthusiasm which followed the elucidation of the structural and functional properties of C_4 photosynthesis (Hatch and Slack, 1970; Björkman, 1973, 1975), righteous confrontations between ecologists and physiologists were not uncommon. "If C_4 plants are so efficient" ecologists asked "why have they not taken over the world?". "Give them 10,000 years and they will" was the optimistic reply of physiologists. For the physiologist, efficiency of C_4 photosynthesis primarily refers to processes of photosynthetic CO_2 fixation in relation to water use, light intensity, and temperature in cells and tissues, response functions which are entirely appropriate within scales 10^{-1} to 10^4. To the ecologist, efficiency has to do with survival or succession at a scale 10^8 to 10^{11} and may be only tenuously related to photosynthetic performance.

The crossed purposes of the above exchange are plainly evident. By virtue of his training, a physiologist interprets the presence of a C_4 plant in a particular climatic-geographic location in terms of performance response functions appropriate to his conception of that habitat. An ecologist interprets the same correlation largely in terms of his experience with the experiments of nature which tend to be survival functions. The presence of a plant in a particular habitat demonstrates "Firstly, that none of the environmental variables, either alone or in combination, have exceeded the tolerance range of the plant at any stage during its lifetime; and secondly, that while growth has presumably varied with environmental fluctuations, the plant has made net growth" (Scott, 1974). These separate analytical problems of defining tolerance limits, and performance within the tolerance limits, are implicit in recent treatments of niche specification (Wuenscher, 1974) and plant-environment relationships (Scott, 1974).

Although the C_4 photosynthetic pathway and the hand lens may have brought physiologists and ecologists together in a remarkable way, such exchanges have not been as effective as they might have been.

Figure 1.1 can serve another important function for evaluating the significance of physiological processes in plant ecology. It facilitates the matching of environmental states and processes in scale with plant states and processes. Logic suggests that if the scales of the plant processes and environmental processes are compatible, then significant interactions between plants and environment are likely to be uncovered (Scott, 1974).

The matching of scales between environmental and plant states and processes is likely to be most meaningful in relation to performance response functions. Significant interactions between photosynthetic performance and temperature are more likely to be discovered by matching maximum and minimum daily temperature (scale 10^3 to 10^4) than monthly means of average daily temperature (scale 10^6 to 10^7). In arid environments, the stochastic nature of precipitation means that plant processes such as germination (scale 10^4 to 10^5) or productivity (scale 10^5 to 10^6) are unlikely to be well interpreted against annual rainfall data (scale 10^7 to 10^8).

On the other hand, tolerance limits may be set by environmental processes operating at scales not in harmony with the plant processes. A single frosty night rather than growth responses to minimum temperature may determine the survival of frost-sensitive species. The demise of perennial vegetation may likewise be determined by a single year of unusually low rainfall rather than age-dependent mortality functions. The intrinsic stochastic nature of environmental processes which exceed the tolerance limits of key plant processes is not easily accommodated at different scales in the simple space–time continuum.

We introduced the space–time scale to describe the continuum of research activities which are part of physiological ecology and noted that various practitioners could be identified by the problem scale at which they most frequently operate. The utility of the space–time scale in delineating the "niche" of particular disciplines or individual researchers should not be underestimated. It can, for example, be a particularly helpful guide for the integration of research activities, as we shall illustrate in this book. De Wit (1970) suggested that individual biologists can work effectively within a scale or two of the continuum shown in Fig. 1.1, and can adequately comprehend states and processes a further scale or two up and down the continuum from this "niche". We three authors have together presumed to span most of the scale and consequently have been acutely aware of these limits of effective comprehension. We have found, however, that one's appreciation, if not total comprehension, is somewhat more elastic.

1.3 The Quest for Adaptive Significance Among Physiological Processes

To understand adaptation is to account for the ways the components of biological systems are fitted together in particular habitats and the ways they respond to particular environmental perturbations. Adaptations may be expressed in terms of survival or performance functions, as discussed above, and in plants these attributes are fitted together to maintain the organism in communities in changing environments over relatively long intervals of time. This is potentially a difficult concept, for it implies that each putative adaptation has a particular space–time scale (or perspective) which must be specified (Eckardt, 1975). It implies that processes which may have survival value at one scale subsequently may be deleterious to performance at another scale.

An outstanding illustration of this shift from performance to survival attributes with time is the change in photosynthetic pathways of *Mesembryanthemum crystallinum* during its life cycle (Winter et al., 1978; Bloom and Troughton, 1979). When adequate water is available during spring, this species engages in normal C_3 photosynthesis and develops a large leaf area which permits high vegetative productivity, enabling the plant to fully occupy its habitat on exposed cliffs near the seashore. With the onset of the dry summer, when these same leaves are exposed to water stress, they begin to engage in dark CO_2 fixation and crassulacean acid metabolism (Osmond, 1978). The induction of this photosynthetic pathway, which is much less productive in vegetative terms but very conservative in water use, evidently allows the organism to extend the available water resource and gain sufficient carbon to set seed, thus ensuring survival of the species.

Another aspect of the complex space–time perspectives implicit in concepts of adaptation is the notion that processes which can be shown to have an adaptive significance appropriate at a small scale may effectively represent little more than molecular noise at a larger scale. This is analogous to the role of individual mutations in the genetic problem of evolutionary mechanisms. Although "it remains true enough to say the ultimate units, the pebbles in the concrete or genes in the organism, have been produced by random processes, this is almost irrelevant to the engineering of a bridge and, in many cases, not much more relevant to the anatomical or physiological construction of the organism" (Waddington, 1972).

Natural selection acts not on the isolated components within the organism, which may respond, within limits, to random mutations, but upon whole integrated physiological processes which are products of the total genetic potential of the organism. Most formulations of biological hierarchies reflect the holistic view that these whole integrated systems are more than the sum of their parts. Paraphrasing Weiss (1972) this implies that the top level operations of the organism are neither structurally or functionally referable to the processes on the molecular level in a steady continuous gradation, but are relayed stepwise from higher levels of determinacy through intermediate layers to next lower levels of again more rigorously ascertainable determinacy. We suspect that the biological hierarchy, although administratively neat, reflects a human passion for hierarchical thinking (Anderson, 1971c) and is not necessarily the best pragmatic avenue for synthesis and integration. The philosophy behind truisms displayed in Fig. 1.1 is based on the premise that the whole is no more than the sum of its parts, and for practical purposes it may be a good deal less. In terms of adaptation, many of the states and processes which are significant at smaller scales may be as irrelevant as the pebbles in the concrete of Waddington's bridge.

1.4 Structure for a Synthesis

As a pragmatic necessity, interactions between single plant processes and isolated factors of the total environment have been investigated and this has emerged as a successful experimental strategy. In the main, performance under favorable conditions, rather than tolerance limits, has been explored. Wild plants such as those in the genus *Atriplex* have been most useful in exploring the ecological and generic limits of adaptation of physiological processes, largely because they occupy a wider than usual range of environmentally extreme habitats. Wild plants offer an additional advantage over cultivated plants as experimental subjects in that man has less consciously selected or altered adaptive characteristics of potential ecophysiological significance.

We have used our familiarity with such studies among members of the genus *Atriplex* as a vehicle to illustrate the ways in which the context and adaptive significance of physiological processes may be assessed in plant ecology. In so doing we have obviously narrowed the range of physiological and ecological processes which can be usefully discussed. Although members of the genus are widespread in many communities, species from arid shrubland habitats have been most extensively studied and this emphasis is evident in our treatment. When appropriate we have considered other species as well but overall the treatment is largely autecological. We acknowledge that deficiencies are introduced by this bias, but hope that they are largely offset by an improved coherence of the present treatment.

The book is divided into two parts. Although we began by attempting to write an account in the simple linear style suggested by Fig. 1.1, it became evident that the bulk of the book, the physiological processes themselves, could not be fully integrated in this way. Chapters 2 to 5, which deal with systematic and ecological states, and the processes of genecology upon which they are founded, form the first part. These chapters are structured so that large-scale states and processes (about 10^{10} to 10^{14}) form a basis for subsequent treatments of smaller-scale states and processes, the scale extending to about 10^6 or 10^5, to the level of the individual plant.

The second part, Chaps. 6 to 10, is almost entirely concerned with interactions between individual plants and the environment. In treating physiological processes we found it convenient to use the conventional structure for discussion, that is, to begin with small-scale states and processes and integrate these upwards to the individual plant. This change in the direction of our progress through the continuum at the point of the individual organism does not necessarily mean that we have found the bridge between synecology and autecology to be impassable. Rather, it reflects our efforts to synthesize a coherent account of an enormous amount of data pertinent to small-scale states and processes which could conceivably be relevant in the larger scale. Perhaps too, "our imaginations have been stirred more by the successes of the biochemist than by the problems of the ecologist" (Passioura, 1976).

The structure adopted in the book thus differs a little from the linear sequence in Fig. 1.1. Effectively, the space–time scale has been arborized, a trend which is common to other pedagogic abstractions in biology, and the branches arise at the scale of the individual plant. Effectively, having established the trunk in the first part, we begin the second by considering the leaves one by one in an attempt to reconstruct our expectations of the canopy as a whole. As the cover picture is intended to suggest, the structure then begins to resemble Niggle's tree "with all of its leaves in the same style, and all of them different".

Chapter 2. Systematic and Geographical State of Atriplex

"Annual and perennial herbs and shrubs, more or less pubescent with inflated scurf-like hairs. Leaves alternate, or the lower opposite, rarely all opposite, sessile or petioled, entire to dentate or irregularly and deeply lobed. Flowers monoecious or dioecious, solitary or in glomerules, the single or clustered flowers in the leaf-axils and often also in terminal spikes or panicles, the staminate and pistillate flowers often mixing in the same cluster, but the staminate usually confined to the upper axils or to terminal inflorescences. Staminate flowers without bracts or bracteoles; perianth 3- to 5-parted, the segments obovate or oblong and obtuse; stamens 3 to 5, inserted on the base of the perianth, the filaments either united at the base or distinct, the anthers 2-celled; rudiment of the ovary conical or wanting. Pistillate flowers each subtended by 2 bracts; bracts accrescent, distinct or usually united at least at the base and inclosing the fruit, entire or the margins variously dentate, sometimes fleshy-thickened or spongious; perianth none or rarely present and then consisting of a 3- to 5-lobed membranous calyx or of 1 to 5 squamellae; disk and rudimentary stamens wanting; ovary ovoid or depressed-globose; stigmas 2, nearly filiform, or slightly thickened or compressed near the base, where also shortly connate; ovule oblique or erect and with a short funicle, or inverted and suspended from the end of an elongated funicle. Utricle inclosed between the bracts, the pericarp membranaceous and usually free from the seed. Seed erect or inverted, rarely horizontal, the coats membranaceous, coriaceous or almost crustaceous; embryo annular, surrounding the farinaceous albumen, the radicle inferior, lateral, or superior".

(Hall and Clements, 1923)

"It's the damnest stinking genus there is, taxonomically"

(M. Nobs, pers. comm.)

There are several important reasons for undertaking a serious, but limited review of the systematics of the genus *Atriplex*. First, many of the species which have attracted ecological and physiological attention in the past are members of large species complexes. Some of these complexes have been examined recently, species have been renamed, regrouped, and sometimes subdivided. We have attempted to translate the species name used in original ecological and physiological reports into the currently accepted species name throughout the book. Second, a reasonably accurate estimate of the number of species is an essential starting point for a discussion of geographical distribution and of the generic processes of evolution and radiation of these plants. Third, because the systematics of the genus is largely based on differences in morphological characters, some of which are functional and significant in physiological and ecological processes, a review of systematics and geography may indicate trends or relationships which are helpful in evaluating these processes. With an agreed and reasonably accurate species list it may even be possible to present these trends or relationships in quantitative terms.

In undertaking a review of the systematic state of *Atriplex* for these purposes we are conscious of our lack of familiarity with large portions of the genus, as well as our limited appreciation of formal taxonomic practice. In spite of this, we hope that the majority of our decisions will be adequately documented and acceptable to our taxonomist colleagues who will, we trust, sense the purpose and the limitations of this exercise.

Atriplex is a large and widely distributed genus of the Chenopodiaceae, and is regarded by most taxonomists as a natural and workable group at the generic level. This view is sustained by the restoration of the genus to its present status after several major revisions. For example, in North America Hall and Clements (1923) restored the previously separated genera of *Obione, Pterochiton,* and *Endolopsis* to the genus *Atriplex* and Australian taxonomists prefer to retain the segregate genera *Theleophyton* Moq., *Blackiella* Aellen, *Senniella* Aellen, and *Morrisiella* Aellen within the genus (Eichler, 1965; L.A.S. Johnson and J. de Nade, personal communication). Whether this acceptance of the genus as a natural group is based on more than the sentiments expressed by our colleague Malcolm Nobs, is not known.

Ideally the classification of a group of organisms should be based on a defined relationship between constant discernible characters. Such an ideal is approached in a very small proportion of classifications, however, and in *Atriplex,* as in most other plant genera, taxonomists have commented on the difficulty of selecting constant characters, and on the intergrading of characters that occurs between even the most widely accepted species. Usually like Anderson (1930), they appeal for "the cooperation of the geneticist and cytologist in order to define adequately the individual species". That is, faced with variation which makes a rigorous state description nearly impossible, the response has been to examine the processes responsible for the variation. This approach has been remarkably successful in recent studies of closely related populations of *Atriplex* in Scandinavia and the United States, as discussed in Chap. 4.

The alternative to seeking a functional definition of the species in genetic terms is to apply numerical or chemical methods to the classification of species, but these have not been widely used in the genus *Atriplex*. Cozic-Trichet and Goas (1969) investigated the amino acid and amine composition of 15–20 day seedlings of the closely related species *A.laciniata, A.glabriuscula, A.hastata, A.littoralis,* and *A. patula*. These species are difficult to separate at the seedling stage and an identification based on chemical analyses would be very useful. The authors concluded, however, that quantitative differences in amino acid and amine composition provided no additional help in the separation of these species and that morphological characters were a more reliable basis for species identification. Jones (1975) found no clear discontinuities in the composition of leaf flavanols among British *Atriplex* species but noted the flavanols of *A. glabriuscula* and *A. triangularis* were more similar than those of any other between species comparison. This correlated with the greater similarity of leaf shapes noted between these two species.

Numerical methods were used by Parr-Smith (1977) in his treatment of the woody group of Australian *Atriplex* species, the predominantly dioecious, perennial species related to *A.vesicaria*. He applied the ordination technique of principal components analysis (see Chap. 5) to a collection of 3 male attributes,

16 female attributes, and 12 vegetative attributes such as erectness, branching, leaf shape, etc, measured on a representative collection from species in this group. Although most species were satisfactorily segregated, those of greatest interest, which were also the most variable, were not clearly delimited by this technique. Parr-Smith concluded that an extended numerical analysis at the species level was not helpful and resorted to a combination of numerical and conventional treatments. However, the numerical method proved invaluable in unraveling the polymorphism of *A. vesicaria,* as discussed below.

The separation of species in *Atriplex* thus rests largely on evaluation of highly variable morphological characters. These characters and their variability will be discussed because together they provide a description of the organisms and their tissues, which is an essential basis for much of the physiological and ecological discussion that follows.

2.1 The Habit and Morphology of Atriplex

The plants grouped in the genus *Atriplex* are herbs and shrubs ranging in size from herbaceous annuals of a few centimeters in height to woody perennials of up to 5 m. The formal generic diagnosis quoted above is weighted by an account of floral characters and is not particularly helpful in conveying a meaningful picture of these plants for our purposes. Much more helpful in this respect are two morphological features used in most Floras to separate *Atriplex* from *Halimione, Chenopodium, Rhagodia* and other genera of the Chenopodiaceae. These are the appendages to the fruit, the fruiting bracts, and the bladder-like hairs of the leaf surface, the epidermal trichomes. The emphasis on these characters in different Floras varies. In the Floras of Britain and Turkey for example, *Atriplex* is separated from *Halimione* by the fact that the fruiting bracts of *Atriplex* are not fused above the middle (Clapham et al., 1962; Davis, 1965). *Halimione* is absent from the floras of South Australia and Texas and the presence of fused fruiting bracts and leaf hairs is used to delineate the genus *Atriplex* in Floras of these regions (Black, 1948; Lundell, 1969). Although both these characters are highly variable, they are also extensively used as convenient characters at the species level. Both morphological structures are functionally significant in physiological processes, as discussed later in this book.

It is a commonly held view among modern biosystematists that great caution must be exercised in applying functionally important characters for purposes of classification. Yet in this genus, in spite of the acknowledged variability, these two functionally important characters have been retained simply on the basis of convenience and practical experience. Too great a reliance on such characters may be responsible for the present uncertain taxonomy of the genus.

2.1.1 The Fruiting Bracts

The fruiting bracts which enclose the seed in *Atriplex* were considered as modified upper leaves by early systematists, a view supported by recent studies on the vascularization pattern in this organ (Bisalputra, 1960). In species such as

Fig. 2.1. Fruit of *A.patula (left)* and *A.hastata (right)* showing leaf-like bracts enclosing the seed. (×4 from Collins, 1901)

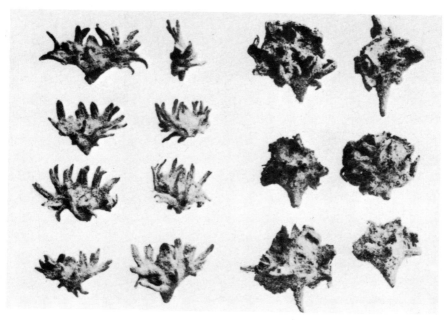

Fig. 2.2. Fruit of *A.polycarpa (left)* and *A.argentea (right)* showing the woody, highly sculptured apendages of the bracts enclosing the seed. (×4 from Collins, 1901)

Fig. 2.3. Fruit of *A. canescens* showing the wing-like apendages to the bracts enclosing the seed. (× 4 from Collins, 1901)

Fig. 2.4. Fruit of *A. spongiosa* showing the bulbous, spongy bract tissue enclosing the fruit. *Dark lines* on the bracts are vascular elements surrounded by a normal "Kranz" complex of mesophyll and bundle-sheath cells. (Approx. × 2)

A. hortensis and *A. patula*, the bracts are very leaf-like indeed and are free at the base (Figs. 2.1, 2.5). In others the bracts are tightly fused and quite woody and may bear highly sculptured appendages *(A.elegans)*, wings *(A.canescens)*, or, in many Australian species, show elaborate development of spongy tissue. Some of the best illustrations of these structures and their variability were presented by Collins (1901) in his survey of the seeds of commercially used *Atriplex* species in the United States, reproduced in Figs. 2.1, 2.2, and 2.3. In species such as *A.spongiosa* (Fig. 2.4) the bract is expanded to an almost spherical mass of aerenchyma-like tissue. The surface of the bract is covered with the remains of collapsed epidermal trichomes, the epidermis contains numerous stomata, and the network of vascular elements is surrounded by a "Kranz" complex indistinguishable from that of regular leaves (Bisalputra, 1960). The vascular network persists through the structural tissue, but the "Kranz" complex is restricted to those vascular elements immediately below the epidermis.

Although variation in bract characters is useful in classification of species, many authors believe undue reliance has been placed on this morphological feature (Hall and Clements, 1923; Van der Meijden, 1970). It may be that these modified leaves are subject to much the same genetic control as the leaves themselves and some aspects of variation in bract and leaf characters are correlated. For example several major edaphic forms of *A. vesicaria* have been recognized (Wood, 1936) which show a gradation between extreme forms, one with long narrow leaves and large spongy appendages to the bracts (characteristic of heavy clay, saline soils) and the other with short broad leaves and bracts with small appendages only (characteristic of calcareous soils). Parr-Smith (1977) in an extensive analysis of 69 *A. vesicaria* populations has further subdivided this species into 8 forms on the basis of leaf shape, bracteole characters, and inflorescence structures. Other authors have correlated the polymorphism of bract characters with those of the seed itself. Thus Kadman-Zahavy (1955) noted that wrinkled bracts of *A.rosea* bore small black seed and smooth bracts contained brown seed. Jones (1975) noted that seed morphology and bracteole weight were correlated. For example, *A.glabriuscula* contained small seeds in small bracteoles and *A.triangularis* contained larger and dimorphic seeds in larger bracteoles.

The great variety of bracts found in different species of *Atriplex* may be correlated with dispersal factors such as wind, flotation, seed burial, and animal vectors. These are considered in more detail in later chapters. Furthermore the bracts appear to play an important role in seed germination, particularly in those species in which the bracts contain high concentrations of salt, and these functional relationships are discussed in detail in Chap. 6.

2.1.2 Seed Morphology

Regarded by Hall and Clements (1923) as the most constant of taxonomic criteria, seed morphology and the position and orientation of the seed within the bracts have been retained as important features of modern species classifications. In *A.hortensis,* the species which is probably close to ancestral stock (Chap. 2.3), seed is produced in two types of flowers; horizontally within a regular five-part calyx, and vertically between two bracts. This is the only instance of horizontal seed

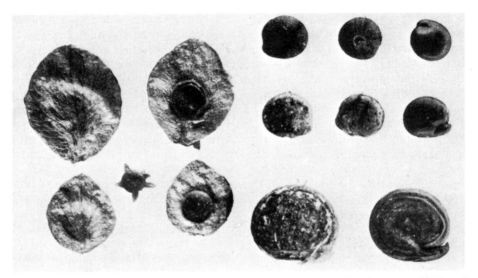

Fig. 2.5. Vertical and horizontal orientation of seed in bracts of *A.hortensis (left)* and examples of small hard, black seed *(right top)* and large, soft, brown seed *(right bottom)*.
($\times 8$ from Collins, 1901)

orientation in the genus and the horizontally orientated seeds are all small and black (Fig. 2.5). Seeds produced vertically between the partly fused fruiting bracts of *A.hortensis* may be of two types, commonly described as small, hard black seed, or large, soft brown seed. The two types are clearly illustrated in Collins' photographs shown in Fig. 2.5.

Seed dimorphism of the type shown in Fig. 2.5 is particularly common in the cosmopolitan species, such as *A.patula, A.triangularis,* and in European species (Jones, 1975). Collins (1901) and Hall and Clements (1923) noted that such dimorphism is not found in native North American species. However, Taschereau (1972) suggested that the relative abundance of the two seed types is a useful taxonomic character and noted only one species of eight in Nova Scotia which did not produce dimorphic seed. Seed dimorphism is common in Australian species (Burbidge, 1945; Beadle, 1952) and so does not appear to be restricted to species of more mesic regions. It is not exclusively associated with other important morphological features such as leaf anatomy or elaborations of the fruiting bracteole. However, since the proportion of black and brown seed is controlled by plant water relations and day length (Chap. 6), dimorphism is therefore of doubtful value in classification. Black seeds are longer-lived, but slower to germinate than brown seed, and seed dimorphism is a morphological feature with important implications for reproductive processes (Chap. 6).

2.1.3 Epidermal Trichomes

The epidermis of leaves and stems of many *Atriplex* species is covered by trichomes which often collapse to form a layer variously described as a grey waxy scurf (Hall and Clements, 1923), whitish mealiness (Frankton and Bassett, 1968) or

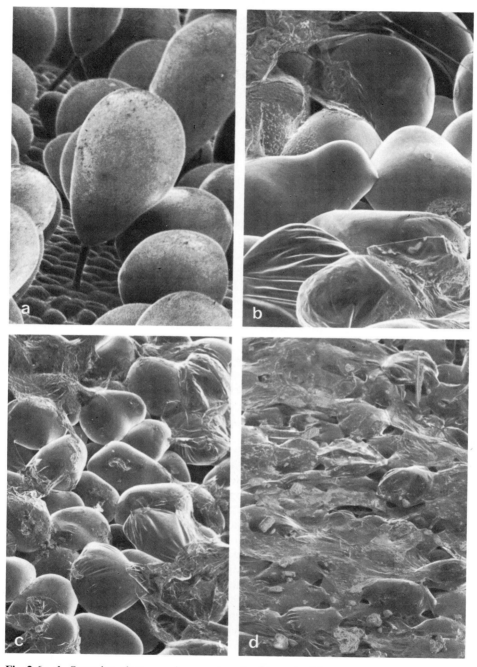

Fig. 2.6a–d. Scanning electron micrographs of epidermal trichomes on *Atriplex* leaves. **a** Young leaves of *A.spongiosa* showing trichomes with stalk cell and nearly spherical bladder cells; **b** young leaf of *A.nummularia* showing bladder cells of trichomes in close contact; **c** collapse of trichomes on epidermis of old leaves of *A.nummularia* showing some crystalline material (**d**). (Photographs courtesy of J.H. Troughton)

in any number of other picturesque terms. Black (1954) reported that the trichomes arise by differentiation and division of an epidermal cell to form a trichome initial cell which later divides into stalk cells and the terminal bladder (Osmond et al., 1969a; Smaoui, 1971). The trichomes of *Atriplex* are essentially similar to those found on the epidermis of *Chenopodium album* (Brian and Cattlin, 1968) and other chenopods. Scanning electron microscopy of young trichomes on *A.spongiosa* leaves (Fig. 2.6a) shows the arrangement of stalk and bladder cells most elegantly. In older leaves, and particularly those on plants raised under natural conditions, the bladders collapse and coalesce to form a dense layer over the entire leaf surface. This sequence of events is traced for the trichomes of aging *A.nummularia* leaves (Fig. 2.6b,c,d). The layer of collapsed trichomes frequently displays complex channeling which may expose a small portion of the epidermal surface.

Species differ in the extent of trichome development. Hall and Clements (1923) noted that species such as *A.hortensis* and *A.patula* had a relatively poorly developed scurf which was lost, at least from the upper leaf surface, during development (cf. *Chenopodium album*). In other species, particularly in the perennial species of semi-arid regions, several layers of trichomes are found. New trichomes are produced as the older trichomes collapse, so that a thick multi-layered scurf results (Wood, 1925; Black, 1954). The surface of the scurf frequently shows crystalline deposits which have been identified as NaCl and the waxy surface is almost nonwettable. Controlled studies of the development of the layered trichomes in different environments have not been undertaken but it is evident that, aside from species differences described above, the development of the layered trichomes within a species responds to seasonal environmental conditions (Mooney et al., 1974). The epidermal trichomes appear to function in salt excretion, as well as playing a role in leaf energy balance, by increasing leaf reflectance (Chaps. 7–9).

2.1.4 Leaf Morphology and Anatomy

The leaves of *Atriplex* species display a wide range in length from about 5 mm in *A.hortensis*. The ratio of length to width, the leaf margin and the leaf base are extremely variable. Polymorphism in these characters of natural populations in the *A.triangularis* complex of northern Europe has been studied by Van der Meijden (1970), Jones (1975), and Gustafsson (1973a,b, 1974) and similar complex relationships are found in artificial hybrids (Chap. 4.3.3). Leaf size and thickness are not very constant characters and respond markedly to nutrition, salinity, water availability, light intensity, and temperature, as detailed in later chapters.

In contrast to leaf shape, the organization of photosynthetic tissues in the leaf is a constant character. In the leaves of the majority of higher plants, chloroplast-containing spongy and palisade mesophyll cells are not highly organized in relation to vascular tissues. However, in many members of the Gramineae and Cyperaceae, two or more rows of cells are frequently found arranged in concentric cylinders about the vascular elements. Termed "Kranz" or wreath-like, this arrangement was clearly recognized by plant anatomists in the 19th century (Haberlandt, 1884), and was also described in several other families and genera of dicotyledons. The basic "Kranz" arrangement consists of an inner cylinder of bundle-sheath cells about the vascular tissue and an adjacent outer cylinder of mesophyll cells. In recent years this

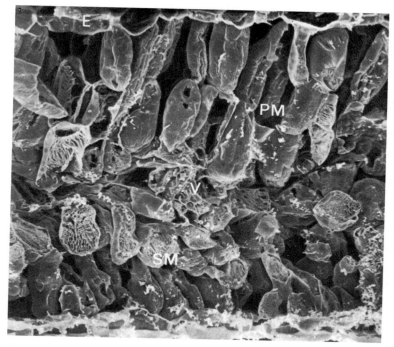

Fig. 2.7. Scanning electron micrograph, showing transverse section of *A. triangularis* leaf with vascular tissue *V*, spongy and pallisade mesophyll cells *SM*, *PM*, and epidermis *E*. (Photograph courtesy of J.H. Troughton)

arrangement of chloroplast-containing cells has been recognized as an essential component of the C_4 dicarboxylic acid pathway of photosynthesis. Plants capable of photosynthetic carbon assimilation by this pathway are known as C_4 plants (Chap. 10).

Moser (1934) was the first to observe that some *Atriplex* species were characterized by a "Kranz" arrangement of the simple type, whereas other species had a conventional palisade and spongy arrangement of the photosynthetic tissues. Scanning electron microscopy provides the most revealing display of chloroplast-containing cells in leaves with and without the "Kranz" complex, and scanning micrographs of *Atriplex triangularis* and *A. spongiosa* are shown in Figs. 2.7 and 2.8. The large hypodermal cells between mesophyll and epidermis contain few chloroplasts. Moser examined 98 *Atriplex* species and found 79 with "Kranz" leaf anatomy. The only contentious determination of leaf anatomy discovered thus far in the genus *Atriplex* is that of *A. phyllostegia*, which was described as "Kranz" by Moser (1934) but has twice been described as non-"Kranz" in recent studies (Welkie and Caldwell, 1970; Frankton and Bassett, 1970). Observation by Nobs at the Carnegie laboratory on specimens from several species populations of this species confirm the absence of "Kranz" anatomy.

The presence or absence of the simple "Kranz" complex is readily determined in leaves of herbarium specimens. In well-preserved specimens transverse sections of

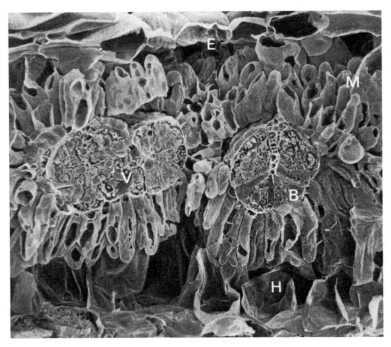

Fig. 2.8 Scanning electron micrograph, showing transverse section of *A.spongiosa* leaf with vascular tissue, *V*, "Kranz" complex of mesophyll *M*, and bundle-sheath cells *B*, hypodermal cells *H* and epidermis *E*. (Photograph courtesy of J.H. Troughton)

the hydrated tissue may be prepared, but failing this, squashing the hydrated material usually reveals a paradermal view of the more robust bundle-sheath arrangement of cells about vascular elements (Osmond, 1974). Staining with iodine frequently reveals the presence of starch in these cells, presumably because plant collectors have been so successful in the rapid drying of their material that respiratory loss of starch has been minimized. The presence or absence of the "Kranz" arrangement in samples of *Atriplex* leaves has been successfully determined by these techniques more than a century after collection.

Leaf anatomy is now recognized as an important taxonomic character at the species level and is included in recent systematic treatments (Frankton and Bassett, 1970; Taschereau, 1972), although it was rarely considered in earlier studies. More complex arrangements of photosynthetic tissues are found in other chenopods related to *Atriplex,* such as *Salsola kali* (Carolin et al., 1975). These have prompted more precise definitions of the key features of the relationship between leaf anatomy and photosynthetic pathway. Histological techniques, for example, have been developed to determine the distribution of enzymes between cells, which is an important part of the functional basis of "Kranz" anatomy (Hattersley et al., 1977). These techniques are difficult to apply to herbarium specimens but conveniently, another character can be invariably correlated with functional "Kranz" anatomy and the C_4 pathway.

Table 2.1. Correlation between the "Kranz" leaf anatomy in leaves of *Atriplex* species and the $\delta^{13}C$ value of leaf carbon

Species	"Kranz" complex	$\delta^{13}C$ value ($^0/_{00}$)	Reference
A. australasica	–	−21.7, −24.2	b
A. triangularis	–	−26.5, −29.6	a, d, e
		−25.1, −27.8	
A. hortensis	–	−24.6, −27.3	d
A. glabriuscula	–	−24.3, −25.2	a, d
A. novae-zealandiae	–	−27.3	d, e
A. patula	–	−27.2, −28.3	d, e
A. argentea	+	−10.5, − 9.6	a, d
A. buchananii	+	−17.3, −18.7	d, e
A. hymenelytra	+	−14.9	c
A. nummularia	+	−14.7, −12.9	a, d
A. lentiformis	+	−16.4	c
A. polycarpa	+	−14.8	c
A. rosea	+	−11.1, −14.3	a, d
A. spongiosa	+	−16.2, − 9.1	a, d
A. sabulosa	+	−10.4, − 8.1	a, d

a Hatch et al., 1972; b Osmond, 1974; c Philpott and Troughton, 1974; d Troughton et al., 1974a; e Troughton et al., 1971

Plants with the C_4 pathway do not discriminate against carbon-13, the naturally occurring heavy isotope of carbon, to the same extent as C_3 plants. First noted by Bender (1968), the measurement of carbon isotope discrimination has emerged as the most unequivocal method of determining C_4 photosynthesis in plant specimens. As discussed in more detail in Chaps.4 and 9, the ^{13}C concentration in plant material, expressed as a $\delta^{13}C$ value relative to a standard, is the most reliable indication of functional "Kranz" leaf anatomy and C_4 photosynthesis (Osmond and Ziegler, 1975). The correlation among *Atriplex* spp. is shown in Table 2.1. It is a particularly useful correlation for systematic and ecological surveys because the carbon isotope discrimination properties of photosynthesis in leaves are preserved in all of the carbon assimilated by the plant. Consequently, small samples of fruit, stem, or root can be used to type a plant specimen. An example of the potential application of this correlation to plant geographical and geobiological questions is given by Troughton et al. (1974b) who confirmed that *Atriplex confertifolia* specimens at least 40,000 years old had a photosynthetic pathway similar to present-day specimens in the same locality. Although there are some qualifications necessary in the interpretation of the data from carbon isotope discrimination measurements (Lerman, 1975), this technique has great potential in delimiting states and processes relevant to problems at this scale. Leaf anatomy is a constant morphological character of tremendous functional significance, as discussed in Chap. 10. Using the best anatomical, histological, and carbon isotope techniques there is no indication of intra-specific differentiation with respect to "Kranz" and non-"Kranz" anatomy in natural populations or in artificial hybrids of *Atriplex* species.

2.2 A Species List for Atriplex

After three centuries of systematic evaluation, 417 species of *Atriplex* have been distinguished on the basis of floral and morphological characters and cited in the *Index Kewensis* (Franclet and Le Houérou, 1971). Using the *Index* as a starting point and as many monographs and floras as were conveniently accessible, we have attempted to unravel the major synonyms in the literature, particularly in instances where the species involved is important in the ecological and physiological contexts which follow. Many of the species listed in Table 2.2 are known to us only from citations in the *Index Kewensis;* others are recently described species which have not yet been cited by the *Index Kewensis*. Table 2.2 contains about twice the number of species recognized by Ulbrich (1934), the last major treatment of the genus, and about twice the number of species usually attributed to the genus in most Floras and texts. This discrepancy may be due to the methods we used to construct Table 2.2, in particular to the inclusion of the *Index Kewensis* listings of species from South America.

In general, when authors have differed on the status of a species, we have given preference to those listed in the most recent regional flora or revision of the genus. An exception to this rule is that we have generally chosen to give Hall and Clements (1923) preference over Ulbrich (1934) because of the comprehensive nature of the former regional work, and Hall and Clement's views have been sometimes amended on the advice of M. Nobs. A most obvious deficiency which became apparent in this exercise was the need for a careful comparison of species from different regions. Aellen (1940a,b; 1960) was responsible for extensive reviews of the European, Australian, Oriental, and South African species, but comparisons with the large species groups of North and South America remain desirable. Aside from the contributions from Cabrera (1957), Pizarro (1959), Pyykko (1966), and Aellen (1968) we have been unable to discover comprehensive reviews of the genus in South America.

There are several areas of major uncertainty in Table 2.2. For example, there are two different species entries under *A. chenopodioides, A. prostrata,* and *A. pusilla,* respectively, and we have been unable to evaluate the synonomy in these cases. In the case of *A.prostrata,* P.G. Wilson (personal communication) comments that *A.prostrata* Boucher ex DC. is the correct name for the plant wrongly called *A.hastata* (Gustafsson, 1976). It includes *A.calotheca* (which is listed as a distinct species) and *A.triangularis. A.prostrata* R.Br. refers to an Australian plant. The name is a later homonym of *A.prostrata* DC., and Wilson recommends that it be deleted and replaced by *A.pumilio* R.Br. which is a synonym. The synonyms listed in Table 2.2 are by no means exhaustive but the more uncertain taxa are indicated by a greater number of synonyms. In the remainder of the book we shall use only the names given in Table 2.2 and indicate in brackets the synonym used in the original report, e.g., *A.glabriuscula (babingtonii)* in Binet (1965). We recognize, of course, that this action is itself potentially misleading in the absence of a thorough taxonomic evaluation in each instance, but this is clearly impossible in the present work. Table 2.2 is thus a listing of convenience which we hope will remove some of the uncertainty and confusion as to the species used in ecological and physiological studies described later.

Table 2.2. A working list of *Atriplex* species showing principal synonyms and regional distribution. Numbered regions correspond to those listed in the text and shown in Fig. 2.7

Species	Synonyms	Reference	Distribution
A. acadiensis Taschereau		t	9
A. acanthocarpa S. Wats.	*pringlei* Standl.	m	9
A. acuminata Waldst. et Kit	*nitens* Schkuhr, many others	d, o	1, 2, 3
A. acutibractea R.H. Anderson		f	1, 5
A. acutiloba R.H. Anderson		f	1, 5
A. alaskensis S. Wats.		z	9
A. albicans (Soland) Ait.	*odorata* Pers.	x	6
A. aldamae Griseb.		p	8
A. amblyostegia Turcz.	*desertorum* (Iljin) Sosn. *aucherii* Moq. *argentea* Pall ex Steud.	o	1, 2, 3
A. amboensis Schinz		a	2, 6
A. andina R. Fries.		h	7
A. angularis Larranaga		IK	7
A. angulata Benth.		f	1, 5
A. angustifolia Phil.		IK	7
A. arenicola Hauman-Merck		c	1, 7
A. argentea Nutt.	*volutans* A. Nelson; *trinervata*, Jepson; *rydbergii* Standl.; *nodosa* Greene; *mohavensis* Standl.; *hillmanii* Standl.; *caput-medusae* Eastw.	m	1, 9
A. argentina Speg.		IK	7
A. asplundii Standl.		IK	7
A. asphaltitis Kasapl.		IK	2
A. atacamensis Phil.		q	7
A. australasica Moq.		g	5
A. axillaris Phil.		IK	7
A. barclayana Hall and Clements	*insularis* Rose; *rosei* Standl.	m	9
A. belanger (Moq.) Boiss		b	2, 3
A. bonnevillensis C.A. Hanson		IK	9
A. boecheri Aellen		y	7
A. braunii Soriano		IK	7
A. buchananii (Kirk) Cheesem.		e	5
A. bunburyana F. Muell.		u	5
A. cachiyuyu F. Kurts		IK	7
A. californica Moq.		m	9
A. calotheca (Rafn.) Rafn. & Fries	*hastata* L.	l, t	1
A. cana C. A. Mey	*halimus* Pall.	o	1, 2, 3, 4
A. canescens (Pursh.) Nutt.	*tetraptera* Rydb.; *odontoptera* Rydb.; *occidentalis* D. Diet.; *macropoda* Rose and Standl.; *linearis* S. Wats.; *garretti* Rydb.; *berlandieri* Moq.; *aptera* A. Nelson.	m	9
A. centralasiatica Iljin		o	3, 4
A. cephalantha Aellen		n	5
A. chapinii I. M. Johnston		IK	7

Table 2.2 (continued)

Species	Synonyms	Reference	Distribution
A. chenopodioides Batt.		u	2
A. chenopodioides Aellen		y	7
A. chilensis Colla.		IK	7
A. cinerea Poir.	isatidea Moq. humilis F. Muell.; capensis Moq.; bolusii C. H. Wright.	f	1, 5, 6
A. clivicola I. M. Johnston		IK	7
A. colerei Maire		IK	2
A. conduplicata F. Muell.		n	5
A. confertifolia (Torr. & Frem and S. Wats.)	subconferta Rydb.; spinosa D. Dietr.; collina Wooton and Standl.	m	9
A. coquimbana Phil.		IK	7
A. cordifolia J. M. Black		n	5
A. cordubensis Gand et Stuckert		IK	7
A. cordulata Jepson		m	9
A. coriacea Forsk.		s	2
A. coronata S. Wats.		m	9
A. corrugata S. Wats.		m	9
A. costellata Phil.		IK	7
A. coulteri (Moq.) D. Dietr.		m	9
A. crassifolia C. A. Mey		o	3
A. crassipes J. M. Black		f	5
A. crenatifolia Chod. Et Wilcz.		IK	7
A. crispa Urb.	halimus Ritter; domingensis Standl.	P	8
A. cryptocarpa Aellen		f	5
A. davisii Aellen		IK	2, 3
A. decumbens S. Wats.	watsoni A. Nelson	m	9
A. densifolia Phil.		IK	7
A. deserticola Phil.		IK	7
A. dimorphostegia Kar. et Kir.	transcaspica Bornm.; bracteosum Trautv.	o	2, 3
A. dioica (Nutt.) Macbride	suckleyana Rydb.; endolepis S. Wats.	k, u	9
A. eardleyae Aellen	campanulata Benth.	f	1, 5
A. eichlerii Aellen		f	5
A. elachophylla F. Muell.	varia Ewart and Davies; cornigera Domin.	f	1, 5
A. elegans (Moq.) D. Dietr.	thornberi Standl.; saltonensis Parish; fasiculata S. Wats.	m	9
A. expansa S. Wats.		N	9
A. espostoi Speg.		IK	7
A. exilifolia F. Muell.		n	5
A. eximia Soriano		IK	7
A. farinosa Forsk	arabicum Ehrenb.	s	2, 3, 6
A. fasciculiflora Aellen		g	5
A. fera (L.) Bunge	lenticulare C. A. Mey	o	3, 4
A. fissivalvis F. Muell.		f	5
A. flabellum Bunge		o	2, 3
A. flavescens Speg.		IK	7
A. flexuosa Moq.		u	6
A. foliolosa Phil.		IK	7

Table 2.2 (continued)

Species	Synonyms	Reference	Distribution
A. frankenoides Moran		M	9
A. franktonii Taschereau		t	9
A. frigida Speg.		IK	7
A. fruticulosa Jepson		m	9
A. glabriuscula Edmonston	babingtonii Woods	d	1, 9
A. glauca L.		d	2, 6
A. glaucescens Phil.		IK	7
A. gmelinii C. A. Mey		o, z	4, 9
A. graciliflora M. E. Jones		m	9
A. griffithii Moq.	stocksii Boiss; persica Boiss	b	2, 3
A. halimus L.	halimoides Tineo; capensis Moq.	a	2, 6, 7
A. heterosperma Bunge	micranthum Kar and Kir; micrantha C. A. May	j, o	1, 3, 9
A. hjertingii Aellen		y	7
A. holocarpa F. Muell.		f	5
A. hortensis L.	microtheca Moq.; lucida Desf.; heterantha Wright; hermannii Vill.; benghalensis Lam.; arieghinoi Speg.; acuminata Bieb.	d, j, o	Cosmop.
A. humilis Phil.		IK	7
A. hymenelytra (Torr.) S. Wats.		m	9
A. hymenotheca Moq.		n	5
A. hypsophila I. M. Johnston		IK	7
A. hystrix Phil.		IK	7
A. iljinii Aellen		a	3
A. inamoena Aellen	rotundifolia Boiss; crystallina Ehrenb.	b	2
A. incrassata F. Muell.		f	2, 5
A. inflata F. Muell.		IK	7
A. integribracteata Lillo.		f	5
A. intermedia R. H. Anderson		u	5
A. jubata S. Moore		m	9
A. julacea S. Wats.		f	5
A. kochiana Maiden		t	1, 9
A. laciniata L.		o	1, 3
A. laevis C. A. Mey		r	7
A. lampa Gill		c	1, 2, 3
A. lasiantha Boiss	pruinosa Sieber; autranii Pest		
A. lentiformis (Torn.) S. Wats.	torreyi S. Wats.; orbicularis S. Wats.; griffithsii Standl.; breweri S. Wats.	m	9
A. leptocarpa F. Muell.		f	1, 5
A. leptostachys L. Chevall		u	2
A. leuca Phil.		IK	7
A. leucoclada Bioss		a, s	2
A. leucophylla D. Dietr.		m	9
A. lilloi Hauman		h	7
A. limbata Benth.		f	1, 5
A. lindleyi Moq.	halimoides Lindl.	f, n	5
A. linifolia Humb. and Bonpl.	polygama Sesse	m	9

Table 2.2 (continued)

Species	Synonyms	Reference	Distribution
A. littoralis L.	*sulcata* Michx.; *serrata* Huds.; *salina* Desf.; *salicina* Pall.; *maritima* Pall.; *marina* L.	o, v	Cosmop
A. lobativalvis F. Muell.	*lobaticarpa* Domin.	f	5
A. longipes Drejer and Fries	(4 sympatric species)	d, l	1
A. macropterocarpa (Aellen) Hj Eichler		g	5
A. madariagae Phil.		a	7
A. malvana Aellen and Sauvage		IK	2
A. matamorensis A. Nelson	*oppositifolia* S. Wats.	m	9
A. megalotheca Popov.		o	1, 3
A. microcarpa (Benth.) Dietr.	*pacifica* A. Nelson	m	9
A. microphylla Phil.		q	7
A. mollis Desf.		d	2
A. moneta Bunge		o	3
A. monilifera S. Wats.		m	9
A. montevidensis Spreng.		IK	7
A. morrisii R. H. Anderson		f	5
A. mucronata Phil.		IK	7
A. muelleri Benth.	*rosea* F. Muell.	f	1, 5, 6
A. multicolora Aellen		a	2, 3
A. myriophylla Phil.		h	7
A. navajoensis Rydb.		IK	9
A. neurivalvis Domin.		v	5
A. novae-zelandiae Aellen		e	5
A. nudicaulis Boguslaw	*validum* Iljin; *praecox* Hulpers	o	1
A. nummularia Lindl.		f	5
A. nuttallii S. Wats	*tridentata* Kuntze; *fruticulosa* Osterh.; *falcata* Standl.; *cuneata* A. Nelson (and others)	m	9
A. oblongifolia Waldst and Mit.	*tatarica* Schkuhr.	o	1, 2, 3, 9
A. obovata Moq.	*sabulosa* M. E. Jones *jonseii* Standl.; *greggii* S. Wats	m	9
A. olivieri Moq.		a	3
A. oppositifolia Vill.		m	1
A. oreophila Phil.		IK	7
A. ornata Iljin		o	3
A. pacifica Nels.		N	9
A. paludosa R. Br.	*reniformis* R. Br. *drummondii* Moq. *moquiniana* Moq.	P	5
A. pamirica Iljin		o	3
A. pamparum Griseb.		r	7
A. papillata J. H. Willis		n	5
A. parishi S. Wats.	*minuscula* Standl.; *depressa* Jepson	m	9
A. parryi S. Wats.		m	9
A. parvifolia H. B. and K. Nov.		IK	9
A. patagonica D. Dietr.		IK	7
A. patens Iljin		o	1, 2, 3

Table 2.2 (continued)

Species	Synonyms	Reference	Distribution
A. patula L.	*spicata* S. Wats.; *procumbens* Jundz.; *dioica* Rafin.; *carnosa* A. Nelson.	d, o, t, v	Cosmop
A. pedunculata L.		o	1, 2, 3
A. pentandra (Humb. and Bompl.) Hall and Clements	*texana* S. Wats.; *muricata* Humb and Bompl.; *glomerata* S. Wats.; *crispa* D. Dietr.; *arenaria* Nutt. (and others)	m, p, N	7, 8, 9
A. perrieri Leandri		i	6
A. peruviana Moq.		w	7
A. philippii R. Fries		IK	7
A. phyllostegia (Torr.) S. Wats.	*draconis* M. E. Jones; *corvillei* Macbride	m, k	9
A. platensis Speg.		c	1, 7
A. plebeja Carmich.		IK	7
A. pleiantha W. A. Weber		k	9
A. podocarpa Phil.		q	7
A. polycarpa S. Wats.	*curvidens* T. S. Bradley	m	9
A. polyphylla Phil.		IK	7
A. portulacoides L.		s	1, 2, 6
A. powelli S. Wats.	*philonitra* A. Nelson *nelsonii* M. E. Jones	km, m	9
A. prostrata Phil.		IK	7
A. pseudocampanulata Aellen		n	1, 5
A. pterocarpa Ewart and Rees		n	5
A. pugae Phil.		IK	7
A. pumilio R. Br.	*prostrata* DC.		5
A. pungens Trautv.	*serpyllifolium* Bunge	o	3, 7
A. pusilla Phil.		IK	7
A. pusilla S. Wats.		m	9
A. quadrivalvata Diels.	*sessilifolia* Ising; *hochreutineri* R. H. Anderson	n	5
A. quinii F. Muell.		f	5
A. recurva D'Urv.		l	2
A. repanda Phil.		q	7
A. repens Roth.	*cristata* Heyne	u	4
A. retusa C. Gay		q	7
A. rhagodioides F. Muell.		f	5
A. robusta Speg.		IK	7
A. rosea L.	*spatiosa* A. Nelson; *monoica* Moench; *laciniata* Bieb.; *axillaris* Tenore; *alba* Scoop.	m, o	Cosmop.
A. rotundifolia Dombey		IK	7
A. rusbyi Britton		IK	7
A. sabulosa Rouy		N	1, 9
A. saccaria S. Wats.	*cornuta* M. E. Jones	m	9
A. sagittifolia Speg.		r	7
A. salaris Phil.		IK	7
A. schugnanica Iljin		a, o	3
A. semibaccata R. Br.	*microcarpa* Benth.; *flagellaris* Wooton and Standl.; *denticulata* Mow.	f	1, 5, 6, 7, 9

Table 2.2 (continued)

Species	Synonyms	Reference	Distribution
A. semilunaris Aellen		n	1, 5
A. serenana A. Nelson	*A. bracteosa* S. Wats.	N	9
A. sibirica L.	*lucida* Moench	o	3, 4
A. sphaeromorpha Iljin		o	1, 2, 3
A. spinibractea R. H. Anderson		u, n	1, 5
A. spinifera Macbride		m	9
A. spongiivalvis Aellen		g	5
A. spongiosa F. Muell.		f	5
A. stipitata Benth.		f	5
A. stylosa Viv.	*rotundifolia* Boiss; *parvifolia* Lowe; *palaestina* Boiss; *ehrenbergii* F. Muell.; *alexandrina* Boiss	s	2
A. subcordata Kitagawa		IK	4
A. suberecta Verdon		g	1, 5
A. subspirata (Nutt.) Rydb.		t, z	9
A. taltalensis I. M. Johnston		IK	7
A. tatarica L.	*tornabenii* Tineo; *procumbens* Less (and others)	d, o, t	Cosmop.
A. tenuissima A. Nelson	*greenei* A. Nelson	m	9
A. thunbergiaefolia Boiss	*aralensis* Iljin	o	2, 3
A. transandina I. M. Johnston		IK	7
A. triangularis Willd.	*hastata* L; *latifolia* Whalenb; *prostrata* Boucher ex DC	l, o, t, U	Cosmop.
A. trigonophylla Phil.		IK	7
A. truncata (Torr.) Gray	*subdecumbens* M. E. Jones	k, m	9
A. tuberculosa Farranaga		IK	7
A. tularensis Coville		m	9
A. turbinata Aellen		g	5
A. turcomanica Fisch.	*leptocarpa* Boiss; *lehmanniana* Bunge	o	2, 3
A. undulata D. Dietr.		IK	7
A. vallicola Hoover		IK	9
A. velutinella F. Muell.		f	5
A. verreauxii Moq.	*farinosa* Moq.	u	6
A. verrucifera Bieb.	*portulacoides* Pall.; *glauca* Pall.	o	1, 2, 3, 4
A. vesicaria Heward		f	5
A. vestita (Thumb) Aellen		b	6
A. vulgatissima Speg.		IK	7
A. wolfii S. Wats.		m	9
A. wrightii S. Wats.	*radiata* Coult.	m	9
A. zahlensis Mout.		IK	2
A. zosteraefolia S. Wats.		u	6

a Aellen, 1940a; b Aellen, 1940b; c Aellen, 1960; d Aellen and Hulme, 1964; e Allan, 1961; f Black, 1948; g Eichler, 1968; h Cabrera, 1957; i Cavaco, 1954; j Frankton and Bassett, 1968; k Frankton and Bassett, 1970; l Gustafsson, 1973a, 1976; m Hall and Clements, 1923; n Johnson and De Nardi, unpublished assessment; o Komarov, 1970; p Leon and Alain, 1951; q Pizarro, 1959; r Pyykko, 1966; s Täckholm, 1956; t Taschereau, 1972; u Ulbrich, 1934; v Van Der Meijden, 1970; w Wiggins and Porter, 1971; x Wright, 1912; y Aellen, 1968; z Bassett and Crompton, 1973; IK Index Kewensis, 1885–1974; M Moran, 1975; N M. Nobs, unpublished assessment; P G. A. Parr-Smith, unpublished assessment; U P. Uotila, 1977

This table reflects modern taxonomic practice in the sense that major treatments, such as that of Hall and Clements (1923) have emphasized species polymorphism and favored the larger species complex. In some instances, however, recent cytological studies have led to the recognition of species earlier lumped together in larger species complexes. An example is the recent separation of hexaploid *A. alaskensis* and *A. subspicata* by Bassett and Crompton (1973) from the tetraploid *A. patula*. All three species were considered by Hall and Clements to be part of the *A. patula* complex. Further accounts of modern genecological research which go a long way to meeting Anderson's (1930) plea for the cooperation of the geneticist and cytologist in order to define more adequately the individual species in this genus are given in Chap. 4. Table 2.2 is by no means an authoritative statement of the systematics of the genus *Atriplex* but with the reservations expressed above we will use it as a basis for discussion of the geography and genecology of these plants.

2.3 Geographical Distribution of Species

Members of the genus *Atriplex* are established on all continents and many islands. They range from at least 70° N (Aellen and Hulme, 1964) to at least 46° S (Troughton and Card, 1974). Nine major regions of distribution have been arbitrarily defined as follows:
1. *European,* including France but excluding other nations bordering the Mediterranean;
2. *Mediterranean,* including North Africa and all nations bordering the Mediterranean (excluding France), as well as Iraq and Saudi Arabia;
3. *Eastern Asian,* including India, China, Japan, South East Asian nations, and Indonesia;
4. *Central Asian,* including Asian U.S.S.R., Iran, Afghanistan, and Pakistan;
5. *Australasian,* including Australia, New Zealand, and Pacific Islands;
6. *African,* including nations bordering the Gulf of Guinea, and of Central, Eastern, and Southern Africa;
7. *South American,* comprising nations south of Panama;
8. *Central American,* including nations between Mexico and Columbia and those of the Caribbean;
9. *North American,* including Mexico, United States, and Canada.

The boundaries are most arbitrary between the Mediterranean and Central Asian regions, but correspond approximately to the Caucasian Mountains and Iran. Mexico is included with the North American region because the deserts and mountains of the southwestern United States do not respect national boundaries. When considered in terms of these nine regions, species of the genus *Atriplex* show several important features of distribution and development which are illustrated in Fig. 2.9. Only seven species are truly cosmopolitan, that is, occur in four or more of the nine geographical regions. Of these, only one species complex, *A. triangularis* is reported for all regions, the remaining cosmopolitan species being largely restricted to the northern hemisphere.

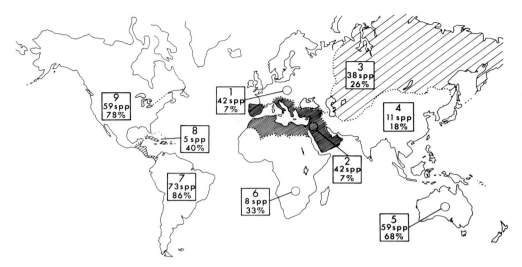

Fig. 2.9. Geographical distribution of *Atriplex* showing species groups in regions *1–9* identified in the text, the approximate number of species in each group (from Table 2.2) is shown, with an estimate of the percentage of endemic species in each group. Small discrepancies between Table 2.2 and this figure have arisen as a consequence of taxonomic advice during final preparation for publication

2.3.1 Diversification of Species in Relation to Aridity

The maximum diversification of the genus *Atriplex* is found in three remote centers, the arid regions of North America, South America, and of Australasia (Fig.2.9). In each region about 60 to 70 species are found, most of them in areas which receive from 100–500 mm precipitation annually. In each center of diversification, the proportion of endemic species is greater than about 70%. This is an underestimate of the real extent of the endemic character of these *Atriplex* floras due to exchanges of species in the course of recent trade. The semi-arid regions dominated by endemic species in each continent have been exploited for the grazing of sheep and movement of species in the wool trade, particularly to Europe, has occurred in the last 100 years. Probst (1949) reported 15 Australasian species of *Atriplex* in the "Wolladventiv" flora of Europe and smaller numbers of South American and Central Asian species have found their way to Europe in the same way (Aellen, 1960; Aellen and Hulme, 1964). At the same time, European and Asian species have been introduced to North America (Frankton and Bassett, 1968; Bassett and Crompton, 1973). Several Australasian species have been introduced into North Africa (Franclet and Le Houérou, 1971) and to North America (Hall and Clements, 1923) for grazing purposes.

In each of these three isolated regions of maximum diversification the genetic resources of the genus have given rise to large, distinct populations of species in arid and semi-arid regions of these continents. As discussed later (Chap.5), there is an astonishing similarity in growth form of the different species occupying similar habitats in Australasia, North America, and South America. Associated with this

Table 2.3. Proportion of *Atriplex* species with the "Kranz" complex leaf anatomy in regional species groups

Region	"Kranz" +	complex −	Percent "Kranz"	Reference
1. Europe[a]	2	11	15	b
2. Mediterranean	8	6	57	b
3. Central Asia	3	7	30	b
4. East Asia	2	2	50	b
5. Australasia	51	3	94	c
6. South Africa	7	2	78	b
7. South America	10	0	100	a, b, d
8. Central America	2	0	100	b
9. North America	36	6	86	c

[a] Excluding wool adventives

a R. Carolin, personal communication; b Moser, 1934; c Osmond, 1974; d Pyykko, 1966; e Welkie and Caldwell, 1970

similarity in growth form is the fact that these isolated species populations in arid habitats are dominated by species with "Kranz" leaf anatomy. In each of the three regions *Atriplex* species with the "Kranz" arrangement of mesophyll and bundle-sheath cells comprise more than 85% of the taxa (Table 2.3). In regions such as Central Asia and Europe in which the degree of endemism is very low, fewer than 30% of the species have "Kranz" leaf anatomy.

The distribution in the European, Mediterranean, and Central Asian regions of *A.sabulosa* and *A.rosea,* species with "Kranz" leaf anatomy, and of *A.littoralis* and *A.acuminata (nitens),* species without "Kranz" leaf anatomy, is shown in Fig.2.10. *A.sabulosa* and *A.littoralis* are coastal species and their ranges overlap for the most part, except that the latter species is found along more northerly coasts. *A.rosea* is common in more arid regions of Spain and the Eastern Mediterranean whereas *A.acuminata (nitens)* is not. Although this closer inspection of the distribution of individual species in greater Europe gives further indication of the relationship between aridity of habitat and the morphological character of "Kranz" leaf anatomy seen at the larger scale (Table 2.3), it also emphasizes that species differing in this character cohabit in a wide range of environments.

Statistical treatment of taxonomic data can be used to test more rigorously the generalizations relating morphological characters to habitat. Teeri and Stowe (1976) and Stowe and Teeri (1978) estimated the percentage of taxa with "Kranz" leaf anatomy among the grasses and among the dicotyledons of the spermatophyte floras of about 30 locations throughout North America. Their primary data are shown in Fig.2.11 and although for both groups there is an increase in the percentage of taxa with "Kranz" anatomy at lower latitudes, there is a distinct difference in the trends observed for monocotyledons and dicotyledons. In the monocotyledons the percentage of taxa with "Kranz" anatomy is far higher than in dicotyledons, and is generally greatest in the east and south-eastern United States, a region characterized by hot, wet summers, although high percentages were found also in the floras of Arizona and the Sonoran desert. In the dicotyledons the

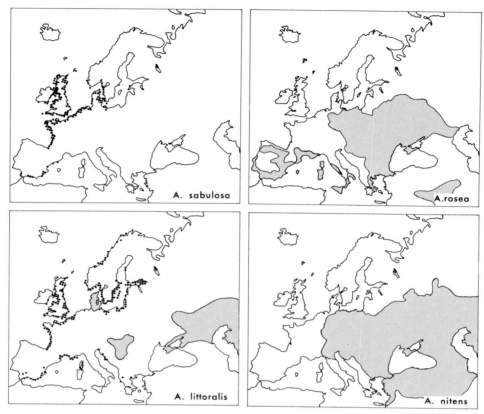

Fig. 2.10. Geographical distribution in Europe of *Atriplex* species with and without "Kranz" anatomy. Of the coastal species, *A.sabulosa* (with "Kranz" anatomy) is confined to lower latitudes than *A.littoralis,* a species with normal leaf anatomy. Of the inland ruderals, *A.rosea* with "Kranz" anatomy is found in the more arid regions of the east and west Mediterranean whereas *A.nitens (A.acuminata),* a species with normal leaf anatomy, is not.
(Redrawn from Hulten, 1958; Meusel et al., 1965)

percent of C_4 taxa was generally lower in regions of high summer rainfall and highest in regions of hot dry summers.

These authors then undertook stepwise multiple regression analyses of correlations between these data and about 20 climatic parameters taken from long-term weather records. They concluded that this morphological character in the monocotyledons of these floras was best correlated with night temperature during the growing season, whereas in dicotyledons the best correlation was found with summer pan evaporation, an index of aridity. This marked distinction between the correlations shown for monocotyledons and dicotyledons may point to significant physiological interactions in these classes, as discussed below. However, this quantitative, statistical approach has several inherent difficulties. Although it is clear from Teeri and Stowe (1976) that the percentage of monocotyledon taxa with "Kranz" anatomy, and hence C_4 photosynthesis, was much better correlated with

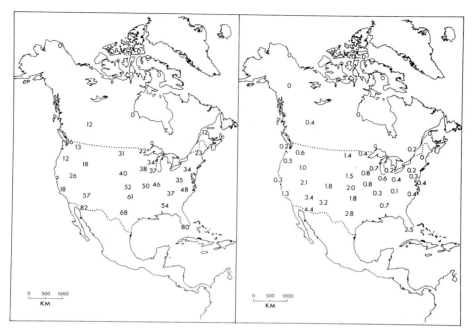

Fig. 2.11. Distribution of C_4 species in the grass floras *(left)* and dicotyledon floras *(right)* of different regions of North America. Figures refer to percentage of total species in each flora thought to have the C_4 photosynthetic pathway and "Kranz" anatomy. (Redrawn from Teeri and Stowe, 1976; and Stowe and Teeri, 1978; with permission)

the complex of environmental factors associated with July minimum temperature ($r = 0.917$ to 0.972) than with irradiance factors ($r = -0.228$ to 0.560), the correlation with dryness factors was also quite substantial ($r = 0.683, 0.697$). It is difficult to evaluate such differences in correlation coefficients, and one wonders whether some other complex of environmental variables, many of which are interdependent, might not give a different picture.

The matrix of correlation coefficients was not given for the dicotyledons, so the relative correlations with other factors are unknown. However, Stowe and Teeri (1978) noted that the C_4 dicotyledon species belonged to only eight families and recognized that these families themselves may be more abundantly represented in arid habitats. They therefore expressed the C_3 species in these same families as a percentage of the spermatophyte floras of the same regions and found that the correlation with summer pan evaporation was only marginally less ($r = 0.886$) than that for C_4 species of these families ($r = 0.974$). The authors concluded that families of dicotyledons with C_4 species are favored in arid areas, regardless of photosynthetic pathway, but that C_4 species of these families are particularly abundant in such areas. Thus, although this detailed statistical approach provides some clear-cut quantitative relationship, the physiological and functional interpretation of these relationships is not much easier than that suggested by the simple gross geographical comparisons given in Figs. 2.9 and 2.10.

2.3.2 The Tropics as Barriers to Migration

Given that the genus *Atriplex* has diversified most spectacularly in arid regions, it is not surprising to find that these species populations are isolated from each other by the tropics. Figure 2.9 shows that the North and South American regions with a high percentage of endemic *Atriplex* species are separated by the Central American tropics, a region containing only five species of *Atriplex* as far as we can ascertain. Only one characteristically North American species (*A.pentandra*) is common in South America, and no species characteristic of the latter region are common in North America. *A.pentandra* occurs in the three American regions but two species, *A.aldamae* and *A.crispa,* are restricted to Central America.

The Australasian group of *Atriplex* species is isolated from that of Eastern Asia by the tropics of the East Indies which seem to have been a most effective barrier to mixing of these populations. Aside from the *A.triangularis* complex there are no species in common to these two regions. Colonial links between Australia and Europe were such that trade by-passed the Asian region and no interchange of species by this route has taken place in recent times. However, in the last 100 years, trade with Europe has led to the recording of 15 Australian *Atriplex* species in that region (Probst, 1949).

The African tropics have been less effective in isolating North African and South African populations of *Atriplex* species. Seven Mediterranean species are found with only six endemics and several others of European, Central Asian, and Australasian origin. This pattern possibly reflects the longer history of trade around the Cape of Good Hope since the 17th century. The resulting low proportion of endemic species and the diverse origins of the introduced species is similar to, but less extensive than that resulting from the much older land trade routes of Central Asia.

The tropics have not been a barrier to the migration of the now cosmopolitan maritime *Atriplex* species. The extent to which natural long-distance dispersal in the sea has contributed to these movements is uncertain. It is likely that a good deal of the movement of these species has taken place with trade in recent times, as suggested in the case of the trans-Pacific migration of *A.subcordata* (Bassett and Crompton, 1973).

2.3.3 Origins of the Genus

Genetic and evolutionary relationships and the processes involved in speciation are discussed in detail in Chap.4. However, it is useful to consider here some questions of the origin of the three similar but isolated populations of *Atriplex* discussed above. Several authors imply that primitive stock of the genus was a form in which the pistillate flower contained an inferior ovary subtended by some form of perianth (Hall and Clements, 1923; Ulbrich, 1934). This hypothesis is based on the general features of other members of the Chenopodiaceae and is most nearly satisfied in the form of *A.hortensis*. Now a commonly cultivated species in many continents, *A.hortensis* is thought to be derived from *A.acuminata(nitens)* of Central Asia (Frankton and Bassett, 1968).

The relationship of the major regional populations of North and South America, and particularly of Australasia, to this primitive stock is tenuous (see Hall

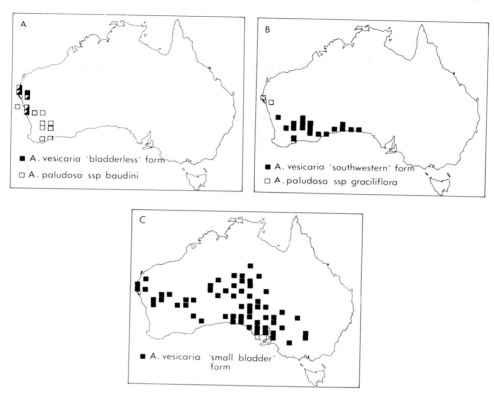

Fig. 2.12A-C. Distribution of closely related subspecies of *A.paludosa* and forms of *A.vesicaria* indicating possible coastal origin of inland *Atriplex* populations in Australia. Populations of the closely related "bladderless" form of *A. vesicaria* overlap with *A.paludosa* ssp *baudini* in coastal central Western Australia (**A**). These populations overlap with *A.paludosa* ssp *graciliflora* and the "south-western" form of *A.vesicaria* (**B**). Both *A.vesicaria* forms overlap with the most widespread "small bladder" form of this species (**C**). (Redrawn from data provided by G.A. Parr-Smith, with permission)

and Clements, 1923). *A.hortensis* and closely related species are all recent introductions to these regions and there are no obvious links between the present populations and the supposed primitive stock. A possible link between *A.hortensis* and most primitive North American species was uncovered by Weber (1950) in the form of a new species *A.pleiantha*. This herbaceous desert annual from southwestern Colorado has a perianth in both male and female flowers, which are mixed in axillary clusters. The female flowers are enclosed in a pair of bracts and the seed has an inferior radicle. For the North American population of *Atriplex* this species embodies the morphological features of Hall and Clements' hypothetical ancestor. Although *A.pleiantha* meets the morphological specifications of the ancestral stock, its very restricted distribution in a remote part of the continent makes speculation as to relationships with Central Asian or with cosmopolitan coastal species a hazardous exercise.

    There is no such link between Australasian *Atriplex* species and primitive forms in Central Asia. The only Australasian species with affinity with European species is

A.australasica, a taxon of very restricted, east coastal distribution which has some affinity with the *A.patula, A.triangularis* complex of Europe. The relationship between *A.australasica,* a C_3 plant, and other coastal species such as *A.cinerea* and *A.paludosa,* which are C_4 plants having some affinities with the more diverse and characteristic arid shrubland species, has yet to be established. It has been suggested that the chenopod shrubland vegetation of arid and semi-arid Australia arose from remnants of a coastal flora which may have extended to central Australia during the Cretaceous (Burbidge, 1960).

Following a detailed numerical analysis of relationships within the highly polymorphic taxon *A.vesicaria,* Parr-Smith (1977) conjectured a close relationship between the "bladderless form" of this species and *A.paludosa* ssp. *baudini.* Overlapping adjacent populations of these two taxa occur near the central coast of Western Australia, and both intergrade with populations of other forms of their species in south-western and southern Australia (Fig. 2.12), including the widespread "small bladder" form of *A.vesicaria.* These studies pave the way for closer cytological and genetic analysis of what may be a major link between coastal and arid shrubland species in this genus. P.G. Wilson (personal communication) considers *A.paludosa* R.Br. to be distinct from *A.moquiniana* Moq., *A.drummondii* Moq., and presumably *A.reniformis* R.Br. It is a decumbent plant found around the coasts of southern Australia, usually in water. He considers the inland and more northern collections referred to *A.paludosa* to represent a distinct species, i.e., those at present called *A.paludosa* subsp. *cordata* (Benth.) Aellen, subsp. *baudinii* (Moq.) Aellen, and subsp. *graciliflora* Aellen.

The links between *A.paludosa* and coastal species such as *A.australasica* and *A.triangularis* remain obscure. However, there is no evidence for such a flora until the late Pleistocene. Pollen deposits at the margins of the present chenopod shrublands show several periods of alternating domination by chenopod and myrtaceous pollen in the last 100 to 200,000 years (Martin, 1973; G. Singh, personal communication) and the stomach contents of *Diprotodon,* the fossil marsupial found in South Central Australia, contained fragments of chenopod shrubs (Stirling and Zietz, 1900).

2.4 Perspectives from the Systematic and Geographical States

The expressed purposes of this chapter were to clarify the systematic state of the genus *Atriplex* to the extent of a working list of species, and to note whether any of the characters used to establish this list of species, some of which may be functionally significant, were in any way related to major geographical and environmental parameters. Table 2.2 shows the genus *Atriplex* to consist of about 245 species that are generally considered to be closely related, in spite of their diverse habits and habitats. The genus shows greatest species diversification in the isolated semi-arid to arid regions of three continents. Before this description of the systematic and geographical state of the genus is considered in more detail it is important to recognize the limitations which may be imposed by the data base available at this scale of enquiry.

The systematic and geographical record discussed here is an account of the presence or absence of species in broad geographical regions. The recording of a species in a flora indicates only that this taxon has survived to be collected by a field botanist. Rarely do herbarium sheets provide quantitative or detailed comments of abundance of the species or of habitat factors. The systematic and geographical data base thus indicates only survival of taxa in broadly identified geographical regions. It does not distinguish between survival and performance limits of the species under consideration. The detailed correlations of Teeri and Stowe, for example, evaluate the survival of C_4 plants in relation to environment, not their performance. There are very few physiological data on the survival of these plants in relation to environmental factors because physiological experiments, by their nature, are concerned with performance, not survival. There is a great deal more than photosynthetic performance involved in determining survival of plants in different habitats. This point was most emphatically made in Stowe and Terri's analyses of dicotyledons from which they concluded that families containing C_4 species appear to have adaptive properties other than C_4 photosynthesis which favor their presence in arid regions. Moreover, the strong correlation between grass taxa with the C_4 pathway and night temperature during the growing season also indicates that processes other than the C_4 pathway are involved in adaptation.

Thus great care must be exercised in extrapolating from systematic and geographical data, which outline the survival limits of a species, to the functional significance of morphological characters, some of which may be related to performance limits of the species. That this is so should be obvious if the scales of the exercise are kept in mind. Thus the systematic state and geographical comparisons are based on scales of about 10^{12} in Fig. 1.1. The states are the product of processes with correspondingly large relaxation times such as speciation and dispersal, and it is likely that the significant environmental interactions are those with similarly large relaxation times. In the North American floras for example, the distribution of C_4 taxa described by Teeri and Stowe may be more a product of northward migration of a tropical grass flora and the migration of C_4 dicotyledon families from a Sonoran Desert refuge since the last glaciation than a product of the performance of C_4 plants in present-day environments.

"Kranz" leaf anatomy and the C_4 pathway are states and processes with scales of about 10^3 in Fig. 1.1. The significant environmental factors relevant to these states and processes are likely to be effectively operational over correspondingly small scales of space and time. Microhabitat data at these scales are not available from long-term weather records. An experimental approach to assess the significance of this morphological character in relation to the performance of plants in controlled environments is more effective. Subsequently, by the integration of these performance characteristics in the growth response of the plants, it may be possible to assess the advantage of this character and the associated photosynthetic pathway to the life cycle of the plant. Only then can we hope to validate these integrated studies of physiological and environmental processes with correlations based on systematic and long-term climatic data.

The systematic and geographical record thus serves two important purposes in physiological ecology. It is a resource which can be used, in general terms, to identify potentially significant relationships between functional morphological characters and the environment. Our treatment of the systematics and geography of *Atriplex* is an example of this general approach. The suggested relationships are only useful, however, in the extent to which they serve to erect testable hypotheses upon which the adaptive significance of the character can be assessed in ecological, physiological, or biochemical terms. Physiological ecology has an enormous catalog of such suggested relationships, of which few reach the stage of testable hypotheses and even fewer relate to the survival or performance of the plant in its habitat. The systematic and geographical record can also be used in a more precise way to validate the adaptive significance of processes associated with specific characters when these have been placed in the appropriate perspective. The treatments of Teeri and Stowe, discussed above, are examples of this second approach and will become much more valuable when the specific contributions of C_4 photosynthesis to the performance and survival of plants under natural conditions have been more accurately assessed. We have attempted to achieve this in Chaps. 8, 9, and 10.

In the next chapter we consider the wider questions of ecotypic differentiation and its relationship to polymorphism, which becomes only too evident when the systematics of the larger species complexes in the genus are considered. This more detailed evaluation of genecological differentiation leads naturally to a consideration in Chap. 4 of the genetic processes and evolutionary relationships in the genus, the product of which is the systematic state described in Table 2.2. The ecological state, comprising the numbers, abundance, and distribution of *Atriplex* and other organisms in specific habitats, is discussed in Chap. 5. This completes the discussion of the larger-scale states and processes which identify the main questions of functional significance and adaptive value examined in the latter part of the book.

Chapter 3. Genecological Differentiation

"The prevailing practice in the segregation of genera and species is chiefly detached analysis, with little consideration of relationship and practically none of evolution ... To be both comprehensive and thorough, taxonomy must draw its materials from all other fields, just as it must serve them in turn. While it leans most heavily upon morphology, it can not afford to neglect histology and physiology, and it must learn to go hand in hand with ecology and genetics in the future. Indeed, if it is to reflect evolution as accurately as it should, it must regard physiological adjustment as the basic process, and morphological and histological adaptations as the measurable results. This means that the taxonomist of the future will think in terms of evolutionary processes, and will learn to treat his morphological criteria as dynamic rather than static."

(Hall and Clements, 1923)

The above statement was made over half a century ago in the introduction to a book entitled *The Phylogenetic Method in Taxonomy: The North American Species of Artemisia, Chrysothamnus, and Atriplex*. This work by Hall and Clements marked a new epoch in North American taxonomy, as did the pioneering work by Göte Turesson (1922a,b) and Jens Clausen (1922) in Europe. A new line of taxonomic inquiry, that of genecology, had begun in which genetic relationships and the importance of distinctions between environmentally induced modifications and heredity variations in ecologically important characteristics were emphasized. Since then much progress has certainly been made in the field of biosystematics (or experimental taxonomy) and the necessity for a genecological approach is generally, in theory at least, accepted by modern taxonomists. However, it is still too often ignored in practice. This is not to say that we do not consider classification for its own sake an important and sufficiently respectable activity; in fact we agree with Heslop-Harrison's (1964) view that redefinition of biosystematics which effectively makes biosystematics synonymous with genecology, is undesirable. One can certainly understand the reluctance of the taxonomist to adopt the genecological approach, realizing the tremendous task of experimentally determining genetic and ecological relationships in genera with worldwide distribution, composed of hundreds of recognized species and thousands of taxa at the subspecific level. Yet, there is no substitute for this approach if the objective is to gain an insight into evolutionary and functional relationships.

In the context of this monograph, evolutionary and genecological aspects are obviously much more important than descriptive classification, and for this reason we will attempt in this and the following chapter to bring together available information on genetic, cytological, and functional relationships within the genus, even though this information is both scattered and sketchy.

3.1 Intra-Specific Variation and Ecotypic Differentiation

Atriplex species were the first experimental plant materials used by Turesson in his classical studies on intra-specific variation in relation to habitat that led to the development of the concept of ecotypic differentiation (Turesson, 1919, 1922a,b).

One of Turesson's first studies was made on *A. littoralis L.,* a cosmopolitan annual species commonly occurring along the Swedish coast where it occupies the lower portion of sandy beaches. On the protected coast along the Sound, separating Sweden and Denmark, this species occurs in sheltered places on the beach lines bordering bays and coves. The forms growing there are tall, erect, and robust; prostrate forms are only very rarely found. Further up along the exposed North Sea Coast the plants are of low stature and more spreading. Turesson asked the question whether these differences in form were only modifications directly induced by the differences in the environment or whether they were due to hereditary differences. To answer this question he collected seed and small plants from various places along the coast and from these he raised several generations of plants in a uniform transplant garden, taking the necessary precautions to prevent cross-pollination. He found that although noticeable variations existed within a given population, plants derived from exposed locations of North Sea Coast were exclusively procumbent or prostrate types (Fig. 3.1, top) whereas those derived from sheltered locations in the Sound region were predominantly tall and erect types (Fig. 3.1, bottom). The former types were also found to flower earlier.

Turesson (1922b) also obtained similar results with regard to growth habit and earliness in other species of *Atriplex,* now known to belong to the *A. triangularis* complex, as well as with species of a number of other genera. He concluded that although in certain cases the observed characteristics were found to be modifications, brought about by a direct influence of the environment, in the majority of cases they were caused by a differentiation of the species-population into different hereditary variations. It was clear that these hereditary characteristics were related to the particular native habitats. This and other evidence strongly supported his conclusion that the habitat type is a result of a genotypic response to a definite habitat. Turesson coined the term *ecotype* to denote such habitat types and the underlying evolutionary process was named *ecotypic differentiation.*

A mass of evidence accumulated subsequently by numerous investigators, notably Turesson (e.g., 1923, 1925, 1930) in Sweden, and Clausen and Hiesey and their coworkers in North America (Clausen et al., 1940, 1945, 1948; Clausen and Hiesey, 1958; Hiesey et al., 1971); Gregor in Britain (1944, 1946, 1956), as well as many other workers in different parts of the world, have established that ecotypic differentiation must be largely responsible for the ability of species to occupy a wide ecological range of habitats. (For an excellent review on this subject see Heslop-Harrison, 1964). Thus, ecotypic differences have been shown to exist with regard to complexes of climatic factors, such as occur along altitudinal gradients (e.g., coastal to alpine), latitudinal gradients (e.g., warm temperate to arctic), precipitation gradients (moist coastal to desert), to mention only a few. Ecotypic differentiation in response to edaphic factors including soil type, nutrient deficiency and metal toxicity has also been demonstrated. Just as major habitat factors may vary spatially in a discontinuous or continuous manner, so ecotypic variation may be either discontinuous or more or less continuous.

Fig. 3.1. Erect and prostrate ecotypes of *A. littoralis* from a sheltered habitat on The Sound (*bottom*) and an exposed location on the coast of Halland (*top*). Both ecotypes were grown in a garden at Malmö, Sweden. (From Turesson, 1922b)

3.2 Morphological Aspects of Genecological Differentiation

In the vast majority of studies on genecological differentiation, morphological characteristics have served as indicators of ecotypic variation. In some instances these habitat-correlated morphological differences could be shown to have an adaptive value in themselves, but often this is not so; it is rather implied that these morphological traits are genetically linked to physiological characteristics that enable the plant to succeed in its particular habitat. Evidently Turesson (1922b) was already aware of this problem:

"It is a generalization of the facts to maintain that the prostrate habit of growth enables the prostrate forms of *Atriplex sarcophyllum* (= *A. triangularis* in current usage) to live on the exposed Swedish west coast while erect forms of the same species, because of their erectness, are expelled from this coast strip. Comparative physiological experiments with individuals differing genotypically from each other only as to the factor or factors responsible for the difference of growth would probably yield some information as to this point ... Thanks to their genotypical constitution – and not necessarily on account of certain phenotypical characteristics – these prostrates support life where the erects do not."

In this particular case comparative micrometeorological and physiological studies may well have provided evidence that the prostrate growth habit does indeed constitute an adaptive advantage in a cool, windy habitat, in part because of the more favorable temperatures and lower windspeeds that occur close to ground. The steep vertical gradient of salt spray is also an important factor (Chap. 5.2.1).

Another morphological trait that can be considered to have a direct ecological significance is the presence on the upper leaf surfaces of a dense and highly reflective pubescence layer, or, as is found in many *Atriplex* species, several layers of bladder-like trichomes (Chaps. 2.1.3, 8.1.3, 9.2.3). These reflective layers have been shown substantially to decrease the heat gain by the leaves, thereby avoiding excessively high leaf temperature and transpirational water loss in habitats with extreme loads and limiting water supply. However, these layers would be disadvantageous in cool mesic habitats, primarily because they inevitably also decrease the light available for photosynthesis (Ehleringer, 1977; Mooney et al., 1977b; Smith, 1978).

Ehleringer's (1977) studies on two closely related nonpubescent and pubescent species of *Encelia* occupying mild coastal and hot desert habitats, respectively, are exemplary in their comprehensiveness and the rare combination of approaches used. *E. californica* is a nonpubescent plant restricted to coastal areas along the coasts of southern California and upper Baja California, whereas the pubescent species *E. farinosa* is found in hot, dry habitats of the Sonoran desert extending into portions of the Mojave desert. The two species tend to be allopatric, but hybridization often occurs along the distribution margins. The taxonomic differences between the two species are slight; the distinguishing characteristics are the difference in pubescence and a small difference in the color of the disk flower; only a single gene seems to be involved to produce this latter difference. Controlled crossing experiments indicate that there are no genetic barriers to hybridization between these two taxonomic species and that from a genecological viewpoint, they can be considered to belong to the same biological species.

Field studies showed that *E. californica* remains nonpubescent and the leaves absorb 83% of the incident sunlight throughout the year with negligible variation among different sites. In *E. farinosa* the pubescence increases from winter to summer with the mean corresponding solar absorptance values of plants from a number of habitats, ranging from 72% in December to 50% in July. In this species there was also a strong negative correlation between pubescence and the precipitation received at each site. In extreme cases, such as found during the summer in Death Valley, the heavily pubescent leaves appear almost white, and the leaves absorbed, on the average, only 29% of the incident light; as much as 70% was reflected by the pubescence layer. In *E. californica* leaves, or *E. farinosa* leaves from which the pubescence layer had been removed by shaving, 84% of the incident light was absorbed and only 14% was reflected by the upper leaf surface (Fig. 3.2). Other experiments showed that the degree of pubescence in *E. farinosa* is largely controlled by the plant water status at the time of leaf initiation. *E. californica* totally lacks the ability to develop pubescent leaves.

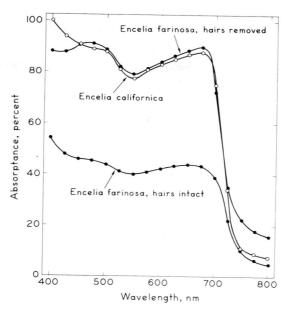

Fig. 3.2. Spectral leaf absorptance of intact leaves of glabrous *Encelia californica*, pubescent *E. farinosa*, and of *E. farinosa* leaves from which the leaf hairs have been removed. (From Ehleringer, 1977)

The presence of dense leaf hairs on *E.farinosa* typically reduced the absorption of solar energy to about one-half of that absorbed by *E.californica* leaves. Energy balance determinations showed that this has a most important effect on the temperature which a sunlit leaf will assume. A heavily pubescent leaf can maintain its temperature at least 10 °C below that of a nonpubescent leaf under the same conditions of radiation, windspeed, air temperature, humidity, and stomatal conductance. This has two consequences: transpirational water loss is much reduced and leaf temperature becomes considerably more favorable for photosynthesis during the warmer part of the year. In the desert habitat, it also minimizes the risk of the leaves reaching excessively high temperatures that cause irreversible injury or even death.

An important potential disadvantage of pubescence is that it inevitably also reduces the light available for photosynthesis. As shown in Fig.3.3a, the photosynthetic efficiency of utilization of incident light is strongly reduced as the pubescence increases. That the difference in this efficiency is entirely attributable to the difference in pubescence is shown in Fig.3.3b where photosynthesis for these same leaves is plotted as a function of absorbed instead of incident light. This indicates further that there are no detectable differences in intrinsic photosynthetic characteristics between *E.californica* and *E.farinosa,* and this was confirmed in detailed measurements of the response of photosynthesis to temperature and CO_2 concentration. All observed differences in the photosynthetic responses could be directly and quantitatively accounted for by pubescence effects alone.

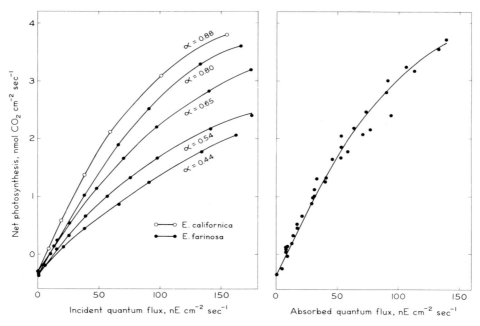

Fig. 3.3. Light dependence of photosynthesis in glabrous *Encelia californica* leaves and *E. farinosa* leaves with varying amounts of pubescence. The absorptance (α) varied from 0.44 for most pubescent leaf to 0.88 in the glabrous leaf. The graph on the *left* shows the response of photosynthesis to the incident light intensity (quantum flux density); in the graph on the *right* differences in leaf absorptances have been taken into account and photosynthesis is thus plotted as a function of absorbed light. (Data from Ehleringer, 1977)

Using a "trade-off" computer simulation model in which leaf parameters determined in the laboratory (pubescence, stomatal conductance, and intrinsic photosynthetic characteristics) were entered as inputs together with seasonal environmental parameters (radiant energy load, air temperature, and evaporation demand) for coastal and desert habitats of *Encelia,* Ehleringer was able to predict the relative advantages and disadvantages of pubescence. He found that in the coastal habitat the disadvantages of pubescence by far outweighed any advantage (which was slight). In the desert habitat, the optimum amount of leaf pubescence predicted by the model varied greatly with the season, ranging from extreme pubescence in the hot, dry summer to light pubescence in the coolest period of winter. These predictions are thus in excellent agreement with the actual situations found in nature. Ehleringer speculated that coastal *E.californica* is very similar to the ancestral species which occupied relatively mesic habitats and were unable to invade and survive the desert environments as these developed because this physiology could not cope with the high temperatures. In place of an evolutionary change in biochemical and physiological characteristics, a change in a morphological characteristic occurred. A small genetic change enabled the cells of the rudimentary leaf hairs, already present on the glabrous leaf, to expand and form the reflective pubescent surface of *E.farinosa* leaves, allowing this new form to occupy the desert habitats.

In the case of *Encelia* there appears to be little doubt that pubescence is an important genetically based adaptation to a particular habitat type and that it represents an excellent example of ecotypic differentiation of a morphological characteristic. The question was asked: in what way is this feature adaptive? The appropriate experiments were made, and satisfactory answers were obtained.

It seems likely that large differences in leaf reflectance frequently observed in plants of the same or closely related species occurring in cool, mesic, and hot, dry environments have a similar adaptive significance as in *Encelia,* but direct experimental evidence is lacking and in most cases it is not known whether or not the differences are genetically based. In *Atriplex,* species occupying cool-temperate environments generally tend to have low leaf reflectance, whereas those occurring in hot, arid environments predominantly possess highly reflective leaves. In certain *Atriplex* species, ecotypic differences in these characteristics are indicated. For example, like other arid-region Australian *Atriplex* such as *A.vesicaria* and *A.nummularia, A.paludosa* is characterized by leaves covered with a highly reflective dense mat of collapsed trichome bladders. However, two forms, *A.paludosa* ssp. *paludosa* and *A.hypoleuca* (which is probably conspecific with *A.paludosa*) entirely lack these reflective layers and possess green, glabrous leaves. The first of these forms is found in salt marshes on the south of east Australia, the other occurs across the continent in the south of Western Australia where it is restricted to coastal swamps (Parr-Smith and Calder, 1979).

The great intraspecific morphological variation observed in many widespread *Atriplex* species has frequently been attributed to ecotypic differentiation. From statistical studies on extensive field collections and examination of most Australian herbarium material, Parr-Smith (1977) recognized eight morphological forms in *A.vesicaria* (Chap.2). Four of these are restricted geographically and the others appear to be differentiated on an edaphic basis. Some of these forms had previously been considered ecotypes (Wood, 1936; Jessup, 1951). The distinctions among these forms were based on the morphological characters of leaf, fruit, and growth habit. Such differences point to the possibility that these forms may be different ecotypes and they should therefore provide a suitable basis for genecological studies. However, at present neither the genetic nor the ecological components of the observed variation are at all known and the use of the term ecotype is therefore unfounded.

3.3 Genetic Coherence Between Functional and Morphological Characteristics

The extensive work on genecological differentiation by the Carnegie group for almost forty years, under the leadership of Jens Clausen and William Hiesey has provided much of our present knowledge of the genetic structure of natural populations and the genetic basis of ecotypic differences. A general treatment of these subjects matters is beyond the scope of this monograph, but we would like to use a few examples from their work to illustrate some basic approaches and principles. The common practice used to distinguish between environmental modifications and genotypically based characteristics is to grow the plants in a common "neutral" environment as was done by Turesson. While this method may be reasonably satisfactory in determining the genetic and environmental components of certain morphological features, it is highly unsatisfactory as far as the detection of functional differences is concerned. By eliminating environmental differences, we may well fail to detect genetically determined differences in the *ability to respond adaptively* to particular environments. This is well illustrated by the early work by the Carnegie group along an altitudinal transect in California, extending from the coast to over 3,000 m elevation in the Sierra Nevada. When cloned individuals of the same or closely related species were collected from alpine and lowland habitats, they often performed equally well (or poorly, depending on the species) when tested at the mid-elevation transplant garden. By extending these transplant studies to include the extremes occupied by the respective populations (Stanford, 30 m altitude and Timberline, 3,050 m altitude) the tremendous differences in their ability to grow and survive in these contrasting environments were clearly expressed, demonstrating a genecological differentiation within these species in characteristics that are of paramount importance to their success. Subsequent studies by the Carnegie group on many genera and species have provided much information on the mode of inheritance of the factors underlying the different abilities of these ecotypes, or climatic races, to succeed in contrasting environments. Although in many instances these climatic races differ in a number of heritable morphological characters, these workers concluded that the functional basis of the observed ecotypic differences in terms of performance and survival cannot be explained by these differences in morphological features but must be largely physiological. Examples taken from the work on the *Mimulus cardinalis – M. lewisii* complex (Hiesey et al., 1971) may serve to illustrate these points.

These taxa are diploid with $n=8$ chromosomes and extensive crossing experiments show a very close genetic relationship between them. First-generation hybrids between *M. cardinalis* and *M. lewisii* from the same geographical region are completely fertile and the mild genetic barriers found between certain geographically distant populations of the two taxa are similar to those present within each taxon. From a genecological viewpoint these taxa can be considered to belong to the same biological species, but for practical reasons they will be referred to here by their binomials. *M. cardinalis* occurs on the Pacific coast from 32° to 43° N and inland to Nevada at elevations extending from sea level to 2,500 m. *M. lewisii* tends to complement the distribution of *M. cardinalis*, being widespread at higher altitudes (1,500 to 3,200 m) and latitudes (35° to 62° N) and extending into climates with more severe winters. Both taxa are in turn composed of a large number of ecological races. Many of these, and numerous F_1, F_2, and F_3 progenies of crosses between them, have been tested for growth and survival at the different transplant stations.

Two of these climatic races are *M. cardinalis* from Los Trancos Creek at 30 m elevation, only a few kilometers from the Stanford garden, the other *M. lewisii* from the Timberline garden. Several hundred cloned individuals of these plants and their first-generation hybrid were transplanted into each of these two gardens and

Table 3.1. Growth[a] and survival[b] of *Mimulus cardinalis*, Los Trancos, *M. lewisii*, Timberline, and their F_1 hybrid in transplant gardens at Stanford, Mather, and Timberline. (After Hiesey et al. 1971)

Culture	Stanford (30 m)		Mather (1,400 m)		Timberline (3,050 m)	
	Growth	Survival	Growth	Survival	Growth	Survival
	(g)	(%)	(g)	(%)	(g)	(%)
Mimulus cardinalis Los Trancos	152	100	14	0	0	0
Mimulus lewisii Timberline	4.4	0	0	0	14.9	100
F_1 hybrid	72	100	96	50	8	100

[a]   Growth expressed as mean dry weight yield per plant (g) of above ground material per growing season of cloned transplants. Values are means over a three- to five-year period at each station
[b]   Survival of cloned transplants in percentage over the same period

also into the mid-elevation Mather garden (1,400 m). Table 3.1 shows the dry matter yield and the percentage of plants that were able to survive for more than two years. *M. cardinalis* grew well and 100% of the plants survived in the Stanford garden whereas all were killed in the Timberline garden. The exact opposite was observed with *M. lewisii*. Interestingly, neither of them was able to survive in the mid-elevation Mather garden, but their F_1 hybrids performed well and all survived in all three gardens. Several hundred F_2 hybrids, obtained by selfing F_1 hybrids, were then cloned vegetatively and ramets of each clone tested at the transplant stations. A great variation in the ability to perform and survive at the three stations was observed among the different clones. Some behaved like the *M. lewisii* parent, surviving only at Timberline, others behaved like the *M. cardinalis* parent, surviving only at Stanford, and yet others behaved like the F_1 hybrid, performing well at all three stations.

An analysis of up to 24 morphological features distinguishing *M. lewisii* and *M. cardinalis* showed that there was a remarkably high correlation between the survival pattern and morphological characteristics among the F_2 hybrids. Those surviving only at Stanford, although showing some of the traits of *M. lewisii*, were morphologically similar to the *M. cardinalis*, those surviving only at Timberline closely resembled *M. lewisii*, while those surviving at all three stations were morphologically intermediate between parental types although a wide range of different combinations was evident.

Table 3.2 shows a partial analysis of the correlation between the inheritance in second-generation hybrids of two functional characteristics distinguishing *M. lewisii*, Timberline and *M. cardinalis*, Los Trancos. The functional characteristics, frost tolerance at Timberline and winter-activity at Stanford, are highly correlated with and causally related to the capacities for survival at Timberline and Stanford, respectively. F_2 hybrids having a high frost tolerance, similar to that of *M. lewisii*, were scored a low number and those having a frost susceptibility similar to that of *M. cardinalis*, were scored a high number on an

Table 3.2. Partial analysis of the correlation[a] between essentially nonmodifiable morphological and two functional characteristics in F_2 hybrids between *Mimulus lewisii* and *M. cardinalis*. (Data from Hiesey et al., 1971)

Morphological character	Winter activity at Stanford	Frost resistance at Timberline
Petal reflexing	3	3
Petal width	3	0
Corolla aperture	3	3
Yellow upper epidermis in petal lobes	2	2
Yellow lower epidermis in petal lobes	2	3
Anthocyanin in petal lobes	3	2
Unpigmented bands in petal lobes	2	2
Brilliant magenta in corolla throat	2	2
Pistil length	3	3
Leaf width to length ratio	1	3
Number of leaf dentations	3	2
Average of all characters	2.5	2.3

[a]   0 Correlation values not significant, indicating free recombination
      1 Correlation values significant between 5% and 1% levels
      2 Correlation values significant above 1% level
      3 Correlation values very highly significant

arbitrary scale from 1 to 9. The converse ranking was made for winter activity at Stanford so that hybrids possessing a high winter activity similar to that of *M. cardinalis* were scored a high number on a 1 to 9 scale. Similarly, for each of the morphological characters, the F_2 hybrids were ranked so that those closely resembling *M. lewisii* were scored a low and those resembling *M. cardinalis* a high value on the 1 to 9 scale. The scored values for each character were averaged on individual F_2 plants over a period of three to five years.

It is important that the morphological traits chosen had been shown to be essentially *nonmodifiable* by the growth environment and that in this experimental design they could not have played any *functional* role in the different abilities of the clones to grow and survive at the three stations. Character combinations likely to be pleiotropic were also excluded as far as possible. It should further be noted that meiosis in the F_1 and F_2 hybrids was perfectly regular with complete chromosomal pairing forming eight bivalents. As shown in Table 3.2, nearly all of the morphological characteristics analyzed are strongly associated with the two functional characteristics with 20 of 22 possible combinations showing a very highly significant correlation. The correlation values obtained among the various morphological characters were even higher (not shown).

These results demonstrate that the inheritance of parental character combinations was very much more frequent than would be predicted on the basis of free random recombination – a phenomenon termed *genetic coherence*. Evidence pointing to genetic coherence as a general principle in the evolution of ecological races and species was also obtained in previous work by the Carnegie group on the genera *Achillea* and *Potentilla*. A discussion of the genetic mechanisms underlying coherence and its role in speciation is beyond the scope of this book; the reader is referred to Grant's (1964) extensive review on these subjects. In the present context it is sufficient to make two points: morphological character constellations, unrelated to performance or survival, cohere genetically with physiological properties which determine survival. It follows that such genetic coherence within the characters of any ecotype tends to ensure its perpetuation as an evolutionary entity which may result in morphologically distinct ecotypic subspecies. *This indicates that great caution is advisable in attributing adaptive functions to morphological traits because of their ubiquitousness in a certain ecological race unless it is also clear in what way they are adaptive.*

3.4 Physiological Aspects of Genecological Differentiation

The striking ecotypic differences in the ability to grow and survive in contrasting environments such as those demonstrated in species of *Mimulus, Potentilla, Achillea,* and many others, raise the question of the nature of the underlying physiological characteristics. The demonstration that differences in important physiological properties must exist between ecotypes such as those growing in warm temperate lowland regions and those inhabiting alpine habitats does not disclose which habitat factor or complex of factors determines the differential performance or survival of these ecotypes. Neither does it reveal which physiological mechanism or mechanisms responded differentially in the two ecotypes.

To provide a sufficient data base for a reasonably realistic assessment of the relative importance of the multitude of differing habitat factors and of the great number of potentially differing physiological characteristics it would be necessary to conduct a factor-by-factor analysis of the respective environments and a process-by-process analysis of the physiological responses of the plants. In environments differing in a complexity of interacting variables at the same time and which also show great seasonal variations, such habitat factor analyses become very difficult and time-consuming tasks in themselves. The task of unscrambling the response patterns of a multiplicity of interacting physiological processes is beset with formidable difficulties especially when timing adjustments of the developmental cycle such as induction of seasonal dormancy (e.g., to avoid winter cold or summer drought) are also involved. Moreover, the analysis of physiological characteristics requires that the response of each particular process to each particular factor be determined under a range of growth and measurement conditions since it is the ability of an individual to adjust its physiological response to a certain factor in an adaptive manner which is inherited, not necessarily the final adaptation itself.

This does not mean that we consider it futile to attempt to determine the major selective environmental factors and to provide an insight into the physiological bases for ecotypic adaptation to contrasting climates. Valiant attempts to do so even in very complex cases have met with considerable success (e.g., the extensive studies by Billings and coworkers on arctic and alpine ecotypes of *Oxyria* and by the Carnegie group on *Mimulus*). However, the chances of obtaining information on the direct causal relationships between differentially acting habitat factors and physiological processes increase tremendously in situations where differences in a single habitat factor are large and distinct while differences in other habitat factors are small. This is especially true where a difference in this dominant factor primarily affects a single physiological process and the functional relationship between the factor and the process is known. *Adaptive differentiation is still likely to be a holistic process, but the system lends itself more easily to experimental studies and interpretations of the adaptive significance of observed physiological characteristics.*

3.4.1 Light Intensity

Examples of such ecotypic differences are provided by studies on the photosynthetic response to light intensity in sun and shade forms of *Solidago virgaurea* (Björkman and Holmgren, 1963; Björkman, 1968b; Holmgren, 1968) and *Solanum dulcamara* (Gauhl, 1969, 1976). Both of these widespread species occupy a wide range of ecologically diverse habitats including open, sunny heaths and meadows and densely shaded beech-oak forests where the light intensity reaching the forest floor during the growing season is only a few percent of that found in open habitats (Chap. 9.2). Inevitably, such sun and shade habitats differ to some extent also in other correlated factors, but the one to two orders of magnitude difference in light intensity can undoubtedly be considered of paramount importance especially in regard to photosynthesis since light provides the sole force for this process. In a comparison of *Solidago virgaurea* populations from several sun and shade habitats, including a dense forest in south Sweden and an open meadow in the same general region, Björkman and Holmgren (1963) demonstrated that under controlled conditions the shade clones were able to grow under considerably lower light intensities than the sun clones. Conversely, at high light intensities the sun clones grew considerably faster than the shade clones and the latter showed evidence of chloroplast damage in the upper palisade cell layers. This situation is similar to that often found between "sun species" and unrelated "shade species". Comparative measurements of photosynthetic characteristics of the sun and shade clones of *S. virgaurea* showed that the rate of photosynthesis as a function of light intensity during measurement was strongly influenced by the light intensity under which the plants were grown but the response was strikingly different between shade and sun clones (Fig. 3.4). Two expressions of photosynthetic activity are especially important in this context: (1) the efficiency by which light of low intensities can be utilized in photosynthesis (indicated by the initial slope of the light-dependence curve) and (2) the photosynthetic capacity at high, saturating light intensities. The values for six shade clones and six clones of *Solidago* are shown in Table 3.3. It is important that in each case the photosynthetic efficiency in utilizing light of low intensities was determined on plants grown in weak light, whereas the photosynthetic capacity in saturating light was determined on plants grown at a high light intensity. The shade and the sun clones differ in both respects. The shade clones have a markedly higher efficiency in using weak light than the sun clones, whereas the converse is true of photosynthetic capacity of high light intensities. Obviously these genotypically based differences in photosynthetic characteristics can be expected to be of great importance to growth and survival in the respective

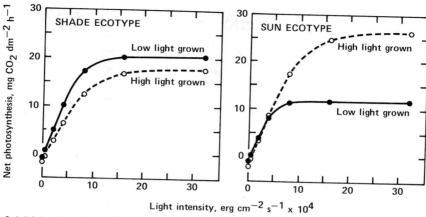

Fig. 3.4. Light dependence of photosynthesis in a shade and a sun clone of *Solidago virgaurea*. Each clone was grown under a low and a high light intensity regime. (Data from Björkman and Holmgren, 1963)

habitats. Like most species from open habitats, including *Atriplex*, the sun ecotypes of *Solidago* are able to markedly increase their capacity for photosynthesis at high light intensities when they are grown in strong light. This adjustment involves a number of changes in the intrinsic properties of the photosynthetic apparatus at the chloroplast level and is also associated with changes in leaf morphology. The shade ecotypes lack the ability to increase their photosynthetic capacity when grown in

Table 3.3. Light-related photosynthetic characteristics of sun and shade ecotypes of *Solidago virgaurea*. (After Björkman and Holmgren, 1963)

	Shaded habitat clones				Exposed habitat clones			
	A	B	C	D	A	B	C	D
Habitat I								
Clone 1	3.1	18.9	0.63	0.95	2.4	31.8	1.12	1.86
Clone 2	3.0	18.6	0.63	0.92	2.2	27.3	0.98	1.65
Clone 3	2.6	15.2	0.67	1.04	2.3	27.6	1.06	1.59
Habitat II								
Clone 1	2.9	18.6	0.65	0.88	2.3	24.6	1.11	2.14
Clone 2	3.0	17.0	0.66	0.86	2.3	23.8	0.94	1.59
Clone 3	3.0	19.2	0.61	1.19	2.4	27.8	1.00	2.19
Mean	2.93	17.9	0.64	0.97	2.31	25.5	1.04	1.84
Standard error	±0.08	± 0.6	±0.01	±0.05	±0.04	± 2.0	±0.03	±0.11

Column A: Efficiency of low light utilization of plants grown at low light intensity
Column B: Capacity of light-saturated photosynthesis of plants grown at high light intensity
Column C: Ratio of low light utilization efficiency of a clone grown at high light intensity to that of the same clone grown at low light intensity
Column D: Ratio of light-saturated photosynthetic capacity of a clone grown at high light intensity to that of the same clone grown at low light intensity

strong light, but do show the same changes in leaf morphology as the sun ecotypes. The molecular and physiological mechanisms underlying photosynthetic adaptation to different light intensities are discussed in Chap. 9.2.

Intraspecific differentiation similar to that in *Solidago* was also found between sun and shade clones of *Solanum dulcamara* by Gauhl (1969, 1976). The existence of differences in the response of light-saturated photosynthesis to light intensity during growth between populations is also indicated by the studies of Eagles and Treharne (1969) in *Dactylis glomerata* and by Björkman and Holmgren (1966) in *Rumex acetosa*. These findings suggest that ecotypic differentiation in the efficiency of light utilization by the photosynthetic apparatus is a widespread phenomenon in species which occupy habitats with greatly contrasting light intensities.

3.4.2 Temperature

This is another factor which shows great differences among different habitats occupied by the same species. However, the range of seasonal temperature variation in a given habitat with a long growing season is often very wide, and the temperatures thus often overlap with those prevailing during the period of active growth in a much colder climate with a short growing season. Moreover, microclimatic variation is often substantial and the temperature of the various plant organs may also be substantially different from meteorological air temperatures, further obscuring the difference between habitats. Analyses attempting to determine the relationship between the differential habitat factor (temperature) and a given physiological process (such as photosynthesis) are further complicated by the fact that nearly all physiological processes are highly temperature-dependent and also highly modifiable by the growing conditions. Nevertheless, when study habitats and experimental plant materials are carefully and appropriately chosen and the specific effects of temperature on the physiological process studied are well understood, it becomes feasible to conduct meaningful investigations on the physiological basis for temperature-related ecotypic differences.

Frost resistance is undoubtedly an extremely important aspect of temperature-related differences which affect all metabolic processes and play a key role in the survival of plants. Resistance to freezing injury has long been the subject of extensive investigation and a great complexity of underlying mechanisms has been proposed. The reader is referred to Levitt's (1972) comprehensive treatment of this subject.

Intraspecific variation in frost resistance has been observed in many plants, especially tropical and subtropical grasses and legumes (e.g., Edye and Kiers, 1966; Jones, 1969b; Hacker et al., 1974; Clements and Ludlow, 1977). For example, accessions of the perennial leguminous species *Centrosema virginianum*, ranging in distribution from 40° N.L. in New Jersey, U.S.A. to 31° S.L. in Argentina, show marked differences in their ability to survive controlled frosts, and the differences were correlated with the latitude of origin (Clements and Ludlow, 1977). These workers also found that the differences in ability to survive frosts were primarily due to frost *avoidance* (the ability to avoid low temperatures or to avoid ice formation) rather than frost *tolerance* (the ability to tolerate ice formation in tissues without injury). It seems that the height of the lowest growing points was an especially important factor in determining frost resistance. A strong negative correlation was obtained between frost survival and cotyledon node height. Crossing experiments between a number of accessions from different latitudes showed that this characteristic is under strong genetic control. Although it is conceivable that the observed correlation between cotyledon node height and frost resistance could be due to genetic coherence (discussed in Sect. 3.3), other experimental evidence strongly suggests that cotyledon node height is causally related to frost avoidance. It seems likely that intraspecific differences in frost resistance are largely due to frost avoidance rather than tolerance, although avoidance of ice formation in the tissues by freezing point depression may be at least as important as avoidance of low temperatures.

Resistance to chilling injury, i.e., tolerance to low, but nonfreezing temperatures, is quite distinct from frost resistance. Differences in chilling resistance are well documented between species from mild tropical and cool-temperate regions. Recently, differences in resistance to chilling injury have also been found within the same or, at least, among closely related species. Patterson et al. (1976) compared the chilling sensitivity in a series of species and hybrids of *Passiflora* from natural and cultural habitats with different temperature regimes. Among the taxa studied were *P. edulis* forma *flavicarpa*, which is grown in tropical areas such as Hawaii, *P. edulis* which is grown in cooler climates (to 34° S.L.) and *P. caerulea*, the only *Passiflora* which can be grown outdoors in cool-temperate climates such as south England (46° N.L.). A number of F_1 and F_2 hybrids between these and other *Passiflora* species were also studied. All plants were grown in a greenhouse at a 25°/18 °C day/night cycle; isolated leaves were then kept at 0 °C for varying periods. The main criterion of chilling injury was the rate of electrolyte leakage from leaf strips into a bathing medium. Potassium was the principal cation lost from the leaf tissue and its leakage was proportional to that of total electrolyte during the course of the experiments. Those taxa which had originated from tropical lowlands had the fastest rate of electrolyte leakage and those which are successful in colder climates had the slowest leakage rate. For example, *P. edulis flavicarpa* lost 50% of its electrolyte content in two days, whereas no significant leakage could be detected from *P. edulis* in the first two days and less than 20% of the electrolyte was lost after 20 days at 0 °C. Hardly any leakage could be detected in cold-resistant *P. caerulea*. Hybrids which involved *P. caerulea* were chilling-resistant even in those crosses where the other parent (e.g., *P. edulis flavicarpa*) exhibited extreme chilling sensitivity. The number of F_2 individuals studies was much too small to indicate the mode of inheritance of chilling resistance but the data show convincingly that chilling resistance is under strong genetic control.

Lyons (1973) and Raison (1973) have provided evidence that chilling injury phenomena derive from effects of temperature on the lipo-protein matrices of the various cellular membranes, including the chloroplast membranes (Chap. 9.3.3). Using an electron spin resonance technique Patterson et al. (1978) were able to show that phase-transition temperatures of phospholipids, isolated from the *Passiflora* taxa mentioned above, correlated with their chilling sensitivity. The temperature dependence of the motion of the spin labels showed a change at 1 °C for lipids from

P.caerulea and at 9 °C for lipids from *P.edulis flavicarpa*. Lipids from other *Passiflora* taxa showed changes at intermediate temperatures and the greater the chilling sensitivity of the taxon, the higher was the temperature of the spin label motion change. This clearly shows that differences in chilling-sensitivity are related to physical differences in the membrane lipids. However, the fatty acid compositions of eight polar lipid classes from *P.caerulea* were generally similar to those of the corresponding lipids from *P.edulis flavicarpa*. Thus, while in *Passiflora* the evolution of resistance to chilling has been accompanied by proportionate changes in the temperature characteristics of the polar lipids, no marked change has taken place in the proportion of unsaturated fatty acids.

Another example of intraspecific differentiation in chilling sensitivity was reported by Raison (1974). *Themeda australis* is a perennial grass which is found mainly in tropical regions of Australia, but populations of presumably the same taxonomic species extend into temperate regions of Tasmania where they grow successfully. Examination by spin-labeling of chloroplasts isolated from races of *Themeda* from tropical New Guinea and temperate Tasmania showed that phase transition occurred at 14° and 7 °C, respectively.

Unusually high heat tolerance is frequently found in species occupying hot desert environments (Chap.9.3). Significant varietal differences in heat tolerance have also been shown in certain crop species such as *Sorghum* (Sullivan, 1972), but little is known about underlying mechanisms or relationship to physiological processes or habitat factors. The most comprehensive information on intraspecific adaptive differentiation in regard to high temperature tolerance is provided by Pearcy's (1976, 1977) studies on desert and coastal populations of *Atriplex lentiformis*.

Atriplex lentiformis is an evergreen perennial shrub native to phreatic habitats at low elevations in the Sonoran, Mojave, and Colorado deserts of south-western North America. A slightly different form of this species, recognized as a subspecies (ssp. *breweri*) by Hall and Clements (1923) occurs along the Pacific coast of southern and central California. Both the desert and the coastal plants are diploid with n = 9 chromosomes (Nobs, 1975, 1978) and possess "Kranz" anatomy and the C_4 photosynthetic pathway. Desert forms of this species, growing around the Saratoga Spring on the floor of Death Valley and close to Thermal by the Salton Sea in southern California, remain active all year even though the mean daily maximum temperature varies from a low of about 18 °C in January to 43 °C in July. Field measurements of photosynthesis, using a mobile laboratory, showed that in spite of the striking differences in prevailing leaf temperatures in July (17° to 23 °C in the coastal and 40° to 46 °C in the desert site), the diurnal photosynthetic rates were very similar in the two sites. This was made possible by a large difference in the temperature dependence of photosynthesis so that the coastal plants were able to photosynthesize twice as fast as the desert plants at 20 °C, whereas the opposite was true at 43 °C, and the optimum temperature for photosynthesis was shifted from 32° to 43 °C (Pearcy and Harrison, 1974). When cloned individuals of these coastal and desert forms were transplanted into well-watered experimental gardens on the Death Valley floor and at the coastal site of Bodega Heads, it was found that both forms grew equally well in the coastal garden and also during the cooler part of the year in Death Valley. However, during the extremely hot summer, growth rates

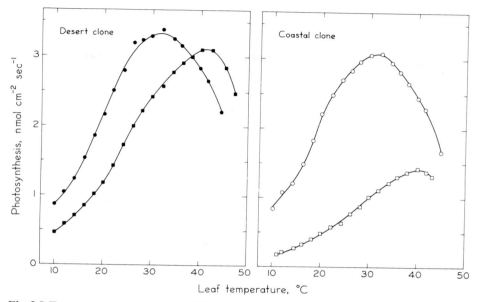

Fig. 3.5. Temperature dependence of light-saturated photosynthesis in a hot-desert clone and a cool-coastal clone of *Atriplex lentiformis*. Each clone was grown at a daytime temperature regime of 23 °C (○, ●) and 43 °C (□, ■). (Data from Pearcy, 1977)

were considerably higher in the desert plants and the coastal plants suffered visible injury (Björkman et al., 1974b). Growth experiments under controlled conditions at a series of temperatures yielded similar results (Pearcy, 1977). No difference in relative growth rates was detected between the desert and the coastal clones at 23°/18 °C or 33°/25 °C day/night regimes but at 43°/30 °C the growth rate was significantly higher in the desert clone.

Striking changes in photosynthetic characteristics occurred in response to increased growth temperatures in both types of plants (Fig. 3.5). In the desert clones, growth at high temperatures resulted in an upward shift of the temperature response curve with little change in the maximum photosynthetic rate. The photosynthetic characteristics of the desert and the coastal plants are very similar when the plants have been grown at low temperature, and like the situation in the desert clones, the optimum temperature for photosynthesis in the coastal clones is shifted upward when the plants are grown at a high temperature regime. However, in contrast to the situation in the desert clones, growing the coastal clones at high temperature results in a strong inhibition of photosynthetic capacity. This and other studies (Pearcy et al., 1977a) indicate that the inhibition is attributable to thermal inactivation of key components of the photosynthetic machinery itself and is not caused by the general effect on cellular integrity. The mechanisms involved are discussed in Chap. 9.3.

Pearcy's studies provide convincing evidence that desert clones of *Atriplex lentiformis* possess a higher thermal stability of its photosynthetic apparatus and are therefore capable of photosynthetic acclimation over a wider temperature range than are coastal clones of the same species, a feature that must be of great importance in coping with the great seasonal changes in temperature that occur in the desert environment. In the coastal habitat, the ability to cope with high temperatures would of course be of little consequence and the coastal clones lack this ability. There is, however, no evidence that the coastal clones possess any physiological characteristics that would enable them to perform better than the desert clones in the coastal environment.

Striking adaptive differences in temperature-related physiological characteristics are found between different species of *Atriplex*, native to habitats with contrasting thermal regimes. These characteristics and the adaptive significance of the C_4 pathway of photosynthesis in relation to temperature and other habitat factors are discussed in Chap. 9.3.

3.4.3 Drought

The availability of water is undoubtedly the major factor limiting the distribution of plants in the large areas of the world where rainfall is low and the potential evapotranspiration is high. There is little doubt that profound differences exist among plants in their abilities to cope with drought (Chap. 8). But in spite of the paramount importance of drought resistance both to the fundamental and the applied plant sciences and the wide interest in the water relations in plants, alarmingly little is known about ecotypic adaptation or varietal differences in response to drought. This is largely also true for adaptive differentiation among different species regardless of their relatedness. It is of course fully recognized from field observations that certain species of wild plants, as well as species and varieties of agricultural plants, exhibit differences in their performance or endurance in response to aridity. In many cases increased drought resistance appears to be related to various forms of drought evasion rather than desiccation tolerance, or improved efficiency of water use, and it is frequently believed that drought resistance is inversely related to productive capacity under conditions of ample water supply. However, only in a very limited number of cases have analyses of these factors been made.

Three examples of such investigations on intraspecific differences will be briefly mentioned here. The first is a study of Fanous (1967) who compared five varieties of *Panicum typhoideum*, ranging from regular commercial varieties to a collection from Rajastahn, India, where it is grown at the edge of the desert without irrigation with less than 250 mm annual rainfall. The results showed that the response of growth to decreased water potential in the soil followed the same pattern in all varieties, indicating that there were no significant varietal differences in physiological drought resistance. However, the Indian variety reached maturity earlier, which may enable it to complete its life cycle before the onset of severe drought. A somewhat similar situation was found by McKell et al. (1960), who compared two races of the grass *Dactylis glomerata*, one native to the coast of Portugal, the other to an inland semi-arid area in Israel. Following establishment of the plants in cans with equal amounts of soil water present, the cans were allowed to dry out and soil water potential and water use by the plants were continually recorded. It was found that the inland races used water at a markedly lower rate than the coastal race. However, in both cases the plants ceased to grow when the soil water potential declined to -30 bar, and the final leaf weight was somewhat lower in the inland race. The authors suggest that beside any role that the slower rate of water use may have, the principal means of the inland race for survival during the period of water stress in its native habitat is its ability to pass into a state of dormancy.

A different situation was found by Sullivan and coworkers in different *Sorghum* varieties. In one series of experiments, two varieties, one from the United States and the other from India, were grown in mixed culture (to provide the same root moisture availability), and following establishment of the plants, water was withheld until the leaf water potential of both varieties reach -33 bar. Upon rewatering, all plants of the Indian variety, but none of the American variety, recovered, indicating that the former had a much higher desiccation tolerance (Sullivan, 1972). Tests on other *Sorghum* varieties, using different laboratory techniques to determine desiccation tolerance (such as the water potential at which ion leakage from the leaves and reduction in photosynthetic rate occurred), indicated substantial varietal differences (Blum and Sullivan, 1972). In general, a correlation was also found between the desiccation tolerance determined in the laboratory and performance under drought conditions deduced from field experience. These results point to the possibility that intraspecific differences in intrinsic physiological characteristics which determine drought tolerance may prove an important mechanism in adaptive differentiation to aridity.

One aspect of morphological features, leaf pubescence, which can serve to improve water use efficiency was mentioned in a preceding section (3.1.1). The role of stomatal regulation and the significance of the C_4 photosynthetic pathway in improving the efficiency of water use will be discussed in Chaps. 8 and 9.

3.4.4 Edaphic Factors

Heavy Metal Contaminated Soils. No other aspect of genecological differentiation in response to habitat factors has been so extensively studied as the evolution of resistance to potentially toxic levels of heavy metals in contaminated soils. This may seem paradoxical since such soils constitute only a very small component of the environment, as they are usually restricted to small areas in the immediate vicinity of man's mining activities. The subject of metal tolerance can therefore be regarded to have only marginal relevance in comparison with such factors as temperature extremes, drought, and salinity. However, studies of the genecology of heavy metal tolerance have probably produced the best available examples of natural selection in plants and have shed much light on the mechanisms of speciation.

The advantages of metal tolerance as a subject for studies of intraspecific differentiation and adaptation to habitat are obvious. A single selective factor, metal concentration, is of overriding importance in determining plant establishment and survival in metal-contaminated habitats and these habitats are usually spatially distinct, easy to define, and relatively constant with time.

Moreover, most areas contaminated with heavy metals are of recent origin, usually less than a century old, so the history of the vegetation is well known; this permits a direct study of evolution in action.

There is a vast literature demonstrating the recent evolution of ecological races that are resistant to very high concentrations of one or several heavy metals contaminating the soils of their respective habitats. The existence of heavy metal-tolerant races has been reported for a great number of species in many families. Antonovics et al. (1971) list about 30 species of higher plants in which heavy metal-tolerant races have been demonstrated. Since that time new studies must have added at least another score of higher plant species as well as a great number of microorganisms. The evolutionary, physiological, and genetic aspects of heavy metal tolerance are discussed in detail in the excellent and comprehensive review by Antonovics et al. (1971). We will limit our treatment on this topic to a few additional brief comments.

Because of the common occurrence of metal-tolerant races, one is tempted to conclude that most, if not all, plant species may possess the capacity to evolve heavy metal tolerance. Moreover, the evolution of metal tolerance can be extremely rapid. For example, Abbot and Misir (cited by Antonovics et al., 1971) showed that when seed from nontolerant plants of *Agrostis tenuis* were sown on a copper-mine soil whose toxicity had been lowered somewhat, about one plant in 7,000 survived, demonstrating that selection for tolerance could be achieved in a single generation. In spite of this, it appears that heavy metal tolerance is genetically complex. Physiologically, the most common tolerance mechanism evidently involves the chelation of the metal in the cell wall thereby keeping it away from the active sites of metabolism. However, tolerance to one metal does not confer tolerance to another; independent complexing systems are used for different heavy metals such as zinc, copper, lead, and nickel. Moreover, tolerance to even a single metal is not an all-or-nothing effect, but the level of tolerance is related to the concentration of metal in the soil. Thus, in natural populations the tolerance level may be a more or less continuous variable. For example, the inheritance of resistance to zinc may appear to be determined by a single dominant gene at low levels of the toxic metal, but at higher concentrations it may appear to be determined by several recessive genes. This and many other studies support the conclusion that, in general, metal tolerance is of a polygenic nature. Interestingly, there is much evidence that metal-tolerant races have evolved a requirement for an increased level of the metal. The simplest interpretation of these results is that trace amounts of the metal are normally needed in metabolism and because of the presence of a powerful complexing mechanism in the tolerant plants these demand a higher external level of the metal to satisfy the metabolic requirements.

The presence of nontolerant populations immediately adjacent to tolerant populations has also provided a unique and nearly ideal natural laboratory system for studies of gene flow and population dynamics. These studies have shown increasingly that genetic divergence within a species can evolve and be maintained in the absence of geographical isolation. They also support the view that in the absence of such isolation, genetic breeding barriers tend to evolve in its place and may eventually isolate the populations. Other evidence supporting this view has also been obtained in *Atriplex* species (Chap. 4.2.2).

Serpentine Soils. In comparison with heavy metal contamination that is caused by mining activities, serpentine soils represent a great complexity of natural environmental factors which may also differ greatly from one serpentine habitat to the next. Common denominators in these soils which have developed on serpentine rock in various parts of the world are a predominance of magnesium over calcium, a short supply of major nutrients, and in many cases a high concentration of heavy metals such as nickel and chromium. Serpentine soils are often also characterized by low soil moisture levels.

It is well documented that ecotypes capable of occupying serpentine soils have evolved within many species in different parts of the world, and that their success in these habitats depends at least in part on their unusual ability to cope with low calcium levels (Kruckeberg, 1969; Proctor, 1975). In plants from serpentine soils containing high heavy metal contents this ability is also combined with an increased tolerance to these metals. No information is available on the mechanisms that enable serpentine plants to cope with a general deficiency of major nutrients.

Aluminium Toxicity. Extensive areas in the warm-temperature, subtropic and humid tropic zones of the world are characterized by acid soils which contain sufficiently high levels of soluble aluminium to be toxic to many plants. The high aluminum level is a direct result of low pH which causes the normally insoluble soil aluminum to dissolve. The low pH in turn results from a high rainfall which leaches out bases as the water percolates through the soil. Estimates show that about 40% of the world's arable soils and perhaps as much as 70% of potential new lands that can be brought under cultivation are acid, and thus have an aluminum toxicity problem (World Food and Nutrition Study, 1977). In spite of the importance of the problem, work on intraspecific differences in aluminum tolerance has been started only recently and has so far been limited to screening of certain crop plants. Wide genetic diveristy in aluminum tolerance has been found among varieties of wheat, barley, and sunflower (Foy et al., 1967, 1974; Lafever et al., 1977). Interestingly, newer, high-yielding varieties of many species are generally sensitive and much of the tolerance is found in primitive varieties indigenous to acid soils. To date very little is known on the mode of inheritance or the physiological mechanisms of aluminum tolerance and studies of ecotypic differences in wild plant species occupying soils of different levels of soluble aluminum are notably lacking.

Metal Tolerance in Atriplex. No studies of ecotypic differences with regard to metal tolerance in *Atriplex* are known to the authors. However, the investigations by Rees and Sidrak (1956) on the colonization of *A. triangularis* in areas covered with "fly ash" from the combustion of coal in power stations are noteworthy in the present context. These fly ash deposits contain high levels of metals that are toxic to plant growth, especially aluminum manganese and boron, while the nitrogen level is very low.

Interestingly, the fly ash becomes covered with *Atriplex triangularis* var. *deltoida* which often forms an almost pure stand. Other common weeds which are prominent in the early plant succession on many other industrial waste materials appear to be almost entirely excluded from the ash. Controlled growth experiments with a number of species not growing in the ash and *Atriplex* seedlings, raised from seed collected from plants occupying such sites in nature, showed that the high aluminum and manganese content of the ash caused severe toxicity symptoms, and the low nitrogen content caused nitrogen deficiency symptoms, in all species except *Atriplex,* which grew vigorously on the ash. Like the other species, *Atriplex* accumulated high quantities of aluminum and manganese in the leaves but showed no toxicity symptoms. Neither did *Atriplex* show any symptoms of nitrogen deficiency. Unfortunately, no comparisons were made with *A. triangularis* plants from habitats outside the ash area so it is not possible to say whether the remarkably high tolerance is the result of ecotypic adaptation or whether it is a general characteristic of the species.

Salinity. This is probably by far the most important edaphic factor limiting the distribution of plants in natural habitats and it is also becoming an increasingly severe problem confronting agriculture in many parts of the world. Although it has been recognized since the beginning of this century that large differences in salt tolerance exist among species of different genera, it is only recently that attention has been given to intraspecific differences. Much of this work has been concerned with screening of varieties of agriculturally important plants for increased salt tolerance. Significant intraspecies differences in the response of dry matter yield to salinity level have been observed in several species (Dewey, 1960, 1962; Greenway, 1962; Tarimoto, 1969).

While these screening studies indicate that there exist considerable genetic variation in salt tolerance in agricultural species and that breeding for more salt-tolerant plants therefore has much promise, they do not provide any information on the relationship between salt tolerance and habitat factors. In a recent study, Rush and Epstein (1976) compared the salt tolerance of the cultivated tomato, *Lycopersicon esculentum* CV.VF36, with that of a wild tomato species, *L. cheesmanii,* from a saline coastal habitat on the Galapagos Islands. The two taxa show a high degree of genetic compatibility and can perhaps be regarded as belonging to the same species complex. Representatives of four populations of *L. cheesmanii* showed 40% to 90% survival when tested in nutrient solutions with full seawater salinity, although the variability among them may be indicative of ecotypic differences within this taxon. In contrast, cultivar VF36 was killed by this high salt treatment. Similarly, although all of the plants tested reached maximum growth rates in nonsaline solutions, cultivar VF36 and the montane populations of *L. cheesmanii* showed a considerably greater inhibition of growth in a 20% of full strength seawater solution than the coastal population. The tolerance of salinity in *L. cheesmanii* was correlated with a controlled accumulation of salt in the shoot and differences in K^+ and Na^+ discrimination, features which distinguish halophytes generally from other plants (Chap. 7).

The first genecological study on adaptation to salinity to natural populations was made by McMillan (1959). Examination of a *Typha* (cattail) population distributed in a saltflat near Lincoln, Nebraska, showed that clones of *T. angustifolia* occupied the drier (more saline) sites and those of the closely related species *T. latifolia* occupied the moister (less saline) sites. Clones with intermediate morphological characteristics, probably hybrid, were distributed with both these taxa. As the saltflat dried during subsequent years, many *T. latifolia* clones died or were much reduced in area of occupancy whereas the *T. angustifolia* clones remained vigorous. The intermediate clones increased their coverage primarily in areas that had previously been occupied by *T. latifolia.* This pattern indicated that the spatial adjustments within the population resulted from the selective action of increased salt concentration accompanying the drier conditions. Laboratory studies on clones obtained from the field site confirmed that large differences in salt tolerance existed within the population, with *T.angustifolia* possessing the greatest, *T.latifolia* the least, and the morphologically intermediate clones an intermediate salt tolerance. Similar but somewhat less pronounced differences in salt tolerance have also been reported by McNaughton (1966) for *T.domingensis* clones from habitats with different salinity levels and in populations of *Phragmites communis* (Waisel, 1972). In contrast to the results obtained by Rush and Epstein (1970) with *Lycopersicon*, only small differences were found between the *Phragmites* ecotypes in their abilities to take up sodium, potassium, and calcium, suggesting that other modes of adaptation to different salinity levels must operate in these plants.

Habitats occupied by *Atriplex* species are often characterized by moderate to high salinity. Chatterton and McKell (1969) investigated the responses of three Californian populations of *Atriplex polycarpa*, a common saltbush in the deserts of south-western North America. They found that all three populations were able to grow and survive without any visible toxicity symptoms in NaCl concentrations of 670 mM which is approximately equal to an osmotic potential of -27 bar. However, there were very marked differences in the response of growth to salt concentration among the populations. In one of them, total yield fell gradually with increasing salinity, reaching about 20% of the control (1 mM NaCl) at 670 mM NaCl. In another population, the yield at 670 mM NaCl still remained over 60% of the control; moreover, optimum yield was obtained not at the lowest NaCl concentration but at about 220 mM NaCl. In a third population, the relative yield was reduced to about 34% at the highest salt concentration (670 mM) but the absolute yield was two- to fourfold higher than in the other populations. While these results clearly demonstrate that large population differences in salinity response exist with *A.polycarpa*, it is unknown whether or not these differences are ecotypic since no information is available on the salinity levels of the native habits.

Conspicuous local intraspecific variation in the morphology of *Atriplex* is often found along salinity gradients. In Australia, morphologically distinct forms of *A.vesicaria* can be separated by as little as 10 m across boundaries between highly saline and nonsaline, calcareous soils. Osmond (unpublished) compared the growth of such forms under a range of calcium and magnesium chloride concentrations in the laboratory. He found that the distinguishing morphological traits, such as leaf shape and the development of bladders on the fruit remained constant irrespective of the salinity of the culture medium. The seedlings which were native to the saline soil showed a much greater tolerance to high chloride concentrations than those from the nonsaline soil. In the former there was no evidence of inhibition of growth at 160 mM Cl^- the highest concentration used, whereas the latter died at a concentration of 80 mM. These results provide strong evidence for the existence of ecotypic differentiation in the salinity tolerance of *A.vesicaria*. The mode of inheritance and the physiological mechanisms of salt tolerance are unknown, however.

Similar salinity gradients extend from the beaches of the highly saline Great Salt Lake of Utah to the surrounding non-halomorphic soils. *A.nuttallii* extends across this entire gradient ranging from the most saline community, dominated by

Table 3.4. Shoot dry weight, relative leaf growth rate, and nitrate reductase activity of *Atriplex nuttalli* plants, collected from zones with differing soil conductivity, and subsequently grown in a common environment[a]. (Data from Goodman and Caldwell, 1971)

Site	Soil conductivity (mmhos)	Shoot dry weight (mg/shoot)	Relative leaf growth rate	Nitrate reductase activity (mmol $NaNO_2$/g fr.w./h)
Atriplex-Curatoides Zone	1.4	10	0.39	29
Sarcobatus Zone	5.2	17	0.60	106
Sarcobatus-Suaeda Zone	96	21	0.60	173
Salicornia-Allenrolfea Zone	121	14	0.42	9
Standard error	—	3.2	0.05	24

[a]   All plants were grown in nutrient culture having a conductivity of 50 mmhos, equivalent to a salinity of 10 mM

Salicornia and *Allenrolfea,* to the least saline community, dominated by *Eurotia* and *Atriplex* species. Distinct morphological forms of *A.nuttallii* are observed at the extremes of this gradient. Goodman and Caldwell (1971) transplanted individuals of *A.nuttallii* along the gradient and grew them in the same greenhouse environment in nutrient culture with a 10 mM salt concentration approximately equal to the mean concentration of the gradient. On the basis of observed differences in growth rate and nitrate reductase activity, these authors concluded that three genetically distinct populations, i.e., ecotypes, of *A.nuttallii* existed along the salt gradient and that this may account for the wide salt tolerance of this species (Table 3.4). However, the differences in the three measured characteristics between the plants collected at the two extremes of the salinity gradients were quite small, whereas the differences between populations collected from adjacent zones were much larger despite the fact that the differences in salinity were only slight. Furthermore, nitrate reductase activity responds remarkably rapidly to the nitrate status of the leaf tissue (Chap. 7) and would seem to be an unreliable indication of metabolic differentiation between populations. In a subsequent study, Goodman (1973) grew three of these populations in sand-soil cultures with addition of three different salt concentrations: 0.001, 0.015, and 3.4 mM. According to the author the salt responses of the different *A.nuttallii* clones were correlated with the salinity of their natural habitats. However, the response difference was small and marginally significant statistically. This may not be surprising since it appears that even the highest salinity level used is very low and less than one-fifth of the salt concentration found in the more saline natural sites from which *A.nuttallii* was collected. It would certainly be very interesting to know how these different plants would respond if they were grown in a wider range of salt concentrations. Such experiments need to be done before a definite conclusion can be reached regarding the involvement of ecotypic differentiation in the salinity tolerance of *A.nuttallii*.

3.5 Concluding Remarks

In spite of the scarcity of investigations which at the same time provide sufficient information on pertinent habitat factors, genetic relationships, and functional characteristics, it seems safe to conclude that intraspecific variation resulting from genecological differentiation in *Atriplex* species is probably a common phenomenon. This must certainly be kept in mind in any attempts to characterize an entire species on the basis of experiments conducted on a single biotype. Future comparative studies in which the ecological, genetic, and physiological components are carefully measured and analyzed promise to provide a much deeper insight into the processes of speciation and plant adaptation to the great diversity of habitats that exist on earth. *Atriplex* should be one of the genera well suited for such studies, especially in regard to the mechanisms underlying adaptation to the stresses imposed by high temperatures, limited water supply, and high salinity.

Chapter 4. Genetic and Evolutionary Relationships in Atriplex

In this chapter we will review the available information on genetic relationships among various taxa of *Atriplex* as determined by cytogenetic methods, and the attempts to hybridize these taxa both at the intra- and inter-specific levels. Such studies are of utmost importance in understanding the evolutionary processes operating in speciation. In *Atriplex,* they also serve as a unique tool in elucidating the evolutionary and genetic aspects of an especially interesting example of adaptive differentiation, namely the C_4 pathway of photosynthesis which is present in certain taxa but absent in others.

4.1 Chromosome Numbers in Atriplex

Table 4.1 lists the chromosome numbers of all *Atriplex* species for which counts have been found in the literature. Only 11 species are listed in chromosome atlases. In terms of the classification used here, our list contains 56 species from Europe, Asia, North America, and Australia, representing a great diversity of ecological habitats. This number would increase to about 62 if some of the taxa that are here delegated to subspecific rank were raised to the species level as they usually are in the taxonomic literature. In all, the total number of populations on which these counts are based is in the order of one thousand.

With no exception, the basic chromosome number in the genus is $x = 9$. The great majority of species (about two-thirds) are diploid with $2n = 18$ chromosomes although tetraploidy ($2n = 36$) is not uncommon. Higher ploidy levels than tetraploid are found in less than 10% of the species. Four of these species are hexaploid ($2n = 54$) and one species is octaploid ($2n = 72$).

Polysomaty or endomitosis, i.e., the occurrence of large cells containing multiples of the normal chromosome number in the somatic tissue of a plant, is commonly encountered in the tips of primary roots of *Atriplex* and other members of the Chenopodiaceae (Witte, 1947; Hulme, 1957; Gustafsson, 1973a). Although it is sometimes possible to mistake polysomaty for polyploidy, it is clear that endomitosis is not responsible for the high ploidy levels given in Table 4.1, especially since most of these counts were also confirmed in meiotic divisions. One reported count of $2n = 36$ in root tips for *A. hastata* ($= A. triangularis$ Willd.) by Heiser and Whittaker (1948) has been excluded from the list, since it is probable that it was either made on endomitotic cells (Gustafsson, 1973a) or that the plant used

Table 4.1. Chromosome numbers in *Atriplex* (species name used in reference given in parentheses)

Species		Origin	Reference	Chromosome number	
				n	2n
A. acadiensis		Quebec	Taschereau, 1972		36
		Nova Scotia	Taschereau, 1972		36
A. alaskensis		Alaska	Bassett and Crompton, 1973		54
A. argentea		Saskatchewan	Frankton and Bassett, 1970	9	18
		Alberta	Bassett, 1970		18
A. barclayana	(*A. barclayana* ssp. *palmeri*)	Baja	Bassett, 1969		36
		California	Nobs, 1978	18	36
A. californica		California	Nobs, 1975	9	18
A. canescens		Alberta	Stutz et al., 1975	18	36
		Montana	"	18	36
		Wyoming	"	18	36
		Colorado	"	18	36
		Utah	"	18	36
		New Mexico	"	18	36
		Old Mexico	"	18	36
	(*gigas* form)	Utah sand dune	"	9	18
A. confertifolia		Nevada	Nobs, unpublished		18
A. coronata		California	Nobs, 1975	18	36
A. coulteri		California	Bassett, 1969		18
A. dioica		Saskatchewan	Frankton and Bassett, 1970		18
		Wyoming	Bassett, 1970	9	18
A. elegans	(*A. elegans* ssp. *fasciculata*)	California	Nobs, 1975	9	18
A. expansa	(*A. argentea* ssp. *expansa*)	California	Nobs, 1975	18	36
A. franktonii		Nova Scotia	Taschereau, 1972		18
A. fructiculosa		California	Nobs, 1975	9	18
A. glabriuscula	(*A. babingtonii*)	Europe	Wulf, 1936	9	18
	(*A. hastata-glabriuscula*)	Britain	Hulme, 1957	9	18
		Britain	Nobs, 1975	9	18
		Nova Scotia	Taschereau, 1972		18
		Sweden	Gustafsson, 1973a	9	18
		Iceland	Löve and Löve, 1956		18
A. gmelinii		Alaska	Bassett and Crompton, 1973	27	54
		Japan	Nobs, 1975	18	36

Table 4.1 (continued)

Species	name used	Origin	Reference	Chromosome Number	
				n	2n
A. halimus		S. Europe	De Castro and Fontes, 1946		18
		Israel	Nobs, unpublished		18
A. heterosperma		California	Nobs, 1975	18	36
		British Columbia	Frankton and Bassett, 1968	18	36
		British Columbia	Mulligan, 1965		36
		North Dakota	Frankton and Bassett, 1968		36
A. hortensis		Wisconsin	Frankton and Bassett, 1968		18
		California	Nobs, 1975	9	18
		Europe (?)	La Cour, 1931		18
A. hymenelytra		California	Bassett, 1969		18
A. inflata		Australia	Nobs, 1978	9	18
A. julacea		Baja California	Nobs, 1978	18	36
A. lentiformis		California	Nobs, 1978	9	18
	(A. lentiformis ssp. breweri)	California	Nobs, 1975	9	18
		California	Bassett, 1969		18
A. leucophylla		California	Nobs, 1978	18	36
A. littoralis		Britain	Hulme, 1957	9	18
		Holland	Van der Meijden, 1970	9	18
		Europe (?)	Pólya, 1948		18
	(A. litorale)	Europe	Winge, 1971	9	
		Nova Scotia	Taschereau, 1972		18
A. mohavensis		California	Nobs, 1978	18	36
		Nevada	Nobs, 1978	18	36
A. nitens		Europe (?)	Wulff, 1936		18
A. nummularia		Australia	Nobs, 1978	18	36
A. nuttallii	(var. *falcata*)	California	Nobs, 1978	9	18
A. pacifica		Baja California	Nobs, 1975	9	18
A. parryi		California	Nobs, 1975	9	18

Table 4.1 (continued)

Species		Origin	Reference	Chromosome Number	
				n	2n
A. patula		Europe	Winge, 1917	18	36
		Britain	Hulme, 1957	18	36
		Holland	Van der Meijden, 1970	18	36
		Ohio	Nobs, 1975	18	36
	(A. patula ssp. (?))	California	Nobs, 1975	18	36
		Nova Scotia	Taschereau, 1972		36
		Iceland	Löve and Löve, 1956		36
A. pentandra		Baja California	Nobs, 1978	9	18
A. phyllostegia		California	Nobs, 1975	9	18
A. polycarpa		Baja California	Nobs, 1978	18	36
A. powelli		Alberta	Frankton and Bassett, 1970		18
A. recurva		Greece	Gustafsson, 1970		18
A. rosea		California	Nobs, 1975	9	18
		Europe	Wulff, 1936		18
		British Columbia	Mulligan, 1957		18
A. sabulosa		Sweden	Nobs, 1975	9	18
		France	Nobs, 1975		
	(A. laciniata)	Nova Scotia	Taschereau, 1972		18
	(A. arenaria Woods)	Europe	Wulff, 1936		18
A. saccaria		Arizona	Nobs, 1978	9	18
A. semibaccata		Australia	Darlington and and Wylie, 1956		18
		Australia (?)	Wulff, 1936		18
A. serenana		California	Nobs, 1975	9	18
A. spongiosa		Australia	Bassett, 1969		18
A. stocksii		Pakistan	Bassett, 1969		36
A. suberecta		California introduced from Australia	Nobs, 1978	9	18
A. subcordata		Japan	Bassett and Crompton, 1973		54

Table 4.1 (continued)

Species		Origin	Reference	Chromosome Number	
				n	2n
A. subspicata		Quebec	Taschereau, 1972	27	54
			Bassett and Crompton, 1973		
		Nova Scotia	Taschereau, 1972	27	54
			Bassett and Crompton, 1973	27	54
		British Columbia	Bassett and Crompton, 1973	27	54
		Alberta	''	27	54
		Saskatchewan	''	27	54
		New Brunswick	''	27	54
		Prince Edward Island	''	27	54
		British Columbia	Crompton, 1973		36
		Alberta	Crompton, 1973		36
A. tatarica		Germany	Bassett, 1969		18
		Rumania	Tarnavschi, 1948		18
		Hungary (?)	Pólya, 1949		18
A. triangularis	(*A. longipes* ssp. *longipes*)	Sweden	Gustafsson, 1972	9	18
	''	Denmark	''	9	18
	(*A. longipes* ssp. *preacox*)	Sweden	Gustafsson, 1972	9	18
	''	Finland	''	9	18
	''	Norway	''	9	18
	(*A. calotheca*)	Germany (?)	Wulff, 1936		18
	''	Sweden	Gustafsson, 1973a	9	18
	''	Denmark	''	9	18
	(?)	Europe	Winge, 1917		18
	(*A. calotheca*)	Europe	Wulff, 1936		18
	(*A. hastata*)	Rumania (?)	Tarnavschi, 1948		18
	(*A. hastata-glabriuscula*)	Britain	Hulme, 1957		18
	(?)	Holland	Gadella and Kliphuis, 1966		18
	(*A. patula* ssp. *hastata*)	California	Nobs, 1975	9	18
		Finland	Gustafsson, 1973a	9	18
		Sweden	Gustafsson, 1973a	9	18

Table 4.1 (continued)

Species	Origin	Reference	Chromosome Number	
			n	2n
A. triangularis	Denmark	Gustafsson, 1973a	9	18
	Norway	Gustafsson, 1973a	9	18
	Nova Scotia	Taschereau, 1972		18
(A. hastata)	Holland	Van der Meijden, 1970	9	18
A. truncata	British Columbia	Frankton and Bassett, 1970		18
	Utah	"		18
	California	Nobs, 1975	9	18
	Nevada	"	9	18
A. vesicaria	Alice Springs Australia	Nobs, 1978	18	36
A. watsonii	Baja California	Nobs, 1978	18	36
A. wrightii	Arizona	Nobs, 1978	9	18

was a misidentified *A.patula* which is tetraploid. Two different counts are listed in Table 4.1 for *A.gmelinii*, with 2n = 54 and 2n = 36 for the plants of Alaskan and Japanese origin, respectively. However, these may well prove to be different species (Bassett and Crompton, 1973). These authors also reported two different chromosome counts for *A.subspicata*. Most populations of this species are hexaploid with 2n = 54, but two tetraploid populations with 2n = 36 chromosomes were also found. The only morphological difference observed between the hexaploid and the tetraploid populations in the field is that pollen size was somewhat larger in the hexaploid. In greenhouse-grown plants another distinguishing characteristic was observed: the leaves were directed sharply downward in tetraploids and were horizontal or ascending in hexaploids. Of the remaining species, only one, *A.canescens*, has more than one ploidy level; this interesting case will be discussed at some length below.

The relative constancy in chromosome number within a single species holds true even for those species which have a very wide geographical range. There appears to be no relationship between chromosome numbers among the different species and their geographical or ecological origin, nor the type of photosynthetic pathway (C_3 or C_4). Neither are such relationships to be expected on theoretical grounds. Higher ploidy levels may seem to have some association with weediness (e.g., *A.patula*, *A.expansa*, *A.heterosperma*, *A.subspicata*), but if so there are several exceptions and many diploid species are weedy (e.g., *A.rosea*, *A.semibaccata*, *A.serenana*).

There seems to exist, however, a clear relationship between chromosome number and breeding system. While the monoecious *Atriplex* species can be either diploid or polyploid, it appears that the strictly dioecious species are diploid (see the following paragraphs for the special situation in *A.canescens*). Such lack of polyploidy in dioecious species is commonly observed in higher plants and animals although exceptions are known, e.g., species of *Salix, Rumex,* and *Melandrium* (Stebbins, 1950). One reason for this is considered to be that if a tetraploid individual emerged from a dioecious diploid this would be a dead end evolutionarily, since being unisexual, the tetraploid would have to backcross with the diploid, and the progeny from such a cross would be triploid and hence completely sterile.

A particularly interesting case of different ploidy levels within an *Atriplex* species is found in *A.canescens.* This is probably the most widespread species of all perennial *Atriplex* in North America, extending from Alberta in Canada south to central Mexico and from the Pacific coast east to the Missouri River. Stutz et al. (1975) counted the chromosomes in 67 populations of this species, extending over the full distribution range (some of these are listed in Table 4.1). All populations were tetraploid with one single exception: a diploid *gigas* form, growing on the Little Sahara Sand Dunes in central Utah. A search for other diploid populations besides the one restricted to the Little Sahara Dune area (about 800 km²) was made in many other dune habitats of *A.canescens* in the south-western United States and even included the plants growing in sandy soils and on drifting dunes contiguous to the Little Sahara Dune site but these plants were all tetraploid. Just as the diploid is excluded from all areas outside this dune, so are the tetraploid plants excluded from the dune itself, although they do occur on its sandy periphery.

The diploid *gigas* form is distinct from the tetraploid forms in a number of characteristics, some of which are listed in Table 4.2. In nature, tetraploid forms of *A.canescens* usually reach about 1 m in height and diameter whereas the diploid *gigas* often grows to a height of 3 to 4 m and a spread of 4 to 5 m. Growth of the plants in the same environments confirmed that the *gigas* habit is genetically controlled. The germination rate and percentage of germination were very much higher in the diploid. Seedling growth rate was also much more rapid and this

Table 4.2. Characters distinguishing normal tetraploid forms and the *gigas* diploid form of *Atriplex canescens*. (After Stutz et al., 1975)

Character	Diploid	Tetraploid
1. Chromosome number	2n = 18	2n = 36
2. Stature	Giant	Normal
3. Seed germination	High	Low
4. Growth rate of seedlings	High	Low
5. Growth rate of new twigs	High	Low
6. Distribution	Restricted	Wide
7. Polysomaty	Common	Rare
8. Root sprouting	Common	Rare
9. Time of flowering	Late	Early
10. Pollen fertility	Higher	Lower
11. Pollen size	Dimorphic	Monomorphic
12. Sexuality	Strictly dioecious	Often monoecious
13. Meiotic chromosomes	Bivalents only	Multivalents
14. Fruit size	Larger	Smaller

continued into maturity. A most striking difference is the unusual disposition of the *gigas* plants for extensive elongation of growing tissues. New twigs grow as much as 100 cm in one season compared with 20 to 40 cm in normal plants and hypocotyl elongation is twice that of normal plants. The *gigas* character is also reflected in unusually large and heavy fruits. Flowering occurs 5 to 6 weeks later in the *gigas* than in the normal form, resulting in a complete isolation barrier in nature, thus preventing the formation of sterile triploid hybrids that would be formed if cross pollination between the diploid and tetraploid forms could occur. Such triploids also could not be found in the border areas. In addition to these and other distinguishing features, the *gigas* form is strictly dioecious with an almost exact 1:1 sex ratio, whereas the tetraploid forms can be either monoecious or dioecious.

Stutz et al. (1975) considered the characteristics of the *gigas* form as highly adaptive in their sand dune habitat. The plants are confined to the leeward side of the dunes. As dunes shift position, plants are often buried or unburied and left exposed. Under these conditions the plants must grow fast enough to match the rate of sand deposition, and the rapid growth habit of the *gigas* plants is therefore probably of great adaptive advantage. The deep roots and the rapid rate of root elongation keeps them anchored and in contact with moist soil at all times. Exceptionally high rates of root elongation are also required because of the low water-holding capacity of sandy soils. The high propensity of the *gigas* plants for forming adventitious roots and shoots permits them to root in any available strata.

According to Stutz et al. the main selective factor against the diploid outside the dune habitat is the much heavier grazing pressure. The remarkably luxuriant growth may make the *gigas* plants too attractive to grazing animals. Inside the dunes this is not a problem, since the lack of drinking water and the difficulty of walking in the loose drifting sand almost completely exclude grazing animals. One might perhaps also speculate that the rapid growth rate of the *gigas* form is likely to increase the demand for water and that this demand might not be met in the relatively flat areas outside the dunes. In the dune habitat, the steep pitch of sand deposits on the leeward side of the dunes, where the plants grow, greatly facilitates water percolation, providing a locally more favorable water supply and thus permitting the higher growth rates of the *gigas* form.

Stutz et al. favor the view that the normal tetraploids were derived as auto-tetraploids from diploid ancestors and that the diploid form, now found in the Little Sahara Dunes, is a relic of those ancestors although it may subsequently have evolved certain characteristics which make it uniquely adapted to the sand dune habitat. Other diploids presumably have been subject to a very rapid demise, probably within the past 5,000 years, and have been replaced by tetraploids. The alternative possibility that the *gigas* form is a polyhaploid (cf. Dewet, 1971), derived from tetraploid *A. canescens*, cannot be completely ruled out but the evidence speaks against this. The diploid has regular meiosis with nine bivalents forming in almost every cell. This is also accompanied by a very high pollen and seed fertility. Together with the unusual vigor these features are suggestive of a well-established ancestral diploid rather than a recent polyhaploid derivative. The tetraploids, on the other hand, show several quadrivalents during meiosis which is strong evidence for an autoploid origin. In addition, pollen fertility is reduced and seed fertility is much lower than in the diploids, probably reflecting incompatible combinations that one would expect to segregate from autoploid, but not diploid, parents. All of this evidence supports the view of an autopolyploid origin of the tetraploid plants and the diploid being an ancestral form. Moreover, the possibility that the diploid is a polyhaploid derivative appears highly improbable in view of the fact that it is strictly dioecious; a polyhaploid origin of the diploid would therefore require simultaneous repetitions of highly unlikely events.

For reasons given in the beginning of this section the origin of a highly successful tetraploid from a strictly dioecious species also appears very unusual. A possible explanation is given by Stutz et al. (1975). If the sex in diploid *A. canescens* is determined by X-Y chromosomes so that one sex is XX and the opposite sex is XY, then an autotetraploid derivative of an XY plant would be XXYY. This could give rise to plants that are both male and female, thus capable of self-fertilization and the production of fertile progeny.

Another possibility would be that the ancestral diploid plants were monoecious. In this connection, it is noteworthy that other *Atriplex* species which are normally dioecious in their native habitats may become monoecious when the environmental conditions are changed. For example, *A. hymenelytra,* a diploid perennial shrub native to the deserts of south-western North America, is dioecious when growing in its native habitats, such as Death Valley, California, where it flowers during the winter. When cloned individuals of these plants were grown in a cool-climate transplant garden at Bodega Head on the California coast they flowered during the summer instead and all of the plants were now monoecious (Björkman et al., 1974b). It is tempting to speculate that the ancestral *A. canescens* plants might have responded in a similar manner so that an environmental change induced a monoecious habit in normally dioecious plants and these gave rise to autotetraploid populations.

It is especially noteworthy in this connection that tetraploid *A. canescens* may be capable of changes in sex expression in response to environmental stress. McArthur (1977) conducted a study of the year-to-year floral trait constancy of a half-sib family of *A. canescens* transplanted into an experimental station in Utah. The seed that produced the family was collected from a single pistillate plant from New Mexico. This half-sib family showed a preponderance of pistillate plants with an approximate ratio of 56 pistillate: 34 staminate: 5 monoecious: 5 nonflowering individuals. This ratio was generally consistent for the years 1972, 1974, and 1975. However, a dramatic change occurred between September 1972 and September 1973, presumably in response to unusually unfavorable weather in the late fall and winter of 1972–1973. Of 372 pistillate individuals 28 changed to staminate, 58 to monoecious, 36 to nonflowering, and 29 were killed. Of 228 staminate individuals only one changed to pistillate, 11 to monoecious, 22 to nonflowering, and 14 were killed. Of 34 monoecious individuals 7 changed to pistillate, 10 to staminate, 5 to nonflowering; none was killed. Finally, of 31 nonflowering individuals, 4 changed to pistillate, 12 to staminate, 2 to monoecious, and 6 were killed. This resulted in ratio 38 pistillate: 37 staminate: 14 monoecious: 11 nonflowering. Thus the most striking change was a reduction in the number of pistillate plants. The ratios found in 1972 were essentially restored in 1974 and were sustained in 1975.

McArthur (1977) proposes that the XXXX state is invariably pistillate, the XXYY state invariably staminate, the XXXY state pistillate, staminate or monoecious, and the XYYY state not viable. Such a mechanism allows consistent interpretation of available data from the half-sib family and natural tetraploid populations.

It is certainly most remarkable if a doubling of the chromosome number alone is sufficient to explain most of the drastic differences in the whole battery of characteristics that distinguish the *gigas* diploid from tetraploid populations of *A. canescens*. It is of course well known that polyploidy often results in larger cell size; but it is probably unprecedented that a diploid of a species with a relatively low basic chromosome number not only possesses *gigas* characteristics in comparison with its autotetraploid but also shows profound differences in other respects. It would certainly be most interesting to know the extent to which tetraploids produced artificially by chromosome doubling of the *gigas* diploid would resemble the natural tetraploid forms of *A. canescens*.

Many *Atriplex* species are noted for their exceptionally high degree of polymorphism. In view of the situation in *A. canescens* one might ask whether variation in chromosome number may in part be responsible for this. *A. triangularis* and the very closely related taxa *A. glabriuscula*, *A. longipes,* and *A. calotheca* are especially notorious for being highly polymorphic in characters such as habit, height, and the size and shape of leaves and bracteoles. This morphological variation is often pronounced even within populations. In the course of his very extensive cytogenetic and hybridization studies of the *A. triangularis* complex, Gustafsson (1972, 1973a,b, 1974, 1976) must have determined the chromosome numbers of at least one thousand individuals from a great number of populations but without a single exception the chromosome number was $2n = 18$. It is thus eminently clear that variation in chromosome number plays no part in the polymorphism found in the *A. triangularis* complex. (The underlying causes will be discussed in the following section.) Fairly extensive chromosome counts are also available for populations of *A. patula,* and as mentioned above, also for 67 populations of *A. canescens*. With the notable exception of the *gigas* form of the latter species no variation in chromosome number could be detected within any of these polymorphic taxa.

4.2 Genetic Structure and Evolutionary Trends in the Atriplex triangularis Group

By far the most extensive and comprehensive investigations on the genetic structure in *Atriplex* have been made in the *A. triangularis* complex and closely related taxa. Because of their pertinence to evolution and speciation in *Atriplex* we will consider these studies in some detail.

The *A. triangularis* group is a polymorphic aggregate of taxa with worldwide distribution. In northern Europe the group is represented by five taxa which are variously recognized as species or subspecies. In addition to *A. triangularis*[1] (= *A. prostrata* Boucher ex DC.), *A. glabriuscula,* Edmonst., *A. calotheca* Rafn et

---

1 It seems probable that what is commonly named *A. hastata* L. in the literature in most instances is *A. triangularis* Willd. However, the typification of *A. hastata* L. by Taschereau (1972) indicates that it corresponds to *A. calotheca*

Fries, and *A.longipes* Drej. have usually been given specific rank. Two subspecies are distinguished within *A.longipes:* ssp. *longipes*, and ssp. *praecox* (Hülph.) Turess. The geographical distributions and habitats of these taxa are given in Table 4.3. *A.triangularis* itself is by far the most widely distributed, occurring in most continents and occupying habitats ranging from coastal strand to inland alkaline flats, including saline marshlands in the interior of certain deserts. *A.glabriuscula* is limited to coastal strands and marshes but has a wide geographical distribution, ranging from the coast of western Scandinavia to the north-east coast of North America. The remaining taxa have very limited distribution in certain parts of Scandinavia. *A.longipes* ssp. *praecox* is of special interest since it is evidently of Arctic origin while the other taxa have probably reached Scandinavia from the south-west (Gustafsson, 1976).

Prior to the cytogenetic and hybridization studies by Hulme (1957) it was thought by some authors that two additional taxa, *A.patula* and *A.littoralis*, might also belong to this species complex (Bentham and Hooker, 1924; Hall and Clements, 1923). *A.patula* has since been found to be tetraploid, whereas *A.littoralis* and all taxa in the *A.triangularis* group are diploid. Hulme (1957) reported success in producing many vigorous F_1 hybrids by crossing *A.patula* × *A.littoralis*, *A.patula* × *A.triangularis*, and *A.patula* × *A.glabriuscula*, and reciprocally. However, as expected, the F_1 hybrids were all triploid and largely sterile. On this ground Hulme considered *A.patula* a biologically distinct species from *A.littoralis* as well as from *A.triangularis* and *A.glabriuscula*. Hulme did not succeed in her attempts to obtain artificial hybrids between *A.littoralis* and *A.triangularis* and although Turesson (1925) reported such hybrids it now appears doubtful that the progeny obtained in his crossing experiments were of hybrid origin (Gustafsson, 1976). Further controlled crossing experiments are therefore needed before any conclusions regarding the genetic relationship of *A.littoralis* to the *A.triangularis* group can be made.

According to Gustafsson all of the taxa listed in Table 4.3 are annual, diploid, and monoecious with both male and female flowers; bisexual flowers have not been observed. With the exception of *A.longipes* the taxa show little protogyny or

Table 4.3. Distribution and habitats of members of the *Atriplex triangularis* group

Taxon	Distribution	Habitats
A. triangularis	Most continents; in Europe along all coasts but rare in the Arctic	Coastal strands and marshes; also inland alkaline or saline areas
A. glabriuscula	Western Europe including British Isles, west Scandinavia and Iceland; north-east North America	Coastal strands and marshes
A. calotheca	South-west Scandinavia; rare in the Baltic region	Mainly restricted to coastal marshes; rare on coastal strand
A. longipes ssp. *longipes*	South Scandinavia; coasts of North Sea and Baltic Sea	Exposed and protected coastal marshes
A. longipes ssp. *praecox*	Arctic Sea and Baltic Sea coasts	Seashore meadows and marshes

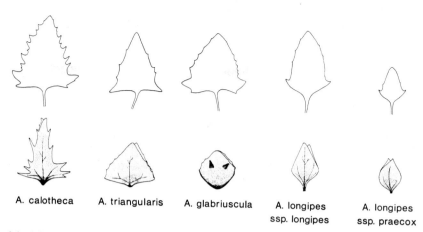

A. calotheca   A. triangularis   A. glabriuscula   A. longipes
                                                      ssp. longipes

A. longipes
ssp. praecox

Fig. 4.1. Morphological appearance of lower leaves and bracteoles of the taxa within the *Atriplex triangularis* complex. (Redrawn from Gustafsson, 1973a)

protandry and the degree of self-fertilization is high, probably even dominant. *A. longipes* is markedly protogynous, which favors outbreeding in this taxon. Pollen is wind-dispersed in all taxa. Leaf and bract shape characteristic of the five taxa are shown in Fig. 4.1.

The mechanisms that reduce or prevent gene exchange between taxa can be divided into two principal groups: prezygotic and zygotic (Dobzhansky, 1970). Prezygotic isolation is caused by external barriers to hybridization such as limited or nonexistent overlap in the geographical and ecological distribution or in flowering time and pollinating agents of the different populations. Zygotic isolation is caused by internal barriers which reduce the fertility or viability of hybrid zygotes.

4.2.1 Prezygotic Hybridization Barriers

In the *A. triangularis* group, external barriers to gene exchange between taxa are generally weak but vary considerably with region. The external isolating mechanisms between the taxa in different regions of Scandinavia are illustrated in Fig. 4.2. No real spatial isolation exists although the degree of overlapping between populations of *A. longipes* ssp. *praecox* versus *A. calotheca* and *A. glabriuscula* is small. In general, there is also little ecological isolation *between* taxa, although ecological differentiation is very common *within* most of them. Seasonal isolation must completely prevent hybridization between *A. longipes* ssp. *praecox* and the other taxa in the Baltic region but in northern Norway, this taxon and *A. triangularis* flower at the same time. Interestingly, the arctic form of *A. triangularis* probably represents a special ecotype characterized by a rapid rate of development. In western Scandinavia seasonal isolation between the early flowering *A. longipes* ssp. *longipes* and the other taxa is only partial as there is some overlapping of their flowering periods. No significant seasonal isolation exists

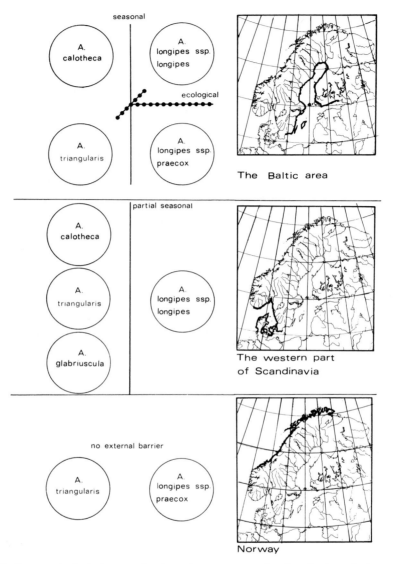

Fig. 4.2. Prezygotic isolating mechanisms between taxa in the *A. triangularis* group in different regions of Scandinavia. (From Gustafsson 1973b, with permission)

between the other taxa. *A.triangularis*, *A.glabriuscula*, and *A.calotheca* flower at the same time in areas where they occur sympatrically (Gustafsson, 1973b).

Hybrid derivatives between the different taxa of *Atriplex* commonly occur along the western and northern parts of Scandinavia. The frequency with which such hybrids occur is inversely correlated with the extent of partial seasonal isolation. The absence of external barriers evidently results in the formation of hybrid swarms, especially between *A. longipes* ssp. *praecox* and *A. triangularis* in the northern part of Scandinavia. Hybrids between *A.longipes* ssp. *longipes* and *A.triangularis, A.calotheca* and *A.glabriuscula,* are less commonly found in nature; this is consistent with the observation that flowering occurs earlier in *A.longipes* than in the other taxa. Such hybrids are even less frequent in the Baltic area where the seasonal isolation between *A.longipes* ssp. *longipes* and *A.calotheca* and *A.triangularis* is especially strong. No hybrids have been found between *A.calotheca* and *A.longipes* ssp. *praecox* in the Baltic area. The former taxon flowers late in the season and inhabits marshes with tall vegetation, whereas the latter flowers early and is restricted to open habitats in the lower parts of seashore meadows. Evidently, in this case, the combination of very strong seasonal and ecological isolation effectively prevents any hybridization.

That ecological isolation alone is not quite sufficient to prevent hybridization is indicated by the formation of hybrids between *A.longipes* ssp. *longipes* and ssp. *praecox.* The former subspecies occurs in the same habitats as *A.calotheca* but its flowering time overlaps with that of ssp. *praecox.* Conversely, even though seasonal isolation is usually absolute between *A.calotheca* and *A.longipes* in the Baltic region, hybridization can still occur under unusual circumstances in the absence of ecological isolation. Exceptional environmental stresses can cause strong relative shifts in the flowering time so that the reproductive periods of the two species will overlap. For example, Gustafsson observed instances in which grazing of flowering shoots of *A.longipes* resulted in a later reflowering. Presumably, this displacement was sufficient to allow the occasional formation of hybrids between *A.longipes* ssp. *longipes* and *A.calotheca.*

4.2.2 Zygotic Hybridization Barriers

The correlation between prezygotic isolating mechanisms and the natural occurrence of hybrids between the various taxa strongly suggests that zygotic barriers to hybridization are generally weak in the *A.triangularis* group. This is strongly supported by experimental evidence obtained by Gustafsson (1973a) in his extensive crossing and cytological studies. The controlled crossing experiments included all combinations of the taxa listed in Table 4.3, and also represented most of the different geographical regions in which the taxa occur in Scandinavia. All taxa were found to cross easily giving rise to vigorous, vegetatively well-developed F_1 hybrids, some of which displayed heterosis in comparison with their parents. Development of somewhat deformed leaves was occasionally observed in crosses *within* taxa but the plants were normal in other respects. Chromosome counts made on a large number of hybrid individuals gave $2n = 18$ with no exception. Seed with deviating chromosome numbers are thus either not formed or they fail to germinate.

Figure 4.3 summarizes the results of Gustafsson's studies on the male fertility in first-generation hybrids in crosses within and between the taxa of the *A.triangularis* complex in Scandinavia. With the exception of *A.longipes* ssp. *praecox* all taxa form viable F_1 hybrids with very high male fertility and good seedset. Most F_2 hybrids are also vigorous and well developed and in crosses *between* taxa many of the F_2 hybrids show increased fertility compared with the F_1 hybrids. There was no

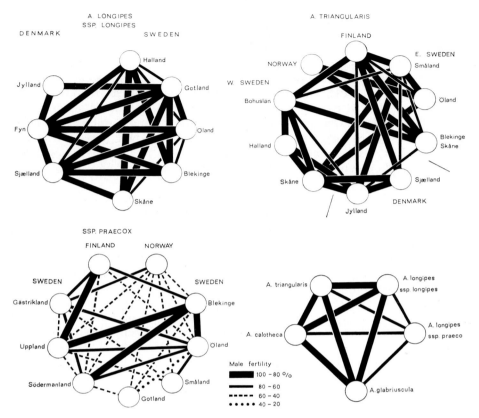

Fig. 4.3. Crossing polygons depicting the mean male fertility values of F₁ hybrids in crosses within and between different taxa of the *A. triangularis* complex from different regions of Scandinavia. The polygon of crosses between taxa *(lower right)* is based on the mean fertility values of all crosses carried out in each combination. (From Gustafsson, 1973a, with permission)

general correlation between fertility and the geographical distance between the parent populations in crosses within or between taxa. However, in certain instances where two taxa occur sympatrically and prezygotic barriers to hybridization are weak, the frequency of hybrids with reduced fertility is greater than in crosses between populations of allopatric origin. These results are in accordance with the view that maintenance of closely related sympatric species in the absence of strong prezygotic barriers requires that internal barriers develop instead (Stebbins, 1950; Dobzhansky, 1970). It thus appears that the reduced fertility in certain hybrids between sympatric taxa in the *A. triangularis* complex reflects selection for reproductive isolation (cf. Grant, 1966).

The strongest zygotic crossing barriers have evolved between *A. longipes* ssp. *praecox* and the other taxa; only rarely are any significant zygotic barriers found between *A. glabriuscula, A. longpipes* ssp. *longipes, A. triangularis,* and *A. calotheca.* Gustafsson's (1972, 1973a) cytological studies demonstrate that reduced fertility in

F_1 and F_2 hybrids results from the presence of bridges and paracentric inversions at anaphase I. He interprets these to result from heterozygosity for paracentric inversions in the chromosomes, involving a few small inversions rather than a single large one. The principal evidence on which these conclusions are based is as follows. Chromosomal pairing in crosses between, as well as within, taxa is usually quite normal even in hybrids that have low fertility; nine ring bivalents are usually formed. Configurations comprising more than two chromosomes have never been detected. In most of those hybrids that show meiotic irregularities and reduced fertilities, the chromosomal disturbances are exclusively of two kinds: formation of bridges and laggard univalents at anaphase I, presumably caused by a single crossover within the inverted segment, but in four crosses involving *A. longipes* ssp. *praecox* two bridges have been observed at the same time. However, in all such hybrids the bridges always occur in different chromosome pairs, indicating that the inversion differences involve separate chromosomes. No other anaphase configurations suggestive of other crossover types have been detected. The paracentric inversions in different crosses are probably dissimilar since the crossover frequency within the inverted segments varies greatly. The frequency of meiotic disturbances in various crosses shows a very high correlation with fertility.

It is important that reduction in fertility is at least as common in hybrids between different populations of the same taxon as in hybrids between different taxa. The highest frequency of meiotic irregularities is in fact found in population crosses within *A. longipes* ssp. *praecox* (cf. Fig. 4.3). *This supports the view that reduced fertility in certain crosses between taxa is not caused by insufficient genetic relatedness between them but is rather a consequence of cytological barriers that have evolved to permit a separation between specially adapted sympatric populations* (cf. Chap. 3.4.4, heavy metal toxicity).

4.2.3 Polymorphism

As previously mentioned, the *A. triangularis* complex is noted for its great morphological variation. Gustafsson (1973a, 1976) has conducted a comprehensive study on the degree of morphological variation and its causes within each taxon of this species complex. The variation was examined at five levels: within single individuals, within populations, between populations, regional variation, and total variation within each taxon. Moreover, cultivation experiments enabled him to assess the extent to which the variation observed in nature was genotypic. Typical morphological appearances of lower leaves and bracteoles of the different taxa are shown in Fig. 4.1.

Intrapopulation variation in morphological characteristics such as erectness vs. prostrateness, base angle of lower leaves, shape of upper leaves, dentation of leaf margins, bracteole shape and dentation, is most conspicuous in *A. triangularis,* and least pronounced in *A. longipes* ssp. *praecox.* In the other taxa the intrapopulation variation is wide in some populations and narrow in others. In *A. glabriuscula* the intrapopulation variation is more or less continuous. Differences between populations generally are relatively small in all taxa and the overlapping in variation is extensive, particularly in *A. triangularis, A. calotheca,* and *A. longipes* ssp. *longipes.* In the other taxa interpopulation differences are present between some populations but not between others. For example, *A.glabriuscula* populations inhabiting exposed locations are prostrate to ascending, whereas those growing in sheltered habitats, typically dominated by *Scirpus maritimus,* are tall and ascending to erect. In mixed habitats, such as sandy shores with marshy parts, the populations are highly variable and erect, prostrate, and ascending types are represented.

Regional differences in morphology are small or absent in all taxa. The variation patterns tend to be similar in all regions and the range of variation of the populations investigated corresponds fairly well with the total variation observed in each of the taxa. Moreover, many morphological characters show extensive overlapping between taxa. In many instances introgression is probably a very important underlying factor leading to a transfer of morphological characters from one taxon to another. Introgression occurs between most taxa, being most frequent between *A. triangularis* and *A. longipes.* For example, the wide variation in the shape and margins of leaves and bracteoles in *A.calotheca* is at least partly due to introgression of genes from *A. triangularis.* In certain populations individuals are found that have all characters in common with *A.calotheca* except for stalked bracteoles, a trait which is probably caused by introgression of genes from *A.longipes* ssp. *longipes.*

In order to determine the extent to which morphological differences observed in nature are genetically fixed and the extent to which they are environmentally induced modifications, Gustafsson (1976) performed cultivation experiments on all taxa in the *A. triangularis* group but most of his work focused on *A. triangularis,* the most polymorphic of the taxa. Characteristics such as leaf and bracteole size, lamina length, and plant height are mostly modifiable although there seems to be some genotypic variation. In contrast, more qualitative traits such as leaf and bracteole shape are largely genotypically fixed. Growth habit can evidently be either genotypically fixed or modifiable by the environment.

Two different prostrate forms of *A. triangularis* were observed by Turesson (1919). One form is erect in moderate light but becomes prostrate in intense light, whereas the other form remains prostrate regardless of the light intensity under which it grows. These early observations are in good agreement with the results of Gustafsson's studies. Progenies from the obligate prostrate form mostly retained the prostrate habit in cultivation. About 73% of the plants were prostrate and 20% ascending. Those individuals which developed an ascending habit may be derived from crosses in which the maternal plant is prostrate and the paternal plant is of the ascending type. About 85% of the progeny from facultative prostrate plants developed an ascending habit in cultivation, while only a few were prostrate or erect. In the offspring of parental types with an ascending habit in nature, 65% were ascending, all the others were erect in cultivation. The genetically fixed prostrate form is entirely restricted to extremely exposed seashore sites whereas in more sheltered areas the plants have an ascending to erect habit, indicating a genecological differentiation. However, all morphological types can be found together within a single population and in many populations the opportunity for gene exchange between these types is very good. There are no prezygotic or zygotic barriers to such gene exchange. Hybridization experiments showed that crosses between different morphological types are as fertile as crosses within either type.

Interestingly, there is no correlation between morphological and cytological differentiation. In *A. triangularis,* numerous different morphological types have evolved but the cytological variation is slight. The same is true of *A. longipes* ssp. *longipes.* Conversely, evolution within *A. longipes* ssp. *praecox* has led to marked cytological differentiation between populations, mainly in the form of paracentric inversion, yet morphological differentiation is slight in this taxon. *This indicates that chromosomal and morphological differentiation have evolved independently and are governed by different evolutionary factors.*

It thus appears that chromosomal differentiation is not an underlying factor in the wide morphological variation found in the *A. triangularis* taxon but that the polymorphism is primarily attributable to differentiation at the genic level and in some characteristics, also to environmental modifications. Introgression between the different taxa is a contributing factor to the morphological diversity in this group.

4.2.4 Evolutionary Trends

According to Gustafsson (1976) the extent of differentiation in the *A. triangularis* group in Scandinavia apparently depends in large degree on evolutionary events that occurred before and during the last glaciation some 8,000 to 10,000 years ago. With the exception of *A. longipes* ssp. *praecox,* the *Atriplex* taxa have probably migrated to Scandinavia from the south-east. *A. triangularis* may have reached Scandinavia at different times and along somewhat different routes. Possibly, some populations may even have survived the glacial period in ice-free areas along the arctic coast. *A. longipes* ssp. *praecox* appears to be of arctic origin. The present distribution of this taxon is disjunct: the arctic coasts of northern Norway and the U.S.S.R., Iceland, and the Baltic coasts of Sweden, Finland, and Estonia. There is strong evidence that the Baltic populations are also of arctic origin. Presumably, ssp. *praecox* survived the glacial period in small ice-free areas along the coast of Norway and migrated to what is now the Baltic area during the early post-glacial period and survived the fresh water lake stage. The chromosomal differentiation in ssp. *praecox* probably evolved during the post-glacial period. This rate of evolution is remarkably rapid, particularly in view of the observation that crossing barriers build up quite slowly in the genus. As was mentioned earlier, selection for reproductive isolation between this taxon and ssp. *longipes* in the Baltic region has probably been, and may still be, an important factor in the rapid evolution of chromosomal differentiation. Genetic drift is probably another important factor.

The other taxa of the *A. triangularis* group show a much slower rate and degree of differentiation. In these taxa the rate of evolution is evidently primarily governed by selection at the genic level. Nevertheless, crosses between populations of *A. triangularis* which have been subject to introgression from other taxa have a frequency of reduced fertility which is about three times higher than in crosses between nonintrogressive populations. The frequency in interpopulational chromosomal differences also appears to be higher in introgressive populations. The establishment of structural chromosomal changes is probably counteracted by strong karyotypic selection in introgressive populations.

The divergence between the taxa in the *A.prostrata* group is evidently strongly counteracted by spontaneous hybridization and introgression. With the exception of ssp. *praecox* populations in the Baltic area isolating mechanisms are poorly developed in all Scandinavian regions and as a result a steady gene flow occurs between most of the taxa.

4.3 Interspecific Relationships and Distribution of C_4 Photosynthesis in Atriplex

As was mentioned in Chap.2, *Atriplex* provides an excellent opportunity for studies of the evolutionary and genetic aspects of a particularly important and interesting functional characteristic, namely the C_4 dicarboxylic acid pathway of photosynthesis. For this reason considerable work has been devoted to comparative ecological, physiological, anatomical and genetic studies on C_3 and C_4 species of *Atriplex*. Soon after the discovery of the C_4 photosynthetic pathway it was found that, contrary to early expectations, this pathway does not represent a deep evolutionary split in the plant kingdom but that its origin is polyphyletic and both C_3 and C_4 plants can be present even within a single genus. *Atriplex* was one of the very first genera in which such a split was recognized (Björkman et al., 1970; Downton et al., 1969; Osmond et al., 1969b) and it was soon found that certain C_3 and C_4 species of *Atriplex* can be sufficiently closely related to permit hybridization between them (Björkman et al., 1970).

4.3.1 C_4 Photosynthesis

The C_4 pathway of photosynthesis is now known to occur in at least 15 diverse families of the more highly evolved orders of angiosperms (Table 4.4). Moreover, at least 16 different genera, four of which are in the Chenopodiaceae, are known to contain both C_3 and C_4 species (Table 4.5). For many good reasons C_4 photosynthesis is generally considered to have evolved as a special adaptation to certain selective factors operating in particular environments. The adaptive significance as well as the biochemical mechanism and physiological aspects of C_4 photosynthesis will be discussed at length in Chap.9. For the present purpose it is sufficient to point out a few features of C_4 photosynthesis.

The C_4 pathway is an addendum to, not a replacement for, the conventional C_3 pathway; and C_4 plants evolved from C_3 progenitors although it is possible that, in rare instances, reversion has occurred. Two sequential CO_2 fixation events occur in C_4 plants by two linked cycles whereas only one of these cycles is present in C_3 plants. The first is the fixation of atmospheric CO_2 by the C_4 cycle to C_4 dicarboxylic acids. The special enzyme requirement for the operation of the C_4 cycle includes very high activities of PEP carboxylase. This and certain other C_4 cycle enzymes are located in special photosynthetic chloroplast-containing cells here called PCA (primary carbon assimilation) cells. The C_4 acids formed in these PCA cells are then transported to another type of specialized chloroplast-containing cells, here called PCR (photosynthetic carbon reduction) cells (Hattersley et al., 1977). These PCR

Table 4.4. Families with C_4 photosynthesis

Acanthaceae	
Aizoaceae	Cyperaceae
Amaranthaceae	Euphorbiaceae
Boraginaceae	Gramineae
Capparidaceae	Nyctaginaceae
Caryophyllaceae	Portulacaceae
Chenopodiaceae	Scrophulariaceae
Compositae	Zygophyllaceae

Table 4.5. Genera known to contain both C_3 and C_4 species

Cyperaceae	Boraginaceae	Euphorbiaceae
Cyperus	*Heliotropium*	*Euphorbia*
Scirpus		
	Chenopodiaceae	Nyctaginaceae
Gramineae	*Atriplex*	*Boerhaavia*
	Bassia	
Alloteropsis	*Kochia*	Zygophyllaceae
Panicum	*Suaeda*	*Kalistromia*
		Zygophyllum
Aizoaceae	Compositae	
Mollugo	*Flaveria*	
Amaranthaceae		
Alternanthera		

cells contain all the components of the conventional Calvin-Benson (C_3) cycle with its CO_2-fixing enzyme, RuP_2 carboxylase, in addition to the enzymes needed to release CO_2 from the C_4 acids. This internally released CO_2 is refixed by the C_3 cycle. In C_3 plants atmospheric CO_2 is fixed directly by the C_3 cycle and all photosynthetic cells are of the same kind. The first CO_2 fixation cycle in C_4 photosynthesis acts as a very efficient trap for CO_2 and serves to increase the CO_2 concentration at the site of CO_2 fixation by RuP_2 carboxylase in the PCR cells. This enables the C_4 plant to utilize limiting CO_2 concentrations in the intercellular air spaces more efficiently than the C_3 plant.

The cellular compartmentation provided by the two different types of photosynthetic cells is evidently required to prevent short-circuiting of the two sequential CO_2 fixation cycles. In C_4 plants possessing classical "Kranz" anatomy, which is the case in all C_4 species of *Atriplex*, the PCA (or "mesophyll") cells form a concentric cylinder around the PCR (or "bundle-sheath") cells which in turn form a concentric cylinder around the vascular bundle. Cross-sections of such cells are shown in Fig. 4.4 (and also in Fig. 2.8).

It may be recalled from Chap. 2 that in *Atriplex*, well over one-half of the species possess C_4 photosynthesis and most, but by no means all, of these occur in semi-arid or arid environments in many parts of the world. With one possible exception, all *Atriplex* indigenous to the Australian continent have C_4 photosynthesis, whereas in

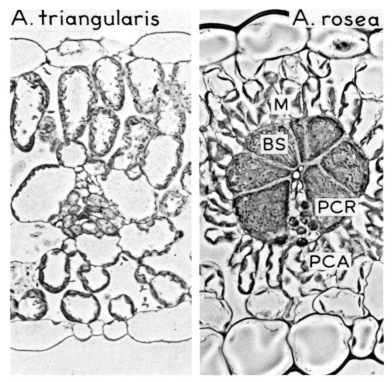

Fig. 4.4. Phase contrast micrographs of cross sections of leaves of C$_3$, *Atriplex triangularis* and C$_4$, *A.rosea*. *BS* bundle-sheath (or *PCR*) cells. *M* mesophyll (or *PCA*) cells. (Micrographs by courtesy of Dr. John Boynton)

North America, Europe, and Asia both C$_3$ and C$_4$ plants are common in certain regions. C$_4$ species are extremely rare at latitudes higher than 50° N. No member of the *A.triangularis* group or any other species native to Scandinavia has C$_4$ photosynthesis, although *A.sabulosa* has succeeded in invading the Scandinavian west coast in this century.

4.3.2 A Phylogenetic Scheme for Atriplex

The taxonomic distribution of C$_4$ species in *Atriplex* is a very difficult problem and depends to a large extent on the taxonomic scheme one follows. It can be said with some certainty that C$_3$ and C$_4$ plants do not occur within the same species complex or even in the same section, regardless of the scheme used. However, if one groups the species according to the scheme of Hall and Clements (1923) which divides the genus into two separate subgenera, one finds that both C$_3$ and C$_4$ plants are represented in both subgenera. Hall and Clements' chart (Fig.4.5), which attempts a classification according to phylogenetic relationships, is based primarily on the position of the embryo in the seed. The development of dioecism and shrubby habit are used as further important distinguishing characters. "Kranz"

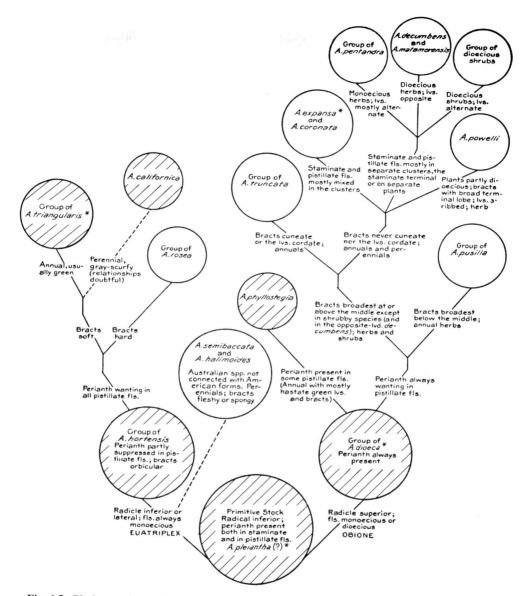

Fig. 4.5. Phylogenetic relationships among *Atriplex* species according to Hall and Clements (1923). Certain modifications of the original scheme have been made here. The names of some species (indicated by an *asterisk* after the name) have been changed to conform to the nomenclature used in the present monograph. *Shaded circles* indicate C_3 species, *open circles* indicate C_4 species. *A.halimoides* is now regarded as a synonym for *A.lindleyi* (Table 2.2)

anatomy, which we now know to be directly and causally linked to C_4 photosynthesis and which would therefore introduce a drastic bias and render our comparison useless, was not included as a character in constructing the scheme. According to this scheme the two subgenera, *Euatriplex* and *Obione*, are derived from a common primitive stock in which the essential organ of the pistillate flower was subtended by some sort of perianth. The inferior position of the radicle is so common in the Chenopodiales that Hall and Clements considered any modification of this as indicative of a divergent phylogenetic line. The absence of a perianth was also looked upon as a case of suppression and thus a mark of advance.

On the basis of these criteria Hall and Clements found it impossible to select any one *Atriplex* species as a potentially common ancestor. However, Weber (1950) reported the discovery of a new *Atriplex* species, *A.pleiantha,* which embodies the characters of the primitive stock postulated by Hall and Clements. This species, a herbaceous desert annual, native to a barren clay mesa slope in south-western Colorado, has an inferior radicle; a perianth is present both in staminate and pistillate flowers, and the flowers of each sex are mixed together in small axillary clusters. The characters of *A.pleiantha* suggest a close relationship with the *A.hortensis* group on the one hand and with the *A.dioeca* group on the other.

Of the species known to Hall and Clements, only *A.hortensis* and the closely related *A.acuminata* (*nitens*) have an inferior radicle and a true perianth in the pistillate flowers; and were placed at the beginning of the subgenus *Euatriplex.* This subgenus thus becomes the branch in which the radicle is always inferior and the perianth either present or absent in the pistillate flowers. No dioecious species are present in *Euatriplex.* The section *Teutliopsis* (all C_3), which includes the *A.triangularis* complex *(A.glabriuscula, A.calotheca, A.longipes, A.triangularis)* and *A.patula, A.littoralis,* and *A.spicata,* has a perianth only in the staminate flowers and has developed distinctive features in the bracts. On these grounds, Hall and Clements placed the *A.triangularis* group next to *A.hortensis* but did not consider it to be a direct derivative. *A.californica* (C_3) was placed near to *A.triangularis* (C_3) in the chart, but its connections are probably much more primitive and its position very doubtful. *A.rosea,* together with the annuals *A.tartarica* and *A.sabulosa* and the perennials *A.recurva* and *A.halimus* form the section Sclerocalymma (all C_4). Hall and Clements considered the group of annual Sclerocalymmas to be close to the *A.triangularis* group, with particularly close connections between *A.sabulosa* (C_4) and *A.glabriuscula* (C_3)[1]. *A.rosea* (C_4) was considered to have an intermediate position between *A.sabulosa* (C_4) and *A.tartarica* (C_4). However, in certain characteristics such as the degree of induration of the fruiting bracts the *A.rosea* group resembles some forms of *A.expansa* (C_4) and other members of subgenus *Obione.*

Subgenus *Obione* is characterized by an inverted embryo so that the radicle is superior and a much reduced perianth is found only in a few primitive species. According to Hall and Clements, the most primitive American species of *Obione* are *A.dioeca* and *A.monolifera* (C_3). A much reduced but evident perianth is present in

---

1   According to Gustafsson (pers. communication to O.B.) *A.sabulosa* is not more closely related to *A.glabriuscula* than the other annual species of Sect. Sclerocalymma. The close connection that Hall and Clements (1923) and later also other authors have considered to exist between *A.sabulosa* and *A.glabriuscula* is mainly based on a certain similarity in the bracteoles but this similarity is analogous, not homologous

the pistillate flowers of these species. This is also true of *A.phyllostegia* (C₃) but in this species the perianth sometimes is completely suppressed as it is in all of the succeeding species in the subgenus *Obione*. In their further phylogenetic arrangement of species in *Obione* the evolution of dioecism and perennial shrubby habit were used as the main criteria. In *Obione*, a tendency toward a separation of the sexes is noted at a number of places but complete dioecism is attained only in the truly shrubby species (*A.barclayana, A.acanthocarpa, A.obovata, A.nuttallii, A.corrugata, A.polycarpa, A.julacea, A.hymenelytra, A.lentiformis, A.confertifolia, A.spinifera, A.parryi,* and *A.canescens*) and in two herbaceous species (*A.decumbens* and *A.matamorensis*). All of these dioecious species possess C₄ photosynthesis. A transition between the group of monoecious herbs and dioecious shrubs is found in the *A.pentandra* group (*A.leucophylla, A.pentandra, A.elegans, A.microcarpa, A.wrightii, A.serenana, A.linifolia, A.fruticulosa,* and *A.coulteri*). All of these are C₄ plants and this is also true of the remaining groups of *Obione*. These are the group of *A. pusilla* (*A.cordulata, A.tularensis, A.pusilla, A.tenuissima,* and *A.parishi*), the group of *A.truncata* (*A.wolfii, A.gracifolia, A.saccaria, A.truncata*), the *A.expansa* and *A.coronata* pair, and *A.powelli.*

Finally, Hall and Clements did not consider the introduced Australian species *A.semibaccata* and *A.lindleyi (halimoides)* to be connected with American or Eurasian forms. These species have highly spongious bracts which are not found in American species but are characteristic of several other Australian species, notably *A.holocarpa, A.vesicaria* and *A.spongiosa*. All of these species have C₄ photosynthesis. Because of its lateral position of the radicle and the nearly distinct fruiting bracts, these authors considered *A.semibaccata* one of the more primitive forms.

It is evident that according to Hall and Clements' phylogenetic chart all of the more primitive forms of both *Euatriplex* and *Obione* are C₃ plants. The more highly evolved forms in *Obione* are all C₄ plants, whereas in *Euatriplex* they may be either C₃ or C₄ plants. If these phylogenetic relationships are correct, then one would arrive at the conclusion that not only have the C₄ *Atriplex* species evolved from C₃ progenitors, but also that the origin of C₄ species is polyphyletic even within this single genus. In order to cast further light on this question, investigations by the Carnegie group have been initiated along two different lines. The first involves classical cytogenetic and crossing studies between different C₃ and C₄ taxa; the second attempts to establish phylogenetic relationships by the modern approach of comparing the sequences of single copy DNA in certain C₃ and C₄ species of *Atriplex*. Both types of study also serve to aid in the search for C₃ and C₄ species with sufficient genetic compatibility to permit the production of highly fertile segregating diploid populations in the second and subsequent generations. This would allow more detailed and precise analyses of the genetic aspects of C₄ photosynthesis than have been possible so far.

4.3.3 Inter-Specific Crossing Experiments in Atriplex

The first artificial hybrid ever to be produced between a C₃ and C₄ species was obtained in 1968 between *A.rosea* (C₄) and *A.triangularis* (C₃) (Björkman et al., 1970), demonstrating that C₃ and C₄ plants can be relatively closely related. Much

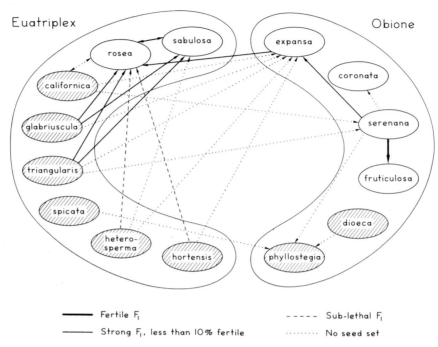

Fig. 4.6. Summary of attempted crosses between *Atriplex* species. *Open ovals* indicate C₄ species, *shaded ovals* indicate C₃ species. *No seeds set* means that the cross failed completely. (From Nobs, 1976, with permission)

additional important information was obtained from subsequent studies of the progeny of this cross and these experiments will be discussed in detail in following pages. During the ensuing years, 22 additional interspecific hybridizations have been attempted in *Atriplex* (Nobs, 1976). The crossing diagram (Fig. 4.6) summarizes the results of these experiments. The attempted hybridization includes crosses in a number of combinations: $C_3 \times C_4$, $C_3 \times C_3$, $C_4 \times C_4$, within, as well as between, species of either subgenus. Within *Euatriplex* eight out of nine attempted crosses yielded hybrid seed and six of these combinations gave rise to strong F_1 plants while two were sublethal. Seven of the nine attempted crosses in this subgenus were $C_4 \times C_3$, one $C_4 \times C_4$, and one $C_3 \times C_3$. Of the two crosses that failed completely in *Euatriplex* one involved *A.californica,* the phylogenetic position of which is doubtful according to Hall and Clements. The other attempted cross that failed was between the tetraploid *A.heterosperma* and the diploid *A.sabulosa.* Similarly, the two crosses that yielded sublethal F_1 hybrids involved *A.heterosperma* and the closely related diploid species, *A.hortensis* which according to Hall and Clements is fairly far removed from the *A.triangularis* and the *A.rosea* groups. It thus appears that the results of these crossing experiments within *Euatriplex* are in substantial agreement with Hall and Clements' scheme.

    To date, only five crosses have been attempted within *Obione* and only one of these was a $C_3 \times C_4$ cross. Two of the $C_4 \times C_4$ crosses were successful and one was unsuccessful. Both the $C_3 \times C_3$ and the $C_3 \times C_4$ cross failed. The fact that *A.fruticulosa* $(C_4) \times$ *A.serenana* (C_4) yielded a strong fertile F_1 hybrid, having regular meiosis with nine bivalents and 98% normal pollen, confirms Hall and Clements' suggestion that these species are closely related. The other successful hybrid combination within *Obione* is *A.serenana* $(C_4) \times$ *A.expansa* (C_4). The former is diploid, the latter tetraploid, and the F_1 is triploid. In the hybrid nine pairs of chromosomes (probably formed by autopairing of the *A.expansa* parent) and nine univalents are formed at metaphase I of meiosis. A close relationship between the two species is also indicated in Hall and Clements' scheme. Both of these $C_4 \times C_4$ hybrids appear to posses normal "Kranz" anatomy and properly functional C_4 photosynthesis. The attempted cross between *A.phyllostegia* (C_3) and *A.dioeca* (C_3) yielded aborted embryos but no hybrid seed matured. This is surprising since these species closely resemble one another in several key characteristics. Both are diploid, they have very similar inflorescences with a rudimentary perianth present in the female flowers, and succulent leaves with conventional C_3 anatomy. Hall and Clements even considered them sufficiently closely related that they might be united into a single species. It nevertheless appears that they have diverged considerably from a genetic standpoint.

Nine crosses were attempted between the two subgenera. Of these, eight failed to produce seed. However, one cross, *A.rosea* (diploid C_4) × *A.expansa* (tetraploid C_4) yielded strong hybrids. Like the *A.serenana* × *A.expansa* hybrid the *A.rosea* × *A.expansa* hybrid was triploid ($2n = 27$) with nine chromosome pairs and nine univalent at metaphase. Also this F_1 hybrid possesses normal "Kranz" anatomy and functional C_4 photosynthesis. The success of *A.rosea* × *A.expansa* cross suggests that the Eurasian *A.rosea* group of C_4 species is more closely related to *A.expansa* and certain other North American herbaceous C_4 species than indicated by Hall and Clements' phylogenetic scheme.

4.3.4 Inter-Specific Hybridization in Natural Populations

As mentioned earlier (Chap.4.2.2), spontaneous hybrids between different taxa in the *A.triangularis* group are frequently found in the western and northern parts of Scandinavia. Such absence of strong zygotic barriers to hybridization is also evident among a number of woody perennial *Atriplex* species of south-western North America. According to Hanson (1962), Plummer et al. (1957), and Drobnick and Plummer (1966), *A.canescens* readily hybridizes in nature with a number of perennial *Atriplex* species with which it is sympatric. Perhaps the most common natural hybrids involving *A.canescens* are those with *A.confertifolia* and *A.cuneata*. Hybrids between *A.canescens* and *A.confertifolia* are similar to *A.canescens* in the possession of pedicellate bracts and slender branchlets, but they are similar to *A.confertifolia* in general stature and habit and intermediate between the parents in fruiting bracts, spines and foliage characteristics. Hybrids between *A.canescens* and *A. cuneata* are highly variable with individual plants combining parental characteristics in various ways. According to Blauer et al. (1976), several natural populations suggest the existence of local introgression. Drobnick and Plummer (1966) and Blauer et al. (1976) also reported that artificial pollination of pistillate

A.canescens with pollen from *A.confertifolia, A.corrugata,* and *A.gardneri* resulted in the production of viable seed which gave rise to hybrid plants.

   A.confertifolia is known to hybridize not only with *A.canescens* but also with *A.corrugata* (Hanson, 1962), *A.cuneata* (Plummer et al., 1957), *A.gardneri* and possibly *A.hymenelytra* (Hanson, 1962). Plummer et al. (1957) found a number of hybrids between *A.confertifolia* and *A.cuneata* in roadside sites in Utah and Blauer et al. (1976) state that this hybrid apparently is fairly common in nature. Hansen (1962) examined 420 *Atriplex* plants growing along a road near Price, Utah. Of these, 51% were *A.confertifolia*, 44% were *A.corrugata*, 7% were *A.cuneata*, and 1.7% were apparent hybrids between *A.confertifolia* and *A.corrugata*. No hybrid was found in adjacent stabilized vegetation away from the roadside. The hybrids plants were nearly sterile; out of 500 large utricles only two had seeds. *A.cuneata* apparently hybridizes not only with *A.canescens* and *A.confertifolia* but also with all other sympatric *Atriplex* species except *A.obovata*. Included are *A.corrugata, A.garrettii, A.tridentata,* and *A.welshii* (Hanson, 1962).

   It is noteworthy that several of the woody perennial *Atriplex* species of the intermountain regions of the Western United States also appear to hybridize with certain species of related chenopod genera such as *Grayia* and *Sarcobatus*. For example, the occurrence of natural hybrids of *Grayia brandegei* with *A.confertifolia* and *A.cuneata* has been reported by Drobnick and Plummer (1966) and viable seed have been obtained by artificially pollinating pistillate *A.canescens* with pollen from *Grayia brandegei, G.spinosa, Sarcobatus vermiculatus,* and *Ceratoides lanata* (Blauer et al., 1976). Since *C.lanata* is a C_3 species, hybrids between it and the C_4 species *A.canescens* would be particularly valuable material for further study, but unfortunately none of the hybrids seedlings produced by Blauer et al. (1976) have survived. It should be emphasized that it is not definitely known that the seed produced in the above-mentioned artificial pollination experiments are genuine intergeneric hybrid seed in all cases since such pollination may stimulate the development of seed from unfertilized egg cells, or from associated somatic cells.

4.3.5 DNA Sequence Comparisons in Atriplex

   Since DNA is the basis of genetic differentiation during evolution, comparison of DNA sequences should reflect the degree of evolutionary divergence between different taxa. The accumulation of base substitutions of DNA during evolution has frequently been used to measure divergence in animal systems (Kohne, 1970). Recently such measurements have been extended to include higher plants (Stein and Thompson, 1975; Belford and Thompson, 1976, 1977). Relationships determined by such molecular hybridization techniques have the important advantage that they are not complicated by either prezygotic or zygotic barriers to hybridization as is often the case in crossing experiments at the organismic level. Chromosomal differentiation, such as the development of inversions and translocations, does not affect the results obtained in molecular DNA hybridization since the DNA used is sheared to fragments containing only a few hundred base pairs. Moreover, this method completely avoids the problem of similarities resulting from convergent evolution which often complicates determinations of relationships that are based on morphological or functional characteristics. DNA sequence comparisons should thus provide an independent tool which is complementary to classical methods involving taxonomic, cytogenetic, and crossing techniques in determinations of phylogenetic relationships.

A major limitation to DNA sequence comparisons in establishing genetic relationships is the considerable time and effort required to isolate sufficient quantities of uncontaminated DNA from plant material and to perform the hybridization analyses. In comparisons of higher plant DNA this problem is frequently compounded because in most higher plants, including *Atriplex,* a majority of the DNA is made up of repetitive sequences (Flavell et al., 1974; Belford and Thompson, 1976). Comparisons using total DNA are complicated by possible addition or differential amplification of repeated sequences in the different taxa and by the fact that related but nonidentical sequences existing in the same genome frequently reassociate with one another in vitro. The use of single-copy DNA in determinations of phylogenetic relationship has the advantage that only identical single-copy DNA sequences can reassociate to form precisely paired duplexes, so that almost all the mispairing observed in the DNA hybridization studies can be attributed to the evolutionary divergence between the species being compared. Another advantage is that the homology of single-copy DNA can be more directly related to the evolutionary divergence time. The disadvantage with using single-copy DNA is that it is more difficult and time-consuming (Belford and Thompson, 1976).

To date, Belford and Thompson have hybridized radioactively labeled single-copy DNA from *A.fruticulosa* (C_4), *A.serenana* (C_4), *A.truncata* (C_4), *A.phyllostegia* (C_3), *A.hortensis* (C_3), *A.triangularis* (C_3), *A.sabulosa* (C_4), and *A.rosea* (C_4) with a 2,000- to 9,000-fold excess of total unlabeled DNA from *A.serenana* or *A.rosea.* Two somewhat different criteria for phylogenetic diversity are used by these authors. The first is cross-reactivity of single copy DNA between each species and the reference species. For example, Fig. 4.7 presents the results of cross-reaction experiments in which single-copy DNA sequences from *A.phyllostegia, A.truncata,* and *A.triangularis* were hybridized with *A.rosea* or *A.serenana* unlabeled DNA. Self reactions between *A.serenana* and *A.rosea* labeled tracer DNA's with the homologous unlabeled DNA were also included as controls. Data from these and other cross-reaction experiments are summarized in Table 4.6, center column. Obviously, the lower the cross-reactivity, the fewer are the DNA sequences homologous between species. While this measure should provide a good estimate of the difference in their single-copy DNA, it does not reflect the evolutionary divergence time between species, since deletion of blocks of DNA sequences can occur abruptly and independently in the different taxa, which would have a considerable effect on cross-reactivity.

The second measure, ΔT_m, does reflect evolutionary distance between taxa by quantifying base sequence divergence within sequences able to cross-react. ΔT_m is the difference in melting points of homologous DNA hybrids (e.g., *A.rosea* tracer × *A.rosea* DNA) and heterologous DNA hybrids (e.g., *A.rosea* tracer × *A.serenana* DNA). The T_m of the heterologous hybrid is lower than that of the homologous hybrid by an amount directly proportional to the base sequence divergence between taxa. Examples of such experiments are shown in Fig. 4.8 and

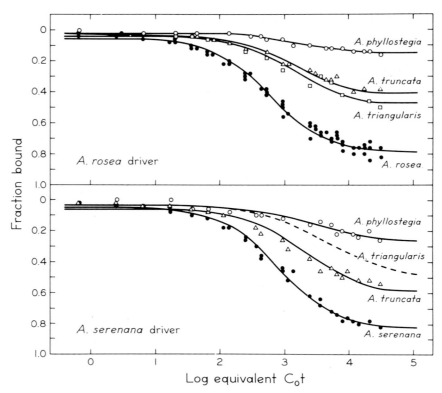

Fig. 4.7. ^{3}H-labeled single-copy sequences from *A.phyllostegia, A.truncata,* and *A.triangularis* were reacted with excess total, unlabeled DNA from *A.rosea (top panel)* and *A.serenana (bottom panel)*. ^{14}C-labeled single-copy sequences of *A.rosea* or *A.serenana* were included in each mixture as a homologous reaction control. After reassociation for various times, aliquots of the reaction mixtures were fractionated on hydroxylapatite under conditions in which only double-strands bind to the column. The fraction bound was calculated for each tracer as a function of C_0t (molar concentration of DNA nucleotides × incubation time in seconds). The *A.triangularis* curve *(bottom panel)* is actually the theoretical single copy component from a reaction including a repeated sequence component. (Redrawn from Belford and Thompson, 1979)

the results obtained to date are summarized in Table 4.6, right column. A ΔT_m value of 1 °C is interpreted as a mismatch in DNA base pairs of approximately 2%. The smallest degree of base sequence divergence as well as high DNA cross-reactivity was found between *A.serenana* and *A.fruticulosa* (C_4) and between *A. rosea* (C_4) and *A. sabulosa* (C_4). These species were shown by Nobs to form fertile F_1 and vigorous F_1 of reduced fertility, respectively. Hall and Clements recognized each pair as closely related species within the genus. The sequence divergence between the remaining taxa and *A.rosea* or *A.seranana* is greater. The majority of ΔT_m values cluster between 5 and 6 degrees, indicating that several taxa (*A.truncata, A.phyllostegia,* and *A.hortensis*) separated from *A.serenana* and *A.rosea* ancestors at about the same time that the *A.serenana* and *A.rosea* lines were

Table 4.6. Divergence in single-copy DNA homology between *A. serenana* and *A. rosea* and certain other species of *Atriplex*

Unlabeled DNA	Labeled DNA	Labeled DNA photosynthetic type	Cross reactivity, %	Corrected ΔT_m, °C[b]
A. rosea	*A. rosea*	C_4	100	0
	A. sabulosa	C_4	72	4.5
	A. triangularis	C_3	57	5.0
	A. truncata	C_4	53	5.2
	A. phyllostegia	C_3	29	5.5
	A. hortensis	C_3	58	6.2
	A. serenana	C_4	61	6.5
	A. fruticulosa	C_4	55	7.2
A. serenana	*A. serenana*	C_4	100	0
	A. fruticulosa	C_4	92	4.4
	A. truncata	C_4	72	5.0
	A. sabulosa	C_4	81	5.0
	A. hortensis	C_3	46	5.2
	A. phyllostegia	C_3	28	5.7
	A. rosea	C_4	46	6.7
	A. triangularis	C_3	56	8.2

[a]   Cross reactivity equals the fraction of a labeled tracer able to react with heterologous unlabeled DNA divided by the fraction of that labeled tracer able to hybridize with homologous unlabeled DNA × 100

[b]   A ΔT_m of 1 °C is interpreted to indicate about 2% base pair mismatch under the conditions used in these experiments

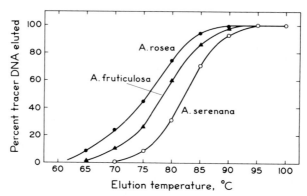

Fig. 4.8. ³H-labeled single-copy sequences from *A. rosea* and *A. fruticulosa* were reacted with excess total, unlabeled DNA from *A. serenana*. ¹⁴C-labeled single-copy sequences from *A. serenana* were included in each mixture as a homologous reaction control. After incubation to a C_0t value at which the majority of homologous tracer sequences had reacted, duplex molecules were isolated by binding to hydroxylapatite columns at 60 °C. Thermal elution was performed by raising the temperature of the column in 5 °C increments and collecting the DNA which became single-stranded at each temperature. The results are presented as the cumulative percent of tracer duplexes eluted at increasing temperatures. From these curves the T_m (the temperature at which 50% of the duplex DNA has "melted" into single strands) and ΔT_m (the difference between T_m's for inter- and intraspecific duplexes) can be determined. (Figure by courtesy of Belford and Thompson)

diverging from each other. This radiative scheme of speciation is in contrast to the gradually reticulating form of the Hall and Clements phylogeny and suggests that all C_4 lines arose during the same time period. Thus the DNA data do not support the concept of the polyphyletic origins for C_4 photosynthesis within *Atriplex*. If it is supposed that C_4 photosynthesis did evolve more than once in this genus, the different evolutionary events must have occurred within a period corresponding to the resolution of the hybridization technique and at about the same time as divergence of the main species lineages was occurring. The DNA hybridization data also do not support the view that separation of the *Euatriplex* and *Obione* subgenera occurred much earlier than the divergence of the branches leading to the main species groups in each subgenus.

In future studies concerning the evolution of C_4 photosynthesis within *Atriplex*, it would be particularly interesting to include representatives of woody perennial Australian, American, and Eurasian species in the comparisons. In this connection it is interesting to note that all evidence points to a polyphyletic origin of C_4 photosynthesis within the Chenopodiacaeae. It may be recalled that in addition to *Atriplex*, at least three other genera in this family, *Suaeda, Kochia (Maireana)* and *Bassia* contain both C_3 and C_4 species.

4.3.6 Intraspecific C_3/C_4 Differentiation?

Work in other genera such as *Alloteropsis, Panicum,* and *Mollugo* also indicates that C_3 and C_4 species can be closely related. Ellis (1974) reported that some individuals of the grass *Alloteropsis semialata* possess "Kranz" anatomy, while other individuals, occurring in the same region, do not. Smith and Robbins (1974) reported two distinct groups both within this taxon and also in *Mollugo cerviana* with regard to the carbon isotope discrimination ($\delta^{13}C$) values. The presence or absence of "Kranz" anatomy and the $\delta^{13}C$ value serve as excellent indicators of whether an individual is a C_3 or C_4 plant. However, the taxonomies of both *Alloteropsis* and *Mollugo* are very difficult and nothing is known about the genetic relationships between the suspected C_3 and C_4 forms in these taxa. Crossing experiments as well as cytogenetic and further taxonomic studies are therefore needed before any conclusions can be reached regarding the existence of an intraspecific differentiation with regard to C_3/C_4 photosynthesis in these taxa.

In virtually all cases the distinction between C_3 and C_4 photosynthesis is absolute, i.e., a given plant either possesses fully functional C_4 photosynthesis or not at all. However, there are some reports that certain taxa of *Panicum* and *Mollugo* possess features that are intermediate between C_3 and C_4 plants. Artificial hybrids between C_3 and C_4 *Atriplex* species also exhibit such characteristics (Sect. 4.4). *Panicum milioides* and certain other related taxa in the *laxa* group, such as *P. hians* and *P. laxum,* exhibit leaf anatomical features which are intermediate between those of C_3 and C_4 species of *Panicum* and they also possess certain photosynthetic characteristics of an apparently intermediate nature, although the overall functional performance is much more similar to a C_3 than to a C_4 plant (Brown and Brown, 1975; Brown, 1976). These authors also reported that the $\delta^{13}C$ value for *P. milioides* was typical for a C_3 species, implying that there is no significant net carbon fixation by two linked CO_2 fixation pathways in series (one using PEP carboxylase, the other RuP_2 carboxylase) as in C_4 plants. Nevertheless, *P.milioides*, *P.hians*, and *P.laxum* may well prove to be true C_3/C_4 intermediates. Whether they represent a stage in the evolution of C_4 photosynthesis, a reversion of C_4 to C_3 photosynthesis, or derivatives of spontaneous C_3/C_4 hybrids is completely unknown.

A situation somewhat resembling that in *P.milioides* is found in *Mollugo verticillata* (Kennedy and Laetsch, 1974). This taxon also exhibits a leaf anatomy which appears to be intermediate between C_3 and C_4 species. Early products of carbon fixation show a pattern intermediate between those found in C_3 and C_4 species and certain other photosynthetic characteristics also suggest an intermediate nature of *M.verticillata*. Sayre and Kennedy (1977) reported that interpopulational differences in the extent of C_4 resemblance exist within this taxon and even suggested that this variation is the result of ecotypic differentiation. However, the $\delta^{13}C$ values obtained in *M. verticillata* so far are typical for C_3 plants, indicating that no significant amount of carbon net fixation occurs via C_4 photosynthesis in this taxon (Troughton et al., 1974a; Smith and Robbins, 1974). This does not exclude the possibility that *M.verticillata* represents a link in the evolution of C_4 photosynthesis but weakens the evidence. The polyploid nature of this taxon ($2n = 64$ compared with $2n = 18$ for the C_4 species *M.cerviana*) suggests that it may rather be a hybrid derivative. Again, genetic studies are required to resolve this problem and to elucidate the evolutionary and genetic aspects of C_4 photosynthesis in general.

4.4 Inheritance of C_4 Photosynthesis in Atriplex

The ideal materials for investigation of the inheritance of C_4 photosynthesis would obviously be C_3 and C_4 individuals that are capable of producing freely segregating F_2 hybrid populations. Unfortunately, such materials, if they exist, have not been found so far. The only crossing experiments between C_3 and C_4 plants of any taxon that have been successful in producing fertile F_1 hybrids are those obtained in the crosses between *A.rosea* (C_4) × *A.triangularis* (C_3) and *A.rosea* (C_4) × *A.glabriuscula* (C_3). Although these two crosses produce strong diploid F_1 hybrids, chromosomal pairing is irregular, resulting in F_2 populations with varying chromosome numbers thus preventing a precise and detailed genetic analysis. Nevertheless, these crosses have provided all of the information that is available on the inheritance of C_4 photosynthesis and will therefore be discussed in some detail. The cross *A.rosea* × *A.triangularis* is the more thoroughly studied and most of the examples will be taken from this cross. On the whole, the results obtained with the two crosses are quite similar.

A.rosea, the C_4 parent, is a summer annual native to the Mediterranean region but it is widely naturalized in North America and very common in California, especially in the interior valleys. It is also frequently found in milder regions such as along the San Francisco Bay where it often grows in the vicinity of the cosmopolitan *A.triangularis*, the C_3 parent used in this cross. No spontaneous hybrids have been

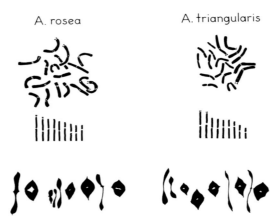

Fig. 4.9. Somatic chromosomes *(top)* and meiotic chromosomes *(bottom)* in *Atriplex rosea* *(left)* and *A. triangularis (right)*. (From Nobs et al., 1971)

detected in areas where the distribution of the two species overlap, but this could be due in part to an incomplete coincidence of their flowering periods. As was shown in Fig. 4.4, *A.rosea* has classical "Kranz" anatomy, whereas *A.triangularis* has conventional leaf anatomy. Both species invariably are diploid with $2n = 18$ chromosomes. The chromosomes are small, the largest having a length of approximately $5 \mu m$. The karyotypes of the two species are very similar as shown in both mitotic and meiotic divisions (Fig. 4.9, top). For example, two of the chromosome pairs have satellites in both species. As shown in the lower part of Fig. 4.9, meiosis is perfectly regular with nine bivalents forming in metaphase I (Nobs et al., 1971).

Although the flowers are unisexual, all of the *A.triangularis* inflorescences contain both male and female flowers and since the species is highly self-fertile, controlled crossing experiments with *A.triangularis* as the maternal parent would therefore require emasculation. Because of the small size of the flowers (less than 1 mm at anthesis), this is a very difficult task. A similar situation exists in *A.rosea* but in addition to the capitate bisexual inflorescences this species also possesses axile flowers that are female only and can readily be used as the maternal parent in controlled pollination experiments, after removal of the capitate inflorescences. In experiments where axile flowers of *A.rosea* had been repeatedly pollinated with *A.triangularis* pollen, approximately 10% of the flowers yielded seed. Almost all seeds germinated readily and gave rise to strong F_1 seedlings (Nobs et al., 1971). The F_1 hybrids were very uniform and intermediate between the parents in a number of characteristics such as habit (Fig. 4.10), leaf shape and dentation (Fig. 4.11, left column). More importantly, they were also intermediate in several characteristics relating to C_4 photosynthesis. One of these is leaf anatomy (Fig. 4.11, center column). Like the C_4 parent the F_1 hybrids have a distinct bundle-sheath surrounded by a radiate mesophyll and both of these cell types contain chloroplasts. Electron microscopic examination (not shown; see Boynton et al., 1971) reveals that the chloroplasts and mitochondria contained in the bundle-

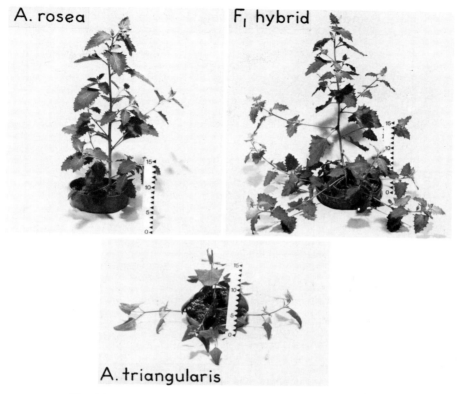

Fig. 4.10. *Atriplex rosea, A.triangularis,* and their F_1 hybrid

sheath cells are larger and more numerous than those of adjacent cells in the radiate mesophyll, resembling the situation in *A.rosea*. The cell walls separating bundle-sheath and mesophyll cells in the F_1 hybrids are traversed by plasmodesmata as they are in *A.rosea*. However, on the average, the bundle-sheath cells are somewhat thinner and the cells more vacuolated than in *A.rosea* and the radiate mesophyll is also less distinct in the F_1 hybrid than in *A.rosea*. Moreover, in the hybrid, chloroplast-containing cells are also found outside the two concentric layers of bundle-sheath and mesophyll cells, a feature apparently contributed by the *A.triangularis* parent which completely lacks the leaf anatomical specialization found in *A.rosea* and other C_4 plants possessing "Kranz" anatomy. Thus the gross leaf anatomy as well as the fine structure of cells and organelles of the F_1 hybrid is truly intermediate.

The intermediate nature of the F_1 hybrids is also expressed in the activities of several enzymes known to be necessary for C_4 photosynthesis, such as PEP carboxylase, pyruvate dikinase, and alanine and aspartate amino transferases (Pearcy and Björkman, 1971; Hatch et al., 1972). All of the enzymes known to be essential for C_4 photosynthesis are present in the F_1 hybrid but the activities are lower than in the C_4 parent. For example, PEP carboxylase activity in *A.rosea* was

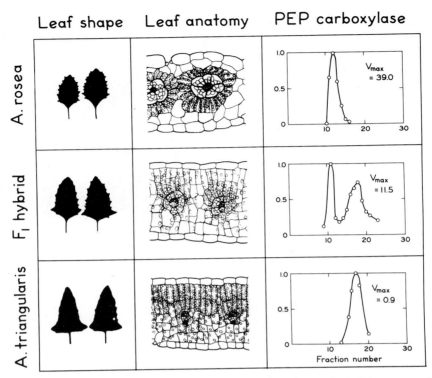

Fig. 4.11. Leaf shape, leaf anatomy and PEP carboxylase of *Atriplex rosea* (C_4), *A. triangularis* (C_3), and their F_1 hybrid. The *center column* shows camera lucida drawings of leaf cross sections (after Boynton et al., 1971). *Curves* shown in the *right column* are elution profiles of PEP carboxylase on DEAE-cellulose using a phosphate gradient, and the V_{max} values shown are substrate-saturated PEP carboxylase activity in μmol CO_2 mg^{-1} chlorophyll min^{-1}. (Redrawn from Ting and Osmond, 1973b)

approximately 42 times that of *A. triangularis*. In the F_1 hybrids the activity of this enzyme was intermediate between the parents or about 13 times that of *A. triangularis* (activities are given by the numbers shown in the right column of Fig. 4.11). However, in addition to these differences in PEP carboxylase activity between *A. rosea* and *A. triangularis,* this enzyme is qualitatively different in two species, having distinctly different electrophoretic, chromatographic, and kinetic properties (Hatch et al., 1972; Ting and Osmond, 1973a). Examples of elution profiles are shown in Fig. 4.11, right column. In the F_1 hybrids both of the parental forms of PEP carboxylase are present. Similar results have been obtained with other C_4 pathway enzymes (Hatch et al., 1972).

The labeling pattern of early carbon products of photosynthesis, following a 6 s feeding of radioactive CO_2 to photosynthesizing leaves, is also clearly intermediate in the F_1 hybrids (Pearcy and Björkman, 1971). In *A. rosea* more than 90% of the radioactivity was recovered in C_4 dicarboxylic acids aspartate and malate (which is typical for C_4 plants) whereas in *A. triangularis* over 90% was recovered in Calvin cycle intermediates (which is typical for C_3 plants). Partial operation of C_4

Table 4.7. The efficiency of utilization of CO_2 at low concentrations in the intercellular spaces, $\Delta P/\Delta C_i$, in parents and hybrids at 27 °C.[a] (After Björkman et al., 1971)

Plant	Efficiency, $\Delta P/\Delta C_i$	
	21% O_2	1.3% O_2
Atriplex rosea	0.67	0.67
Atriplex triangularis	0.25 ± 0.02	0.36 ± 0.02
F_1 hybrid	0.11 ± 0.01	0.19 ± 0.02

[a] The efficiency of utilization of low CO_2 concentration in the intercellular spaces, $\Delta P/\Delta C_i$ is the initial slope of the curve for photosynthesis as a function of CO_2 concentration in the intercellular spaces. $\Delta P/\Delta C_i$ has the dimensions $cm\ s^{-1}$

photosynthesis in the F_1 hybrid is suggested by the intermediate labeling pattern obtained in these plants: 44% of the radioactivity was recovered in C_4 dicarboxylic acids and 56% in Calvin cycle intermediates.

We have seen that the F_1 hybrids are intermediate between the parental species in number of characteristics pertaining to C_4 photosynthesis, including leaf anatomy, fine structure of cells and organelles, presence of key enzymes and nature of early products of carbon fixation. Therefore one might perhaps expect that the F_1 hybrids would also be intermediate in terms of their photosynthetic performance, but this is not so. For example, *A.rosea* has a considerably higher efficiency of utilizing low CO_2 concentrations for photosynthesis than *A. triangularis* (Table 4.7). Likewise, 21% oxygen causes a 40% inhibition of net photosynthesis at normal CO_2 concentrations in *A. triangularis* and other C_3 plants, but has no significant effect in *A.rosea* and other C_4 species. As was mentioned in Sect. 4.3.1 (also see Chap. 9.1.2), it is now generally accepted that these characteristics are a direct result of the C_4 pathway serving as a metabolic CO_2 pump that increases the CO_2 fixation by the Calvin cycle which in C_4 plants is located in the bundle-sheath (or PCR) cells. It was also indicated that the leaf anatomical specialization of C_4 plants evidently serves to provide spatial compartmentation for the two sequential CO_2 fixation cycles. Lack of such compartmentation would be expected to have serious consequences in terms of the efficiency of C_4 photosynthesis (also see Chap. 9.1.2).

Comparative measurements of photosynthetic CO_2 exchange characteristics show that the F_1 hybrids are not intermediate between its parents. The efficiency of utilization of low CO_2 concentrations by the F_1 hybrids is not higher but lower than in the C_3 parent, *A. triangularis* (Table 4.7), and the inhibitory effect of 21% O_2 on net photosynthesis is approximately the same as in *A. triangularis*. The low efficiency in utilizing low CO_2 concentrations for photosynthesis is also reflected in the growth rate in an atmosphere of low CO_2 (Table 4.8). The capacity for photosynthesis and growth at high, saturating CO_2 concentrations is very similar in *A.rosea*, *A.triangularis,* and their F_1 hybrids, indicating that the inferior photosynthetic performance of the F_1 hybrids is limited to the *efficiency* of CO_2

Table 4.8. Dry matter production of *Atriplex rosea*, *A. triangularis*, and their F_1 hybrid, grown under an atmosphere containing 60 ± 1 ppm CO_2. [a] (After Björkman et al., 1971)

	Dry weight, mg		Relative growth rate, $g\,g^{-1}\,day^{-1}$
	At harvest	At start of experiment	
Atriplex rosea	4032	249	0.19
Atriplex triangularis	1377	597	0.07
F_1 hybrid	395	220	0.03

[a]   Light was continuous, with an intensity of approximately 1.2×10^5 erg cm^{-2} s^{-1}; temperature was held constant at 27 °C; and oxygen concentration was 21%

utilization. The hybrid is fully functional in all other respects. This suggests the possibility that compartmentation of the two CO_2 fixation cycles is incomplete or lacking in the hybrids. Recent work on the localization of the enzyme RuP_2 carboxylase lend strong experimental support for this. Using an in situ immunofluorescent labeling technique, Hattersley et al. (1977) were able to determine the distribution of this enzyme in living leaf tissue of a number of C_4 and C_3 species, including *A. rosea, A. triangularis* and their F_1 hybrid. In all C_3 plants examined, including *A. triangularis*, RuP_2 carboxylase was present in all chloroplast-containing cells throughout the leaf parenchyma, whereas in all C_4 plants examined this enzyme was present in the PCR cells but absent in the PCA cells. In *A. rosea* and other C_4 species with classical "Kranz" anatomy, the PCR and PCA cells are the bundle-sheath and radiate mesophyll cells, respectively. In the F_1 hybrid, RuP_2 carboxylase was found in all chloroplast-containing cells, including those of the radiate mesophyll cells. This is also in agreement with the findings that the total activity of RuP_2 carboxylase in leaves of the hybrids was not intermediate between the two parents but essentially as high as in the *A. triangularis* parent (Pearcy and Björkman, 1971).

These and other results provide strong evidence for lack of a proper intercellular compartmentation of the two CO_2 fixation cycles in the *Atriplex* hybrids. It is not difficult to visualize that in the absence of compartmentation the two CO_2 fixation pathways would compete rather than cooperate. In such a system the C_4 pathway would serve no useful purpose and a reduction rather than an increase in the efficiency of CO_2 utilization over the C_3 parent is the probable result. However, there should be no significant effects at saturating CO_2 concentrations. This is consistent with the experimental data obtained with the *Atriplex* hybrids.

Endeavoring to investigate the recombination of photosynthetic characteristics, the Carnegie workers set out to produce a segregating F_2 population by self-pollinating the F_1 hybrids. This work met with only partial success since chromosomal pairing in the F_1 hybrids is highly irregular (Fig. 4.12). At most, six of nine possible chromosome pairs are formed. In general there are only three to four pairs, and not infrequently no pairing takes place. In those cells where no pairing occurs (Fig. 4.12, center), precocious division occurs at metaphase I resulting in

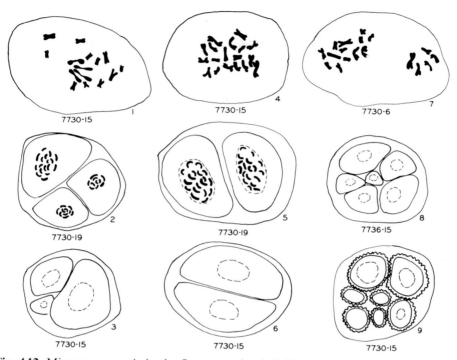

Fig. 4.12. Microsporogenesis in the first-generation hybrid *Atriplex rosea* × *A.triangularis*. For details, see text. (From Nobs et al., 1971)

diploid gametes. In those cells where partial pairing occurs the paired chromosomes undergo normal separation. However, some of the unpaired chromosomes undergo precocious division, and the remaining unpaired ones are randomly distributed between the new cells (Fig. 4.12, left and right columns). This results in gametes with highly variable chromosome numbers; only those containing at least a full complement of chromosomes are presumably viable. On the basis of male fertility in the F_1 hybrid (6% to 7% stainable pollen, compared with 95% in the parents), most of the pollens have incomplete or incompatible genomes. In all probability the meiotic irregularities in the embryo sacs are similar to those in the pollen mother cells, resulting in a very low seed set (<1%) in the self-pollinated F_1 hybrids. However, the seeds obtained were viable, germinated readily and, in most cases, gave rise to strong plants. As can be expected, the F_2 population shows a remarkably wide variation in chromosome number, from slightly above diploid to more than pentaploid (Table 4.9). The median number is $2n = 34$ but almost any number between $2n = 20$ and $2n = 48$ has been obtained.

The variation in seed set in the F_2 hybrids was very much greater than in the F_1 hybrids (Björkman et al., 1971). As shown in Table 4.9, seed set varied from 0% in F_2 hybrid 7736-1 to 86% in F_2 hybrid 7744-4, the latter approaching the seed set in the parental species. One might expect that a tetraploid F_2 hybrid such as 7735-3 would have a better seed set than an essentially triploid such as 7744-4, but there is

Table 4.9. Chromosome numbers and percentage of flowers yielding viable seed in F_2 hybrids between *Atriplex rosea* and *A. triangularis*. (From Björkman et al., 1971)

F_2 hybrid	Chromosome number (2n)	Seed-set, % full seed
7735- 1	34	1
7735- 2	34	28
7735- 3	37	18
7735- 4	44	11
7735- 5	36	10
7736- 1	21	0
7740- 1	30	0
7740- 2	44	0
7740- 4	46	78
7740- 5	31	75
7744- 1	34	12
7744- 4	26	86
7746- 1	29	15
7746- 9	29	40
7746-10	28	0
7746-12	30	30
7746-14	22	85
7746-16	28	20

no indication that the percentage of viable seeds is in any way related to the ploidy level in the F_2 hybrids. It is also noteworthy that the chromosome numbers obtained in F_3 populations varied substantially from those of their F_2 parents (Table 4.10). This shows that apomixis was not a factor underlying the remarkably high seed set of many F_2 hybrids. Unfortunately, however, the observation that highly fertile F_2 hybrids with low chromosome numbers (such as 2n = 22) produce progenies with highly varying numbers (2n = 21 to 32) makes it unlikely that repeated selection for individuals with low chromosome numbers would lead to self-perpetuating diploid genotypes needed for detailed analysis of the inheritance of C_4 photosynthesis.

Table 4.10. Somatic chromosome numbers in meristematic cells among F_3 hybrids between *Atriplex rosea* and *A. triangularis* in relation to those of their F_2 parents. (From Björkman et al., 1971)

Chromosome number of F_2 parent	Chromosome number of F_3 progeny
22	21, 22, 24, 28, 32, 32
26	26, 26, 28, 34, 38
34	28, 30, 32, 32
37	32, 34, 34, 36, 36, 36
44	40, 44, 44, 46, 48, 52

Fig. 4.13. Variation in growth habit and morphology among F$_3$ hybrids between *Atriplex rosea* and *Atriplex triangularis*

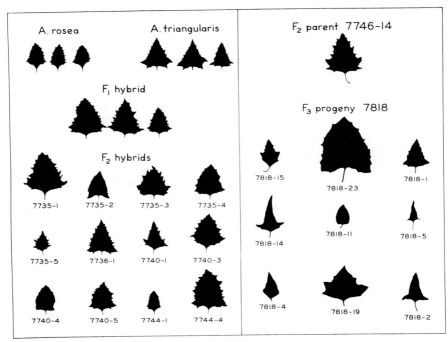

Fig. 4.14. Leaf shape and size in *A. rosea, A. triangularis,* their F$_1$, F$_2$, and F$_3$ hybrids

In spite of the polyploid nature of the second and subsequent generations of hybrids between *A. rosea* and *A. triangularis,* the variation in morphological and leaf anatomical characteristics is at least as great as one would expect to find in a freely segregating diploid hybrid population. This variation is likely to result from recombinations of parental genes together with gene dosage effects caused by varying number of chromosomes contributed by one or the other of the parental species. Examples of the dramatic variation in characters such as growth habit, leaf shape, and leaf size among F$_2$ and F$_3$ hybrids are given in Figs. 4.13 and 4.14. Many of these hybrids widely transgress the characteristics of *A. rosea* and *A. triangularis.*

Wide variations are also present in leaf anatomical characteristics among different F$_2$ and F$_3$ individuals with a complete spectrum ranging from apparently perfect "Kranz" anatomy, very similar to that of *A. rosea,* to the conventional non-"Kranz" anatomy of *A. triangularis,* and almost any conceivable combination of the two types. A random sample of leaf anatomies from a segregating F$_2$ hybrid population is shown in Fig. 4.15. This figure also shows the PEP carboxylase activities in the leaves of the same F$_2$ hybrids. Also this characteristic shows a wide segregation in the F$_2$ population but it does not co-vary with leaf anatomy. Some individuals closely resembling the C$_4$ parent in leaf anatomical characteristics have low PEP carboxylase activities while other individuals, more closely resembling the C$_3$ parent anatomically, have high PEP carboxylase activities. The variation in the relative *amounts* of the two parental types of PEP carboxylase is illustrated in Fig. 4.16, in this case for hybrid individuals from an F$_3$ population. In some

F$_2$ hybrids

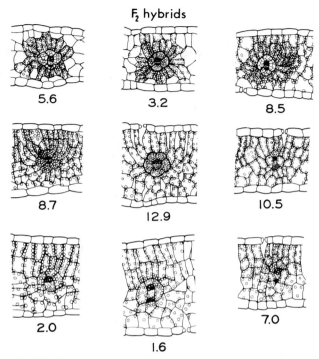

Fig. 4.15. Camera lucida drawings of living leaf cross sections of F$_2$ hybrids between *Atriplex rosea* and *A. triangularis*. The number under each section gives the PEP carboxylase activity of that hybrid, expressed in μmol CO$_2$ g^{-1} fresh wt. min^{-1}. (Redrawn from Boynton et al., 1971, and Pearcy and Björkman, 1971)

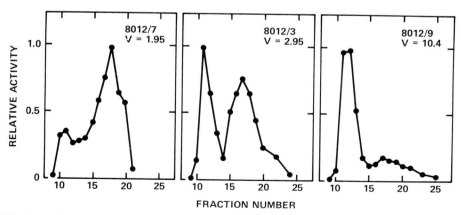

Fig. 4.16. Elution profiles and PEP carboxylase activity of three different F$_3$ hybrids between *Atriplex rosea* and *A. triangularis*. See Fig. 4.12. (Redrawn from Ting and Osmond, 1973b)

individuals, e.g., 8012-7, the C_3 parental enzyme form is dominant, in others, e.g., 8012-4, the C_4 parental enzyme form is dominant, and in yet others, e.g., 8012-3, both forms are present in approximately equal amount. These and other results show that the inheritance of two key components of C_4 photosynthesis, "Kranz" anatomy and PEP carboxylase, is not closely linked but largely independent. If this is also true for other components of the C_4 pathway (and there is some evidence that this is so) then the frequency in the second and subsequent generations of segregating individuals with complete and fully functional C_4 photosynthesis would be expected to be very low. The frequency may, indeed, be even lower than expected on this basis alone since not only is it necessary that all biochemical and anatomical components of C_4 photosynthesis be present but evidently each biochemical component must also be confined to its proper anatomical compartment. The presence of RuP_2 carboxylase in the radiate mesophyll cells of the F_1 hybrids, mentioned above, exemplifies the lack of proper compartmentation. The uniformly high level of this enzyme among the approximately 30 F_2 and F_3 hybrids examined, regardless of their leaf anatomy or PEP carboxylase level, suggests similar lack of proper compartmentation of RuP_2 carboxylase in these hybrids (Pearcy and Björkman, 1971; Björkman, unpublished).

Whether such lack of compartmentation in hybrids also applies to PEP carboxylase and other key enzymes is not yet known. Nevertheless, it is highly probable that lack or deficiency of any C_4 enzyme, or any aberration in the intra- or intercellular location of C_3 or C_4 enzymes in a hybrid individual will result in a photosynthetic performance inferior to that of the C_4 parent. Unless the inheritance of the various component enzymes, as well as their compartmentation, is linked the segregation of a hybrid possessing complete and fully coordinated C_4 photosynthesis would therefore be a very rare event. The crossing experiments indicate that such linkage is weak and none of the approximately 300 F_2, F_3, and F_4 hybrids screened so far has been found to have fully functional C_4 photosynthesis with the highly efficient use of low CO_2 concentrations that characterizes the C_4 parent, although several of these hybrids have approached their C_4 parent in such characteristics as leaf anatomy, PEP carboxylase activity and photosynthesis CO_2 compensation point (Björkman, 1976). Interestingly, the presence of a low CO_2 compensation point is not necessarily an indication of a high efficiency of CO_2 utilization even though these two characteristics always go together in C_4 plants. Evidently, high efficiency of CO_2 utilization requires a complete and fully coordinated C_4 photosynthetic system; the requirements to give a low CO_2 compensation point appear less stringent. This situation closely resembles that found in the "C_3/C_4 intermediate" species *Mollugo verticillata* and *Panicum milioides,* mentioned in a previous part of this section.

Perhaps the best indicator of normal C_4 photosynthesis in nonsucculent plants is the carbon isotope discrimination ratio expressed as $\delta^{13}C$ value (see Chap. 9.1.2). The $\delta^{13}C$ value of parental C_4 *A.rosea* is about $-11^0/_{00}$ and that of parental C_3 *A. triangularis* is about $-27^0/_{00}$. As shown in Fig. 4.17, all the hybrid *Atriplex* individuals so far examined show $\delta^{13}C$ value similar to those of the C_3 parent, indicating that normal integrated C_4 carbon assimilation does not take place in the hybrids. To our knowledge, all the so-called "intermediate forms" of C_3 and C_3 metabolism reported in other species result in $\delta^{13}C$ values similar to those of C_4

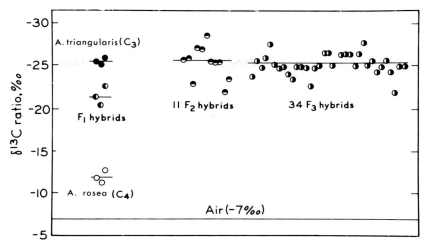

Fig. 4.17. Carbon isotope ratios ($\delta^{13}C$) in leaves of *Atriplex rosea*, *A.triangularis* and their F_1, F_2, and F_3 hybrids. *Horizontal bars* indicate means for each group. (From Björkman et al., 1974c)

plants. These data may be interpreted to indicate that even if some CO_2 is directly assimilated by PEP carboxylase in these plants, this carbon is not further metabolized in the PCR cells as it is in C_4 plants.

Any conclusive statement regarding the mode of inheritance of C_4 photosynthesis and the number of genetic loci involved must await further hybridization experiments and genetic analysis on materials capable of producing fertile and diploid segregating hybrid populations. Whether or not C_3 and C_4 plants that meet these demands do indeed exist is still an open question. Further research on the phylogenetic relationships between C_3 and C_4 plants in *Atriplex* and certain other genera should aid in providing an answer to this question and point to the most suitable C_3 and C_4 plant materials for such genetic analysis. Crossing experiments between different C_4 taxa may also prove valuable in elucidating certain genetic aspects of C_4 photosynthesis if the genes determining the inheritance of a certain component of the C_4 pathway happen to be nonallelic in the two parental taxa.

Keeping in mind the limitations of the hybridization experiments reviewed here we will conclude by making a tentative assessment of the genetic and evolutionary aspects of C_4 photosynthesis in *Atriplex*. C_4 photosynthesis represents an advanced evolutionary state in the genus; all primitive *Atriplex* probably were C_3 plants, although some present-day C_3 species may have diverged as much from this ancestral stock as certain C_4 species. Thus, the maximum genetic divergence between C_3 species may equal or exceed the minimum divergence between C_3 and C_4 taxa. It is possible but far from certain that the origin of C_4 photosynthesis within the genus is polyphyletic. The demonstration that certain C_3 and C_4 species of *Atriplex* can be crossed and form vigorous hybrids, some of which are highly fertile in the second and subsequent generations, shows that the genetic divergence between C_3 and C_4 species need not be great. On the other hand, neither is the inheritance of C_4 photosynthesis simple. The absence of a maternal mode of inheritance in the *A. rosea* × *A. triangularis* crosses shows that C_4 photosynthesis is not simply transmitted by the chloroplast genome. In addition to any role that chloroplast genes may have, at least several nuclear genes must be involved. It is also evident that different components of C_4 photosynthesis, such as the specialized leaf anatomy and certain key enzymes of the pathway, are not inherited together. Thus it is likely but not certain that at least some of the genes determining the inheritance of these different components are located in different chromosomes and hence different linkage groups.

Chapter 5. Atriplex Communities: Regional Environments and Their Ecological Analysis

"A plant community is understood to be a more or less stable combination of naturally occurring species, which are in an ecological equilibrium with one another and their environment."

(Walter, 1971)

Judged against the variable background of evolutionary and systematic relationships of *Atriplex* discussed in the previous three chapters, and the very diverse viewpoints adopted by different authors who have prepared monographs describing regional floristic and vegetation types, there are many ways of preparing a coherent account of *Atriplex* communities on a global basis. Just as we have taken a rather liberal and pragmatic approach to the systematics of the genus, we shall in this chapter explore some of the more generally recognized vegetation types ("communities") and environments in which species of *Atriplex* commonly occur. The difficulties of generalization are exaggerated if one attempts to categorize *Atriplex* communities in terms of the definition offered by Walter, since it seems unlikely from our field experience that many species of *Atriplex* are found in "stable combinations" or indeed in "ecological equilibrium" with their environments. Initially we will consider some of the apparent, qualitative relationships between the vegetational and environmental features of these communities, and attempt to distinguish in very general terms the principal environmental factors that appear to be most significant in each vegetation type. In the latter part of this chapter we shall examine some approaches that have been made toward a quantitative analysis of these relationships, approaches that seek to provide a preliminary coupling between states and processes of the vegetation and environment appropriate to nodes in the span 10^6–10^{11} of our space–time scale.

If we ignore then the problems inherent in attempting to match formal (and perhaps unduly idealized) definitions of communities, our experience suggests that we can usefully recognize three major vegetation types to which *Atriplex* species contribute significantly: arid shrublands, littoral-maritime, and weed-ruderal communities.

5.1 Communities of Arid Continental Interiors

A substantial number of *Atriplex* species are found in semi-arid regions of the world (Chap. 2). In south-western United States, in the pampas region of

Argentina, in the Karoo region of South Africa, in southern Australia, in the steppes of Central Asia and parts of the arid hinterland surrounding the Mediterranean, *Atriplex* species frequently contribute a significant floristic component to two major vegetation types. These are the shrub steppes usually dominated by perennial species, and the less well-defined halophytic communities associated with saline soils, scalds, washes, or sinks, in which both annual and perennial species are well represented.

5.1.1 Arid Shrub Steppe

Shrub steppes are the most visually uniform and most extensive communities in which the genus *Atriplex* is represented. There is a striking similarity in both growth-form and community physiognomy between *Atriplex* shrublands developed in different continents. This similarity is well exemplified by the saltbush steppes of southern Australia dominated by *A. vesicaria,* and the shadscale vegetation of the inter-mountain regions of the western United States, dominated by *A. confertifolia.* The extent of these arid shrubland steppes is shown in maps of Fig. 5.1 and the similarity of growth form and community physiognomy in Fig. 5.2. These particular vegetation types are probably the most intensively researched of the arid shrublands. The ecological states have been documented in early studies (Wood, 1936; Beadle, 1948; Billings, 1949) while more intensive research on both states and processes has been conducted under the auspices of the Desert Biome of the International Biological Program in the United States and by the CSIRO Division of Land Resources Management in Australia.

The structure of these communities reflects a remarkable degree of convergent "community" evolution in that different species from presumably distantly related stocks have come to occupy similar and extensive niches in two continents. Both *A. vesicaria* and *A. confertifolia* can form almost pure stands of low (<1 m), *apparently* regularly spaced plants (Fig. 5.2), although not infrequently they may share regional co-dominance with other shrubs of similar growth-form such as the bluebush *Maireana (Kochia)* in Australia (Wood, 1936) and sagebush *(Artemisia)* in the United States (Billings, 1949).

Within Australia some *Atriplex* species that characterize shrub steppe are also found as part of a shrubby understorey in open woodlands dominated by species of *Eucalyptus* and *Acacia* (Fig. 5.3), particularly in Western Australia. One of these species, *A. nummularia,* was originally the dominant shrub over much of the heavier, slightly undulating (gilgaid) soils of the western Riverine Plain of New South Wales until its populations were decimated during the nineteenth century by a combination of overstocking by sheep and associated ring-barking by rabbits during drought periods (Williams and Oxley, 1979). This period of intensive pastoral utilization led to the replacement of this deeply rooted species by the now ▶

Fig. 5.1. a Range of *A. confertifolia* (shadscale) communities in south-western United States, showing the location of Curlew Valley *1* and Desert Experimental Range *2* sites discussed in the text. **b** Range of chenopod shrublands in Australia showing the location of Koonamore *1,* Riverine Plain *2* and Barrier Range *3* sites discussed in the text. (Redrawn from Billings, 1949; Woodwood, M.T., unpublished)

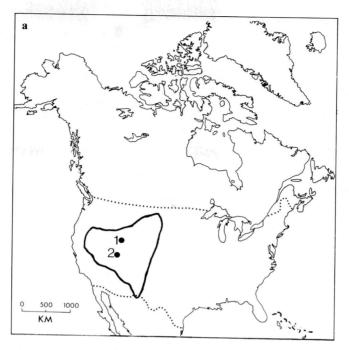

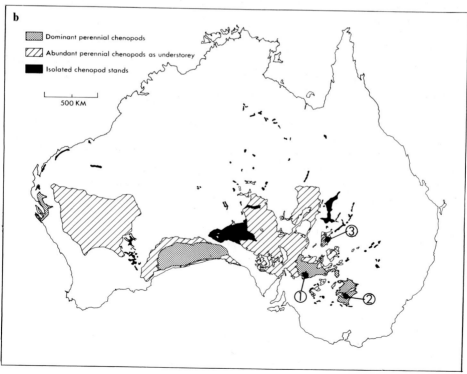

Fig. 5.2. Arid shrublands dominated by *A. confertifolia* in Curlew Valley, Utah (*upper*) and *A. vesicaria* on the Riverine Plain, New South Wales (*lower*). (Upper photograph courtesy M.M. Caldwell)

widely dominant, but shallow-rooting *A.vesicaria*. *Atriplex nummularia* in Australia, *A.lentiformis* in North America, *A.lampa* in South America, and *A.halimus* in the Mediterranean region, comprise the largest and longest-living members of the genus, growing to 3 m in height and in the case of *A.nummularia*, producing individuals that exceed 100 years in age.

Fig. 5.3. Open mixed shrubland with *A. nummularia* and *Eucalyptus*, Riverine Plain, New South Wales

Some climatic characteristics of the extensive arid shrublands of Australia and North America are indicated in the climatic diagrams of Fig. 5.4. These diagrams, constructed essentially in the format developed by Walter and Lieth (1961), show some climatic properties of two representative sites from each area. In the Great Basin, the Curlew Valley, a few kilometers north of the Great Salt Lake, Utah site 1 (Fig. 5.1a) has been studied intensively during the U.S. IBP program and many of the ecological processes operating in the *A. confertifolia, C. lanata* communities at this site will be examined in Chap. 10. The Desert Experimental Range in south-western Utah site 2 (Fig. 5.1a) is the site of long-term studies of productivity and grazing trials which will also be discussed in Chap. 10. It has a lower annual precipitation than Curlew Valley where much of the precipitation falls as snow in winter months. In south-eastern Australia site 1 (Fig. 5.1b) Koonamore has been the site of long-term observations of regeneration and succession of arid shrubland vegetation dominated by *A. vesicaria,* the details of which are presented in Chaps. 6 and 10. In average rainfall it corresponds approximately to the Desert Experimental Range in Utah. The Riverine Plain somewhat further east on the Australian continent site 2 (Fig. 5.1b) is a region which, in terms of rainfall, more nearly corresponds to the Curlew Valley site of Utah.

These climate diagrams emphasize the chief difference between the environmental conditions experienced by the two communities. The Great Basin is

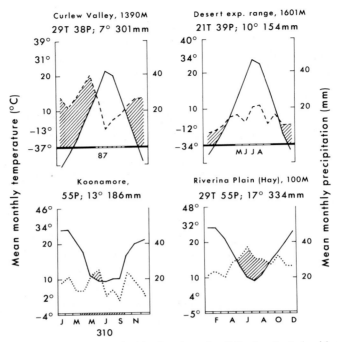

Fig. 5.4. Climate diagrams for two sites in the *A. confertifolia*-dominated arid shrublands in south-western United States (cool desert) and for two sites in the *A. vesicaria*-dominated arid shrublands in south-eastern Australia (warm desert) (cf. Fig. 5.7)

a cool desert in which a large proportion of the precipitation is received at a time when biological activity is limited by low temperature. The average winter minima are about 15 °C lower in this desert than in Australia and absolute minima are very much lower. In southern Australia a larger proportion of the precipitation falls in much hotter months when evaporation is much greater. Although the average summer maxima are similar to those in the Great Basin desert, the absolute maximum temperatures are somewhat higher. The climate diagrams do not show data for the ratio of precipitation to evaporation but the vast difference between the so-called "humid" portion of the year in Curlew Valley, compared with Koonamore, is obvious. This difference is exaggerated because the "humid" period, in the terminology of Walter and Lieth, corresponds to a low temperature period at Curlew Valley.

   The surface soils beneath these communities vary between heavy clays and light-textured sands. They are alkaline and not usually excessively saline, although marked salinity deep in the profile is not uncommon. Data cited by Jessup (1969) for *A. vesicaria* soils in the Koonamore region are comparable to those cited by Billings (1949) for *A. confertifolia* soils in Nevada. Australian arid-zone soils are particularly deficient in both nitrogen and phosphorus in the upper part of the profile compared with either humid zone soils or arid-zone soils elsewhere (Charley and Cowling, 1968; see Fig. 7.2). Although the availability of mineral nutrients may potentially limit the duration and extent of plant growth in these communities, either directly or indirectly (by way of microbial activity in the rhizophere), neither it nor the temperature regimes seem to be the primary limiting growth factors in the shrublands. Water is probably the predominant factor limiting growth in these communities which are characterized both by low and extremely variable precipitation. Commonly, when temperature is optimal for growth, precipitation is minimal and evaporative demand is at its maximum. Thus these communities are typical of the arid ecosystems defined by Noy-Meir (1973) as "water-controlled ecosystems with infrequent, discrete and largely unpredictable water-inputs". In both semi-arid and arid regions, low values of mean annual rainfall are associated with a high percentage variability of falls in time, and this variability is equally marked in spatial terms. The frequency and amount of water inputs to *Atriplex* communities of the Great Basin and southern Australia are substantially unpredictable on either a seasonal or annual basis. The three-year cumulative rainfall records kept at Koonamore in South Australia suggest that a series of low rainfall years are interspersed with higher rainfall years (Hall et al., 1964), but even on this expanded time-scale these sequences are essentially stochastic and there appears to be no statistical basis for the legendary cycles of good and bad years (Noy-Meir, 1973).

Although both the Curlew Valley and the Koonamore communities share these properties of low, stochastic precipitation, there are important differences. Winter snow and low temperature in the Curlew Valley communities effectively restrict biological activity to the spring and the growth response may be interpreted in terms of one single annual pulse of water. This is shown by the detailed phenology of *Atriplex confertifolia* given in Fig. 5.5. In the Koonamore and other similar warm desert communities of South Eastern Australia, biological activity can be initiated in response to precipitation at any time of year. As shown in Fig. 5.5, *Atriplex vesicaria* on the Riverine plain flowered in response to summer, winter, and autumn rainfall in 1962–1964 (Jones, 1969a). One might surmise that ecological processes, such as population dynamics and plant-to-plant interactions, may therefore be much more complex in these communities than in the cool-desert communities of the Great Basin (Chap. 6.3.5).

Because of the essentially stochastic nature of rainfall in arid regions, simplistic predictions of relationships between plant responses and data based on mean precipitation can be seriously misleading (Anderson, 1971a). The effectiveness of a certain amount of water will vary with factors such as season, rainfall intensity, and topography, as well as physicochemical soil conditions which determine the extent of local infiltration, retention, and evaporation, and hence availability of water for physiological processes. Thus it is useful to qualify the term effective precipitation in relation to the process under consideration. Effective precipitation for seed germination, for example, may be less than that required for seedling establishment or more than that required to initiate mineralization of nutrients in litter. For a particular process, the effective precipitation may vary seasonally and when temperatures are too low for that process to continue, any precipitation will be ineffective except to recharge the soil water reservoir.

The arid shrub steppe dominated by perennial *Atriplex* species also supports a rich flora of annuals, including species of *Atriplex* and other chenopodiaceous

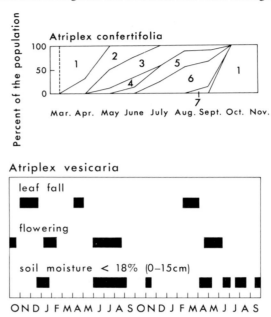

Fig. 5.5. Phenological observations on *A. confertifolia* and *A. vesicaria* communities. Estimates of the percentage of *A. confertifolia* shrubs at each phenological stage are given. The stages correspond to *1* vegetative buds closed; *2* buds opening; *3* buds open shoot elongation; *4* floral buds developing; *5* flowering; *6* fruit developing; *7* twigs woody and yellow (M.M. Caldwell; unpublished). Qualitative observations on *A. vesicaria* redrawn from Jones (1969a)

genera. These ephemerals are much more immediately sensitive to inputs of precipitation and may, in fact, represent a major productive portion of the community. Regrettably, the relationships between the persistence of the perennial vegetation and the performance of these short-lived plants has received scant attention so far.

5.1.2 Arid Halophytic Communities

Although the soils of the shrub steppes described in the previous section are not excessively saline, the dominant perennial species are halophytes in that their distribution frequently extends to much more saline regions. *A.confertifolia* and *A.nuttalli* both extend to the saline margins of the Great Salt Lake of Utah, and *A.vesicaria* can be found in highly saline soils of the Riverine Plain, New South Wales. However, the largest proportion of the species in the genus *Atriplex* are halophytes associated with saline and alkaline soils of continental interiors. The ecological notes accompanying the taxonomic account of North American species of Hall and Clements (1923) indicate more than 50% of the *Atriplex* species are common in these habitats.

Table 5.1. Generalized sequence of halophytes indicating the position of *Atriplex* spp. along salinity gradients in arid regions. (After Chapman, 1960)

Approximate total soluble salts (%) (0–30 cm)	North America (Great Salt Lake)	Mediterranean	Central Asia (Turkestan)
2.5	*Salicornia* *Allenrolfea* *Sarcobatus* *Suaeda*	*Arthrocnemum* *Halocnemum* *Suaeda* *Salicornia*	*Salicornia* *Halocnemum* *Halimione*
0.5	*Atriplex* spp.	*Atriplex halimus*	*Atriplex cana* *Anabasis*
0.02	*Artemisia*	*Tamarix*	*Artemisia* *Kochia*

Some species such as *A. phyllostegia* and *A. lentiformis* are extreme halophytes, occurring on soils with 3%–4% total soluble salts (at depths of 0–30 cm). Most, however, are found on soils of intermediate salinity. Table 5.1 shows three generalized descriptions of the position of *Atriplex* spp. in relation to other species along gradients of decreasing salinity in the vicinity of arid salt pans in three regions. Constructed from the descriptions collated by Chapman (1960), Table 5.1 indicates that *Atriplex* spp. are usually confined to the moderately saline soils, containing less than 0.5% total soluble salts. Floristic and ecological data on the inland halophytes of the central western United States were recently reviewed by Ungar (1974). He reported *A. triangularis* as widespread, co-dominant with *Hordeum* in saline soils containing less than 1% total soluble salts (0–30 cm).

On a local scale, these communities tend to be banded in parallel with the gradient of salinity away from a salt pan, water course, or other source of salt. This banded structure is characteristic and is similar to that encountered in coastal (strand or saltmarsh) communities. On a larger scale, the halosere on the banks of the Great Salt Lake, Utah displays an extensive gradient of halophytic vegetation. With increasing salinity the *Atriplex* comprised a smaller proportion of the total vegetation and Goodman (1973) noted that productivity of the *A. nuttallii* increased with decreasing salinity.

An occasionally striking feature of *Atriplex*-dominated shrublands in the continental interior of Australia is the occurrence of banded vegetation, either as concentric zones reflecting the pattern of depression and shelf components of gilgai mosaics (Charley, 1959), or as interspersed arcs of vegetated and nonvegetated soil usually aligned along topographical contours in the interfluves of rolling, hilly country (Fig. 5.6). Shelf surfaces of the mosaic are highly saline (> 1.5% chloride) whereas the depressions are only moderately saline (< 0.1% chloride), at least in the absence of grazing (Chap. 5.8.2). As a result of this spatial variation in salinity, *Atriplex* and associated species are mainly found in the depressions, whereas the peripheral shelves are only lightly vegetated. Banded concentric patterns found in some parts of the *A. vesicaria* shrubland are thought to arise as a result of slope and

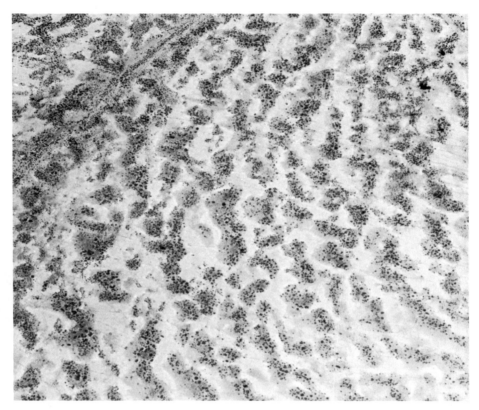

Fig. 5.6. Aerial view of patterned *A. vesicaria* community in the Barrier Ranges, New South Wales. Individual shrubs appear as *dark spots on the gray areas* which correspond to the nonsaline depression zones which follow the contour to the drainage channel at the *top left*. The *white areas* between the depressions are the highly saline shelf regions which are devoid of vegetation. (Courtesy J.L. Charley)

drainage interactions. The ability of these halophytes to modify the micro-environment and to impose and maintain complex patterns of ion and nutrient distribution in the soil is discussed later (Chaps. 5.6.2; 5.8; 7.6).

In later chapters we will also discuss the details of plant processes in one of the most arid, and most saline, habitats in which higher plants are found. The work of the Carnegie Group has included extensive studies of plants in natural habitats on the floor of Death Valley, California, as well as in cultivation at the garden adjacent to the National Park Headquarters near Furnace Creek. Figure 5.7 shows a modified climate diagram illustrating the extremely high temperature and extraordinarily low rainfall of this habitat. The perennial vegetation of this region includes *Atriplex canescens*, found on sandier, elevated saline soils, *Atriplex hymenelytra*, which forms extensive pure stands in the gravel fans of moderate salinity (0.5% total soluble salt), and *Atriplex polycarpa*, which is found in similar but less saline habitats (Hunt, 1966). Another shrub species of particular interest in

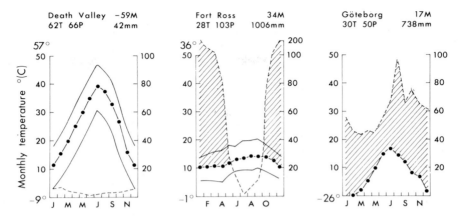

Fig. 5.7. Modified climate diagrams showing features of a hot desert region, Death Valley California and cool coastal regions at Fort Ross, 30 km north of Bodega Bay, California, and at Göteborg, Sweden. In addition to mean monthly temperature (●——●) the mean daily maximum temperatures and mean daily minimum temperatures are also shown for Death Valley and Fort Ross. Monthly precipitation is shown by a *broken line*

this extreme habitat is *Larrea divaricata* which extends from the bottom of Death Valley to well into the surrounding mountains (-80 to $1,2000$ m elevation), frequently associated with *Encelia farinosa*. In the valley bottom the herbaceous shrub *Tidestromia oblongifolia* is frequently found in mixed stands with *Larrea* and *A.hymenelytra* along the margins of washes in regions of relatively low salinity. Extremely salt-tolerant species of these habitats, such as the grass *Distichlis spicata* and the succulent *Allenrolfea occidentalis,* have not yet been studied in detail.

The xerophytic perennial plants in Death Valley are evidently established in particularly favorable seasons and maintain contact with adequate ground water throughout their long lives (Gulmon and Mooney, 1977). The patterns of plant activity throughout the year are not much regulated by rainfall, but seem to be primarily controlled by temperature. Figure 5.8 shows the activity of three perennial species, *A.hymenelytra, L.divaricata,* and *T.oblongifolia* as indicated by their photosynthetic capacity in situ at different seasons. *A.hymenelytra* and *L.divaricata* are both active throughout the year, the former being most active during spring, the latter about equally active at all seasons except at the peak of summer heat. *T.oblongifolia* is of particular interest, showing maximum activity during the hottest period of the year and none at all during the more favorable winter and spring.

Although the pattern of activity in perennial vegetation of Death Valley shows marked temperature dependence, the activity of annual plants in arid saline habitats is probably principally controlled by rainfall. It is also likely that the effective precipitation required for a specific process in these saline habitats is greater than in less saline habitats because salinity modifies the water relations of soils (Chap. 7). First, saline soils are usually finely textured and owing to the high concentration of Na^+ ions, clays are highly dispersed, leading to slow penetration of water. Second, the high concentrations of dispersed clay and soluble salt

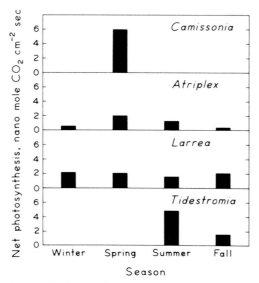

Fig. 5.8. Pattern of photosynthetic activity of different perennial species in Death Valley, California. (From Mooney et al., 1976)

contribute to substantial matric potential and osmotic pressure so that the water potentials even of wet saline soils are low. Thus plants growing in these soils are subjected to "physiological drought" even when the soil is wet. We will show later that to a large measure, tolerance to salinity in halophytes can be functionally equated with tolerance to water stress. Thus although plant distribution in arid saline communities is conveniently described in relation to soil salinity, it is probable that for annual species at least, water remains the "master input".

5.2 Coastal Communities

Although the largest number of *Atriplex* spp. is found in arid and saline regions of continental interiors, the most widely distributed species of the genus are found in plant communities near the sea. Species of the *A. triangularis* complex are common on the foreshores of most continents.

Some workers have proposed that the halophytic vegetation of arid hinterlands has been derived from coastal communities (Burbidge, 1960; Chap. 2.3.3). The inland halophytes discussed by Ungar (1974), for example, may have entered into the Great Lakes of North America during the post-glacial period, from Europe or Asia either via polar regions or by the migration of coastal species. These communities certainly display an impressive similarity of growth forms and generic composition. However, the environment of these coastal communities differs greatly from that of the arid communities and it is perhaps more profitable to explore these differences rather than the apparent similarities.

Marine habitats differ from those of the continental interior in that plant growth in them is not usually water-limited; indeed, precipitation patterns in most maritime environments are more predictable and rainfall is usually more substantial. *Atriplex* species of two coastal regions, western Scandinavia and northern California have been studied extensively and the climates of these regions are shown in Fig. 5.7. The Californian coast shows a typical Mediterranean climate with a rainless summer. This is in marked contrast to the more nearly uniform precipitation encountered in coastal northern Europe. The rainfall pattern for Göteborg, on the Swedish west coast, shown in Fig. 5.7, is more or less an average for north-west Europe. However, although the growing season at Bodega Bay is much longer than at Göteborg, the temperatures during the growing seasons are similar. Thus, well-irrigated plants in the Bodega Bay transplant garden are exposed to temperatures similar to those prevailing in the habitats of the many annual *Atriplex* species of coastal northern Europe, referred to in Chap. 4. The temperature regime in the habitats of *A. novae-zealandiae* on the single beaches of New Zealand is also similar to that of coastal north-west Europe.

In the natural communities of perennial *Atriplex* species on the southern Californian coasts, however, the low winter rainfall and high summer temperatures create much more arid habitats than those of northern California (De Jong, 1977). The extremely arid western coastlines of Australia, Africa, and South America are nearly devoid of *Atriplex* species.

Physical disturbance, rather than climatic, edaphic or biological factors, has been identified by Chapman (1964) and Ranwell (1972) as the overriding environmental constraint in the sand dune and salt marsh communities. The sand or mud is subject to short- and long-term predictable disturbance from wind and tide on a daily and seasonal basis. However, there is also an element of unpredictability associated with high seas which is comparable to that of unpredictable precipitation patterns in arid regions. Although the sand dune and salt marsh may co-exist in close proximity to each other and to the sea, they are subject to a very different complex of environmental factors over and above that of physical instability. Not surprisingly, closely related but different species of *Atriplex* may be associated with these two habitats.

5.2.1 Sand Dune Communities

The coastal sand dune provides a complex of micro-habitats which support a frequently banded plant community parallel to the coastline. Wind and tide as associated influences are the principal sources of instability in the sand dune building and deflation, and are therefore responsible for the formation and differentiation of these habitats (Chapman, 1964; Ranwell, 1972). We shall confine our brief comments to three habitats colonized by *Atriplex* spp.: the drift line, the foredune and the dune slack. All three habitats are comparatively deficient in nutrients and soil moisture, particularly in the surface 30 cm, but water is usually available at depth and a distinct water table is usually present.

On the drift line, the accumulation of tidal litter is commonly held to provide sufficient stability, nutrients and additional water-holding capacity for germination and establishment of annual species. Species of the *Atriplex triangularis* complex are frequent pioneer plants in the drift line (Fig. 5.9). Hulme (1957) followed the advance and retreat of *A. triangularis (hastata)* seedlings down the drift line of an

Fig. 5.9. Coastal *Atriplex* communities. *A. triangularis* seedlings establishing among debris on the drift line (*upper*) and *A. cinerea* on the foredune (*lower*), south-east coast, Australia

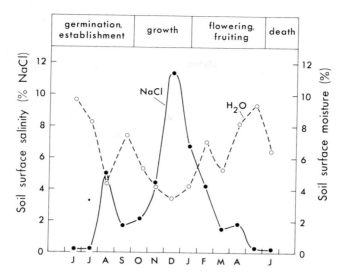

Fig. 5.10. Phenology of *A. novae zealandiae* in relation to changes in soil water and salinity content of a shingle beach in New Zealand. (Data of Ward, 1967a, b, c)

East Lothian dune in response to seasonal changes in high tide. The band of *Atriplex* seedlings was approximately 11 m wide before it was removed by the return of the fall tides. Similar banding of *A. triangularis (hastata)* in the debris at high tide line has been noted in South Australia, and members of the *A. triangularis* complex are found on the drift lines of Scandinavian coasts, as well as along the Pacific and Atlantic coasts of North America (Gustafsson, 1973a; MacDonald and Barbour, 1974; Duncan, 1974). On the Atlantic coast of North America *A.pentandra* and *A.arenaria* are found in similar drift line habitats (Duncan, 1974).

Ward (1967a,b,c) provided comprehensive descriptions of the habitat and vegetation of shingle or shell barrier beaches in New Zealand on which *A.novae zealandiae* is established on drift lines and temporarily stabilized crests. She records that the shallow roots of this species were not in contact with the water table. The water and nutrient relations of *A.glabriuscula* and *A.sabulosa,* relatively shallow rooting species found together in the drift lines of northern European shores (Gimingham, in Chapman, 1964), are equally obscure. Species in all of these sites are presumably exposed intermittently to sea water. As will be emphasized later (Chap.7.1.1) the high salinity levels which are encountered late in the growing season (Fig.5.10) may be a significant factor in the survival of annual species in the driftline. On the Pacific coast of North America Barbour et al. (1976) found *A.leucophylla* present at the leading edge of 13 of 34 beaches surveyed. They concluded that this perennial species must be particularly well adapted to substrate movement, salt spray deposition and seawater immersion. Seedlings of this species were later proven to suffer least mortality of 17 driftline species when subjected to salt spray and seawater immersion experiments in the laboratory (Barbour and DeJong, 1977).

Further up the beach from the drift line it is common to find *Atriplex* spp. amongst the earliest perennial vegetation of the fore dune. In eastern and southern Australia, *A. cinerea* (Fig. 5.9) occupies this habitat (Specht, 1972) corresponding to that occupied by *A. leucophylla* in California (Martin and Clements, 1939), *A. pentandra* and *A. arenaria* on the Atlantic coast of North America (Duncan, 1974), and *A. halimus* on Mediterranean sand dunes. Although the surface sand is usually very arid, a permanent water-table relatively close to the surface is a feature of the sand dune community which sets it apart from the arid communities discussed above. The soil water is quite low in salts and other nutrients. Martin and Clements (1939) report that within 2.5 m of the high tide mark, soil water in the root zone of *A. leucophylla* contained less than 25 mM NaCl, compared with sea water at about 500 mM NaCl. The salinity and water relations, as well as the microclimatology of the foredunes, have been well characterized for two sites on the Californian coast by DeJong (1977) and are discussed later (Chap. 7.1.1; Fig. 8.3). Although soil salinity in the rooting zone was usually less than that of seawater, salt spray rather than soil salinity is a major factor in the zonation of foredune vegetation.

Beyond the dune crest the comparatively level areas of the dune slack provide a much more stable habitat. Dune slacks range from dry slacks, which are little different in water status from the fore dune, to wet slacks which are almost marsh-like (Ranwell, 1972). About 20% of the North American species of *Atriplex* are found in the dune slacks of the eastern and western seaboard. Species such as *A. triangularis* may be found both in the drift line and in the slack, particularly in wet, saline slacks. These sand dune habitats beyond the drift line are usually low in nutrients and salt. There is a noticeable decrease in NaCl content of the sand from the drift line to the slack, and a large increase in the organic carbon and nitrogen content toward the vegetated soils of the slack (Gorham, 1958; Willis et al., 1959). A water table is found in the slack but its depth is closely correlated with seasonal precipitation (Ranwell, 1972).

5.2.2 Salt Marsh Communities

Along the western Pacific coast of the United States, plants of the *A. triangularis* complex are found on the drift line and in salt marshes ranging from about 60° to 30° N. Figure 5.11 prepared from the summary of MacDonald and Barbour (1974), shows, however, that two sand dune species (*A. leucophylla* and *A. julacea*) coexist with the *A. triangularis* complex from about 40° to 30° N, but do not enter the marsh. Two other species in the same region (*A. semibaccata,* introduced from Australia, and the native *A. watsonii*) are found in the salt marsh but not on the sand dunes. In Southern Australia, *Atriplex triangularis* and *A. paludosa* are common in the upper regions of the marsh, behind mangroves and *Arthrocnemum* (Specht, 1972). Table 5.2 shows that, in Scandinavia, only *A. triangularis* is found in all these types of coastal habitats, but that *A. glabriuscula* may be found in the driftline, on the foredune and in sheltered marshes.

The salt marsh habitat has been extensively described by both Chapman (1960) and Ranwell (1972). Low in the marsh, the habitat is exceedingly unstable, tidal influences being the principal source of instability. In the upper marsh where

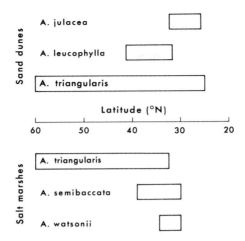

Fig. 5.11. Distribution of *Atriplex* species in sand dunes and salt marshes along the Pacific Coast of North America. (Data of MacDonald and Barbour, 1974)

Table 5.2. Maritime habitats for species of the *Atriplex triangularis* complex in Scandinavia. (After Gustafsson, 1973a)

Habitat	Vegetation type	Substrate	Distribution of *Atriplex* spp[a]			
			A. calotheca	*A. glabriuscula*	*A. longipes*	*A. triangularis*
Drift line[b]	Low	Stones Gravel		+		+
Foredune[b]	Low	Sand	(+)	(+)		+
Exposed salt[b] marshes, Baltic	Low	Stones Gravel Mud			+	+
West coast	Low	Stones Gravel Mud		+	(+)	+
N Norway	Low	Stones Gravel Mud			+	+
SW Denmark	Low	Sand Mud				+
Sheltered salt marshes	Tall (*Scirpus*) (*Phragmites*)	Sand Mud Sand Mud	+	+	+ (+)	+ +

[a] Common + ; rare (+). *A. longipes* sub-species shown together
[b] These communities subject to immersion in seawater

Atriplex spp. are most commonly encountered this effect is minimal and, aside from the unpredictable incursions of high tides or storms, this habitat is physically more stable. It is a mesic, saline environment and soil water, although abundant, usually has a salt and nutrient composition similar to that of seawater (Chap. 7.1.1).

5.3 Weed and Ruderal Communities

As indicated in Chap. 2, many species of *Atriplex* are weeds indigenous to Europe and North America, and while it is fair to regard *Atriplex* as a cosmopolitan genus in its global distribution (Good, 1974) it is also likely that a large and certainly indeterminate part of its present range has resulted from the adventive and weedy characteristics of some species.

The palynological record for some of these weedy species, while neither continuous nor complete, indicates that their "weediness" has developed in association with human activity. Pollen of the *Atriplex triangularis* complex, for example, is known from British Interglacial, Full and Late-Glacial, early Post-Glacial and Early Zone VII deposits, occurrences that were clearly independent of human activity. It has also been recorded from Full-Glacial deposits of the Lea Valley, again pointing to its ancient native status (Godwin, 1956). Later pollen records, however, are mostly associated with human occupation sites, with records from Roman times predominating.

The more recent historical record provides evidence, not only on movement of adventives associated with the European-Australian wool trade and the deliberate introduction of nonindigenous species for fodder or fuel resources, but also of direct human utilization of some species for food. *Atriplex hortensis,* a tall annual known as the garden orache, was cultivated as a vegetable from early times through to the Renaissance, when it was replaced by spinach (a related chenopod, *Spinacia oleracea*). Other species have become weedy in relatively recent times; a closely related species to *A. hortensis, A. acuminata (nitens)* had an indigenous distribution ranging from Western Asia through to eastern Europe, but after World War II it began to spread westward as a colonist of debris on bombed sites. Similarly, *A. rosea* appears to have established itself effectively among the debris of the automobilized culture of California (Fig. 5.12). Some species, such as *A. triangularis,* have become effective colonizers of disturbed habitats created by industrialization: for example Hulme (1957) records this species as common on slag heaps in mining areas. In Britain (Jones, 1975) and in Finland (P. Uotila, personal communication) *A. patula* is principally a weed of nutrient-rich waste ground and arable land which is disturbed by agricultural practice. *A. latifolia* has become a common weed in sugar beet fields of Finland, seed having been introduced along with beets imported for processing from other Scandinavian areas (Uotila, 1975).

It is interesting to note that in Australia and North America endemic species of *Atriplex* have readily occupied habitats disturbed by man, the regularly graded verges of roads being a good example (Fig. 5.12).

Introduced species of the same genus are now found in similar habitats. There are several species, such as *A. subspicata,* endemic to north-western North America, which are now common roadside weeds and weeds of cultivation. The distribution

Fig. 5.12. *Atriplex* spp. as ruderals in disturbed habitats. *A. rosea* (*top*) 4 months after invasion of an abandoned drive-in movie theater in Palo Alto, California. *A. conduplicata* (*bottom*) *A. angulata* and other annual species carpeting a recently cleared roadside following spring rains, western New South Wales

Fig. 5.13. Distribution of endemic and introduced ruderal species in western United States and Canada. The distribution of endemic *A. subspicata*, which is now a widespread ruderal species, is *outlined* and *closed circles* show recordings of introduced *A. hortensis*. (Redrawn from Frankton and Bassett, 1968; Bassett and Crompton, 1973)

of these species, and that of introduced *A.hortensis* overlap to a large extent (Fig. 5.13).

Against this historical background it is not easy to define weed or ruderal communities. If we accept Baker's (1965) definition of a weed as one whose "populations grow entirely or predominantly in situations markedly disturbed by man" then certainly some species of *Atriplex* can be said to be weeds. *A.rosea*, *A.serenana*, and *A.expansa* are typical weeds of open ground. These species are very common weeds in the intensively cultivated Great Valley of California and are rarely, if ever, found in undisturbed habitats. However, most *Atriplex* species which occupy disturbed situations also inhabit undisturbed communities. For example, *A.heterosperma*, a fairly common weed along ditches and fence-cuts in California's Great Valley, is quite common in alkali sinks and flats in deserts far removed from human influence. Similarly, *A.triangularis*, although a colonizer of several kinds of disturbed ground, is very common in undisturbed communities such as coastal strand and salt marsh (Jones, 1975). Figure 5.14 contrasts the distribution of this species in the British Isles, where it is widely distributed in lowland, cultivated areas, with that of *A.glabriuscula*, another maritime species which, although sharing the coastal habitat, is not usually found in disturbed habitats. Thus, not only is it difficult to specify the environment of a disturbed habitat in very precise terms, but

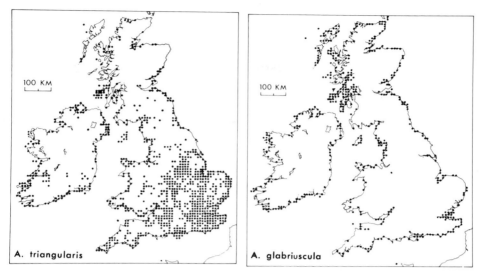

Fig. 5.14. Distribution of *A. triangularis* (*hastata*), a weed of openground which extends into cultivation and coastal habitats in the British Isles, and of *A. glabriuscula*, a coastal species which shares many habitats with *A. triangularis* but is not a ruderal. (Redrawn from Perring and Walters, 1962)

it is also difficult to circumscribe the physical, climatic, and edaphic limits for weedy and ruderal species of *Atriplex* and the communities to which they contribute.

In fact, this difficulty serves to highlight an important element in our understanding of the biology of *Atriplex*: the weedy and ruderal species of this genus occupy niches well within the tolerance limits on the individual species. Baker (1974) places a considerable emphasis on the concept of the "general purpose genotype" as an essential genetic base for a successful weed. Furthermore, he suggests that a plant is more likely to be independent of environmental restraints if its physiological processes can be adapted to changed environmental conditions. In this brief survey we have seen that many species of *Atriplex* occupy environments which are subject to considerable environmental "disturbance", whether it be climatically, edaphically, or biotically induced. It seems then eminently reasonable to suppose that, in studying *Atriplex,* we are probably studying a range of populations that have evolved from a broadly based genome. That broad basis has allowed the development of modern species stocks which have radiated into a wide range of "disturbed" and apparently often inclement habitats, meanwhile exploiting a gamut of strategies associated with growth-form, perenniality, reproductive behavior and physiological processes.

5.4 The Major Atriplex Habitats – a Resumé

From this short summary of the major habitats in which *Atriplex* spp. are found as a significant floristic component, it will become apparent that in this genus salt- and water-relationships, combined with a more general ecological ability to utilize effectively variously disturbed habitats, are important ecological determinants. In order to achieve a fuller interpretation of these relationships, we will have to subject these physiological and behavioral processes to experimental analysis. Before we move into an experimental mode, however, it may be worthwhile to see if natural variations in the field distribution and abundance of *Atriplex* plants can be analyzed in relation to corresponding variation in the habitat; that is, using the existing patterns of communities and associated environments as a "natural" experiment which may be subjected to appropriate analysis.

Using the philosophy – that naturally occurring field "experiments" can be analyzed usefully if appropriate quantitative techniques are employed – we shall in the next section concentrate our attention on approaches that have been used by ecologists to describe and interpret regional environmental variation in *Atriplex*-dominated shrublands. Quantifying the variation in these communities and assessing it in relation to the major environmental constraints should allow us another and somewhat differently based opportunity to analyze and interpret the factors and processes that underlie and determine the distribution of plants and vegetation.

5.5 Regional Variation in Atriplex Shrublands

Many techniques of quantitative vegetational analysis have been developed and advocated during the last 25 years, but for all the debate there has been on the utility and precision, robustness, and applicability of these various methodologies, it now seems fair to state that no one approach is self-sufficient or ideal in the solution of a particular ecological problem. This state of affairs is not surprising when it is recalled that many quantitative methods, including for example variance, principal component and factor analyses, although in widespread use in ecological contexts now, originally were developed for use in the agricultural or behavioral sciences. Their adaptation, development, and refinement in ecological contexts, as a vehicle for data-structuring or hypothesis-generation, has in no way removed the basic need for a clear formulation of a problem and the scale or scales on which that problem is to be tackled (Greig-Smith, 1957; Williams, 1968; Anderson, 1971c).

In the *Atriplex* shrublands we are concerned with a state description of vegetation within the span 10^9 to 10^{11} of the space-time scale referred to in Chap. 1. It is on this scale that the methodological strategy known as ordination has been particularly useful in structuring and analyzing regional vegetation variation (Whittaker, 1973) and we shall begin with a description of this technique.

5.5.1 Ordination of Large-Scale Variation in Plant Communities

On a regional scale – over an area covering several hundred square kilometers for example – the number of factors operating singly or in combination to affect the distribution and abundance of plants and therefore vegetation structure and dynamics is likely to be immense, and the initial problem faced by an investigator in this situation is to identify those factors which are most likely to be of overriding importance in determining these distributional patterns.

Ordination is one methodological strategy for examining such a potentially multivariate field situation. Defined by Goodall (1954) as "an arrangement of units in a uni- or multi-dimensional order", ordination is used to associate information concerning species and habitat (stand) occurrences in such a way that the arrangement provides some ecological insight into the processes that have generated the field distributions of species/stand occurrences. Thus the aim of such a "state" analysis is to reveal more precisely those relationships between plant and environment which are most likely to repay closer ecological and physiological investigation.

The preliminary field data used in ordination usually consist of estimates of the composition of a number of vegetation samples in terms of the occurrences or abundances of the species represented in those samples. Such field data can then be tabulated in a two-dimensional matrix, with the rows occupied by estimates of species abundances and with each of the columns representing a particular vegetation sample.

Since biologists are well used to displaying their data in a graphical form, we might say that a conventional two-axis graph is in fact a two-dimensional ordination. In an ecological context, however, we may have a data matrix made up of records of the occurrences of 50 species in 100 vegetation samples, and the graphical presentation of such multivariate relationships in an essentially three-dimensional world is not possible. However, while there may be 50 different species in our example, our ecological experience would suggest it is unlikely that they are responding to 50 different environmental parameters; indeed our experience from a qualitative appraisal such as that presented earlier in this chapter suggests that only a few such factors are particularly important. Thus we might expect to find a great deal of cross-correlation in our data matrix, and can therefore surmise that it should be possible to reduce the variability present in the original matrix to a handful (rather than 50 or a 100) dimensions.

The notion of a "dimension" in this context is merely an extension of our two-dimensional example; just as it is possible to calculate a line of "best fit" in the two-dimensional case, so analogously it is possible to find the lines – or dimensions – of "best fit" in a multivariate matrix, so that these dimensions progressively account for the greater part of the information (= variance) in the original matrix.

This then is essentially what ordination is about: a geometric appraisal of species distribution and abundances in a coordinate system determined by the correlations present in the data themselves, and hopefully displayed so that the major dimensions are likely to represent some meaningful floristic or environmental gradient, which we can then subject to further analysis or experiment.

Principal components analysis is one very useful and widely employed technique for providing this secondarily derived coordinate system, the dimensions or axes so derived being referred to as principal components or principal axes. The principle of the method embodies the requirement that the first component be located in the scatter of species-in-site data so that the sum of squares of the perpendicular projection of points onto it is the maximum possible; that is, it must account for as much as possible of the total matrix variance (Fig. 5.15a). The second component is then positioned so that it accounts for as much as possible of the

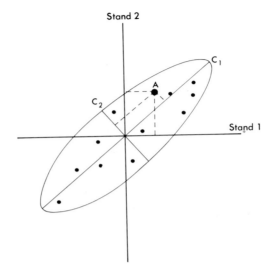

a

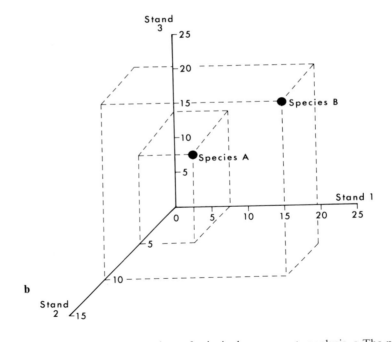

b

Fig. 5.15a, b. Geometrical representations of principal components analysis. **a** The positions of the point (representing species) are determined by reference to the original coordinate axes (relative abundance scaled for stands). The principal components (C_1, C_2) provide a simplified frame of reference. The loading of species A is in relation to these component axes. **b** Point representation of two species in a three-dimensional sample space $(N = 3)$. The *coordinate axes* correspond to stands, the *scales* representing the relative abundance of species. Corresponding vectors can be obtained by drawing lines from the origin to each of the points

remaining variance, and subsequent components are similarly located until all the matrix variance is accounted for (Fig. 5.15b). The lengths of successive components (axes) are proportional to that part of the total variance (i.e., the "latent root") each has accounted for, and in practical terms it has been found that the extraction of three or four components is usually sufficient to account for a major part of the total variance. In these circumstances the component analysis has achieved a simplification of the original correlation matrix by reducing the originally N-dimensional space to one of a few dimensions, which may be more readily interpreted in ecological terms.

The ordination is complete when the coordinates of points (their "loadings") are specified with respect to the principal components.

5.5.2 Ordination of the Australian Arid Shrublands

In an ecological study of the shrub-steppe vegetation of the Western Riverine Plain in south-eastern Australia, Williams (1972) utilized this multivariate approach to provide an economic description of variation in presence and cover of 41 species of shrubby perennials. His results for the preliminary principal components analysis are summarized in Table 5.3, which shows for each

Table 5.3. Principal components analysis of presence data (standardized by site norm) for chenopodiaceous and other perennials. (Data from Williams, 1972)

Component	λ	% Efficiency	Species
1	202	42.2	A. vesicaria B. brachyptera Sporobolus caroli Chloris truncata
2	51	10.6	B. brachyptera B. divaricata A. inflata Sporobolus caroli A. vesicaria
3	33	6.8	Disphyma B. brachyptera Sporobolus caroli Minuria B. divaricata
4	31	6.4	B. quinquescuspis B. tricuspis Sporobolus caroli A. inflata A. angulata
5	23	4.7	K. pyramidata Chloris truncata A. angulata B. tricuspis Disphyma

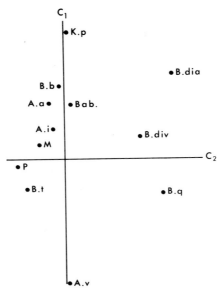

Fig. 5.16. Species loadings of some chenopod perennials on principal components C_1 and C_2 of species standardized presence data. (From Williams, 1972). *A.a*, *Atriplex angulata*; *A.i*, *Atriplex inflata*; *A.v*, *Atriplex vesicaria*; *Bab*, *Babbagia*; *Bb*, *Bassia biflora*; *B. dia*, *Bassia dicantha*; *B.div*, *Bassia divaricata*; *B.q*, *Bassia quinquecuspis*; *B.t*, *Bassia tricuspis*; *M*, *Malacocera*; *P*, *Pachycornia*

component the relative importance (in terms of a so-called eigenvalue, λ) of each component, the percentage efficiency in accounting for the total variance, and the species with extreme positive and negative loadings on that component (and therefore those species "reacting" most strongly to that component). The loadings of the chenopodiaceous species are plotted on the first two components in Fig. 5.16 to provide a species ordination.

Although the first component accounts for 42.2% of the total matrix variance, and the first five components cumulatively account for 70.7% of the variance, Williams had to extract no less than 20 components to account for 96.6% of the total variance in the analysis. The difficulties of providing an intelligent ecological interpretation of every extracted component are immense, and consequently he shared the pragmatic view of other authors describing comparable studies when he attempted an ecological interpretation of only the first few components.

However, for our present purpose the first two components proved to be of considerable interest, since they served to contrast the field distribution of *Maireana pyramidata* vs. *Atriplex vesicaria* (component 1) and contrasted those sites containing *Bassia divaricata* and *Atriplex inflata* with those sites containing neither species (component 2). Williams was able to demonstrate from an associated ordination of soil variables that the first component was related to those soil features which indicate a greater depth of wetting of the *Maireana* soils compared with the *Atriplex* soils, whereas the second component was more closely related to the invasion by annuals or short-lived perennials (*B.divaricata*,

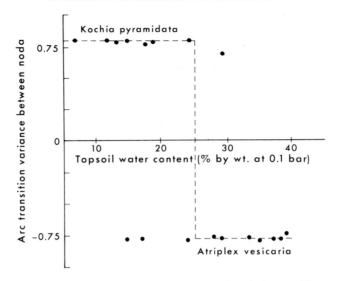

Fig. 5.17. Stepwise dominance transition between *Maireana* (*Kochia*) *pyramidata* and *Atriplex vesicaria* in relation to percentage of available topsoil water (extracted at 0.1 atm). (Redrawn from Noy-Meir, 1970)

A. inflata) of perennial chenopod-dominated pastures which had been subjected to periodic overgrazing by sheep.

Thus despite the large area (25,000 km²) he had surveyed, with its complex floristic, environmental, and historically determined variation, Williams was able to focus his detailed (and later experimental) attention on two central ecological questions: how two economically important chenopod shrubs were physiologically "tuned" to variations in the underlying soil mosaic of the Riverine Plain, (Chap. 7.1.1) and how the direct and indirect roles of the sheep grazing (Chap. 6.3.3) were secondarily superimposed on this primary axis of ecological variation.

Noy-Meir (1970) used component analysis similarly in his investigation of semi-arid vegetation covering an area of some 240,000 km² in south-eastern Australia – an area roughly equivalent in size to Great Britain. He followed his original component analysis of floristic variation (which included woodland, mallee, heath, as well as bluebush and saltbush vegetation types) with multiple stepwise regressions of 37 environmental variables on the original components defining floristic "noda". In our present context Noy-Meir's study of the dominance transition between bluebush (*Maireana*) and saltbush (*Atriplex*) stands in the Riverina is of particular interest, since it demonstrates (Fig. 5.17) the importance of water availability in the topsoil (0.1 atm pressure-extracted water content per weight) in discriminating between the shrubland types.

In general terms, saltbush occurs on soils of heavier texture than bluebush (Beadle, 1948; Carrodus and Specht, 1965) and it has been shown experimentally that *Atriplex vesicaria* can reduce soil moisture to a lower level than can *Maireana sedifolia* (Carrodus and Specht, 1965). Putting together the field and experimental data, Noy-Meir (1970) was able to propose a texture–moisture hypothesis (clearly related to our earlier emphasis on effective precipitation as an overriding ecological factor in these shrublands). This hypothesis was used to interpret the woodland-bluebush-saltbush floristic continuum, with the depth of soil moisture decreasing and the ratio of marginal to available moisture decreasing. At the same time the ratio of marginal to available water was increasing. The associated increasing soil salinity thus operated through the marginal water effect. This provided the ecological rationale for the observed transitions between the floristic noda of the continuum.

This brief synopsis of two investigations that have been based on a multivariate analysis of large-scale floristic and environmental variation of course does less than justice to the detailed and careful work conducted by Williams and Noy-Meir. It does, however, serve to emphasize one methodological approach to the interpretation of the complex patterns of plant variation and abundance in the field that has been used to generate hypotheses relating to the larger-scale features of the ecological state. It is these hypotheses (which at this scale are not only state descriptions but may also invoke process descriptions), rather than any detailed causal explanation and interpretation, that provide useful insights into the key environmental variables which must be assessed when smaller-scale field studies are undertaken. One methodological approach to such smaller-scale ecological appraisals of variation in plant distribution and abundance is described in the next section.

5.6 Local Variation and Pattern in Atriplex Shrublands

There are within the regional framework of variation discussed in the previous section other sub-sets of relationships or patterns that may be related to what we may call local plant/environment interactions. This variation corresponds approximately to the scales 10^6 to 10^9 in space (meter to kilometer) in Fig. 1.1, just as the phenological and demographic processes potentially relevant to this variation have a relaxation time corresponding to 10^6 to 10^9 s. One approach which has been used consistently to investigate these local variations in *Atriplex* field populations is that of pattern analysis (Greig-Smith, 1952, 1957) or what has been called block size variance analysis (Noy-Meir and Anderson, 1970). In order to provide an idea of the utility of this approach we shall describe the method briefly before examining the results of such analyses from particular field stations.

5.6.1 Pattern Analysis

Just as the primary purpose of ordination is to identify ecological relationships between species occurrences and floristic or environmental gradients, so the major objective of pattern analysis is to identify *recurrent* patterns among otherwise similar relationships. Thus, this approach comes into its own at scales of 10^6 to 10^9, where some of the grosser environmental variables, such as rainfall or temperature regime are for all practical purposes uniform, thereby approximating in the field a "natural" experimental situation in which we can think of some of the environmental factors at least as being "held constant".

Once again the basic data available to us are records of species occurrences or abundances (usually some measure of density, frequency, or cover) in defined sites (usually quadrats or parts of a transect). The object of the exercise is to partition the variability (variance) in the data so that we can identify the scale(s) at which such variation is at a maximum, indicating thereby a significant departure from a random distribution. Thus the essence of the approach is a search for significant departures from a random distribution, the underlying assumption being that a random distribution is evidence of a lack of overriding environmental determination, whereas a nonrandom or patterned distribution is evidence of a patterned part of the environment at a similar scale.

5.6.2 Pattern Analysis in Arid Atriplex Shrublands

In the case of pattern analyses conducted on *Atriplex* populations in Australia, the striking feature is their overall consistency, not only within a single area but also between geographically separated areas. If we first review the pattern analyses that have been conducted on one perennial species of saltbush that occurs widely throughout southern Australia – *Atriplex vesicaria* – we can explore the elements of this consistency further, and be in a position to assess two important aspects of the pattern of establishment and persistence of these populations in the field.

In a preliminary reconnaissance of *A. vesicaria* populations located in a range of environments from western New South Wales into the south-eastern corner of South Australia, Anderson (1967) found evidence for a consistent two-scale contagious pattern in cover data (see Table 5.4). Although the actual *scales* of contagion varied from site to site, and not surprisingly so considering the diversity of topographical, edaphic (coarse sands to fine clays) and effective rainfall conditions (ca. 150–350 mm per annum) encompassed by the range of sampled populations, the consistency in the *kind* of pattern found is somewhat surprising.

This consistency has been further emphasized by more detailed analyses of pattern in the density of individuals on sites carrying *Atriplex vesicaria* located in

Table 5.4. Summary of *Atriplex vesicaria* stands examined in south-eastern Australia showing scales, in terms of block size, at which significant patchiness (*) has been detected

Location	Soil type	Mean % cover	Block size (1 = 5 cm)								
			1	2	4	8	16	32	64	128	256
Hay Plains	Gray-brown clay	28.3			*			*			
Hay Plains	Red sandy loam	19.1				*				*	
Boolcamatta	Red sandy loam	27.8				*				*	
Koonamore	Red clayey loam	6.0				*					*
Koonamore	Loam over travertine	37.6				*					*
Koonamore	Loam over travertine	26.9				*					*
Koonamore	Loam over travertine	10.8		*							
Hay Plains	Red sand over clay	28.8							*		*
Koonamore	Gray-brown sand over clay	53.7							*		*

Table 5.5. Summary of peaks in density distributions of "young" and "old" plant categories for seven *A. vesicaria* (AV) populations in south-eastern Australia

Location	Age category	Mean density (plants/m^2)	Block size (1 = 1 m^2)					
			1	2	4	8	16	32
Riverine Plain, N.S.W.								
AV 1	Young	1.32	1.72		2.95		4.36	
	Old	1.28	←		All random		→	
AV 4	Young	4.00			2.89		7.89	
	Old	1.25	←		All random		→	
AV 5	Young	2.19		2.70			26.50	
	Old	1.98	←		All random		→	
AV 9	Young	2.06		1.99			3.67	
	Old	0.97	←		All random		→	
Brewarrina region, N.S.W.								
AV 6	Young	0.50	2.94			4.48		9.44
	Old	0.60	←		All random		→	
AV 7	Young	1.78	5.75				12.41	14.85
	Old	0.89	←		All random		→	
AV 8	Young	1.69						
	Old	0.68	←		All random		→	

the central Riverina of New South Wales (Anderson, 1967; Malik, 1970) in the Barrier Ranges north of Broken Hill (Anderson, 1970) and in northern New South Wales close to the Queensland border (Malik, 1970).

The original (and in retrospect, simplistic) interpretation of the small-scale pattern was that it was due to the intrinsic patchiness of this species and that the larger-scale pattern was due to edaphic and/or microtopographical variation in the associated soils. In pursuing this assumption it became clear from more detailed analyses of populations in the Riverine Plain that these "model" patterns were in fact a good deal more complicated than they appeared at first sight.

As an example of this subsequent approach to "successive approximation" (Poore, 1962), Malik (1970) found that the "standard" two-scale pattern was determined by the pattern exhibited by the younger individuals in the population only, and that the older elements of the population showed a progressive tendency toward a random distribution as their overall density fell (see Table 5.5). Furthermore, it is clear that both the intensity (i.e., degree of departure from randomness) and scale of pattern change as the individuals pass from a seedling to a mature phase in the mosaic of community phases (Fig. 5.18), the youngest individuals alone exhibiting a bi-modal pattern and subsequent phases showing a unimodal or completely random distribution.

From the data presented in Fig. 5.18 it is clear that the greatest shifts in mean density are related to those phases which show the highest intensity of pattern. An explanation of this phenomenon can be offered in terms of the varying selective action of the external environment on different stages – and particularly the earlier stages – in the life cycle of *Atriplex* (Chap. 6.2).

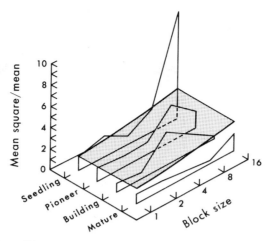

Fig. 5.18. Perspective diagram showing relationship between pattern intensity (mean square/mean), block size and age categories of *Atriplex vesicaria* populations from the Riverine Plain of Australia. The *rising horizontal hatching* represents the upper 95% confidence band. (Redrawn from Malik et al., 1976)

Table 5.6. Peaks (*) in graph of mean square against block size for soil variables estimated in one *Atriplex vesicaria* site in the Riverine Plain. The equivalent scales of contagion for density of young plants were block sizes 1 (2 m) and 8 (16 m) respectively

Soil variable	Block size				
	1	2	4	8	16
Sodium (Na)		*			*
Potassium (K)	*				*
Calcium (Ca)	*				*
Magnesium (Mg)	*		*		
Na/Ca ratio	*		*		*
(Na + K)/Ca ratio	*				*
(Na + K)/(Ca + Mg) ratio	*			*	
pH	*			*	
Moisture (%)	*			*	
Topography	*			*	

Further exploration of the pattern of edaphic variables (Table 5.6) indicated that the detected scales of plant patterns frequently corresponded in scale with some of these variables. It is not therefore surprising that in his preliminary interpretations of these likely time-dependent patterns Malik (1970) invoked the operation of microtopography, run-off and differential water penetration, patterning in soil characteristics, and grazing pressure as important factors that were likely to be involved in pattern determination.

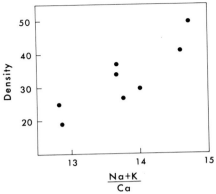

Fig. 5.19. Relationship between density of young plants and the ratio of soil Na + K/Ca at block size 16

Further reflection, however, reveals that the demonstration of correlated patterns of plant and soil parameters do not in themselves provide unequivocal evidence for the operation of ecological processes. For example, while Malik (1970) found that initial germination and establishment of *Atriplex* seedlings were closely related to periods of significant winter rainfall, Anderson (1970) presented field evidence to show that among an established seedling population, and within two days of a substantial rainfall, the distribution of young plants was itself the determinant of the subsequently measured pattern of (gravimetric) soil water. Secondly, although the density of young plants was significantly and positively related to a monovalent/divalent cation ratio (Fig. 5.19) at block size 16 in these analyses, the density of old plants at this scale was significantly and inversely related to this ratio. This fact was emphasized by the negative correlation ($r = 0.9958$, $P = <0.001$) between young and old plants at this scale. Thirdly, while the origins of the small-scale microtopographical variation were obvious in the field, the origin of equivalent variation at a much larger scale was not nearly as obvious.

This last point deserves further amplification before the significance of the detected pattern in soils cation values is discussed. Moore (1953) has deduced that the largest saltbush native to Australia (*A. nummularia* – old man saltbush) was an abundant species on the gray, heavy-textured soils of the eastern Riverina (and now carrying *A. vesicaria*) prior to the introduction of sheep to the Western Riverina in the 1860's (see also Buxton, 1967). *A. nummularia* was subsequently lost over large areas through excess sheep stocking and by the ring-barking activities of rabbits in particularly dry years. In localities still carrying *A. nummularia*, the associated soils exhibit a well developed gilgai structure, with puffs of the order of 20–40 m across. The pattern in cover of these communities also exhibits a single peak at a block size coresponding to the larger scale of pattern in *A. vesicaria* (Fig. 5.20). Furthermore, Sharma (1973b) has demonstrated the existence of a significantly higher sodium absorption ratio[1] in Riverina clay surface soils beneath bushes of *A. nummularia* compared with values for this ratio in soils developed between bushes (Table 7.1).

---

1   Sodium absorption ratio $= \dfrac{Na^+}{\sqrt{\dfrac{Ca^{2+} + Mg^{2+}}{2}}}$

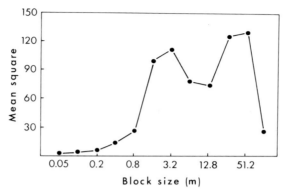

Fig. 5.20. Graph of mean square against block size for projective cover estimated in a relict stand of *A. nummularia* (Block size 1 = 5 cm). (Redrawn from Malik et al., 1976)

Taking these three lines of evidence together, it seems very likely that both the large-scale microtopographical pattern and the pattern in soil cations in modern *A. vesicaria* sites are relict state features imposed by an earlier plant community. Thus, much of the large-scale variation in pattern of both young plants of *A. vesicaria* and associated soil features is a reflection of earlier patterns imposed by *A. nummularia*. In turn, it is apparent that the mature phases of an *A. vesicaria* community are presently imposing a smaller-scale patterning of soil conditions, a feature reflected not only in the small-scale regeneration pattern of *A. vesicaria* but also in annual or semi-perennial populations of *Atriplex* that occasionally replace degraded areas of perennial saltbush country (Malik and Anderson, 1971).

These latter workers reported on the pattern present in four geographically separated populations of even-aged *A. inflata* located on the Riverine Plain, and demonstrated a strikingly similar two-scale contagious pattern for measures of density, cover, and height at scales of 2 to 4 m and 16 m in these four sites. In seeking an explanation for this consistent large-scale pattern they found a significant negative regression between survival (density) and variation in the microtopographical mosaic, whereas properties related to performance (cover and height) were more closely related to the cation balance of the soil in the same mosaic. This latter statistical correlation was subsequently tested in an experiment in which variation in dry weight yield of *A. inflata* plants was clearly related to a varying cation ratio in the experimental soils, although this major response was also mediated by soil-water availability (Fig. 5.21).

In the present context a significant feature of this report was the demonstration, in terms both of circumstantial field evidence and experimental data, that the soil patterns imposed by an *A. vesicaria* community (which presumably had in turn been influenced by the patterns established in an earlier stand of *A. nummularia*) were, two years later, still influencing the development of pattern in the *A. inflata* population which succeeded it. It seems clear that, once a population of *Atriplex* has become established (i.e., having successfully negotiated the most critical environmental selection connected with the phase of establishment; Chap. 6), its presence and persistence in turn impose various patterned features on the external

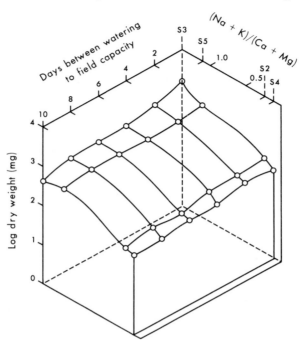

Fig. 5.21. Relationship between dry weight yield of *A. inflata* as a function of soil Na + K/Ca + Mg ratio and watering regime. (Redrawn from Malik and Anderson, 1971)

(and particularly the soil) environment. That such plant-induced patterns are a common feature of *Atriplex* populations is amply demonstrated by the work of Roberts (1950) and Fireman and Haywood (1952) on *A. confertifolia* (shadscale) in the United States and by Jessup (1969), Sharma and Tongway (1973) and Sharma (1973b) working on *A. nummularia* and *A. vesicaria* in Australia. The processes of nutrient cycling which establish these patterns are discussed in Chap. 7.

These analyses of local scale ecological state descriptions provide an elegant illustration of the point that persistence and performance are separate properties which determine states and processes in the ecological structure. In this case the niche for successive species populations of the same genus (*A. nummularia*, *A. vesicaria* and *A. inflata*) in a deflected successional sequence determined by grazing, is conditioned first by the establishment and persistence of the preceding species, and second by the performance of that species. The latter modifies the soil properties to the extent that these determine the state description of the next successional species.

5.7 Regularity in Arid Vegetation

So far in this review of pattern in *Atriplex* shrublands we have drawn attention to the repeated occurrence of either contagious or random distributions exhibited by the sampled populations, the implication being that regular patterns have not

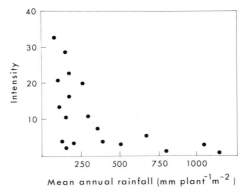

Fig. 5.22. Relationship between intensity (mean square/mean) of large-scale pattern and volume of water available annually on a per plant basis for 19 populations of *A. vesicaria* on the Riverine Plain

been detected. To some students of dryland ecology the lack of demonstrated regularity might appear surprising, since it has been argued frequently (e.g., MacArthur and Connell, 1966) that desert shrubs are generally distributed in regular arrays so as to effect a reduction in inter-plant competition for a limited water-supply. The reasoning that lies behind this assumption is illustrated by a quotation from Woodell et al. (1969), who provided evidence for some regular pattern in populations of *Larrea divaricata* (Zygophyllaceae) established in a California desert:

"... it is not altogether surprising that a shrub growing under such conditions of low water availability is regularly spaced. If water is scarce, plant cover will be sparse, and the result will be that as far as root spread is concerned, the available space *is almost fully occupied ... It is likely, then, that the regular spacing of Larrea* in many sites of low rainfall is the result of root competition for available water."

If we accept this generalization at its face value, it is surprising not only to find no evidence of regularity in the (Australian) *Atriplex* populations that have been assessed quantitatively, but also to find that, in contrast, this presumed relationship between pattern and rainfall availability is reversed in Australian conditions.

Data are presented in Fig. 5.22 to illustrate the relationship between mean square/mean (i.e., intensity) values for the larger of a two-scale pattern in 19 samples of *A. vesicaria* populations, plotted against the mean annual water volume available on a per plant basis. The trend in these data suggests strongly an inverse relationship between pattern intensity and available (rain-)water. Put another way, the drier the average soil environment, the more likely is the pattern contagious rather than regular – the reverse of the more traditional hypothesis. Furthermore, it can be seen from Table 5.7 that individual plants of *Atriplex* are, again "on average", subject to a more arid soil environment, despite a mean annual rainfall in excess of 330 mm, than the *Larrea* individuals growing in the climatically drier (indeed, extremely arid) Californian desert.

In the conventional dogma attaching to the interpretation of pattern analyses, this apparent anomaly is not difficult to justify. Rather it reflects the thesis

Table 5.7. Average volumes of water available annually to *Larrea* shrubs in four Californian desert populations and four *Atriplex* populations from the Riverine Plain in Australia. Calculations based on data of Woodell et al. (1969) and of Malik (1970)

Location	Liters water/shrub/year		Mean annual rainfall
	Larrea	*Atriplex*	
Inyokern	3,464.1		9.1
Emigrant Spring	4,226.1		11.1
Cantil	4,812.4		10.0
Yuma	8,453.8		9.1
Toopuntil		112.3	29.2
Oxley		59.6	31.3
Hay		78.2	32.6
Maude		91.8	27.8

Table 5.8. Relation between mean density in high and low density phases, pattern intensity and diameter size categories for *Atriplex vesicaria* at block size 16 in site AV 9 (Riverine Plain, Australia)

Phase	Diameter size categories					
	5–10 cm		10–20 cm		20—25 cm	
	Mean density	Pattern intensity	Mean density	Pattern intensity	Mean density	Pattern intensity
High	35.5	6.32	20.0	1.16	8.5	0.40
Low	22.0		20.0		10.0	

expounded by Greig-Smith (1957), that the more intense the environmental pattern, the more intense will be those plant patterns that are controlled by that environment. In a real environment where topography is rarely uniform, differential infiltration of water is quite commonplace. Therefore, we might assume the population also varies in physiological vigor of "performance". It is thus likely that contagion would be more likely to develop than regularity. This is not to deny the potential importance of competition between plants for available water resources, but rather to suggest that this process is more likely to operate within the higher density phases of a patterned mosaic, a feature that was demonstrated (from post-hoc evidence) by Anderson et al. (1969) for a population of *A. vesicaria* (see Table 5.8).

It is at this point that we see evidence of some of the innate limitations of ecological methodology. While studies of pattern provide intrinsically interesting data subject to an appropriate interpretational *milieu*, it is apparent from the paradoxical situation just described that pattern studies alone are unlikely to provide by themselves the ultimate interpretation of key ecological *processes*. What they can do is provide field evidence for the construction of appropriate hypotheses relating pattern state to processes. In this particular case they can also offer guidelines to an investigator in setting up appropriate experimental strategies that seek to link these key ecological interactions between observed pattern and the processes that play an important role in initiating and maintaining these patterns.

For example, the origin of contoured patterning of *Atriplex vesicaria* communities (Fig. 5.6) has been a source of interest both to botanists and geomorphologists. Recently Valentine and Nagorcka (1979) have attempted to simulate the processes underlying the formation of these vegetated bands. Developing a computer model in which plant distribution was initially random, and imposing various environmental constraints involving rainfall events, topography, and soil-water absorption, these workers have shown that a consistent run-on/run-off model is sufficient to cause the initiation of banded vegetation. Earlier studies indicated that the pattern is maintained by salt cycling in the halophytic plants. Removal of saltbush by grazing induces a stronger salinity mosaic, wherein the depressions become virtually nonsaline while the shelves become extremely saline. Removal of the saltbush enhances the growth of less salt-tolerant and less drought-tolerant species, the usual floristic progression being from *Atriplex vesicaria* to *Bassia* spp., to the grass *Enneapogon*. The result of this sequential change in species dominance is a rapid decline in productivity and increased susceptibility to loss of all vegetation following drought (Charley, 1959).

We conclude then that rigorous vegetational analysis is most usefully employed if it initiates complementary analysis of environmental factors and if these two approaches can lead to experimentally testable hypotheses using modeling and analytical techniques. The analysis of vegetational and environmental pattern is likely to highlight, as illustrated above, the importance of many factors which may otherwise not attract the attention they deserve. For example, determination of the water availability to plants involves more than the mere computation of precipitation and evaporation and quantitative analysis of vegetation is likely to highlight the significance of topography and soil type which drastically modify the effectiveness of precipitation. The complementary nature of vegetational analysis, which effectively reduces the holocenotic concept of vegetation and environment to manageable proportions, and of the otherwise simplistic component process studies described in the following chapters cannot be over-stressed. Too often the two approaches do not meet, but if vegetational analysis is extended to considerations of biomass or nutrient distribution and if the profile dimension is included, then the significance of these relationships is likely to become more evident.

5.8 Patterns of Biomass and Nutrient Distribution

In the preceding sections we have concentrated on describing the state of vegetational units in a community in two dimensions only, and on their relation to each other. Before we can begin to discuss relevant physiological and ecological processes this state description must be extended to include quantitative measures of biomass and nutrient capital and the patterns of distribution of these in the community. Biomass and nutrient distribution will be considered on both the regional and local scales used in the vegetational analysis described in previous sections (about 10^9 to 10^{11} and 10^6 to 10^9, respectively). Within each scale the distribution of biomass or nutrients will be considered both horizontally and vertically. There are limitations in these data, for when based on harvest methods, they are usually derived by extrapolations from relatively small samples and thus can be subject to large errors. Other estimates based on photographic or other nondestructive methods are equally approximate. Nevertheless, they provide useful "order of magnitude" comparisons for our later discussions of population dynamics and productivity (Chaps. 6, 10). The only *Atriplex* communities for which there are data are the arid shrublands and comparison with other communities such as the sand dune or ruderal *Atriplex* vegetation is desirable.

5.8.1 Biomass Distribution in Arid Shrublands

According to Rodin and Bazilevich (1965) arid shrublands in general are characterized by low total biomass values (approximately 5,000–20,000 kg ha^{-1}) with a low proportion of this in above-ground parts and with only 1%–10% of the aboveground mass devoted to assimilatory tissues. The *Atriplex*-dominated shrublands seem to conform to these generalizations only to a limited extent. The variability in the data available may reflect either substantial differences between otherwise similar communities in different regions or differences in precision and season of sampling and biomass estimation.

In an unpublished survey of *Atriplex* shrublands in the Riverine Plain of New South Wales, J. Marshall (personal communication) has measured aboveground biomass values of perennial shrubs in a mixed *A. nummularia* – *A. vesicaria* stand as high as 5,250 kg ha^{-1}, and as low as 870 kg ha^{-1} in a community composed mainly of *A. vesicaria*. These figures were estimated from sites subjected to grazing, the impact of which is highlighted by data presented by Crisp (1975) from grazed and ungrazed mixed chenopod *(Atriplex vesicaria, A. stipitata, Maireana georgei)* shrublands on the Koonamore Reserve in South Australia. Crisp found aboveground biomass values were 36 times greater in an ungrazed paddock, compared with similarly estimated values in an adjacent but floristically similar paddock subjected to grazing.

The inherent variability in biomass estimation when estimates are made over a period of time is illustrated in data obtained by Shinn (personal communication) for a mixed stand of *Artemisia* – *Atriplex* – *Sitanion* from Curlew Valley in northern Utah. From August 1972 through August of the following year, the total aboveground biomass of *A. confertifolia* rose from 1,269 to 1,884 kg ha^{-1}, an increase of 32.6%, due mainly to an increased growth response of the *Atriplex* to a significantly higher rainfall in 1973.

On a local scale, Rixon's analysis of the spring vegetation of an *A. vesicaria* community on the Riverine Plain of south-eastern Australia shows that the dominant shrub comprised 61% of the total biomass (Table 5.9). Most of the litter was presumably derived from the *Atriplex* individuals so that a further 24% of the total biomass could be attributed to these shrubs. This community is probably one of the denser arid shrublands, falling in the upper range of values recorded by Rodin and Bazilevich (1965). Most of the biomass is clustered in the vicinity of the

Table 5.9. Composition and distribution of surface biomass (t ha^{-1}) in an arid shrubland dominated by *Atriplex vesicaria* on the Riverine Plain of south-eastern Australia. (Data of Rixon, 1971)

Component	Associated with[a] shrubs	Between shrubs
A. vesicaria	7.720	0
Maireana spp.	0	0.006
Legumes	1.313	0.324
Composites	0.155	0.086
Grasses	0.047	0.021
Others	0.009	0.059
Litter	2.390	0.660
Total	11.634	1.056

[a]   Shrubs approximately 50 cm diameter

Table 5.10. Biomass estimates in arid shrubland communities (t ha^{-1})

Location	Great Basin, USA[a]		Eastern Australia	
Dominant	*A. confertifolia*	*C. lanata*	*A. vesicaria*[b]	*E. gilesii*[c]
Shoots	4.17	2.41	2.23	2.37
Litter	8.49	4.0	0.14	0.79
Roots	13.13	16.07	0.91	0.78
Total	25.79	22.48	3.79	3.94

[a]   N.E. West, personal communication (3 yr average)
[b]   Charley and Cowling (1968), Fowler's Gap, N.S.W.
[c]   Burrows (1972), Charleville, Qld.

dominant shrubs, with less than 10% of the total vegetation between shrubs. The individual shrubs clearly provide a favored habitat for most of the ephemeral vegetation. Similar comparisons for other *Atriplex* shrublands are not available. When the vertical profile of biomass distribution is considered, the different communities show substantially different patterns. Table 5.10 shows the biomass profile for the dominant shrubs in four arid shrublands in the Great Basin Desert of the U.S.A. and in eastern arid regions of Australia. Two of these communities are dominated by *Atriplex* species and the table shows that nearby communities of similar life form, but dominated by different species, show a similar distribution of biomass. The cool desert communities of the U.S.A., sampled at the end of the growing season in three consecutive years, show a remarkably high proportion of total biomass associated with the root system. The much lower root biomass in the Australian communities is obvious. The lower litter component in the Australian data presumably reflects the sampling time or the rapid litter turnover (Table 5.10).

An evaluation of the tremendous differences in root biomass in these two community groups is difficult. The cool desert shrubs conform to the general picture of arid shrublands evident in the data presented by Rodin and Bazilevich

(1965), whereas the low root biomass of the Australian communities is clearly at variance with the data presented by these authors. This apparent anomaly persists in studies of the biomass distribution in *Atriplex* spp. grown under irrigated plantation conditions on the Riverine Plain in south-eastern Australia. Basically similar root distributions were recorded for ten-year old *A. vesicaria* and *A. nummularia* plants under the same conditions (Jones and Hodgkinson, 1970).

In these studies the techniques of root biomass estimation were comparable so that the data evidently reflect very substantial differences in root biomass in the two habitats. The proportion of living roots in the fragments isolated from the soil is never easy to estimate and studies by Hodgkinson et al. (1978) suggest that less than 10% of the root biomass in the Great Basin communities may be alive. In this context then it may be that the bulk of the root biomass in these communities should be better considered as belowground litter. We shall return to this question later.

5.8.2 Nutrient and Salt Distribution

The local scale variation in nutrient and salt distribution in arid shrubland communities has long been recognized (Fireman and Haywood, 1952; Gates et al., 1956). This distribution is maintained by the processes of nutrient cycling through litter, as discussed in a later chapter, but has key significance in relation to other processes such as germination and seedling establishment. Because the local variations in nutrient and salt distribution are dependent on plant activity they are closely correlated with the vegetation patterns, as described earlier (Chap. 5.6.2). The distribution of nutrients is likely to be of greatest significance both to seedling establishment and to the retention of available nutrients within the community. Charley and Cowling (1968) described the surface concentration of nutrients in soils of *Atriplex vesicaria*-dominated communities in south-eastern Australia. Similar results were obtained by Charley and West (1975) who examined thirteen sites in arid shrublands of the Great Basin Desert. They found that soil N, available P, and organic carbon were concentrated in the surface few centimeters at all sites (Fig. 5.23). Horizontal patterning of N, organic carbon, and organic phosphorous was evident at most sites.

The patterning and profile of salt in soils supporting the *Atriplex* shrublands is similar to that of the nutrients. These soils are, in the main, not highly saline, but high salt concentrations are found in the surface under the shrubs. Figure 5.23(d) shows the soil chloride profile under *A. vesicaria* shrubs as well as the chloride profile between shrubs (Jessup, 1969) and the concentration of chloride in the soil surface under the bushes is evident. Only at a depth of about 60 cm does the profile match that of the unvegetated soils. Similar reports from sites in the Riverine Plain (Sharma and Tongway, 1973) and elsewhere in south-eastern Australia (Charley, 1959) indicate that this is a regular feature of soil salinity patterns associated with vegetation patterns in these communities. Sharma (1973b) showed that the high cation concentration in the surface soils under *A. nummularia* shrubs was correlated with measurable differences in soil physical characteristics. Indeed there was a significantly poorer surface soil structure under the shrubs, as indicated by lower aggregate stability, reduced drainage, and lower hydraulic conductivity.

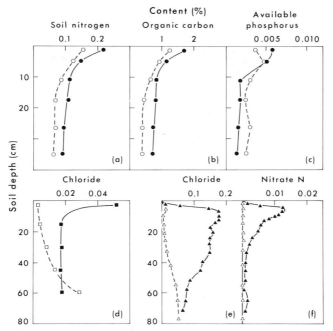

Fig. 5.23. Profiles with depth of nutrients, salt, and organic matter in arid shrubland soils: **a–c** in the Great Basin Desert (redrawn from Charley and West, 1975); **d** in South Australia (data of Jessup, 1969) and **e, f** in the patterned communities of the Barrier Ranges, southeastern Australia. *Open symbols* refer to profiles between bushes, *closed symbols* to profiles under bushes, except in **e, f** where *closed symbols* refer to the barren shelf. (Redrawn from Charley and McGarity, 1964)

The local differences in nutrient distribution and soil properties in the more or less randomly vegetated communities discussed above are small compared with those found in some highly patterned arid shrublands. The nutrient and chloride profiles of soils supporting island communities of *A. vesicaria* on the gilgai habitats of the stony tableland soils of south-eastern Australia and the banded communities of the same species on nearby gently sloping soils (Chap. 5.6.3) display dramatic vertical patterning. The gilgai of the stony tableland soils consists of a vegetated depression a few centimeters lower than the nonvegetated, stone-covered shelf. Charley and McGarity (1964) examined the profile of Cl^- and NO_3^- in the shelf and the depression of these soils. Figure 5.23 (e, f) shows that the surface soils of the shelf are extraordinarily rich in both ions whereas the depression soils show little NO_3^- at any level and are relatively saline only below about 30 cm. These patterns of salt and nutrient distribution evidently reflect the lateral movement of soluble ions as water more readily penetrates the depression and is evaporated from the shelf. Again, the pattern of nutrient and salt distribution is associated with the pattern of vegetation, but in this instance the ionic distribution may depend rather more on soil physical properties than on plant activity. Similar salt distribution profiles have been described by Valentine and Nagorcka (1979) in banded *A. vesicaria* communities. The significance of these states or patterns of ionic distribution for ecological processes is discussed in the following chapters.

5.8.3 Distribution of Microbiological Activity

The local patterns of nutrient and salt distribution in arid shrublands are closely associated with the distribution of the perennial vegetation and are the product of nutrient cycling processes (Chap. 7.5.1) which depend in part on soil microbiological activity. Microbiological activity, particularly that involved with the mineralization of nutrients, is thus also associated with the distribution of the perennial vegetation. In the *Atriplex vesicaria* shrublands of the Riverine Plain in south-eastern Australia Rixon (1971) found that the respiration rate of soil and the rate of nitrogen mineralization were very much greater in surface soils associated with the shrubs than in soils between shrubs (Table 5.11). Furthermore, he noted that this pattern of surface soil microbiological activity persisted for at least two years following removal of the shrub tops. On a soil carbon basis, the microbiological activity was similar both between and under shrubs, indicating that the pattern is perhaps due to litter enrichment under the shrub. Alternatively, it may reflect the activity of a rhizosphere flora and the distribution of roots.

The rhizosphere flora is most evident in the nutrient-poor and dune communities. *A. glabriuscula* and related species which are among the first plants to be established on the foredunes are usually established in the debris of the drift line. Webley et al. (1952) examined the rhizosphere flora of sand which clings to the roots of these and other foredune species and found an enriched microbial population (Table 5.12). The bacterial population associated with *A. glabriuscula* evidently diminishes and fungal population increases as the dunes are established. Whether these micro-organisms function in the mineralization of nutrients on sand dunes remains to be determined.

Table 5.11. Distribution of surface microbiological activity in relation to *Atriplex vesicaria* shrubs on the Riverine Plain of south-eastern Australia. (Data of Rixon, 1971)

Parameter	Associated with shrubs	Between shrubs
Respiration rate (μl O_2 h^{-1} g^{-1} soil)	3.2	1.2
Nitrogen mineralisation rate (μg N h^{-1} g soil N)	74	14

Table 5.12. Rhizosphere flora associated with sand adhering to roots of plants on the foredunes. (Date of Webley et al., 1952)

Site and species	Microbial population (organisms g^{-1} sand)	
	Bacteria ($\times 10^6$)	Fungi ($\times 10^3$)
Open sand, fordune	0.02	0.25
Atriplex glabriuscula	23.29	1.67
Ammophila arenaria	3.58	2.64
Agropyron junceum	3.56	72.47
Stable dunes	1.70	69.47

5.9 Ecological States in Perspective

Past attempts to interpret the ecology of arid regions in process terms based on ecological state descriptions have been criticized in recent years (Anderson et al., 1972; Barbour, 1973). It is possible that these attempts have been singled out for criticism because they stand out against a rather barren literature, much in the manner of arid vegetation against the desert landscape. In fact these attempts are vulnerable to criticism because, although they seek to interpret large-scale state descriptions in terms of small-scale physiological processes, they depend on untenably simplistic approximations.

This is exemplified by the frequently stated view that the shrubs of arid lands are uniformly distributed. This particular description of the ecological state may be derived from the high root/shoot ratios of desert perennials reported in some studies and the extrapolation that "the apparently open plant growth may be biologically closed due to root competition" (Kassas, 1966), presumably for limiting water. But as Anderson et al. (1972) and Barbour (1973) have emphasized, crude state comparisons of root/shoot ratios may well obscure the more important functional ratios of effective absorbing and evaporating surfaces in a water-limited environment. It is likely, for example, that a large proportion of belowground biomass in arid communities is structural in character, and not available for water-uptake. In process terms, it is dead wood. Until we can prune the dead wood from such state descriptive comparisons, we will not be able to focus clearly on the processes on which such hypotheses are in fact based. As will be shown later (Chap. 8), water relations of desert shrubs depend on the dynamics of initiation of new roots following rainfall; processes at a scale of about 10^4. Whether competition for water among these roots prevents seedling establishment or hastens the demise of established species remains conjectural.

Another example of a conventional wisdom which seems to be based on comparisons at inappropriate scales is evident in Shreve's comment that the impact of arid vegetation upon the landscape was almost totally lacking (Shreve, 1942). Our brief discussion of the ecological states of salt, nutrients, and other soil properties shows that this view is clearly untenable. In fact, in arid shrublands the impact of processes of salt and nutrient cycling on the smaller local scale (10^6 to 10^9) is possibly more dramatic, and certainly no less important, than in other communities. Possibly Shreve was comparing the magnitude of impact at larger regional scales ($>10^{10}$), between communities with much greater biomass and greater continuity of biological activity. Comparisons at this scale will certainly obscure the marked patterning and intricate processes which are found in low biomass arid vegetation.

The ecological state description, based on plant density, biomass distribution, and relative abundance of species, provides better indications of performance in response to habitat than is available from the taxonomic or geographical state description. Such ecological data can give further insights into adaptive significance of morphological characters. Observations of phenology and microclimate provide further refinements which may be vital to process evaluations. These observations permit even closer correlation between the environment and key processes in the establishment, performance, and survival of an organism.

Descriptive and quantitative ecology thus represent means of developing more specific hypotheses for mechanisms of adaptation. Figure 5.8, for example, which showed the seasonal patterns of photosynthetic activity among species in Death Valley, indicates that both C_3 and C_4 photosynthetic pathways of carbon assimilation display similar peak rates of photosynthetic performance. However, high rate C_3 photosynthesis is found when mean maximum temperature is about 26 °C, and high rate C_4 photosynthesis was found when mean maximum temperature is about 45 °C. With such evidence, broad correlations between the presence (survival) of species with different photosynthetic pathways in different habitats, based on the systematic and geographical data discussed in Chap. 2, can be developed into specific hypotheses addressing specific physiological processes.

If we return to our analogy of Niggle's tree, it is already evident from the foregoing treatment of generic and ecological states that the form of the tree is interesting and in some places, clearly defined. In the next chapters we shall more closely examine these physiological process which are its individual leaves. While appreciating their sparkle and sheen largely for their own sake, we shall strive to appreciate the place of these leaves in the context of the whole tree. There remains a large gulf between the imaginative interpretation of the most careful analysis of ecological states and testable hypotheses as to the significance of physiological processes in determining the persistence and performance of plants in particular habitats. It is our belief, however, that within the range of habitats occupied by *Atriplex* spp., and the range of physiological processes studied in these plants, this gulf is narrower than in many other realms of physiological plant ecology.

Chapter 6. Germination and Seedling Establishment

"The germination of a seed is the emergence of an embryo from rest ... Getting into and out of rest is a crucial developmental process for the continuity of the life cycle of an organism. We are basically ignorant of the physiologic and histologic conditions that constitute rest."

(Rearranged from Jann and Amen, 1977)

In this chapter we are concerned for the first time with detailed discussions of states and processes as they relate to individual plants. Although the systematic state (Chap. 2) is deduced from analysis of individual plant specimens; although selective pressures leading to the survival of a new species (Chaps. 3, 4) act upon the individual; and although the survival of individual plants through a severe drought determines the response of a plant community (Chap. 5), the states and processes discussed previously were primarily abstractions based on populations of organisms. Birth rate is a key parameter of population ecology, yet in many ecological studies emphasis is given to interactions between adult plants. Similarly, during germination one encounters many fundamental problems of plant physiology in their most simple form, yet ecophysiological studies also concentrate upon the more accessible adult plants. By considering germination and establishment in the present context we hope to draw attention to the neglect of these processes by physiologists, as well as to assess the significance of germination and establishment in larger scale ecological processes. It is true that studies of metabolism in germinating seeds and of the control of morphological development in seedlings have been responsible for major advances in plant physiology. It is also true that understanding of these processes in other than idealized laboratory conditions is negligible. There is no doubt, however, that the period between successful seed set and seedling establishment is the most precarious in the life cycle of angiosperms. Stebbins (1971) suggests that to understand and integrate these states and processes "one must become at the same time a comparative morphologist, a comparative physiologist and an ecologist".

The principal reason for the vulnerability of seed set, germination, and establishment is that these processes involve an extraordinarily complex series of structural and metabolic transitions which often take place at a time when the external environment is also undergoing rapid change. Seed set often involves transition from a fully hydrated, dividing embryonic tissue to a desiccated and dormant tissue as the rest of the plant becomes senescent, frequently as the environment deteriorates. Germination involves the orderly activation of heterotrophic metabolism from dormancy. Establishment involves the transition from heterotrophic to autotrophic metabolism and associated changes from nonvacuolate to vacuolated cells. Integration of these physiological processes during germination and establishment is greater than at most other times in the life of the plant. At this time selective pressure from a changing environment is likely to be greatest and may preserve fascinating and complex physiological processes. To assume that they are functionally significant because they have been selected is one thing; to demonstrate this significance for the survival of the organism in complex environments is another.

6.1 Germination Processes

There is almost limitless scope for philosophical and semantic discourse on the question of "what is germination?" (Jann and Amen, 1977) but for our purposes a simple morphological and physiological definition will be sufficient. Seed germination begins with the imbibition of water by the seed and is completed with the emergence of the radicle and cotyledons from the testa. Germination processes are broadly of two types, the physical processes of water uptake and the complex biochemical processes which are activated as the dormancy of the seed is broken following rehydration. Throughout germination the biological activity of the embryo is sustained by the seed reserves and the living plant remains a heterotrophic system until cotyledons or leaves become capable of net photosynthesis. Although metabolism remains heterotrophic throughout, germination processes involve degradative metabolism in the endosperm and synthetic metabolism in the embryo. Both are accompanied by large changes in cell structure.

Both the physical and biochemical germination processes are conveniently represented on a time scale modeled after Fig. 1.1, the relevant relaxation times being 10^2 to 10^5 s. Figure 6.1, adapted from Mayer (1977), illustrates an idealized time scale sequence of degradative and synthetic metabolism during germination. The space scales relevant to germination processes (seed size and soil volumes)

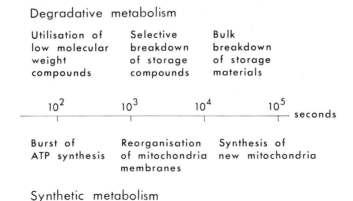

Fig. 6.1. Approximate relaxation times of biochemical processes involved in degradative and synthetic metabolism of germinating seeds. (Redrawn from Mayer, 1977)

correspond, as one would expect, on the basis of the arbitrary matching selected for Fig. 1.1. In this treatment we will first consider the primary physical processes of water movement and uptake by seeds in soil and ways in which water uptake may be modified by seed and soil factors. We will then briefly consider the interaction of environmental factors with the metabolic processes which are initiated by water uptake.

6.1.1 Water Relations of Germinating Seeds

The imbibition of water by the dry seed initiates germination and terminates the dormant state of the reproductive organ. The seed comprises a dormant but living embryo, and the nonliving reserves in the endosperm. Water must be absorbed by both of these parts of the seed before metabolic activity can commence in the embryo and in the cells which mobilize seed reserves. Prior to the imbibition of water, the seed is relatively insensitive to extremes of temperature, radiation, and gaseous environment. After imbibition, these physical aspects of the environment become much more important to successful germination.

The uptake of water by seeds has been reviewed recently by Bewley and Black (1978). Water uptake by seeds is a relatively simple process compared to water uptake by other plant parts (Chap. 8) and it can be used to illustrate some of the more important principles of water state and water movement processes. The most suitable parameter for specifying the state of water is its chemical potential, and it is usual to describe the water potential of the soil or plant as the difference in chemical potential of water in the system relative to that of free water at the same temperature (Chap. 8.1). Water potential is also a useful measure because it can be determined directly by vapor phase equilibration in sealed chambers (Barrs, 1968).

The water potential of seeds is readily measured by these means because most seeds are hygroscopic and readily equilibrate with the water potential of their environment. However, it is not usual to find seed water relations discussed in terms of water potential. More commonly, water content at equilibrium relative humidities at specified temperatures are cited (Harrington, 1972). The seeds of cereals, for example, are ripe and capable of prolonged storage when water contents range between 5%–10% in equilibrium with relative humidities of about 20% at 25 °C. The simple formulae used to convert these commonly used parameters to corresponding values of water potential are given later (Chap. 8.1.1), and the water potential of seeds under these conditions is in the vicinity of $-2,000$ bar. The only other living parts of higher plants which endure desiccation at such low chemical potentials of water are the spores of some phyla and the vegetative tissues of poikilohydric plants (Gaff, 1977; Jones et al., 1980). Total water potential may be divided into several components (Chap. 8.1), and in the dry seed the major component is a matric potential due to the structure of water around the macromolecular colloids of the embryo and endosperm.

Movement of water will take place from regions of high water potential to those of lower water potential. Thus water uptake by seeds is primarily determined by the gradient of water potential between the seed (ψ_d) and the soil (ψ_s) and the conductance of different parts of the water pathway comprising C, the conductance of the soil, C_c, conductance of the seed coat and C_d, the conductance of the seed itself. The rate of water flow (F) is given by

$$F = (\psi_d - \psi_s) \bigg/ \left(\frac{1}{C} + \frac{1}{C_c} + \frac{1}{C_d} \right).$$

Changes of water uptake as a result of change in the water potential gradient may be due to several factors. The gradient in water potential between a dry seed ($\psi_d = -1,000$ to $-2,000$ bar) and wet soil ($\psi_s = -1$ bar) changes rapidly with water uptake because ψ_d increases markedly before germination. *Atriplex hortensis* seeds are unable to absorb sufficient water to germinate if ψ_s is lower than -20 bar, suggesting that water potentials in the seed during germination must be higher than this value. In all probability these water potentials are in the vicinity of -5 to -10 bar, the water potentials measured in meristematic tissues (apices) of higher plants. The water potential gradient also changes rapidly because the soil in the vicinity of the seed dries rather rapidly. In many circumstances it is likely that volume of soil needed to supply the water absorbed by the seed is rather larger than the volume of the seed itself (Hillel, 1972) and the drying of soil adjacent to the seed can be observed experimentally (Currie, 1973). Under natural conditions it is likely that decreases in ψ_s due to evaporation from the soil surface and to water uptake by other plants may also be significant over the time span of germination, further reducing the gradient of water potential and the driving force of water uptake.

Furthermore, each of the conductances to water movement in soil, seed coat, and seed, changes with change in water content in that phase and each influences water flow from soil to seed. In loamy soil conductance may decrease from 10^{-10} to 10^{-12} cm s^{-1} when water potential decreases from -10 to -100 bar. In *Brassica* seeds the value of C_c increases from 10^{-21} cm s^{-1} at the outset to about 10^{-10} cm s^{-1} following water uptake and rupture (Skaykewich and Williams, 1971). In the seed itself the conductance decreases as the water content of the seed increases. A very important component of the water pathway is the area of contact between seed and sources of soil water. This factor and C_d are particularly important for water uptake in larger seeds.

Water uptake by seeds is thus a complex of competing, and sometimes compensating, changes in the driving force and conductance components of the water transport pathway. Seed properties may control water uptake by influencing one or more of these components. The gradient in water potential between soil and seed can be controlled by enclosing the seed with materials having an osmotic or matric potential lower than that required to permit sufficient water uptake for germination. More commonly, water uptake may be regulated by the conductance of the seed coat. Both of these control processes are evident among seed of *Atriplex* spp. and are discussed in the following sections.

6.1.2 Fruit Morphology and the Soil to Seed Water Potential Gradient

The seed of *Atriplex* spp. are almost always born between bracts (Chap. 2.1.1) and the spongy appendages of the bracts found in many Australian species contain high levels of NaCl. Beadle (1952) demonstrated that the concentration of salt in moistened bracts of these fruits was sufficiently high to prevent germination.

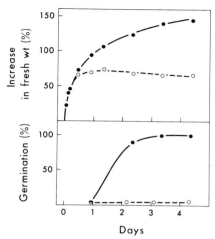

Fig. 6.2. Water uptake, shown by increase in seed fresh weight, and germination percentage in brown seed by *A. hortensis* immersed in distilled water (●) and 0.5 M NaCl (○). (Redrawn from M. Björkman, unpublished)

Leaching of this salt was necessary for germination to occur. The arguments supporting the regulation of water uptake and germination by the salt content of the bracts as proposed by Beadle can be developed as follows.

M. Björkman (unpublished) studied the time course of water uptake by brown seed of *A.hortensis* (Fig. 2.5) in distilled water and in NaCl solutions with a water potential of −21 bar. As shown in Fig. 6.2, water uptake by seeds in distilled water continued for four days. Germination began when the seed fresh weight had increased by 70%–120% due to water uptake. In NaCl solution water uptake proceeded at the same initial rate but ceased when the fresh weight of the seed had increased by 70% and only about 5% germination was recorded at this water content, even after four days. This germination percentage is similar to that found for seeds in distilled water when they too had increased in fresh weight by only 70%. That is, NaCl had little effect on germination other than on water uptake. Other studies (Nicholls, 1972) also rule out specific effects of different osmotica. At the same water potential the inhibition of germination due to NaCl, mannitol, and polyethylene glycol was similar in seed of *A. vesicaria, A. lindleyi,* and *A.nummularia* (Fig. 6.3).

Water potential, rather than specific effects of osmotica, is also implicated in the comparative studies of Sharma (1973a) in which effects of soil matric potential on germination were investigated. Loam soil was brought to equilibrium matric potentials with polyethylene glycol solutions across a semi-permeable membrane. Sharma showed that the final germination percentage for *A. vesicaria* and *A.nummularia* seed was similar, irrespective of whether water potential was controlled by osmotic potential (NaCl, mannitol or polyethylene glycol) or by matric potential in soil. Similar results were obtained for *Danthonia caespitosa.* In Sharma showed that the final germination percentage for *A.vesicaria* and to osmotic or matric potential than was germination of *A.nummularia.*

Germination and Seedling Establishment

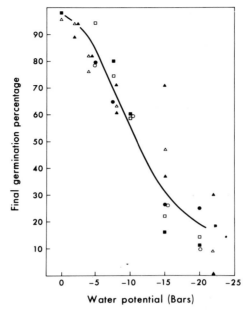

Fig. 6.3. Final germination percentage attained with seed of *A. lindleyi* (□, ■); *A. vesicaria* (○, ●) and *A. nummularia* (△, ▲) in solutions of different water potential adjusted with mannitol (*open symbols*) or NaCl (*closed symbols*). (Redrawn from Nicholls, 1972 and Sharma, 1973a)

    From the above studies it is clear that the water potential of NaCl dissolved in spongy tissue of bracts could be low enough to prevent the absorption of sufficient water for germination. This conclusion was implicit in Beadle's (1952) experiments, but the possibility of toxic effects due to NaCl were not excluded in his experiments. Beadle found that the NaCl concentration in water-saturated bracts of several *Atriplex* species was greater than the NaCl concentration in solution required to inhibit germination of seeds removed from the bracts (Table 6.1). The NaCl concentration preventing germination ranged from about − 16 to about − 32 bar in these experiments.

    Beadle suggested that the high salt content of the bracteoles allows the seed to sense the adequacy or otherwise of the environment for germination. This could function in two ways. First, the seed of freshly fallen fruit would germinate only after substantial rainfall or second, several successive light falls could leach the salt from the bracteole and permit germination in more favorable circumstances. Charley (1959) showed that the equivalent of 100 mm precipitation was sufficient to leach about 95% of the chloride from the bracteoles of *A. vesicaria* in the field, and Beadle (1952) showed that the equivalent of 70 mm precipitation resulted in 100% germination of three *Atriplex* species buried in soil. Several successive smaller falls of rain would presumably be equally effective in leaching salt from the bracteoles and allowing germination to proceed under conditions of optimum water availability.

Table 6.1. NaCl concentration in water-saturated bracteoles of *Atriplex* spp. and the concentration of NaCl required to inhibit germination of naked seeds. (Data of Beadle, 1952; for different seed batches)

Species	NaCl in saturated bracteoles (M)	NaCl preventing germination (M)
A. nummularia	0.72	0.6
	0.48	0.4
A. vesicaria	0.48	0.4
	0.55	0.5
A. semibaccata	0.61	0.4
	0.45	—
A. inflata	0.90	0.7
	0.61	0.5
	0.44	0.4
A. spongiosa	0.62	0.8
	0.69	0.4

Although the removal of chlorides from the bracteoles by leaching stimulates germination of arid shrubland *Atriplex* spp., somewhat contradictory observations have been reported for the coastal species. Gustafsson (1970, 1973a) reported that three weeks treatment of *A. tatarica (recurva)* seed in sea water and up to eight weeks treatment of *A. triangularis* seed did not alter the germination percentage. He observed differences in the response of different populations of *A. triangularis* seed. The germination of seed from the North Sea coast of Scandinavia was insensitive to seawater treatment, whereas germination of seed from populations from the less saline Baltic coast was somewhat inhibited by treatment in 1M NaCl. A further confusing observation relating to the effect of salt on seed germination in coastal *Atriplex* populations has been noted in *A. glabriuscula (babingtonii)*. Binet (1966) observed that the mechanical retardation of germination in this species due to the bracteoles could be eliminated by treatment in seawater. These reports need further investigation, for the NaCl concentrations present at the time of germination are not clear and the effect of salt on water absorption by the seed was not examined. In at least some instances, germination of driftline species is confined to periods of low salinity in the substrate. Ward (1967c) noted that *A. novae-zealandiae* germinated principally during winter when substrate salinity was lowest and water content was highest. It would be most interesting indeed if, in these coastal species, exposure to high concentrations of NaCl had some effect on the permeability of the bracteoles or seed coat which facilitated imbibition.

6.1.3 Seed Dimorphism and the Conductance of the Water Transport Pathway

If the water potential gradient is adequate for water movement from soil to seed then the rate of flow may be determined by the conductance of different parts of the water transport pathway. In many *Atriplex* spp. two types of seed are formed; the

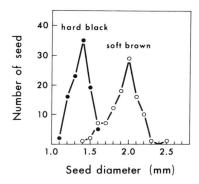

Fig. 6.4. Size of small, hard black seed and large, soft brown seed of *A. triangularis.* (Redrawn from Ungar, 1971)

small, hard black seed and larger, soft brown seed (Chap. 2.1.1). The size distribution of these seed types in *A. triangularis (patula)* is shown in Fig. 6.4 and Ungar (1971) noted about equal proportions of both seed types in this species. The morphological basis for this distinction in seed types resides in the properties of the cell layer immediately below the pericarp. In *A. hortensis* M. Björkman (unpublished) found that this layer in brown seeds was composed of flattened, relatively thin-walled cells 20–40 μm in diameter. In black seeds of the same species this layer was composed of densely packed cylindrical sclereids (Fig. 6.5).

The seed coat of black seeds is extremely resistant to mechanical damage and rather less permeable to water than the seed coat of brown seeds. M. Björkman studied the rate of water uptake by black and brown seed and as shown in Fig. 6.6, uptake was 10^2 times faster in brown that in black seed under the same conditions. When the seed coats were scarified, the initial rate of water uptake by black seeds was similar to that in the brown seed. It should be noted that the percentage increase in weight due to water uptake was much smaller in scarified black seed, yet resulted in a similar germination percentage. The percent increase in weight due to water uptake by untreated black seed remained lower than that in scarified black seed on day 1, at which time there was no germination of the scarified seed. In the first four days, the untreated black seed were simply unable to absorb sufficient water for germination to commence, but after 10 days 22% of the black seed had germinated. In the experiments with *A. hortensis* the absorption of water by black seed was so limited by the conductance of the seed coat that the addition of NaCl to the solution, which reduced the water potential gradient to -21 bar, had practically no effect on water uptake (M. Björkman, unpublished; Nobs and Hagar, 1974) and similar insensitivity to NaCl was observed with the black seed of *A. holocarpa* (Courtice, 1971).

It is possible that the effects of seed dimorphism are more significant in relation to seed longevity than to the regulation of water uptake, although the two phenomena go hand in hand. For example, Beadle (1952) showed that black seed of arid shrubland species remained viable in laboratory storage for longer periods than did brown seed (Fig. 6.7). In *Atriplex semibaccata* and *A. inflata,* germination of

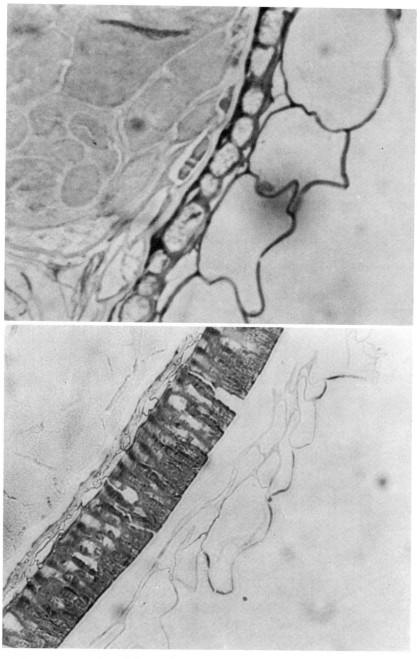

Fig. 6.5. Transverse section of the seed coat of soft brown seed (*top*) and hard black seed (*bottom*) of *A. hortensis*. Below the large pericarp cells the brown seed has a layer of flattened relatively thin-walled cells. The homologous cell layer in black seed consists of closely packed cylindrical sclereids. (× 50, micrographs courtesy M. Björkman)

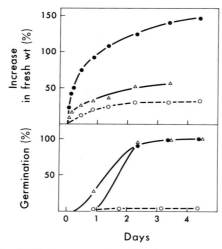

Fig. 6.6. Water uptake in distilled water and germination percentage in brown seed (●) and black seed of *A. hortensis* before (○) or after (△) scarification. (Redrawn from M. Björkman, unpublished)

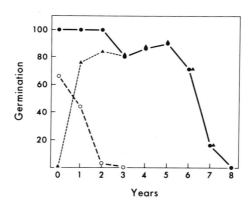

Fig. 6.7. Final germination percentage in distilled water of dimorphic *A. semibaccata* seed during storage in the laboratory. Brown seed (○); black seed before (▲) and after (●) scarification. (Redrawn from Beadle, 1952)

black seed in the first two years was stimulated by scarification, but thereafter scarification had no effect on germination. These observations suggest that the control of water uptake by the seed coat in black seeds is, in these species, a temporary property. In these experiments, the viability of the black seeds remained high even after scarification of black seed ceased to have an effect on germination.

6.1.4 Temperature Responses of Germination

Seed germination is remarkably sensitive to temperature, presumably because of the high temperature coefficients of the complex metabolic processes involved in

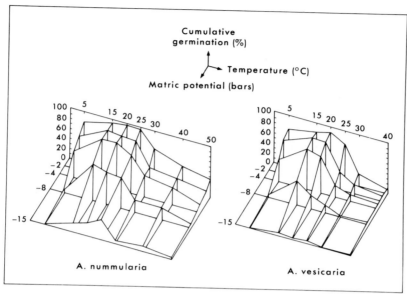

Fig. 6.8. Interaction between temperature and soil matric potential on the final germination percentage for seed of *A.nummularia* and *A.vesicaria*. (Redrawn from Sharma, 1976)

the transition from the resting state. Although it is likely that the principal responses of germination processes to temperature occur after water uptake, and although the physical process of water uptake is likely to have a relatively low temperature coefficient, interactions between water uptake and temperature may be important. These interactions were examined in seed of *A.vesicaria* and *A.nummularia* (Sharma, 1976). The temperature optimum for germination shifted to slightly higher temperatures at low soil water potentials, but the effects of lowering water potential were most marked at temperatures above and below the optimum (Fig. 6.8).

At high soil water potentials, the germination of seeds from arid shrubland *Atriplex* spp. is substantial at low temperatures and declines abruptly between 25° and 35 °C (Fig. 6.9). Similar results were obtained with seed of coastal species (Binet, 1965, 1966) but cold treatment (30 h at 5 °C) of these seeds results in much better subsequent germination (Binet, 1965, 1966; Ward, 1967c). Diurnal fluctuation in temperature under controlled conditions does not alter the pattern of germination (Sankary and Barbour, 1972) and several hours exposure to supraoptimal temperature (35 °C) in the course of a normal low temperature day does not significantly alter the pattern of germination (Beadle, 1952).

Temperature regulation of germination is complex. Koller (1972) emphasized that temperature effects on germination must be placed in a kinetic context and drew analogies with the temperature-induced lags in the rate of enzyme reactions. The sequence of metabolic events in a seed following imbibition of water (Fig. 6.1) involves the reactivation of pre-existing macromolecules and a burst of ATP synthesis (Ching, 1972; Mayer, 1977). Both of these events involve the reactivation

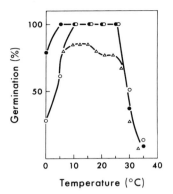

Fig. 6.9. Effect of temperature on final germination percentage of *Atriplex* seed in distilled water. *A. polycarpa* (△); *A. vesicaria* (●) and *A. spongiosa* (○). (Redrawn from Sankary and Barbour, 1972 and Beadle, 1952)

of membrane-bound enzyme systems and it is possible that the temperature sensitivity of these events may account for many temperature effects in germination (Grime, 1979).

In recent years the temperature limits of many metabolic processes have been ascribed to abrupt changes in the structure and stability of organelle membrane lipids (Chaps. 3.4.2; 9.3.3–4). Similar changes, which are reflected in discontinuities in the Arrhenius plot of $\log_{10}$ rate against reciprocal absolute temperature, have been observed with the low temperature limits for seed germination (Simon et al., 1976). However, in these studies respiration of the seed did not show similar discontinuities in the Arrhenius plot. Whether membrane stability is the sensor which determines the upper and lower temperature limits for germination, or whether much more complex interactions are involved, remains to be determined.

6.1.5 Germination and Light

The control of germination by light is readily demonstrated in many seeds. Three aspects of the light environment, light intensity, spectral composition, and periodicity all contribute to the control of germination in some seeds and the effect of light is frequently compounded by modification of the temperature response (Koller, 1972). Of the three principal light responses the control of germination by light intensity may be the most pertinent under natural conditions. The light environment of seeds which germinate on the soil surface is likely to be equal in intensity to the most intense commonly available laboratory light sources. Seeds which germinate while completely buried in soil are likely to be exposed to very low light intensities. Currie (1973) suggested that light was unlikely to penetrate much beyond 1 mm in soils. Light penetration in a silty clay loam and in a sand was less than 1% of incident light at 2.2 mm depth. This was sufficient to induce the germination of light-sensitive lettuce seed at that depth, but did not affect seeds 6 mm below the surface (Wooley and Stoller, 1978).

Some species of *Atriplex* are most sensitive to illumination and other are not. Comparisons between results with sensitive species are difficult because the intensity and spectral composition of light sources used in different studies is infrequently specified. *A.dimorphostegia* has been studied in detail by Koller and serves as a model demonstration of the effects of light intensity, spectral composition, and periodicity in the regulation of germination. Koller (1957) originally reported that germination of the naked seeds from fruits with both "humped" and "flat" bracteoles was inhibited by continuous light, particularly at higher temperatures. However, in the same study, it was shown that a short period of illumination considerably stimulated germination. Subsequently the stimulation of germination by a 15-min period of illumination with red light was shown to be reversed by a 45-min period of far-red light and the light stimulation of germination thus identified as a phytochrome response (Koller, 1970). Germination of seed from the "flat" and "humped" fruits declined when they were exposed to light for longer than 15 min, but photoperiod differences were identified in seed from the two types of fruits. Seed from "flat" fruits declined from 79% to 62% germination when the photoperiod was increased from 15 min to 8 h. Seed from "humped" fruits showed 57% germination after 15 min illumination, whereas those exposed to light for 8 h showed only 22% germination.

Continuous light stimulated germination of *A.rosea* (Kadman-Zahavy, 1955) and *A.glabriuscula (babingtonii)* (Binet, 1966). Courtice (1971) showed that low light intensities stimulated germination of *A.holocarpa* seed, but high intensity light was inhibitory. Whether the light responses of *Atriplex* seeds are modified by the presence or absence of bracteoles is not clear. Binet (1966) and Courtice (1971) showed that the light stimulation of germination was unaffected by the presence of bracteoles. Nicholls (1972) found that continuous light inhibited germination of naked seeds of *A.vesicaria* and *A.lindleyi* but did not inhibit germination of seed remaining within the bracteoles.

Beadle (1952) dismissed light as a factor in the control of seed germination in *Atriplex* under field conditions. Koller (1972) gave considerable weight to this factor. Although these divergent views were formulated on the basis of studies with different species, it is reasonable to conclude that light may regulate the germination of *Atriplex* in some circumstances. Light, like temperature, provides a fine control of germination in these seeds compared with the coarse control exercised by the regulation of imbibition. It is probable that germination of seed exposed to high intensity illumination on the soil surface may be delayed and it is also possible that the bracteoles may protect such seed from the high light intensity to some extent. The germination of buried seed may respond to light quality and photoperiod as proposed by Koller (1970), and the range of responses to light encountered in the polymorphic fruits of *A.dimorphostegia* may add a further degree of seasonal and soil depth control to germination responses in this species. As with temperature responses, it is likely that light responses are sensed by membranes, but via membrane associated pigments such as phytochrome.

6.1.6 Gas Exchange and Seed Germination

The dry seed has extremely low rates of respiratory gas exchange but on imbibition, respiratory O_2 uptake and CO_2 evolution increase dramatically. The exchange of these gases with the atmosphere involves simple diffusion along concentration gradients, as modified by the structure of the seed and soil. The escape of respiratory CO_2 would not normally be a problem, in view of the large concentration gradients away from the seed to the atmosphere, but the pH of the soil solution may be such as to allow the accumulation of respiratory CO_2 as bicarbonate.

The diffusion of O_2 to the seed is of greater importance for although the respiratory oxidases have a high affinity for O_2, the resistance to diffusion of O_2 within the seed increases with increasing seed water content. Other diffusive resistances to O_2 movement in germinating seeds, those associated with the seed coat and water film surrounding the seed and O_2 diffusion through wet soil, are discussed by Edwards (1973).

In spite of the high resistance to O_2 diffusion associated with seed coats, Edwards calculated that the O_2 concentration in more than half the embryo of *Sinapis* was adequate for germination. Although Binet (1965, 1966) failed to stimulate germination of *A. tatarica (tornabeni)* and *A. glabriuscula (babingtonii)* in O_2-enriched environments, it is possible that the water-saturated bracteoles of some species restrict O_2 exchange and inhibit germination (Burbidge, 1945; Beadle, 1952). Thus, although the salt content of the bracteoles may regulate premature imbibition, water retention by this tissue may also inhibit germination in waterlogged conditions (Koller et al., 1958).

6.1.7 Other Factors Regulating Germination

Although the primary effect of the bracts on germination of *Atriplex* seed is that of regulating water uptake and gas exchange, these structures may be important in other ways. Binet (1965, 1966) considered that mechanical restrictions due to the bracteoles should be taken into account and in Beadle's experiments there is evidence that the bracteoles slowed the rate of germination, but did not reduce the final percentage. Chemical inhibitors of germination (saponins) have been identified in extracts of the bracteoles of two North American species *A. canescens* and *A. polycarpa* (Cornelius and Hylton, 1969). These compounds, like salt, are readily removed from the bracteoles by leaching, and presumably it should be possible to separate the effects due to salt from those due to inhibitors by simply measuring water uptake. By reducing the water potential gradient between seed and medium, salt in the bracteoles prevents the absorption of sufficient water for germination to proceed (Fig. 6.2). On the other hand inhibitors which act upon metabolic processes directly are unlikely to interfere with water uptake.

Seed germination is accompanied by changes in many endogenous regulators including gibberellins, cytokinins, abscisic acid and ethylene, all of which may exert control at the gene level, at the levels of information transfer, and protein synthesis, or again, on membrane properties (Khan, 1977). The effects of these internal regulators may be mimicked by external application of naturally occurring or synthetic regulators. The significance of these controls of seed germination under natural conditions is not at all clear. Whether these compounds are the agents by means of which environmentally induced changes in the patterns of seed germination are controlled has yet to be established. Boucard and Ungar (1976) have shown that dormant seed produced by the halophyte *Suaeda* has less gibberellin-like activity than nondormant seed and dormancy can be reversed by application of gibberellic acid. Cytokinin levels are reduced when seeds are imbibed in NaCl solutions. These studies suggest that salinity may directly influence gibberellin levels during seed development and subsequently control germination.

The biochemistry of seed germination and its regulation is a subject of great complexity and it is difficult to evaluate these processes with our incomplete understanding of germination under natural conditions. Although a great many biochemical processes can be isolated and studied in detail under laboratory conditions it is questionable whether some, such as those dependent on light treatments, could have a significant role in natural environments. It is even more difficult to place biochemical processes of seed germination into relevant perspective because by and large, under natural conditions, the germination processes cannot be distinguished from establishment processes. Seedling emergence from soil, a commonly accepted indicator of successful germination under natural conditions, involves rather more than the processes involved in emergence of the radicle and cotyledons from the testa. It can be argued that under most conditions it is the establishment phase rather than germination which is more vulnerable under natural conditions. If this is so the detailed physiology and biochemistry of germination is of somewhat academic interest, and it is appropriate that we should focus attention on the establishment processes discussed in the next section.

6.2 Establishment Processes

The transition from the germinating seed to the established seedling is perhaps the most profound transition in the life of an individual plant. During establishment the physiological states and processes of the plant adjust to a change from the moist, dark soil to the drier, illuminated terrestrial environment. The transition involves the change from an organism composed of largely nonvacuolated meristematic cells to one principally composed of vacuolated cells and increasingly dependent on an elaborately differentiated translocation system for the supply of nutrients and assimilates.

The transition involves complex changes in metabolic patterns, in some cases from a dependence on stored lipids to the utilization of recently assimilated carbohydrates. The whole of the autotrophic machinery of the plant must be developed. This involves the coordination of absorption functions for nutrients in the root and synthetic functions of photosynthesis in the shoot.

Most of the physiological states and processes which develop in the seedling during establishment are retained throughout the life of the plant. Our discussion here could thus form the basis for most of the subsequent examination of physiological structure in this book, but we will confine ourselves to a brief consideration of the more important changes in states and processes which must accompany seedling establishment. For the most part we shall be able to do little more than speculate on the way environmental factors might interact with developmental changes and hope that the study of these processes might find a place in future research in physiological plant ecology.

6.2.1 Water Relations During Establishment

Water relations of seedlings during establishment are made complex by two simultaneously changing factors. First, the structural changes which occur during the transition from nonvacuolated to vacuolated cells are accompanied by important differences in the components of plant water relations. Second, the water environment of the seedling changes dramatically due to the relocation of plant parts in the soil atmosphere interface and to the rapid changes in water content which may occur in the soil surface. The magnitude of the water relations problem can be appreciated when we recollect that germination involves the transition from a dormant system at a water potential lower than $-1,000$ bar to a living system in equilibrium with a soil water potential of about -1 to -10 bar. During establishment the living system moves from the soil environment into the atmosphere where it must maintain the same high water potentials (-10 bar), but in equilibrium with an atmospheric water potential even lower than that of the dormant seed.

Uptake of water by the dry seed is initially driven by the large deficit in the matric potential of seed water and the high potential of water in wet soil (Chap. 6.1.1). Following activation of the embryo and commencement of cell division, cellular function is only possible when cells are turgid; that is when the cytoplasm is maintained at a positive hydrostatic pressure of about $+2$ to $+5$ bar. This positive turgor pressure (T) must be maintained for metabolism and development to proceed and at the same time the total potential of water in the plant (ψ_p) must be lower than soil water potential (ψ_s) if water is to continue to move to the developing seedling. This dilemma is resolved by the accumulation of solutes which exert an osmotic pressure in the cell vacuole (π) and the rigidity of the plant cell wall. More detailed treatment of plant cell water relations is given later (Chap. 8.2) but these components are related as follows

$$\psi_p = T - \pi.$$

In the developing but still heterotrophic seedling buried in the soil the sources of solutes to maintain low osmotic pressure, and hence positive turgor, are limited. These solutes must be spared from reserves otherwise used for synthetic purposes, or absorbed from the soil solution. The complex compromise between the use of seed reserves for turgor generation and extension of existing cells or for new cell synthesis will be influenced by the availability of soil nutrients and electrolytes. Such processes have not been explored and are presumably hidden in simple relationships such as that between seed size and depth (Fig. 6.10). Larger seeds can survive burial at greater depths, as demonstrated by the interspecific comparisons of dune species by Van der Valk (1974), presumably because the greater reserves permit development of larger seedlings, the greater extension growth of these and the more extensive exploration of soil water and nutrients.

Once the cotyledons emerge from the soil surface water uptake by the seedling is predominantly driven by the evaporation of water via the transpiration pathway (Chap. 8.1) but the water requirement of the seedling also increases dramatically at this time. Obviously these complex changes in water relations can only take place

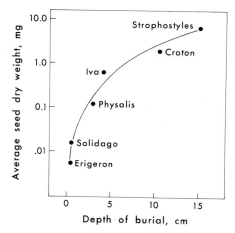

Fig. 6.10. Relationship between seed weight of dune plants and the maximum depth to which seed may be buried for successful seedling emergence. (Redrawn from Van der Valk, 1974)

when soil water potential is high. It is probable that emergence of the cotyledons from the soil takes place mainly at night when the soil to atmosphere water potential gradient and the rate of change of surface soil water potential is minimal. When the surface soil dries to a low water potential subsequent changes in soil water content are proportional to the depth, squared (Hillel, 1972). Successful establishment thus depends on the rate of elongation of the root which must keep ahead of the drying front as it recedes deeper into the soil.

There are few studies of the dynamics of plant and soil water potentials during this critical stage of establishment, largely because the scale of measurement is so small and the seedlings are so fragile. Cowling (1969) examined the drying profiles of clay-loam soils supporting *Atriplex* spp. in southern Australia. Following 25 mm precipitation in summer, the ψ_s in the 0–5 cm portion of this soil declined to -15 bar in less than one day. In the 5–10 cm region, ψ_s declined to -15 bar in 4.5 days. Under winter conditions the time to reach $\psi_s = -15$ bar in the 0–5 cm region was 1.5 days, and in the 5–10 cm region was 10 days. The root growth of *A. vesicaria* and *A. inflata* seedlings planted at the surface of this soil ranged between 2.2 and 2.5 cm day^{-1}. Williams (1972) measured rates of 2.1 cm day^{-1} for *A. vesicaria* in other soils. These studies indicate that, for these plants, seedling root growth can probably keep pace with the drying of soils under winter conditions.

6.2.2 Nutrition During Establishment

Seed reserves of organic and inorganic materials provide the resources for initial seedling development and limit the duration of heterotrophic growth. The size of this reserve is the principal factor which determines the depth to which seed may be buried and still become established (Fig. 6.10). Although the mobilization of inorganic ions during germination and establishment has been studied in detail with cereals and legumes (Sutcliffe, 1976), the relationship between use of internal nutrient reserves and absorption of nutrients from the soil solution has yet to be explored. The rate of mobilization of different nutrients and their sufficiency for the seedling varies greatly. Sufficient quantities of some micronutrient elements, such as zinc, are retained by some seeds to supply most of the requirement of the whole mature seedling so that this element may be transferred from one generation to the next. Seed reserves of other ions, such as potassium, are often sufficient for only a few days normal growth. Potassium is the most rapidly mobilized ion, others such as Ca^{2+} and Mg^{2+} are stored largely as complex organic salts (phytin, myoinositol hexaphosphate) and are released very slowly. The release of nitrogen from storage proteins is delayed until after proteolytic enzymes are synthesized.

The early development of ion absorption systems is as important to the seedling as the switch from heterotrophic to autotrophic metabolism. Although a large part of our understanding of ion absorption processes in plants has been based on seedling roots, the performance of these roots in natural conditions has rarely been examined. It is reasonable to suppose that these roots only function in the absorption of ions and nutrients when soil water potential is high. The intricate connections between nutrient absorption, synthesis of cellular components, turgor generation in the vacuole and expansion growth (Chaps. 7.3; 8.1.1) all determine whether or not the root system of the young seedling is able to explore the soil water and nutrient reserves in such a way that the seedling survives. The principal environmental factor limiting these processes is the rate of change of soil water potential and the distribution of nutrients in the soil surface. The principal plant factors likely to promote survival are seed size and seedling root growth rate.

The significance of nutrient concentration in the soil surface is demonstrated in the "bioassay" experiments of Charley and Cowling (1968). These authors grew *A. spongiosa* seedlings in equal volumes of soil taken from the surface horizons of *A. vesicaria* communities. Figure 6.11 shows the nutrient-limited production of dry matter by *A. spongiosa* seedlings plotted against depth from which the soil was derived. Nutrient limitation was evident in soils below 20 cm and the curve for growth of *A. spongiosa* has the same form as the depth profiles for available phosphorus and nitrogen. The concentration of N and P in the 0–5 cm region of the soil profile ensures that following germination, nutrients are available for these small-seeded species.

The available nutrients are only of significance in seedling establishment if they can be absorbed and utilized in development. In small-seeded species such as *A. vesicaria*, efficient use of nutrients such as phosphorus involves rapid absorption and transport to the most active metabolic regions of the seedling. It may not be surprising that phosphorus absorption begins two days after germination (Cowling, 1969) and that the absorbed ions are rapidly translocated to the shoot. Indeed, the rate of ^{32}P absorption on a root weight basis is higher at four days after germination than later (Fig. 6.12) and the distribution pattern for the absorbed ^{32}P indicates the shifting dependence of metabolism from photosynthesis in the cotyledons to photosynthesis in leaves. Comparisons between seedlings of *Atriplex* spp. and of cereals indicate that the former are much more efficient in transporting phosphorus to the shoot (Cook, 1973; Chap. 7.3.5). This difference may reflect differences in seed size and phosphorus reserves, as well as different developmental patterns between monocotyledons and dicotyledons.

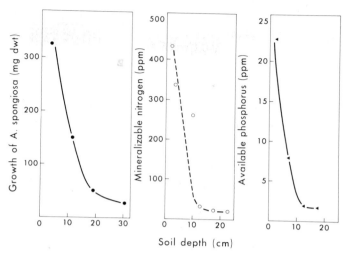

Fig. 6.11. Growth of *A. spongiosa* seedlings in limited volumes of soil from an *A. vesicaria* community. The much larger growth on surface soil is correlated with much higher levels of mineralizable nitrogen and available phosphorous in this part of the profile. (Redrawn from Charley and Cowling, 1968)

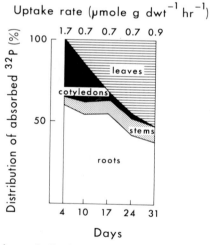

Fig. 6.12. Rate of ^{32}P uptake and distribution of absorbed ^{32}P by young seedlings of *A. vesicaria*. (Redrawn from Cowling, 1969)

6.2.3 Metabolic and Morphological Changes During Establishment

The changes which occur in the water relations and nutritional physiology of the emerging seedling are paralleled by even more significant changes in the metabolic machinery. In those cases where seed reserves are predominantly carbohydrates, the metabolic processes of carbohydrate utilization are retained throughout the life of the plant and the metabolic transition principally involves the development of the photosynthetic apparatus. In fatty seeds, the metabolic machinery involved in conversion of fat to carbohydrate is activated and degraded within the course of a few days and thereafter may play a much smaller role in the metabolism of the plant.

The effect on the developmental biochemistry of environmental factors normally encountered during seedling establishment under natural conditions is almost totally unknown. Recent studies with developing seedlings of *Pennisetum typhoides*, for example, show that low levels of salinity during germination and early seedling growth have marked effects on the rate of development and on the activity of many enzyme systems (Huber and Sankhla, 1976). Whether the change in enzyme activity is due to salt per se or to an effect of salinity on the rate or stage of development is not yet clear, but many of the salinity effects may be matched by treatment with growth regulators. Although environmental conditions may have significant effects on the rate of seedling development and these effects may be ultimately traced to specific metabolic processes, it is unlikely that environmental conditions act directly to determine the presence or absence of major metabolic pathways.

There are many suggestions (Huber and Sankhla, 1976) that a balance between C_3 and C_4 photosynthetic metabolism may exist in some plants and that this balance responds to environmental and developmental factors. The evidence for such claims is most incomplete and, as discussed elsewhere (Chaps. 4.3.6; 9.5.3), even though "intermediate forms" share some properties of both pathways, they remain functionally either C_3 or C_4 plants. In *Atriplex* it is quite evident that the cotyledons have "Kranz" anatomy (Chap. 2.1.4) in those species with C_4 photosynthesis in mature leaves, and conventional leaf anatomy in cotyledons of species with conventional leaf anatomy in mature leaves. Whether the presence of C_4 photosynthesis, which sometimes results in higher assimilation rate, higher temperature optima, and always results in more efficient water use (Chap. 9), confers any advantage during seedling establishment, has yet to be established. In that seedling establishment is probably the most vulnerable stage in the life cycle of higher plants and that many C_4 plants are recorded in water stress habitats, a close examination of the significance of C_4 photosynthesis in seedling establishment may be profitable.

6.3 Physiological Processes of Germination and Establishment in the Context of Ecological Processes

Although it is not difficult to speculate upon the plethora of physiological processes which may determine the establishment of germinated seeds, it is much more difficult to identify the specific processes which are responsible for seedling survival under particular conditions. Few survival studies have been attempted in the *Atriplex* communities or in habitats characterized by similarly extreme environmental change. There are two important limitations which make it difficult at present to assess the significance of the many component processes in germination and establishment.

First, we have yet to gather climatic and edaphic data on a scale, in space and time, which is relevant to that of the germinating seed or the seedling during establishment. That is, although there is a wealth of physiological and biochemical detail relevant to the span 10^2 to 10^5 in our continuum (Figs. 1.1; 6.1), we know relatively little about the environment within this same span. The magnitude of the environmental "noise" on this scale is largely unknown and may well be crucial to seedling survival. The patterns of soil temperature change near the surface as a function of solar inclination, of parent bush shading patterns and so forth are likely to be of crucial importance in seedling establishment. On a smaller scale the processes of water and nutrient movement on scales relevant to the roots of emerging seedlings are poorly understood (Chaps. 7.1; 8.1).

The second limitation relates to field data. The processes of germination and establishment of wild plants under natural conditions are not well enough known in quantitative terms. The significance of the physiological states and processes discussed in this chapter can only be assessed against observations and analyses of individual seedlings under natural conditions, as currently practiced by plant demographers (Harper, 1967, 1977; Grime, 1979). Although the Malthusian philosophy underlying the design of many demographic experiments may be more appropriate to some plant communities and environments than to others, there is no substitute for the methodology employed. Unfortunately, the emphasis in many demographic studies has been on biological rather than environmental interactions. In communities in which *Atriplex* is well represented, it is likely that the latter may be more important.

Some preliminary attempts to analyze the demography of arid shrubland communities and hence to assess the significance of germination and establishment processes in the wider context of ecological processes are discussed here together with laboratory studies under controlled conditions. These allow us to form some appreciation of the relevance of processes with relaxation times 10^2 to 10^5 s in the larger-scale ecological processes with relaxation times of 10^5 to 10^9 s.

6.3.1 Germination Processes and Reproductive Strategy

A good deal of attention has been given in recent years to analysis of the reproductive strategy of angiosperms and the tremendous range in seed size and morphology is taken to reflect a complex of adaptive compromises which have evolved to ensure the survival of these organisms (Stebbins, 1971). Harper et al. (1970) while noting the teleological traps inherent ascribing adaptive significance to seed states, conclude "presumably, some variations in seed shapes and sizes may be of no adaptive value at all, though this can never be proved." The plausibility of such an optimistic proposition comes into question when we examine the relationship between complex seed factors in *Atriplex* and germination processes in natural conditions.

It is commonly thought that seed size and seed number represent alternative strategies in the allocation of production to reproduction. In *A. vesicaria* fruits comprise 50%–65% of the litter dropped by female bushes and litter represents 3–4 times the standing aboveground biomass in this species (Charley, 1959). With about equal rates of stem and root growth (Jones et al., 1970) these observations

Table 6.2. Weight of air dry seeds and bracteoles in *Atriplex* fruits

Species and habitat[a]		Seed weight (mg)	Bracteole weight (mg)	Seed/bracteole
Perennials				
A. vesicaria	A	1.31	13.03	0.08
A. nummularia	A	1.16	7.13	0.16
A. paludosa	C	1.02	3.21	0.32
A. cinerea	C	2.74	8.53	0.32
Annuals				
A. inflata	A	0.70	5.02	0.14
A. spongiosa	A	2.45	7.13	0.34
A. rosea	R	1.49	1.13	1.13
A. triangularis	R	2.37	0.84	2.82

[a]   A arid shrubland; C coastal; R ruderal

imply that about 15% of the total production in this warm desert perennial may be invested in fruit production. Similar values are reported for other herbaceous perennials (Harper et al., 1970). Data are not available for annual species but comparisons with cool desert species such as *A. confertifolia* suggest that less than 5% of the total production is invested in fruits (M.M.Caldwell, personal communication). There are few data which relate to the question of seed size versus seed number in wild populations, most data being based, quite understandably, on seed crops. Whether the principles established in studies of such carefully selected and husbanded species are applicable to wild species remains to be established.

It is not difficult to speculate upon ways in which the functional significance of the seed size parameter might be important to germination processes in *Atriplex*. A large portion of the fruit weight in these species is associated with the bracteole, a structure which is important in dispersal and in germination (Chap. 6.1.2). Table 6.2 shows the average weight of seeds and bracteoles of some perennial and annual *Atriplex* species. In general the annuals have heavier seeds than perennials and the coastal, ruderal species (*A. triangularis, A. rosea, A. cinerea,* and *A. paludosa*) have higher seed-to-bracteole weight ratios. The production of large seed versus large bracteole could itself represent an adaptive compromise with respect to germination processes. Large seeds in small bracteoles with low salt content are likely to germinate more readily in response to smaller precipitation and to germinate from greater depth (Fig. 6.10). Small seeds in large salty bracteoles are likely to germinate only after salt is leached and only when fruits are near the surface. Nutrients concentrated in the soil surface (Fig. 6.11) and nutrient release processes in the surface may compensate to some extent the small reserves of small seeds. The higher rainfall requirement for germination of small seeds in large bracteoles may ensure better soil water reserves for establishment. Such speculation requires further examination.

It is commonly thought that the production of dimorphic seed may represent another compromise in reproductive strategy. The production of hard and soft seed appears to provide the plant with the option of germination in the short and long term in that hard seeds are much slower to germinate and/or much longer lived than soft seed (Fig. 6.7). Nobs and Hagar (1974) noted two germinations in ruderal populations of *A. hortensis* in San Francisco Bay area and speculated that the first germination arose from the large brown seed and the second from the small black seed. Dimorphic seed populations of some *Atriplex* spp. do not show differences in germination behavior in garden experiments. Gustafsson (1973a) reported that the rate and percentage germinations of black and brown seed of *A. triangularis* and *A. longipes* were scarcely different and black seed of the latter species showed greater germination percentages in some experiments. Beadle (1952) concluded that small hard seed of arid shrubland species accumulate in soil as a long-term reserve providing a seed source following prolonged drought periods or following a succession of rainfall which depleted the soft seed reserve but which was ineffective in seedling establishment or for the production of a new seed supply. He noted that hard black seed predominated in the fruit mats under shrubs many months after seeding. However, he was reluctant to conclude that the replacement of soft-seeded perennials by hard-seeded annuals in some communities was largely due to the difference in seed production and germination properties.

Such speculation needs further examination, as does the phenomenon of dimorphic seed production itself. In *A. hortensis* black seed are produced late after the main flush of flowering (Nobs and Hagar, 1974), in *A. heterosperma* they are produced early in the season (Frankton and Bassett, 1968) and in *A. subspicata* the percentage of black seed produced increased with day length (Bassett and Crompton, 1973). In *A. holocarpa* the imposition of water stress during seed set increased the proportion of black seed formed and on irrigation the proportion of black seed produced declined again (Courtice, 1971). Under benign culture conditions Nobs and Hagar (1974) found a slightly slower initial growth of seedlings from hard seed but no other differences in final development or seed production. Plants from both black seed and brown seed each produced 50% black and 50% brown seed under these conditions (M. Nobs and W. M. Hagar, unpublished). Obviously there is room for a great deal of imaginative experimental analysis of the factors determining seed weight versus bracteole weight, of hard versus soft seed. At the same time there is scope for the experimental verification of the functional significance of these factors in germination processes.

6.3.2 Dispersal Factors in Germination and Establishment

Just as we speculated upon the significance of seed factors for germination and establishment and of the allocation of seed resources to bracteole during fruit development, we may also speculate upon the role of the bracteoles in the dispersal of fruit. The light papery bracteoles or wings of the fruit of many arid shrubland *Atriplex* spp. possibly aid dispersal by wind. In *Atriplex* shrublands (Williams, 1972) and in *Artemisia herbalba* (Friedman and Orshan, 1975) 83%–85% of fruit set by female bushes was deposited under the canopy of the parent. Dispersal by wind beyond the canopy of the parent shrub is just as likely to place the fruit near another established shrub, simply because these are the only places in which such mobile fruits can lodge. Any advantages of dispersal by wind must be weighed against the probability that fruit will be lodged close to an established shrub. This may ensure a

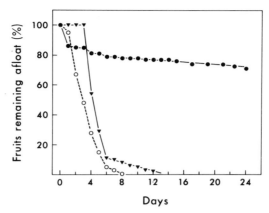

Fig. 6.13. Floatation properties of fruits in seawater from coastal species of *Atriplex*. *A. glabriuscula* (●—●), *A. hastata* (○--○) and *A. tatarica* *(recurva)* (▲). (Redrawn from Gustafsson, 1970, 1973 b)

more hospitable environment during germination but imposes the certain competition for water and nutrients during establishment.

On the other hand, debris can provide the physical stability of the parent shrub environment, without the competing demands for water and nutrients during establishment. In studies of regeneration in the *A. vesicaria* shrubland at Koonamore in south-eastern Australia debris was deliberately placed in some quadrats to trap seed and soil. Hall et al. (1964) noted a substantial increase in the establishment of *Atriplex* seedlings in response to this treatment. Debris serves a similar function in coastal habitats and seedlings are commonly established in debris at the driftline and on the foredune.

The bracteoles may determine the floatation properties of fruits from coastal *Atriplex* spp.. Gustafsson (1970, 1973a) found that fruits of species with quite restricted distribution, such as *A. tatarica (recurva)* in the Aegean and *A. calotheca* in the western Baltic, had limited floatation times in seawater. The fruits of widely distributed species such as *A. glabriuscula* (a species common to both the European and American shores of the North Atlantic Ocean) float for very long periods indeed (Fig. 6.13). There are reports of fruits of the latter species floating for up to five months (Hulme, 1957). It is, however, unlikely that floatation properties alone are responsible for the cosmopolitan distribution of species such as *A. triangularis;* human factors have probably played an important role as well. But once the species has established in different continents its further spread may reflect the floatation properties of the fruit.

The natural dispersal of *Atriplex* fruit by animal vectors is of minor significance, with man's activities being a notable exception. As noted earlier (Chaps. 2.3; 5) long-distance dispersal of seed of many *Atriplex* species in the course of wool, grain, and sugar beet trade is well established. Ridley (1930) records several potentially significant dispersal routes including the Saga antelope of the Siberian steppes which "no doubt swallows and disperses seeds of *Atriplex*". Bison have been observed to eat and to pass the seed of *A. canescens* in North America and in South America the ostrich is thought to distribute in like manner the seed of *A. capensis*.

6.3.3 Herbivory in Germination and Establishment

Although herbivores are unlikely to play an important role in long-distance dispersal in the genus *Atriplex,* they may interact with the reproductive biology of these plants in several ways. The decimation of arid shrublands by exotic herbivores as has occurred in many parts of Australia and North America could be due as much to interference with reproductive biology as to the eating of this relatively unpalatable vegetation. Studies of the grazing preferences of sheep establish that they consume significant quantities of chenopod shrubs only when more palatable vegetation is unavailable such as in times of drought (Leigh and Mulham, 1966). Assuming that the palatability of young seedlings is similar to that of established vegetation and recognizing that when seedlings are abundant, the preferred more palatable vegetation is also likely to be abundant, grazing by exotic herbivores is unlikely to be an important factor in seedling establishment. Indirect effects of grazing such as trampling and compaction of the soil may interfere with seedling establishment. On the other hand, Noble (personal communication) has noted that trampling prior to germination can break up the soil crust and improve conditions for subsequent establishment and the removal of large shrubs by trampling or grazing is likely to open spaces for the establishment of juveniles.

Interference with the reproductive biology of the plant as a result of grazing has been described by Williams et al. (1978). Female shrubs are evidently preferred to male shrubs, for Williams found a ratio of males to females of 1.1 in ungrazed populations of *A. vesicaria* and a ratio of 1.5 to 2.2 in heavily grazed populations. The selective removal of females reduces the seed-producing capacity of the populations and Williams considered that when the ratio males to females exceeded 2.0, production of viable seed declined so that few new seedlings were established. However seed production was limited, not by the number or resources of the females shrubs, but because the spacing between females and males was such that pollination was evidently ineffective. This explanation is based on observations summarized in Fig. 6.14 showing that few seeds germinated and resulted in seedling establishment after 1 year if the distance between male and female plants exceeded 1 m. This conclusion applied irrespective of the vigor of the female shrubs estimated on the basis of foliage and flower quantity and stem diameter. It is clear that the selective grazing of females which reduces their reproductive vigor and their numbers relative to males can reduce the recruitment rate of the population to near zero although the biomass of the population may be relatively unaffected.

The impact of feral herbivores such as rabbits and of native herbivores during seedling establishment has not been studied in detail. Feral rabbit populations have on occasion reached plague proportions in Australian chenopod shrublands, but such population explosions have coincided with periods of abundant vegetation. Again it seems likely that damage to the perennial chenopod shrubs has been greatest during drought periods when rabbits have been observed to remove the moist bark from *A. nummularia* resulting in death of the shrubs. In Australia, native marsupials are very mobile, tending to migrate when herbage is no longer freely available (Newsome, 1971). They are thus unlikely to have a significant impact on adult vegetation but their effects during seedling establishment have yet to be determined.

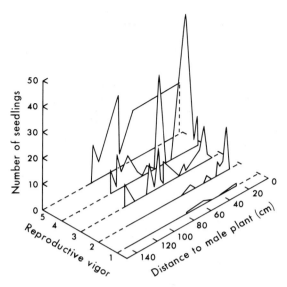

Fig. 6.14. Number of seedlings which had survived after 1 year associated with mature female bushes of *A. vesicaria* as a function of reproductive vigor (indicated by quantity of foliage and flowers and basal stem thickness) and distance from nearest male bush. (Redrawn from Williams et al. 1978)

    Smaller herbivores, such as the longicorn beetle (Cerambycidae) which is a stem and root borer, have been observed to decimate the populations of young seedlings of *A. pseudocampanulata* and *A. lindleyii* (Courtice, 1971). Other insect herbivores, such as locusts and coccids, have been recorded to cause considerable damage to *A. vesicaria* shrubs (Froggatt, 1910). Briese (1974) noted that the sap-sucking coccid, *Pulvinaria maskelli,* is less frequent on small seedlings than on mature shrubs, but the seedlings sustained relatively more damage than adult bushes. These insects may be an important factor in seedling establishment but Briese noted the coccid population was markedly reduced following heavy rainfall, suggesting perhaps that their impact would be minimal under conditions most favorable to seedling establishment. On adult shrubs of *A. vesicaria* this coccid was tended by ants *(Irido myrmex)* which harvested the honey dew it produced. Again, the impact of direct and indirect predators, in this case the ants, seems to be greatest on adult shrubs rather than on juveniles.

    Harvester ants are among the most active predators of arid shrubland communities and could conceivably limit the seed resources for germination processes. Western harvester ants *(Pogonomyex)* are thought to have denuded large areas in the Big Horn Basin, Wyoming, and ant activity results in a marked concentric pattern of *Atriplex nuttallii* in the vicinity of the ant nest (Wight and Nichols, 1966). It is not clear how this pattern is established or maintained but presumably removal of seed and disturbance of the soil in the vicinity of the nest are important factors. Although the ants had an obvious impact on the pattern of vegetation Wight and Nichols considered that they had little effect on the

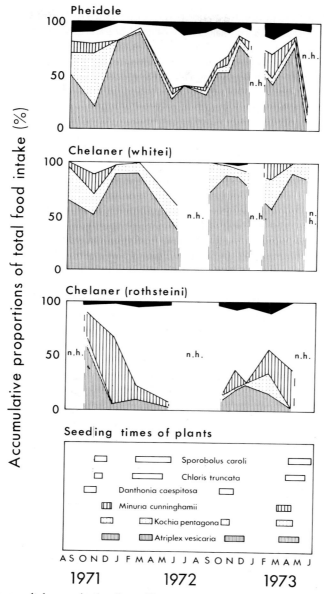

Fig. 6.15. Seasonal changes in the diets of harvester ants in *A. vesicaria* shrublands. *Pheidole* and *Chelaner (whitei)* selectively gather seed of *A. vesicaria* seed and their intake of this seed is greatest at the time of fruit set. (Briese, 1974)

production or stability of the community. Briese (1974) studied the trophic relations, behavioral patterns and effects of ant activity in an *A. vesicaria* dominated shrubland on the Riverine Plains of New South Wales. Seed of *A. vesicaria* predominated in the granaries of the most active species (Fig. 6.15). Over a three-month period Briese estimated that about 20% of the seed production of *A. vesicaria* was harvested by these ants and selectivity of seed harvesting was closely tuned to fruiting time of these shrubs. He concluded, however, that there was scant evidence that seed harvesting and redistribution had any significant impact on the abundance, relative abundance, or pattern of vegetation in this community.

6.3.4 Evaluation of Germination and Establishment Processes

Given suitable dispersal, a suitable site, and the absence of predators, successful seedling establishment depends on the interaction between physiological and environmental processes discussed above. Germination processes can only be separated from establishment processes in laboratory simulations of natural environments and relatively few such evaluations have been detailed in the literature. Williams (1972) asked the question whether germination and establishment processes were likely to be important in maintaining near pure stands of *Atriplex vesicaria* and *Maireana pyramidata* on adjacent different soil types on the Riverine Plain of New South Wales (Chap. 5.1.1). The germination and establishment of seed from fruits of each species was compared on its native soil and on the other soil type. In each case a total of eight different precipitation patterns was compared in a controlled artificial environment which corresponded to springtime temperature and humidity conditions. The results for three of the precipitation patterns are shown in Fig. 6.16. The seed populations used in these experiments showed very different germination behavior under petri dish conditions. The *Maireana* fruit germinated to 85% in five days and *Atriplex* fruit germinated to 56% after 16 days, only 30% having germinated after five days.

This difference in germination behavior is evident in the soil germination results shown in Fig. 6.16. The different precipitation patterns shown in this figure had little impact on the germination percentage of either species except that the time taken to maximum germination was longest in treatments with widely spaced precipitation. In all treatments on both soils the survival of *A. vesicaria* was greater than that of *M. pyramidata* after 25 days. *A. vesicaria* has the C_4 photosynthetic pathway, but whether survival is connected with C_4 photosynthesis has yet to be determined. The survival of *A. vesicaria* after 40 days was greater than that of *M. pyramidata* in all except one treatment and survival of *A. vesicaria* was much greater on its native soil than on the *Maireana* soil. Whether survival of *Atriplex* seedlings was greater on its native soil because of surface nutrient accumulation or the availability of NaCl to this halophyte (Fig. 5.23 d) is not clear. The rate of change in soil water potential in the different treatments was not followed and it is thus difficult to evaluate plant-soil water relations and root growth processes during establishment.

Rates of root growth for each species in each soil at field capacity were compared and these show (Table 6.3) that root growth rate of *Atriplex* was not influenced by soil type. Although *Maireana* roots developed much more rapidly in their native soil, they grow rather more slowly than *Atriplex* roots when *Maireana* was sown in *Atriplex* soil. When the data of Table 6.3 are assessed in relation to Fig. 6.16 they suggest that high root growth rate may be important to the establishment of *Maireana* in its native soil when precipitation is separated by

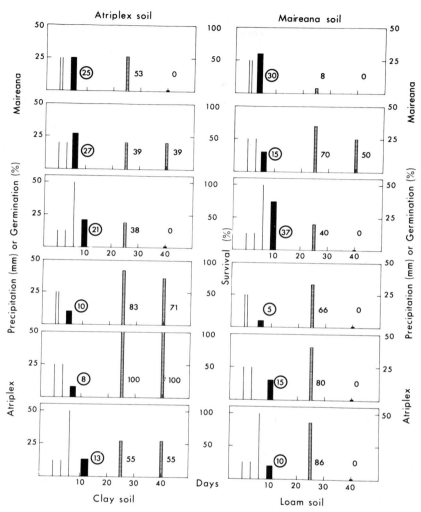

Fig. 6.16. Germination and establishment of *Atriplex vesicaria* and *Maireana* (*Kochia*) *pyramidata* in response to three precipitation regimes on two soils. The extent and separation of the precipitation treatments is indicated by the *thin vertical bars* at the commencement of each experiment. The maximum germination percentage and the day on which it was recorded is shown by the *filled histogram* and the *encircled numeral*. Survival of germinated seed as seedlings was noted after 25 and 40 days and is indicated by the *hatched histograms* and the *associated numerals*. (Drawn from Williams, 1972)

Table 6.3. Mean rates of vertical root growth of seedlings in soils initially at field capacity. (Williams, 1972)

Soil and native species	A. vesicaria	M. pyramidata (cm day^{-1})
Gray clay (*Atriplex*)	2.1 ± 0.1	1.5 ± 0.1
Red sandy loam (*Maireana*)	2.3 ± 0.2	3.6 ± 0.2

several rainless days. Slow root growth rate may prevent the establishment of
Atriplex seedlings on this loamy soil which presumably dries more rapidly than the
heavy clay soil native to *Atriplex*. Slow root growth of *M.pyramidata* on the clay
soil may account for the poor survival of this species in the clay soil. Although
Maireana has a higher germinability, under the conditions used, this did not ensure
a higher survival. This species might have higher survival under natural conditions
on *Atriplex* soil if it produced more seed, but no data on this alternative are
available. Establishment, closely related to rates of seedling root growth and soil
water processes, seems much more significant than germination per se in this
comparison. In pragmatic terms, seed germination under field conditions is more
properly a record of seedling establishment for it is scored by the emergence of
leaves or cotyledons. We can rightly question then whether we are likely to see
evidence of the adaptive compromises and reproductive strategies thought to be
associated with the seed factors discussed in the previous experiments.

6.3.5 Establishment Processes and Population Dynamics in Natural Conditions

The role of germination and establishment processes in the population dynamics
of wild plants is an area of considerable interest at the present time (Harper, 1977).
These processes are most conveniently examined in short-lived species in which the
frequency and the relaxation time of germination and establishment are acceptable
to the investigator or correspond to those of his Ph.D student. Although some
desert ephemerals may approach the ideal of a "seed that germinates to expose a
green flower that immediately proceeds to leave several seeds that germinate
without delay" (Harper and White, 1974), these plants have yet to be examined
systematically (Courtice, 1971).

These interactions could perhaps be most readily and more comfortably
assessed in more salubrious coastal habitats in which the simple dune sand soil
introduces relatively few uncertainties for measurement of water and nutrition.
Binet (1969) studied the establishment of *A.glabriuscula (babingtonii)* on the sand
dunes of the Normandy coast. Plants grew more slowly on the foredune than on the
driftline, presumably because water was less adequate and because they
developed more extensive root systems. The implication from these studies, that a
greater proportion of photosynthetic production is devoted to root production in
water-limited environments, may be very important for establishment. Such
processes need further evaluation under controlled laboratory conditions. In one
such study Williams (1972) found a similar trend to increased root/shoot ratios
when *A.vesicaria, A.nummularia, Maireana sedifolia,* and *Maireana pyramidata*
were grown at controlled low soil water potentials. Obviously, such studies are
needed to further validate the above speculations as to the physiological bases of
germination and establishment processes. Yet they are also important if we are to
understand the significance of germination and establishment in population
dynamics of plant communities.

The records of the Koonamore Vegetation Reserve (page 1, Fig. 5.1 b) provide a
unique account of population dynamics in arid shrubland communities. The
Reserve was established in 1925 on 510 ha of land which had been grazed by

Table 6.4. Observations of *Atriplex* seedling establishment at Koonamore 1926–63. (Average rainfall 186 mm, data of Hall et al., 1964)

Year	Season	Precipitation	
		Seasonal (mm)	Annual (mm)
1926	Winter	45	184
1928	Winter	53	172
1931	Winter	62	205
1933	Spring-summer	70	132
1936–37	Summer	190	337
1937–38	Summer	76	188
1939	Winter	70	318
1941	Summer-fall	101	190
1941	Spring-summer	89	190
1945	Spring	60	162
1946–47	Summer	142	229
1950	Summer-fall	255	366
1950	Winter	60	366
1961	Spring-summer	57	155
1952	Winter	57	324
1952–53	Summer	51	152
1954–55	Summer	89	388
1955	Winter	146	388
1958	Spring-summer	108	300

introduced herbivores since 1863 and sheep and rabbits effectively denuded the landscape. In the period 1925–1962 an average of 50 vegetational records in the form of quadrat chartings or fixed point photographs was made each year. Recent analyses of these data and recent reactivation of some aspects of the project (Crisp, 1975; Noble, 1977) emphasize how valuable an experiment this has been. However, the earlier published reports (Wood, 1936; Hall et al., 1964) contain sufficient detail to allow some assessment of the significance of establishment processes in population dynamics of perennial *Atriplex* spp. in this habitat.

The Koonamore experiment indicates at least 19 separate, significant initial seedling establishment events in the period 1926–1963 and, as shown in Table 6.4, only 42% of these occurred in years of greater than mean annual precipitation. The mean winter precipitation resulting in establishment was 70 mm (n = 7) for the months of June–August. The mean summer precipitation resulting in establishment was 110 mm (n = 5). Of the recorded establishments of *Atriplex* spp., 53% occurred in spring–summer and 78% of these were confined to years with higher than average precipitation. These observations show that in this environment there is no marked seasonality of germination and initial seedling establishment. They show too that establishment in summer is largely confined to years of above average rainfall and that summer establishment is correlated with higher precipitation in this season than that found for winter establishment.

The Koonamore observations accord with the results of field trials reported by Beadle (1952) which showed practically zero establishment of *Atriplex* from seed in soils irrigated with up to 75 mm equivalent precipitation over the summer months. These experiments, conducted in different soils at a different site, can be compared with the observation that summer establishment at Koonamore required an average of 110 mm precipitation. When the experiments were repeated in winter Beadle (1952) recorded 25%–100% establishment of seedlings of five *Atriplex* spp. which persisted for up to 100 days. Woodhouse (1964) reported experiments designed to evaluate different sowing techniques for establishment of *A. vesicaria* in overgrazed communities and eroded soils. On loam soils comparable to those at Koonamore, Woodhouse failed to obtain initial seedling establishment during summer with up to 50 mm precipitation at the time of sowing. The same treatment during a winter sowing resulted in up to 38% initial establishment of seedlings, but none of these survived the following summer.

Interpretation of these empirically established values for "effective precipitation" in terms of the germination and establishment processes discussed above remains complex. Although Charley (1959) established that the equivalent of 100 mm rainfall was sufficient to leach 95% of the NaCl from the bracteoles of *A. vesicaria,* several smaller rainfall events would possibly be just as effective in creating optimal germination conditions near the surface. After germination, the effectiveness of precipitation for establishment will presumably vary greatly with soil type and with season. The high mortality of young seedlings, rather than small germination percentage, is evidently a key parameter in population dynamics in arid shrubland habitats of the Koonamore type.

From quadrat chartings of individual plants during the first 10 years of the Koonamore experiment, Wood (1936) concluded that the longevity of *A. vesicaria* under these conditions was about 12 years and that of *A. stipitata* was about 8 years. Hall et al. (1964) revised the estimated life span of *A. vesicaria* upward to 25 years under more favorable environmental conditions. More recently, analysis of the smaller Koonamore quadrat 10A (Crisp, 1975) indicated that the probability of survival of *A. vesicaria* seedlings established for two years or more was 0.94 and was independent of age. Crisp found that the survival of individual *A. vesicaria* shrubs in this quadrat approximated to the negative exponential model of decay in plant populations (Harper, 1967) and estimated the half-life of *A. vesicaria* shrubs to be 11 years. The co-dominant shrub *Maireana sedifolia* appears to have a half-life of between 150–300 years. Williams (1972) recorded the size distribution of basal stem diameters in an *A. vesicaria* population on the Riverine Plain. His data (Table 6.5) suggest mortality is greatest among the larger plants in this community. However, under field conditions the probability of retaining stems of dead plants which are 1–5 mm in diameter is low and the age-dependent mortality suggested by these data needs further evaluation.

On the basis of the above data we can now assess germination and establishment processes and mortality of established plants in the successional population dynamics of *Atriplex* in these habitats. Quadrat 100 at Koonamore is a 100 × 100 meter enclosure on a desert loam soil and when established in 1926 contained one plant each of *A. vesicaria* and *A. stipitata*. This quadrat was adjacent to a good seed source of *Atriplex* spp. outside the reserve and is flanked on either

Table 6.5. Size distribution of basal stem diameter in a population of
A. vesicaria near Oxley, N.S.W. (Data of Williams, 1972)

Plant status	Diameter (mm)				
	1–3	3–5	5–10	10–20	20+
	Plants ha^{-1} ($\times 10^3$)				
Living	8.12	10.30	9.69	14.90	1.25
Dead	—	—	0.55	5.23	0.39

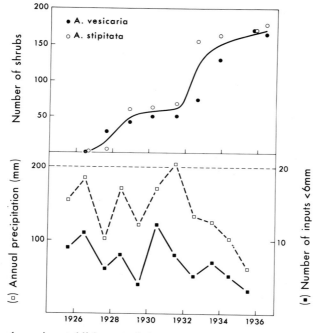

Fig. 6.17. Initial phases in establishment of *A. vesicaria* and *A. stipitata* on quadrat ⧧ 100 at Koonamore, south-eastern Australia, showing annual precipitation and number of precipitation inputs each year of more than 6 mm. Long-term mean precipitation is shown as a *dashed* line. (Data of Wood, 1936)

side by line transects. It provides an excellent illustration of short-term and long-term changes in the population of these two shrub species. Over the first ten years of observation, reported by Wood (1936), the increase in population of both *Atriplex* spp. was remarkably similar and followed the same two-step progression (Fig. 6.17). This period corresponded to a time of below average precipitation, and only once did annual rainfall exceed the mean for the period 1889–1936. The first arrest in population growth (1928–1932) is not associated with abnormally low rainfall, but perhaps with the fact that there were only four falls of greater than

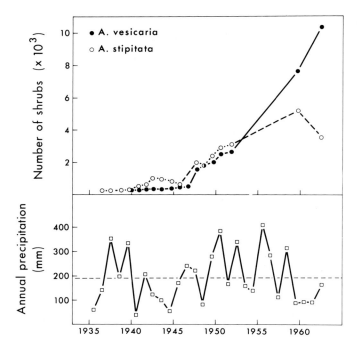

Fig. 6.18. Later phases in the establishment of *A. vesicaria* and *A. stipitata* on quadrat ⧣ 100 at Koonamore, south-eastern Australia, showing annual precipitation. Long-term mean precipitation is shown as a *dashed line*. (Redrawn from Hall et al., 1964)

6 mm precipitation in 1929. This was thought by Wood to be a minimum effective rainfall for germination at Koonamore and compares with 6–11 falls of this magnitude in preceding years. Maintenance of the population in this quadrat is consistent with deductions of age-independent mortality after two years (Crisp, 1975). The second arrest in population growth (1934–1936) is associated with a dramatic decline in annual precipitation and in the number of effective falls during this period. No establishment was recorded 1933–1936/37 (Table 6.4) and the population remained stable throughout this severe drought period.

Longer-term changes in the *Atriplex* populations of quadrat ⧣100 are shown in Fig. 6.18. There was a prolonged period 1941–1945 when no establishment was recorded (Table 6.4) and in which the *A. vesicaria* population remained constant and in which the *A. stipitata* population declined slightly. The establishment of seedlings in 1946–1947 and 1950 took place during summers of higher than usual rainfall and initiated an almost logarithmic increase in the *A. vesicaria* population. As in 1941–1945, the *A. vesicaria* population did not decline in the drought years 1960–1963 whereas *A. stipitata* did. This difference in the stability of the two populations was also evident in line transects flanking quadrat ⧣100. Although some data suggest a more effective establishment of *A. stipitata* in some years (1930–1931, for example), the difference in behavior of these two species

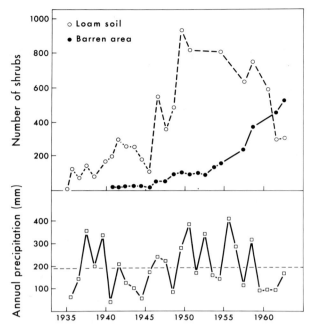

Fig. 6.19. Establishment of *Atriplex* spp. on loam soil and barren areas of quadrat #100 at Koonamore, south-eastern Australia. Climatic data as in Fig. 6.18. (Redrawn from Hall et al., 1964)

populations seems to be a product of mortality factors in plants of more than two years of age.

In some circumstances establishment processes were more obviously responsible for the population responses. Some transects at Koonamore were sited so that they included a soil transition and *A. stipitata* migrated much more rapidly across a loam soil than across a sand plain (Hall et al., 1964). Within quadrat #100, 20 × 20 meter subquadrats on loam soil and barren clay pans show the remarkable difference in population growth in these two habitats (Fig. 6.19). The density of shrubs on loam soil increases more rapidly and declines significantly in drought periods, suggesting density-dependent thinning. On the clay soil over the same period the population did not reach the densities attained on loam soil in good seasons. These observations suggest that although the shrubs may regenerate more rapidly under favorable conditions on favorable soils, these communities suffer substantial mortality when water becomes limiting. On poor soils regeneration is less rapid but, from the view point of population dynamics, the lower density is equally effective in maintaining the community through drought periods. The data in Fig. 6.19 taken from Hall et al. (1964) do not indicate the relative composition of the *Atriplex* population. It may be that the differences between the two plots are due in part to different ratios of the two species, one of which is longer-lived than the other.

6.4 Conclusions

The fascination of, and preoccupation with, different states and processes associated with seed germination in the genus *Atriplex* is as great as in any other widely studied wild plant. In making a distinction between the processes of seed germination and seedling establishment we have focused attention on the similarities and differences of significant component processes such as water relations, temperature relations, and nutrition in a way which may provoke their systematic evaluation. This distinction may remain academic however, for the bulk of presently available evidence suggests that it may be establishment processes rather than the elegant germination control processes which determine survival under field conditions. This conclusion must remain tentative until systematic demographic studies of these processes are made in natural habitats. If we are "basically ignorant of the physiologic and histologic conditions" that underlie germination processes, we must confess to being almost totally ignorant of those underlying establishment. It may be that the intricacies of the physiological processes during establishment in natural habitats arc even more impressive than those associated with germination which are so obviously displayed in the seed states. One thing seems clear, however. In this genus, in the habitats studied, it is the physiological processes of germination and establishment, rather than the ecological processes which follow from seed production (seed and seedling herbivory), which determine the population dynamics of these plants.

Chapter 7. Absorption of Ions and Nutrients

"The amount of salt contained by halophytes is not exclusively determined passively by that of their substratum, but depends chiefly on a craving for salt in the plant itself, since plants that naturally grow in such localities are in ordinary soil also in the habit of storing up larger quantities of sodium chloride than most nonhalophytes".

(Schimper, 1903)

In the previous chapter it was emphasized that, after the utilization of the nutrient and ionic reserves of the seed, the autotrophic activities of land plants depend on the absorption of mineral ions from the soil just as they depend on the harvest of solar energy and atmospheric carbon. This dependence on ion absorption is expressed both in the requirement for nutrient ions, which are incorporated into or complexed with organic molecules in the cytoplasm and metabolic machinery of living cells, and in the requirement for nonnutrient ions which do not participate directly in metabolism, but which play a role in turgor regulation and ionic balance. Some nonnutrient ions may have distinctly unfavorable effects on plant growth due to their interference with metabolism or with the absorption of nutrient ions.

Most of the ions found in higher plants are absorbed from the soil solution by the roots and are redistributed to tissues and cells in other parts of the plant. The absorption processes are much more complex than was supposed by Sachs (1887) who proposed that the mineral ions entered the plant with, and were distributed by water flow during transpiration. Transpiration does play an important role in the movement of ions through the soil to plant roots and from the roots to the shoot. However, the plant is a complex of high and low conductance pathways to the movement of ions and a complex of compartments in which the activity of ions may be similar to that in free solution or may be greatly modified due to electrostatic and other interactions with organic molecules. The soil reservoir of mineral ions is also a complex and dynamic system involving similar differences in ion activity between solution and colloidal phases. Although most ions are absorbed from the soil solution, some may be absorbed by direct contact exchange between root and ions adsorbed on soil colloids (Jenny, 1966).

There is abundant evidence that the ionic composition of different soils and peculiarities of the ion absorption processes in different plants interact to determine the species composition of many communities (Rorison, 1969). One of the more striking examples is the distribution of calcicoles and calcifuges. Calcicoles are plants restricted to soils with a surface pH of 6.0–8.0, and calcifuges are plants restricted to soils with a surface pH 3.0–6.0 as a result of complex interactions between soil pH, aluminum toxicity and iron deficiency. The application of inorganic and organic fertilizers to natural communities is commonly associated with a decrease in the number of species, an increase in vegetation cover, and the replacement of herbs by grasses.

In many parts of the world soil salinity determines the composition of extensive natural communities and is an increasingly important factor in irrigated agriculture (Poljakoff-Mayber and Gale, 1975). Plants growing in saline habitats display a wide range of morphological and physiological properties which enable them to tolerate this ionic environment (Waisel, 1972). Because species of the genus *Atriplex* are well represented in saline habitats and because their distribution is often correlated with salinity factors (Chap. 5) much of the present chapter is concerned with ion absorption processes of significance to halophytes. Considerable progress has been made in the evaluation of these processes since the time of Schimper (1903), and one objective of this chapter will be to assess the "craving for salt" so characteristic of halophytes.

7.1 Ionic Concentration, Distribution, and Pathways in the Soil-Plant System

Activity and charge in aqueous solution are possibly the most significant properties of inorganic ions in relation to processes of ion movement in soils and processes of ion absorption in plants. Ionic activity is commonly equated with concentration, but in most instances, concentration itself is only an approximation, because it is calculated from measurements of ion content and water content. The uncertainty of such approximations becomes a serious limitation when one attempts to evaluate processes and mechanisms. For example, many aspects of salinity tolerance, such as processes of ion uptake and the interaction of ions with metabolism, remain uncertain, simply because it is not yet possible to specify the concentration of ions in the cytoplasm and vacuole.

7.1.1 States of Ions and Nutrients in the Soil

Most ions are absorbed by plants from the soil solution which is in dynamic equilibrium with ions adsorbed to other soil components such as the clay mineral colloids, the organic matter colloids, and with ions contained in the soil parent material. The principal difficulty in determining the ionic concentration of the soil solution is to remove this solution from the soil without changing its composition (Black, 1968). Although we might agree that a *"properly* displaced soil solution is the unaltered, true soil solution" (Adams, 1974), its composition depends on the methods used. Suction and soil paste methods for determining the soil solution composition usually overestimate the concentration of PO_4^{3-} and K^+ compared with displacement methods, but estimates of the concentration of ions such as Na^+ and Cl^- are less influenced by the extraction method used.

Variations with space and time in soil factors such as clay content, organic matter content and water content introduce even larger uncertainties into estimates

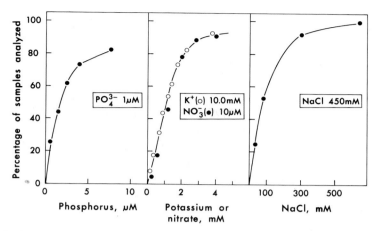

Fig. 7.1. Ionic concentrations in soil solution from analyses compiled by Reisenauer (1966) and Richards (1954). The total percentage of all samples analyzed which fall within arbitrarily chosen ranges of increasing concentrations is plotted against concentration. The concentration corresponding to 50% of the samples analyzed is an indicator of the average soil solution concentration in agricultural or saline soils. The average concentration of these ions in seawater is shown in the *insert* to each graph

of ionic concentration in the soil solution. Figure 7.1 shows the percentage distribution of the concentration of PO_4^{3-}, K^+, and NO_3^- in soil solution analyses tabulated by Reisenauer (1966). Pressure extracts of agricultural soils at field capacity were analyzed and these show that the mean concentration of PO_4^{3-} is about $2\,\mu M$, whereas the mean concentration of K^+ and NO_3^- are about three orders of magnitude greater. The range of K^+ and NO_3^- concentrations found in these soil solutions corresponds to that commonly employed in laboratory nutrient culture solutions. It should be noted that, even in agricultural soils (which may have been fertilized), the concentration of PO_4^{3-} is much lower than that used in culture solutions. Soil solution concentration of PO_4^{3-} can only be reproduced in solution culture experiments by the use of large volumes of dilute circulated solution.

Figure 7.1 also shows the calculated NaCl concentration in saturation extracts of saline soils analyzed by Richards (1954). At soil water-holding capacity, the concentrations of NaCl would be approximately double the values shown, and the concentration of nutrient ions such as PO_4^{3-} may be more than five orders of magnitude lower than Cl^- in saline soils. This relationship is not substantially different from that in sea water, as shown in Fig. 7.1.

Analyses of many arid regions soils supporting *Atriplex* spp. and the probable concentrations of major cations in the soil solution at water-holding capacity for some of these soils are shown in Table 7.1. (See also Beadle et al., 1957; Gates and Muirhead, 1967; Sharma, 1973b). The concentrations of cations are usually in the upper parts of the range shown in Fig. 7.1 and Cl^- concentrations correspond to those of the monovalent ions. The marked differences between the Na^+ content of surface soils under and between bushes of *Atriplex* spp. on the Riverine Plain soils and the high concentrations of Ca^{2+} in the soil solution are

Table 7.1. Soluble and exchangeable cation concentrations in soils supporting *Atriplex* communities in South-Eastern Australia. The concentration of soluble ions is given in mEq l^{-1} and of adsorbed ions (in parenthesis) is given in mEq 100 g^{-1} soil. (Data of Sharma, 1973b; C.B. Osmond, unpublished)

Location and soil		Cation concentration			
		Na^+	K^+	Ca^{2+}	Mg^{2+}
Riverine Plain [a]					
Billabong clay	0– 7.5 cm	18 (2.3)	1 (0.9)	6 (7.8)	9 (6.9)
(Between shrubs)	15–30 cm	100 (5.7)	0.6 (0.9)	10 (10.8)	24 (15.5)
(Under shrubs)	0– 7.5 cm	45 (3.6)	1.4 (1.7)	15 (8.0)	27 (9.9)
	15–30 cm	95 (6.4)	0.6 (0.9)	16 (10.6)	35 (15.0)
Riverina clay	0– 7.5 cm	12 (2.9)	0.6 (1.7)	5 (9.4)	5 (9.8)
(Between shrubs)	15–30 cm	46 (5.4)	0.4 (1.2)	5 (10.6)	8 (14.1)
(Under shrubs)	0– 7.5 cm	38 (3.8)	1.6 (2.4)	11 (7.5)	20 (11.1)
	15–30 cm	60 (5.6)	0.4 (1.2)	6 (10.7)	10 (14.1)
Barrier Ranges [b]					
Saline clay	0– 7.5 cm	150 (1.3)	26 (1.8)	37 (10.0)	25 (2.7)
	15–30 cm	120 (2.1)	6 (0.9)	22 (12.2)	11 (3.1)
Calcareous loam	0– 7.5 cm	170 (0.9)	110 (1.5)	100 (19.5)	32 (0.2)
	15–30 cm	132 (0.4)	15 (0.7)	59 (10.2)	13 (1.3)

[a] Sharma, 1973b; soluble cations in saturation extract adjusted to 50% soil moisture
[b] Osmond, unpublished; soluble cations in saturation extract adjusted to 20% soil moisture

evident. As shown in Fig. 7.2, Australian arid region soils are relatively low in nitrogen and phosphorus, compared with soils of other regions (Charley and Cowling, 1968; Charley and West, 1975). It is probable that the concentrations of NO_3^- and PO_4^{3-} in these soils would be at the lower end of the ranges shown in Fig. 7.1. However, the repeated drying and wetting of arid region soils may result in bursts of biological activity and the mineralization of these nutrients so that these ions may accumulate under some circumstances (Chap. 7.5.1). Very high NO_3^- concentrations have been measured in highly saline soils in *A. vesicaria* communities, presumably because denitrification is inhibited by salinity (Charley and McGarity, 1964).

The interactions between ions in the soil solution and those adsorbed by soil colloids have significant effects on other soil properties, such as dispersion of the soil and its permeability to water. The important interaction between monovalent and divalent cations in saline soils has been extensively studied (Quirk, 1971; Thomas, 1974; Shainberg, 1975). Clay particles carry a net negative charge which is neutralized by adsorbed cations attracted to the colloid by electrostatic forces, but at the same time these counter ions have a tendency to diffuse to the bulk solution where concentration is much lower. From the properties of the ions and colloids, it may be deduced that an increase of Na^+/Ca^{2+} ratio, or a decrease in concentration at constant Na^+/Ca^{2+} ratio in the soil solution will result in swelling and increased dispersion of the soil colloids. Experiments confirm that such changes decrease the hydraulic conductivity of the soil and that alkaline-saline soils can be reclaimed by

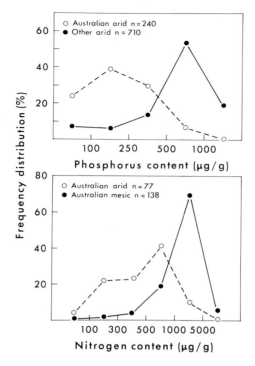

Fig. 7.2. The percentage of soil analyses for total nitrogen and phosphorus which fall within the concentration ranges indicated. Note that the abscissa scale is nonlinear. (Data of Charley and Cowling, 1968)

saline water which serves as a source of divalent cations which flocculate the clay particles and maintain higher permeability (Quirk, 1971).

Theory suggests that to maintain a constant proportion of adsorbed monovalent and divalent ions on the clay colloids, and hence a constant hydraulic conductivity, the ratio $[Na^+]/\sqrt{[Ca^{2+}]}$ in solution must be kept constant. The sodium adsorption ratio $[Na^+]\Big/\sqrt{\dfrac{[Ca^{2+}]+[Mg^{2+}]}{2}}$, calculated from the concentration of ions in the soil solution, is thus an important indicator of water relations of saline soils. In the *Atriplex* communities of the Riverine plain in South-eastern Australia, Sharma (1973 b) showed that the accumulation of Na^+ in the surface soil under bushes as a result of Na^+ cycling in the plants significantly increased the sodium adsorption ratio and substantially decreased the hydraulic conductivity (Table 7.2).

Plant rooting patterns and the dynamics of root development during growth must be taken into account when evaluating the soil nutrient concentrations shown in Table 7.1. During seedling establishment the concentration of salt in the surface soil under bushes is likely to be most important. In the plantations studied by Sharma, two-year-old shrubs of *A. vesicaria* and *A. nummularia* had about 90% of their roots in the 0.30 cm zone (Jones and Hodgkinson, 1970). Mature shrubs of

Table 7.2. Changes in sodium adsorption ratio (SAR) of soils under *A. nummularia* bushes and its effect on soil hydraulic conductivity. (Data of Sharma, 1973b)

Soil type	Sample site	SAR[a]	Hydraulic conductivity[b]
Billabong clay	Between bushes	5.0	6×10^{-2}
	Under bushes	7.0	1.5×10^{-3}
Riverina clay	Between shrubs	4.0	2×10^{-3}
	Under shrubs	6.8	6×10^{-4}

[a] $[Na^+]/\sqrt{\dfrac{[Ca^{2+}]+[Mg^{2+}]}{2}}$

[b] $cm\ min^{-1}$ after 24 h

A. vesicaria retain this predominantly surface rooting pattern, whereas roots of mature *A. nummularia* explore deeper and more saline regions of these profiles.

The concentration of the soil solution in salt marshes resembles that of sea water but is substantially modified by inputs of drainage water from rivers and adjacent land masses. Jefferies (1977) monitored Na^+, NO_3^-, and NH_4^+ concentration in the soil solution at different sites in the Stiffkey marsh, Norfolk, throughout a year and noted that at higher levels in the marsh, Na^+ increased from about 0.4 M in winter to about 0.8 M in summer. Lower in the marsh the variation in salinity was less pronounced (0.5 to 0.7 M). Mahall and Park (1976a, b) noted similar substantial increases in salinity in the upper parts of marshes in San Francisco Bay during the dry summer months. These authors and Jefferies (1977) conclude that this higher salinity during the growing season is an important factor regulating the composition and performance of upper marsh vegetation. Nitrate concentration fluctuated markedly throughout the year, from a minimum of $5\,\mu M$, but NH_4^+ concentration was less variable, commonly in the range of $10\,\mu M$ to $100\,\mu M$ (Jefferies, 1977).

In coastal sand dunes the concentrations of nutrient ions and salinity are unlikely to exceed those of seawater. De Jong (1977) made detailed analyses of the salinity and water relations of two dunes on the Californian coast which support *A. leucophylla* and showed that soil water potential at 30 cm and 100 cm was almost entirely accounted for by the calculated concentration of salt. In these dunes the salt concentration at 30 cm depth was highly variable throughout the year, tending to show both summer and winter peaks but ranging from less than 100 mM up to about 400 mM. At 100 cm depth the concentration fluctuated less markedly and ranged from about 75–200 mM NaCl. *A. leucophylla* at these sites was found to have most roots in the 0–30 cm zone, although at other sites this species has much deeper root systems (Purer, 1936) and access to the lower concentration, more stable ionic environment.

The significant features of the nutrient and ionic status of soils supporting *Atriplex* spp. are that they usually contain lower concentrations of the nutrient ions NO_3^- and PO_4^{3-} and higher concentrations of the nonnutrient Na^+ and Cl^- than other soils. Moreover, because the water status of *Atriplex* soils in arid and coastal regions shows marked fluctuations, the concentration of ions and nutrients in the soil solution fluctuates widely. These fluctuations are greatest and most rapid in the soil surface and hence are likely to have most significant effects during seedling establishment.

Ion absorption processes, which allow *Atriplex* to occupy these habitats successfully, must have some of the following properties: specificity for absorption of adequate PO_4^{3-} from solutions containing 100 times higher concentration of Cl^-; selectivity to absorb sufficient K^+ from solutions containing much higher Na^+; and capacity to absorb copious quantities of NaCl to facilitate osmotic adjustment in the cell vacuole in response to decreasing availability of soil water. The processes which resolve these different and sometimes conflicting requirements form the basis of most of the discussion which follows.

7.1.2 Pathways and Compartments for Ions in Plants

The processes which control the absorption of inorganic ions take place at cell membranes in specific parts of the plant. The structure of these membranes is near the limits of resolution of presently available microscopy and beyond the scope of this book, as outlined in Chap. I. The interested reader is referred to excellent recent accounts in Clarkson (1974), Lüttge and Pitman (1976) and Gunning and Robards (1976). Only a brief outline of larger-scale compartments and pathways is necessary here, as background for discussion of processes likely to make a significant contribution to the ionic status of *Atriplex* spp. described later.

There are two pathways for ion absorption in plants and for ion movement within the plant, the apoplastic pathway (Läuchli, 1976a) and the symplastic pathway (Spanswick, 1976). The apoplast is the space outside the plasmamembrane of each cell and consists of interconnected cell walls and intercellular spaces (Fig. 7.3). There are pathways in the cell walls for movement of ions dissolved in solution (the water free space) and pathways for ion exchange in the charged organic polymers (the Donnan free space, Briggs et al., 1961; Epstein, 1972). The movement of ions in the apoplast is limited by regions in which the cell walls are thickened by hydrophobic deposits of cutin and suberin. These secondary wall substances partly isolate the apoplast of the root from that of the stele and shoot, as shown in the schematic state description of *Atriplex* (Fig. 7.3). In the shoot, similar thickenings may largely isolate the apoplast of leaf cells from the apoplast of glands.

The symplast is the interconnected cytoplasm of adjacent cells in tissues which is bounded by the plasmamembrane on the outside and by the tonoplast on the inside (Fig. 7.3). The key components of the symplast pathway are the plasmodesmata, the cytoplasmic connections through small pores in the wall between adjacent cells. These are lined by the plasmamembrane, and contain cytoplasm and more highly organized structures which may be involved in movement of water and solutes through the pore (Gunning and Robards, 1976). Although the processes involved in water and solute movement through the plasmodesmata are far from being resolved, it is clear that these pores provide a high conductance for flow between tissues. For the movement of ions and water across the root symplast the best estimates suggest that plasmodesmata of the inner wall of endodermal cells allow passage of solutes at a flux equal to that across two adjacent cell membranes and a cell wall, if the cell wall is not suberized (Gunning and Robards, 1976). The presence of suberin in the wall of most cells of the

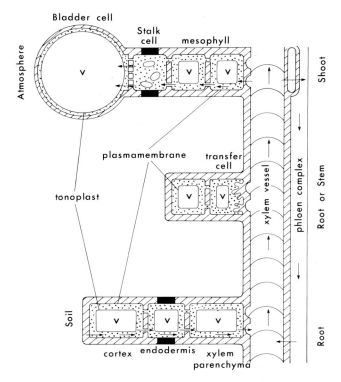

Fig. 7.3. Schematic representation of the compartments and pathways for ion absorption in intact plants. The apoplast (*hatched* cell walls) of different tissues is isolated by suberized regions (*solid black*). The symplast (*stippled* cytoplasm) in different tissues is connected by plasmodesmata and the symplast pathway is indicated by *arrows*. The cell vacuole is identified (*V*) as are wall ingrowths of transfer cells and organelles in cytoplasm of the stalk cells. (Modified after the Darmstadt tradition)

endodermis means that under most conditions the bulk of water flow and the bulk of inorganic ions moving to the shoot pass through the symplast and the plasmodesmata in the endodermis.

In many instances specialized cells are found at the junction between the apoplast and symplast, at the interface between tissues where intensive solute transfer might logically be expected. These "transfer cells" are characterized by wall ingrowths which greatly extend the surface area of the plasmamembrane and which may modify transport properties of the plasmamembrane (Gunning, 1977). They are found in many salt-excreting glands, but not in those of *Atriplex* spp., and also in xylem parenchyma of older roots of some plants where resorption of ions from the xylem stream has been demonstrated. Kramer et al. (1978) have observed similar structures in the epidermal cells and root hairs of *A. triangularis* but transfer cells have not been seen in other tissues.

Figure 7.3 shows diagramatically that the pathways and compartments in leaf tissues are similar to those in roots. The free space of leaf tissues is bathed by ions transported in the xylem stream and these may be absorbed across the plasmamembrane and tonoplast to the vacuole of individual leaf cells or may move through the symplast of leaf cells. Salt glands, if present, function in an analogous manner to the xylem parenchyma cells of the root cortex and serve to transport ions from the symplast to the exterior of the leaf. In many *Atriplex* spp. these glands transport ions to large epidermal trichomes which ultimately burst to release salt onto the leaf surface. The ionic relations of leaves may also be controlled by retranslocation of ions in the phloem as shown in Fig. 7.3.

7.1.3 States of Ions and Nutrients in Plants

The ionic state of plant tissues and cells can be determined, but as discussed previously, the calculation of ionic concentration in these compartments involves many approximations. Table 7.3 shows the concentration of major cations in leaf, stem, and root tissues of *Atriplex* spp. from each of the major habitat types discussed in Chap. 5. The ionic concentration is expressed on the least variable base, tissue dry weight. The fresh weight of plants from coastal areas was about ten times greater than the dry weight, so the concentrations of ions in the bulk tissue water (in mM) is about 10% of those shown in Table 7.3.

The state descriptions in Table 7.3 point to several important processes in ion absorption which will be discussed later. Roots are uniformly low in Cl^- although shoots generally contain about equimolar Na^+ and Cl^-, indicating selectivity in the

Table 7.3. Concentration of ions and the distribution of ions among tissues of *Atriplex* spp. from different habitats. (Unpublished data supplied by A. Läuchli, W.P. Anderson and D. Kramer)

Species and habitat	Tissue	Ion concentration (μ mol g^{-1} dry wt)				
		Na^+	K^+	Ca^{2+}	Mg^{2+}	Cl^-
Arid shrublands						
A. vesicaria	Leaf	2,933	398	130	117	4,252
	Stem	350	108	165	43	272
	Root	850	110	315	78	38
A. leptocarpa	Leaf	3,950	500	148	284	4,200
	Stem	1,500	273	215	187	1,638
	Root	495	170	135	79	75
Coastal sand dunes						
A. cinerea	Leaf	1,075	628	—	67	1,119
	Stem	203	286	—	28	247
A. triangularis	Leaf	2,026	512	—	248	796
	Stem	625	453	—	47	394
	Root	367	321	—	52	173
Coastal saltmarsh						
A. triangularis	Leaf	1,880	450	—	121	1,791
	Stem	762	179	—	25	898
	Root	444	229	—	38	209

Table 7.4. Oxalate concentration and the balance of excess cations in leaves of *Atriplex* spp. grown in culture solutions. (Data of Osmond, 1963)

Species	Ion concentration (mEq g^{-1} dry wt)					
	$\sum$ Cations	Cl$^-$	PO$_4^{3-}$	Oxalate	Total acid	$\sum$ Anions
A. vesicaria	3.11	0.38	0.29	1.51	2.60	3.27
A. inflata	5.20	0.58	0.06	3.12	3.92	4.56
A. spongiosa	5.24	0.45	0.67	3.95	4.55	5.07

absorption and transport of Cl$^-$. In fact, NaCl makes up the bulk of the shoot ionic concentration in these plants from each habitat. The ratio of Na$^+$/K$^+$ increases from roots to stems to leaves, indicating selectivity in the uptake and transport of monovalent cations between each tissue. The ionic content of leaves of *Atriplex* spp. from arid shrublands is higher than that of species from other habitats, and this high concentration of inorganic ions in the shoot is a general feature of vegetation in arid regions (Rodin and Bazilevich, 1965).

Beadle et al. (1957) noted that high salt concentration in the shoot of arid shrubland *Atriplex* spp. was not correlated with variations in the concentration of soluble or exchangeable ions in the soil. Like other halophytes, *Atriplex* spp. are capable of substantial ion uptake even at low concentrations of ions in the soil. Evidently internal plant factors, rather than soil salt concentration, control the "craving for salt" in these halophytes.

Although the analyses shown in Table 7.3 are incomplete, in nearly all species and tissues, total inorganic cations are substantially in excess of Cl$^-$. Some of this cation excess may be balanced by the adsorption of cations to exchange sites in the Donnan free space of the apoplast, but this is unlikely to account for more than about 100 μEq g^{-1} dry weight (Walker and Pitman, 1976). Cation excess of the magnitude indicated by the total cation minus Cl$^-$ concentration in Table 7.3 is a feature common to most tissues of terrestrial plants, and in *Atriplex* spp. it is largely balanced by organic acids. Table 7.4 shows that oxalate is the principal balancing carboxylate in leaf tissues (Osmond, 1963). Whether oxalate is also involved in ionic balance in roots and stem tissues of these plants has yet to be determined. The high oxalate content of the shoots of these and other chenopods is related to the high capacity for NO$_3^-$ uptake and reduction discussed later. *Atriplex* spp. from the arid shrublands have been long recognized as high protein plants (Beadle et al., 1957) and *A. hortensis* was one of the best sources of leaf protein among 24 wild species tested in recent Swedish experiments (Lexander et al., 1970).

Tables 7.3 and 7.4 described the bulk distribution of inorganic ions and carboxylates in different tissues of *Atriplex* spp. but give little indication of the location of ions within the tissues. Ionic concentration may differ greatly in different cells of tissues, as well as within cells. For example, early investigators noted the high Cl$^-$ content of the bladder cells of the epidermal trichomes of *Atriplex* (Wood, 1925). As shown later in this chapter, the gland-like stalk cell is able to transfer more than 50% of the total leaf salt to less than 10% of the leaf water in this external compartment, producing very high salt concentrations in the

Table 7.5. Ionic content in bladder cells of the epidermal trichomes on leaves of *A. halimus*. (Data of Mozafar and Goodin, 1970)

Treatment	Ionic concentration (M)			
	Na^+	K^+	Cl^-	Oxalate
6 mM NaCl control	1.3 ± 0.1	1.0 ± 0.1	1.7 ± 0.3	1.3 ± 0.2
100 mM NaCl + 100 mM KCl	5.5 ± 0.4	6.2 ± 0.6	9.2 ± 0.8	1.2 ± 0.1

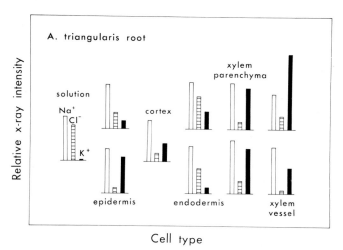

Fig. 7.4. Relative proportions of Na^+, K^+, and Cl^- in different cells of *A. triangularis* roots estimated by X-ray microanalysis of a freeze-fractured specimen. Data were normalized with respect to Na^+ and corrections were applied for the relative intensities of specific emission for K^+ and Cl^-. (Redrawn from Kramer et al., 1978)

bladders (Table 7.5). This great disparity in ion concentration of different tissues in the leaf (cf. Tables 7.3, 7.5) makes it difficult to measure the osmotic pressure of *Atriplex* leaf cells (Chap. 8.2.4; Fig. 8.21).

In the bladders and most other cells the bulk of the ions are presumably located in the cell vacuole and are found in free solution. The ionic concentration within other cell compartments is more difficult to specify but estimates are needed to understand the processes involved in ion uptake and plant responses to inorganic ions. None of the presently available techniques using compartmental analysis, ion specific microelectrodes or electron probe microanalysis, permits confident resolution of the concentration or activity of ions between different cells or within the cytoplasm or organelles of a single cell. Electron probe microanalysis may provide the best resolution.

Using snap-frozen and fractured specimens of *A. triangularis* roots Kramer et al. (1978) observed substantial differences in the ionic content and ionic ratio in different cells. Figure 7.4 shows the relative intensity of the X-ray emission due to the ions Na^+, Cl^-, and K^+ obtained in scans of different cells. These show that all but the endodermal cells of the root contain relatively less Cl^- than the external solution. All cells contain higher K^+/Na^+ ratio than the external solution and the highest selectivity for K^+ (or against Na^+) was found in the root epidermal cells and in the xylem parenchyma. In some cells Kramer obtained sufficient resolution with the microprobe to observe the ion distribution between bulk cytoplasm and the nucleus. The nucleus contained relatively higher Mg^{2+} and the cytoplasm contained relatively low Cl^-. Low Cl^- concentration in the cytoplasm was also observed in freeze-substituted sections of maize leaves (Pallaghy, 1973). Further application of these and other methods is important if we are to verify, directly, many of the key processes believed to control the uptake of ions and their biological effects in plants.

7.2 Ion Absorption Processes in Plants

Our knowledge of ion absorption processes has been largely built upon experiments using plant roots bathed with limited volumes of relatively concentrated culture solutions in which continued stirring is essential to maintain root aeration and ion absorption. In soils the soil solution is not stirred and continued ion absorption is the result of continued root growth and exploration of the soil. In culture solutions roots do not normally produce root hairs; in soil, production of root hairs may be an important means of more complete exploration of the soil, as well as a means of greatly increasing the absorbing surface of the root (Barley, 1970). Furthermore, the roots of many species may form mycorrhizal associations with soil fungi which may increase the absorption capacity of the roots in soil (Barber, 1968). In soils, roots may secrete mucilaginous materials which may facilitate adsorption and transfer of certain ions. Although the bulk of ions is absorbed from the free solution by transfer across a semi-permeable membrane, ion uptake by exchange adsorption or contact exchange may be important for some ions in some circumstances (Jenny, 1966).

The difficulties of integrating the wealth of detailed data on ion absorption processes in model systems (Clarkson, 1974; Baker and Hall, 1975; Lüttge and Pitman, 1976) and assessing their significance in complex systems have been emphasized in several recent treatments (Bowen and Cartwright, 1977; Nye and Tinker, 1977; Scott Russell, 1977). In the present context, we also have to integrate small-scale processes to account for the larger-scale states of ionic distribution discussed in the previous section. It is helpful to adopt a simple terminology, and for our purposes ion absorption processes embrace ion uptake, processes transferring ions across membranes into cells and tissues; ion transport refers to processes moving ions between cells and tissues, and ion excretion refers to processes transferring ions across membrances out of cells and tissues.

Because the soils supporting *Atriplex* spp. are often distinguished by lower concentration of nutrients and higher concentration of nonnutrient ions, it is important to consider the rate of processes in relation to ionic concentration. We will not give much emphasis to mechanisms of ion absorption processes because even in the best-known systems, these mechanisms are poorly understood. The mechanisms which underlie ion uptake in root cells of *Atriplex* at high external concentration, as well as those which are responsible for ion excretion in salt glands of *Atriplex* leaves will be discussed because these may largely determine the ionic status of these plants and may constitute examples of molecular mechanisms of adaptive importance.

7.2.1 Ion Absorption and Concentration

The relationship between external concentration and ion uptake to the vacuole of plant cells, in nearly all tissues examined, is described by at least two hyperbolic functions, one saturating at below 1 mM external concentration and another saturating in the region of 50 mM (Epstein, 1972). Mechanistic interpretation of these hyperbolic functions has far outweighed understanding of the methodological limitations or of the kinetic relationship. The functions describe net flux of ions to the vacuole and the results are therefore a complex function of bi-directional fluxes across two cell membranes and of the activity of ions in the external solution and cytoplasm. As pointed out by Cram (1973a), the hyperbolic relationship between external concentration and uptake to the vacuole is purely fortuitous. As shown in Fig. 7.5 none of the fluxes at the plasma membrane or tonoplast is of this form. Cram found that at low concentrations, influx to the vacuole was limited by the plasma membrane influx and at higher concentrations, influx at the tonoplast limited uptake to the vacuole.

The saturation of ion uptake with increasing external concentration may imply a limited number of sites for uptake, but does not necessarily imply that sites with

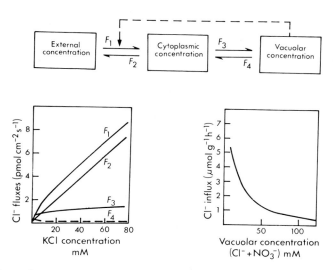

Fig. 7.5. Schematic representation of compartments and fluxes involved in ion uptake to the vacuole of plant cells. The relationship between external concentration and the different fluxes for Cl^- ions and the relationship between internal concentration of $Cl^- + NO_3^-$ and Cl^- influx in barley roots are shown. (Data of Cram, 1973a, b; from Osmond, 1978)

enzyme-like properties are involved (Epstein, 1972; Hodges, 1973; Cram, 1973a). One early alternative interpretation of the saturation phenomena and hence of determination of the rate of process by external ion concentration was that the rates measured represent a balance between influx and efflux across the membrane in question. More recently, evidence that the rate of ion uptake is controlled by *internal* factors such as vacuolar ion concentration has accumulated. For example, in some tissues the Cl^- and NO_3^- influx rates were highly correlated with the $Cl^- + NO_3^-$ concentrations of the cell vacuole (Fig. 7.5) but Cl^- influx was not correlated with internal malate, Na^+ and K^+ concentration (Cram, 1973b).

In this view, control of ion content is independent of the rate of uptake and external concentration but rather is a function of a "set point" corresponding to a particular internal concentration (Cram, personal communication). That is, for the membranes of cells in a particular genotype or tissue there may be a unique relationship between ionic influx across the membranes and the concentration of ions in the vacuole. Ionic state is thus determined by this "set point", a term based on analogies with control systems, and the time taken to attain the "set point" is determined by the rate versus concentration functions described above. Other studies suggest that cell turgor, which is linked to vacuolar osmotic pressure, and therefore a function of ionic concentration in the vacuole, may also be involved in the regulation of ion absorption processes (Cram, 1976).

If these responses can be generalized and if the "set point" is under genetic control, it would provide a simple mechanism for the regulation of tissue ion concentrations in different genotypes. It could be, for example, that the capacity of halophytes and plants from arid regions to absorb ions to much higher concentrations than nonhalophytes reflects a lower sensitivity of ion uptake and transport processes to internal ionic concentration or cell turgor (Osmond, 1979).

Although the hyperbolic relationship between uptake rate and external concentration may be fortuitous, it provides an integration of ion uptake processes which is useful in certain contexts. For example, the apparent affinity for the uptake of some ions in the low concentration range approximately corresponds to the concentration of these ions in soil solutions prepared from agricultural soils (Epstein, 1972). Moreover, in recent experiments the apparent affinities for K^+ uptake by different barley cultivars were closely correlated with growth rates under K^+ limited conditions (Glass and Perley, 1980). However, although arid shrubland *Atriplex* spp. grow in PO_4^{3-} depleted soils, Cook (1973) found no evidence that roots of *A. vesicaria* or *A. conduplicata* had unusually high affinity for this nutrient. The apparent K_m for PO_4^{3-} in *Atriplex* roots (8–25 μM; Fig. 7.6) is similar to that found for cereal roots (Carter and Lathwell, 1967). Comparisons of other species in dilute flowing culture solutions and in sand culture suggest a similar conclusion. For example, Wild et al. (1974) noted that *Anthoxanthum odoratum* had the highest absorption rate of four species tested, but this high rate was principally due to larger root surface area and high root/shoot ratio. It is possible that differences in morphological properties, rather than differences in kinetic properties of ion uptake processes, may underlie the differences in the ability of plants to colonize or exploit differing ionic environments.

Similar conclusions apply to the concentration dependence of Na^+ and Cl^- uptake rates in leaf tissues slices of *Atriplex* and *Citrus* (Fig. 7.6). The high Na^+

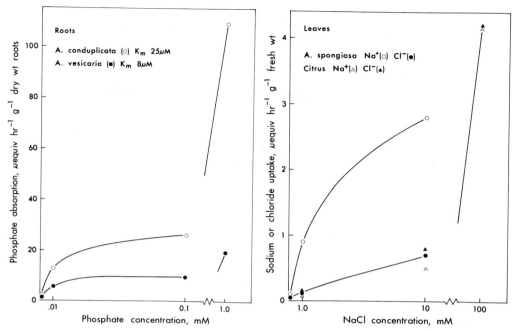

Fig. 7.6. Concentration dependence of PO_4^{3-} uptake in seedling roots of *Atriplex* spp. (data of Cook, 1973) and concentration dependence of Na^+ and Cl^- uptake in leaf slices of *A.spongiosa* and *Citrus* (data of Osmond, 1965; Robinson and Smith, 1970). Note that here and elsewhere PO_4^{3-} is used as an abbreviation of convenience only, and does not imply that phosphorous is usually absorbed as this ionic species.

uptake rates in *Atriplex* at high concentrations were largely balanced by substantial exchange of K^+ from the tissue (Osmond, 1968). There is no indication in Fig. 7.6 that the ability of *Atriplex* leaves to absorb Cl^- to much higher concentrations than in *Citrus* leaves is determined by the relationship between rate of ion uptake and external solution concentration. The higher internal concentration attained by *Atriplex* leaves may imply, however, that ion uptake by leaf cells may continue to a higher "set point" in *Atriplex* than in *Citrus*.

    Uptake versus concentration experiments also show that ion uptake by roots may be quite specific and insensitive to high concentrations of other ions. For example, K^+ absorption from solutions of low concentrations is insensitive to high levels of Na^+ (Epstein, 1972) and PO_4^{3-} uptake is insensitive to Cl^- (Carter and Lathwell, 1967). These data suggest that uptake of essential ions such as K^+ and PO_4^{3-} by root cells may not be directly influenced by soil salinity.

    Rates of ion uptake to the cell vacuole in low salt roots of cereals are very much higher than the rates of ion transport to whole plants. Experiments show ion transport to the shoot to be almost independent of external concentration. Ion transport is a property of the plasmamembrane in the xylem parenchyma at the end of the symplast pathway and the relevant concentration for the rate of the process must be that in the symplast itself. In the same sense, the relevant ion concentration for the rate of ion secretion from a leaf is presumably that in the leaf symplast.

Ion absorption and movement in the root symplast begins with transfer across the plasmamembrane of root epidermal cells into the cytoplasmic compartment. These ions may then move across the tonoplast into the vascular compartment of this cell, move through the symplast and across the tonoplast of the adjacent cells, or move through the symplast to the cytoplasmic compartment of the xylem parenchyma cells abutting the xylem vessels. The plasmamembrane of these cells evidently has different properties from the plasmamembrane of root cortical cells. There is an efflux of ions from the cytoplasm into the free space of xylem vessels and this flux appears to be greater on an area basis than that across the root cortical cell plasmamembrane (Pitman, 1977). The higher fluxes are not unexpected for the symplast in the root cortex gathers ions from the apoplast across a relatively immense area of plasmamembrane compared to the plasmamembrane area of the few xylem parenchyma cells in the stele which transfer these ions to the xylem.

Although ion transport from root to shoot in nonhalophytes is not much influenced by external ionic concentration, it appears to be regulated by many factors in such a way that the ion content of the shoot remains more or less constant. The mechanisms involved in matching ion transport from the root to the demands of the shoot are not precisely known. They may include limitation by availability of respiratory substrates, which presumably fuel active transport (see below), and regulation by specific hormones such as abscisic acid. The hormones may be produced in the shoot in response to local water stress (Vaadia, 1976) which could occur if ions accumulated in the free space faster than they could be transported to the vacuole.

The present accounts of integrated ion transport (Pitman, 1977; Pitman and Cram, 1977), are based entirely on experience with nonhalophytes in which ion transport through the apoplast is thought to be minimal, at least at low external concentrations. There is some evidence which suggests that ion transport through the apoplast may be much more significant at high concentrations, as will emerge in the discussion of halophytes, below. Apoplastic ion transport presumably depends on passive rather than active ion movement across the cell membranes. For this reason apoplastic ion transport is less likely to be subject to fine control by means of metabolites and hormones. The consequences of this for ion relations of the shoot are substantial. The distinction between passive and active ion absorption is a key problem in evaluating processes and mechanisms, as described in the following section.

7.2.2 Passive and Active Processes of Ion Absorption

Passive processes of ion movement are primarily dependent on diffusion and mass flow which depend, in turn, on concentration gradients, electrical gradients, and bulk flow rates. Compared to diffusion and mass flow in simple solutions, these passive processes in soil and plants are modified by the complexity and dimensions of the flow pathways and by the presence of fixed charges on the clay colloids and the cell wall polymers. For example, the rate of exchange diffusion varies inversely with the bond energy between the cation and colloid, so that strongly bound Sr^{2+} diffuses much more slowly than weakly bound Na^+ or K^+. The effective diffusivity of particular ions in the soil is about the same as that in the Donnan free space of plant tissues. Theoretical treatments of the relationship between diffusion and mass flow can be used to predict the effect of these properties on ion distribution about the root, but experimental evidence is needed to validate these predictions (Barber, 1974). Qualitative studies in soil show accumulation of slowly absorbed ions adjacent to the roots (supply exceeds uptake) or the depletion of rapidly absorbed ions (uptake exceeds supply) under different circumstances for different ions. However, the complex state of the soil root interface complicates the prediction and evaluation of these relationships. Within the plant passive processes of diffusion are significant during ion transport through the symplast and mass flow in the transpiration stream is the primary means of ion transport between root and shoot.

The contribution of diffusion to ion uptake or ion excretion processes across cell membranes can be assessed by several methods and used as a criterion for the presence or absence of active uptake. Active uptake is the process whereby ions are moved against an electrochemical potential gradient at the expense of cellular energy. This definition suggests two tests which may be applied to determine if particular ion uptake processes are active or passive. If the distribution of ions across the membrane is due to passive processes, the electrical potential difference across the membrane at equilibrium can be predicted by the Nernst equation (Clarkson, 1974; Findlay and Hope, 1976). Comparison of the calculated Nernst potential and the measured potential indicates whether an active process must be invoked to explain the observed distribution of ions. Alternatively, specific metabolic inhibitors can be used to determine whether the rate of the uptake process is altered by control of energy supply. Two examples of the application of these techniques to ion uptake process in *Atriplex* spp. follow.

The Nernst equation, at 25°, reduces to

$$E_n = \frac{59.2}{z} \log M_0/M_i \, mV,$$

where E_n is the calculated Nernst potential, z is ionic charge and M_0, M_i are ionic concentrations outside and inside the membranes in question. The difference, ΔE, between measured potential (E_m) and the Nernst potential indicates whether the distribution of ions is possible by passive diffusion alone and predicts the direction of active ion movements. When ΔE for a cation is close to zero or some negative value then it may be considered to move passively down its electrochemical gradient. On the other hand, if ΔE for a cation is positive, then movement would be against an electrochemical gradient and active transport would be required. The opposite is true for anions.

Anderson et al. (1977) measured Na^+, K^+, and Cl^- concentration of *A. triangularis* roots from plants treated in a range of NaCl concentrations. As shown in Table 7.6, ΔE for K^+ is -19 to -31 mV, indicating that this ion is near electrochemical equilibrium and the K^+ concentration in the tissue may have arisen by passive processes. Because the ΔE for Na^+ is -130 to -149 mV this ion should also diffuse passively into the tissue until concentrations equal those outside. However, at 300 and 600 mM external NaCl the Na^+ concentration inside is lower than that outside. This indicates that although influx may be passive, there is an outwardly directed active efflux of Na^+. In the case of Cl^-, ΔE is -158 mV

Table 7.6. Comparison of measured potential and Nernst potential (E_n) calculated from measured concentrations in roots of *A. triangularis* in solutions of different NaCl concentration. (Data of Anderson et al., 1977)

Parameters	Added NaCl concentration, mM		
	100	300	600
Measured potential, mV	− 137	− 140	− 130
Tissue Na^+, μmol g^{-1} fr. wt.	132	212	281
Calculated $E_n(Na^+)$, mV	− 7	+ 9	+ 19
$\Delta E\,(Na^+)$ mV	− 130	− 149	− 149
Tissue K^+, μmol g^{-1} fr. wt.	72	71	57
Calculated $E_n(K^+)$, mV	− 107	− 109	− 111
$\Delta E(K^+)$ mV	− 30	− 31	− 19
Tissue Cl^-, μmol g^{-1} fr. wt.	66	140	171
Calculated $E_n(Cl^-)$, mV	+ 10	+ 19	+ 29
$\Delta E(Cl^-)$ mV	− 147	− 159	− 159

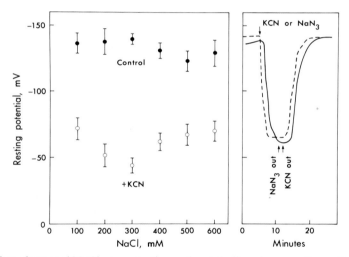

Fig. 7.7. Effect of external NaCl concentration and metabolic poisons on the resting potential of cortical cells in the roots of *A. triangularis* and *A. gmelini*. (Redrawn from Anderson et al., 1977 and R.F. Davis, T.E. Riehl and J.H. Crow, unpublished) (NaN₃ - - -, KCN ——)

showing that the tissue Cl^- concentration is maintained against a strong electrochemical gradient, presumably by an active process.

    The potential of roots of *A. triangularis* (*hastata* var. *salina*) grown in nutrient solutions with 100 to 600 mM NaCl did not respond to external NaCl concentration (Anderson et al., 1977; Fig. 7.7). Similar results were obtained in *A. gmelini* by R.F. Davis, T.E. Riehl, and J.H. Crow (unpublished) and in another halophyte, *Triglochin maritima*, Jefferies (1973) found that the potential was unaffected by high concentration of NaCl. The stability of the potential at high external NaCl

Table 7.7. Chloride concentration, calculated Nernst potential and measured potential in bladder cells of the epidermal trichomes of *A. spongiosa* leaves (Osmond et al., 1969a)

Treatment[a]	Cl concentration (μEq g^{-1} fr. wt.)	Cell potential		
		E_n (mV)	E_m (mV)	ΔE (mV)
Initial	26.2	+43	$- 89 \pm 23$	-122
42 h light	33.8	+49	-105 ± 3	-154
42 h dark	18.3	+33	$- 83 \pm 13$	-116

[a]   Leaf slices treated in 5mM KCl + 0.1 mM CaSO$_4$

concentration distinguishes these halophytes from some nonhalophytes in which the potential is depolarized with increasing external monovalent cation concentration.

Anderson et al. (1977) calculated permeability ratios from their data with *A. triangularis* roots and estimated that the ratio for the permeabilities of Cl$^-$ and K$^+$ was similar to those of other plants. It may be significant, however, that the permeability ratio Na$^+$/K$^+$ was 10 to 30 times smaller than that of nonhalophytes. This low permeability to Na$^+$ accounts for the stability of the resting potential shown in Fig. 7.7, and for the retention of high Na$^+$ concentration in these halophytes.

Figure 7.7 also shows a reversible depolarization of the potential in *Atriplex* roots following treatment with metabolic poisons. These observations support the conclusion from the Nernst calculations that the distribution of the major ions Na$^+$ and Cl$^-$ is due to active processes. They indicate that most of the membrane potential is due to metabolically dependent electrogenic anion influx or cation efflux pumps.

Similar criteria were applied to determine whether Cl$^-$ excretion from the leaf lamina to the bladder cells of epidermal trichomes involved an active or passive process. The potential between the bladder cell vacuole and a 5 mM NaCl solution bathing leaf slices was -80 to -100 mV. The ΔE values obtained after calculation of E_n ranged from -116 to -155 mV (Table 7.7) indicating that Cl$^-$ uptake to the bladder (excretion from the leaf) involved active transport at some step. At the same time, it was shown that the movement of radioactive Cl$^-$ to the bladders was light-dependent and suppressed by the photosynthetic inhibitor DCMU at concentrations which were effective in completely inhibiting photosynthesis (Osmond et al., 1969a). There can be little doubt that active Cl$^-$ uptake from the apoplast takes place at either the plasmamembrane of leaf mesophyll cells or the plasmamembrane of the gland cells. Because the gland cells themselves have little photosynthetic capacity Lüttge and Osmond (1970) favored the view that the active step was probably located at the mesophyll cell plasmamembrane during Cl$^-$ uptake into the symplasm. However, as emphasized by Thompson (1975) and Hill and Hill (1976) many difficulties of experiment and technique remain to be resolved and the above conclusions as to the location and energy dependence of the active excretion process must remain tentative.

One source of the metabolic energy involved in moving inorganic ions across a membrane against an electrochemical potential gradient is the hydrolysis of ATP, and one would expect that this transport might be mediated by an ion-specific ATP hydrolyzing enzyme (ATPase). Many proteins with an ATPase activity which responds to inorganic ions have been identified in plant extracts (Kylin and Quatrano, 1975) but whether these interactions have anything to do with the catalysis of ATP hydrolysis or simply reflect a broad spectrum of enzyme responses to electrolytes (Chap. 7.5.5) has not been evaluated. Plasmamembrane fractions have been separated from other membranes in cell extracts and shown to have ion-dependent ATPase activity in vitro (Hodges, 1973, 1976). A most convincing relationship between active ion movement and a specific ion-activated ATPase is found in the salt gland of *Limonium* leaves (Hill and Hill, 1976). The chloride excretion capacity of this gland is induced by the salt load of the leaf, and as the ion-excretion capacity increases, so does the activity of a specific Cl^--dependent ATPase.

The ATPases associated with the plasmamembrane of nonhalophytes require Mg^{2+} or Mn^{2+}, and are stimulated by monovalent cations with a specificity of activation which reflects the specificity for active ion uptake (Hodges, 1973, 1976). They do not show ouabain sensitivity or Na^+, K^+ synergism characteristic of animal transport ATPases. If the ATPases from halophytes examined by Kylin and Quatrano (1975) are single enzymes and not mixtures, the $Na^+ + K^+$ synergism and sensitivity to ouabain distinguish them from those of nonhalophytes. In line with this possibility R.F. Davis, T.E. Riehl, and J.H. Crow (unpublished) found that the presence of 0.05 mM ouabain in solution cultures stimulated Na^+ uptake and inhibited K^+ uptake in roots, stem and leaves of *A. gmelini,* presumably by inhibiting the Na^+ excretion mechanism. Only much more intensive investigation will establish whether the active extrusion of Na^+ from *Atriplex* roots or the active excretion of Cl^- into the bladder cells of epidermal trichomes involves ATPases with distinctly different properties to those in nonhalophytes.

7.3 Absorption of Nutrient Anions and Interactions with Cell Metabolism

Many *Atriplex* spp. are described as nitrophilous plants, particularly the weedy species, and indeed, the leaf nitrogen content of these plants is among the highest found in wild plants (Lexander et al., 1970). The nitrogen content of *Atriplex* spp. from the semi-arid shrublands has been compared with that of legumes (Beadle et al., 1957), yet the soils on which these plants grow are characteristically depleted in nitrogen and phosphorus (Fig. 7.2). Moreover, *A. vesicaria* and *A. nummularia* growing under irrigation on the natural soils of the Riverine Plain did not respond to nitrogen fertilization (Jones, R. et al., 1970) and Chatterton et al. (1971) noted that growth of *A. polycarpa* was inhibited by an increase in NH_4NO_3 concentration from 15 to 71 mM. Similarly, there is little evidence that the much lower PO_4^{3-} concentration of native *Atriplex* soils limit growth of these species. Jones, R. et al. (1970) found no growth response to phosphorus fertilization of *A. vesicaria* and *A. nummularia,* although the total phosphorus content of the plants increased. Salt marsh species which showed a substantial response to added nitrogen did not respond to phosphorus fertilization (Jefferies and Perkins, 1977).

These studies suggest that *Atriplex* spp. are particularly effective in the absorption and utilization of NO_3^- and PO_4^{3-}. However, it is evident that none of the short-term processes that have been studied accounts for the effective utilization of nutrients in these plants. Processes which interact with development and growth have much longer relaxation times and are poorly understood, but these appear to be the key to efficient nutrient utilization in these species.

7.3.1 Nitrogen Absorption and Incorporation

The absorption of nitrogen has been most extensively investigated in seedlings of cereals and grasses. Nitrate absorption differs from other uptake processes in that the rate of uptake increases with time over the first few hours of exposure to NO_3^-. This interesting phenomenon may be related to the synthesis or activation of a NO_3^- uptake system of a NO_3^- transport system, or to the induction of nitrate reductase, the enzyme which initiates the metabolic incorporation of this nutrient anion (Jackson et al., 1973; Blevins et al., 1974). Uptake is saturated at relatively low concentrations with an apparent K_m of 30–100 μM (Lycklama, 1963; Rao and Rains, 1976) and NO_3^- uptake is insensitive to the presence of 10 mM NaCl (Smith, 1973).

Although NH_4^+ uptake processes have an apparent affinity for this ion which is similar to that for NO_3^- uptake (Lycklama, 1963), subsequent incorporation of NH_4^+ is not dependent on the activity of an inducible enzyme, but rather on the capacity of overall metabolism to assimilate the already reduced nitrogen. The relationships between NH_4^+ uptake rate, metabolism in the cytoplasm, and transport to the vacuole are much less well known than those involved in NO_3^- utilization.

Depending on the tissue and its metabolic status the absorbed NO_3^- may be reduced in the cytoplasm or transferred to the vacuole. The excised roots of some cereals have a low capacity for NO_3^- reduction and thus can be used to study the uptake processes directly (Smith, 1973). In these roots NO_3^- uptake to the vacuole is depressed by high Cl^- or carboxylate concentrations in the tissue, suggesting a control of uptake by internal concentration of the type described for Cl^- previously (Fig. 7.5). However, in other systems most of the NO_3^- absorbed is reduced in the cytoplasm and so the significance of such internal factors in the control of NO_3^- uptake into the cytoplasm has yet to be established.

Nitrate reductase is perhaps the most widely studied of inducible enzymes in higher plants, but the relationship between the uptake of the inducer (NO_3^-) and the induction process itself has only recently been explored (see Osmond, 1976). The enzyme can be induced in roots by 10 μM NO_3^- external to the tissue, well below the concentration required to saturate uptake to the vacuole. Other direct evidence suggests that the enzyme responds to cytoplasmic rather than vacuolar NO_3^- levels and in leaves, the level of the enzyme is determined by the NO_3^- concentration in the xylem stream, not the NO_3^- concentration in the leaf as a whole.

The vacuole may represent a useful storage compartment for NO_3^- when rate of uptake exceeds the rate of reduction in the root and transport to the shoot. Similarly, storage of NO_3^- in the vacuole of leaf cells may occur when transport exceeds reduction capacity in the leaf. For example, when *A. spongiosa* was grown in 50 mM $NaNO_3$ solution NO_3^- accumulated in rapidly growing leaves, but when plants were transferred to cultures with 50 mM NaCl the NO_3^- accumulated in the leaf was reduced within four days following transfer (Osmond, 1965).

Plants differ markedly in their capacity to reduce NO_3^- in the roots and hence to transport NO_3^- to the shoot. Some plants, such as *Xanthium,* have a limited capacity for NO_3^- reduction in the root at all levels of NO_3^- supply, whereas in others, such as *Pisum,* most of the NO_3^- reduction takes place in the root. Analyses of the bleeding sap of *A. spongiosa* grown in 50 mM NO_3^- showed that about 80% of the translocated nitrogen was in the form of NO_3^- (Osmond, 1967a), indicating that these plants are of the *Xanthium* type. This view is sustained by the observation that nitrate reductase activity in shoots of *A. triangularis (hastata)* growing in a salt marsh was 6.4 times greater than in roots, on a fresh weight basis (Stewart et al., 1973). The relationship between uptake, site of reduction, and transport of NO_3^- needs much further clarification. If it is a genetically constant feature of species it may well be an important component of the overall efficiency of nitrogen utilization.

7.3.2 Metabolic Consequences of Nitrogen Incorporation

The implications of the metabolic incorporation of NO_3^- and NH_4^+ for cell metabolism have been reviewed by Raven and Smith (1976). The reduction of NO_3^- produces almost one equivalent of OH^- so the stoichiometric synthesis of carboxylic acids from neutral precursors is required to maintain pH. It has long been recognized that charge balance following the incorporation of NO_3^- or NH_4^+ is maintained by the synthesis or degradation of carboxylic acids (Dijkshoorn, 1962), and that plants grown on NO_3^- have higher carboxylic acid concentrations than those grown on NH_4^+ (Kirkby, 1969). When NO_3^- reduction takes place largely in the shoot, carboxylate synthesis to control pH and to balance the cations which were transported to the shoot with NO_3^- also occurs in the shoot. The metabolic incorporation of NO_3^- in leaves of *Atriplex* spp. is thus associated with substantial carboxylate synthesis and, as shown in Table 7.3, oxalate is the principal carboxylic acid involved. In barley, NO_3^- reduction in the shoot is accompanied by stoichiometric malate synthesis (Blevins et al., 1974).

Figure 7.8 shows the changes with time in the ionic composition of a mature, only slowly expanding leaf of *A. spongiosa* and in a young, rapidly expanding leaf. These plants were grown in culture solution with 50 mM $NaNO_3$ (Osmond, 1965). In the mature leaf the slowing of expansion is associated with a leveling off in the content of reduced nitrogen in each leaf. However, the cation content ($Na^+ + K^+$) and the oxalate content continued to increase while the NO_3^- content remained steady, suggesting that the mature leaf continued to reduce incoming NO_3^- and export the reduced nitrogen to other parts of the plant. In the young leaf, the initial increases in reduced nitrogen were not associated with changes in oxalate content, suggesting that reduced nitrogen may have been imported from mature leaves. However, during the period of maximum leaf growth, changes in reduced nitrogen level were paralleled by changes in leaf cations and oxalate content. Figure 7.9 shows the correlation between change in excess cations and oxalate in expanding

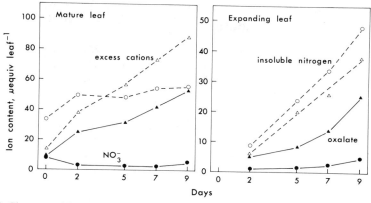

Fig. 7.8. Changes with time in excess cations ($\Sigma\ Na^+ + K^+ - NO_3^-$) oxalate, reduced nitrogen, and NO_3^- in mature and expanding leaves of *A. spongiosa*. (Data of Osmond, 1965)

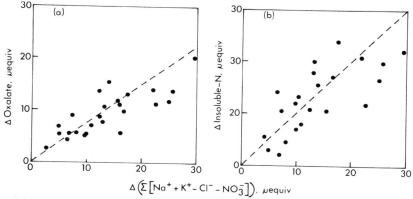

Fig. 7.9. a Relationship between change in oxalate and excess cation content ($\Sigma\ Na^+ + K^+ - NO_3^- - Cl^-$) of expanding leaves of *A. spongiosa* and **b** between insoluble reduced nitrogen and excess cation content of growing leaves of *A. spongiosa*. (Data of Osmond, 1967a)

leaves of *A.spongiosa*. The data are drawn from experiments of the type shown in Fig. 7.8 and demonstrate that oxalate balances 75% of the cations remaining in these leaves after the incorporation of NO_3^-. The correlation between excess cations and reduced nitrogen content is shown in Fig. 7.9.

It will be recalled that about 20% of the nitrogen transported to the shoot in *A.spongiosa* seedlings was already in the reduced form, indicating that some NO_3^- reduction had taken place in the root. It is likely that carboxylates are also synthesized in roots to balance the NO_3^- reduced there and that these balance the excess cations in this tissue (Table 7.2). Whether oxalate is involved, and whether these carboxylates are the source of H^+ which may be actively excreted from roots and account for the inhibitor-sensitive, electrogenic portion of the potential (Fig. 7.6) as proposed by Anderson et al. (1977), remains to be established. The export of some reduced N to the shoot implies that although an equivalence between NO_3^- reduction excess cations and oxalate content may apply during leaf expansion, there is likely to be more reduced N in the shoot as a whole than can be accounted for as excess cations or oxalate. This was found to be the case in whole shoots of *A. spongiosa* (Osmond, 1965) and is commonly observed in monocotyledons (De Wit et al., 1963). The cations entering the root with NO_3^-, which was reduced in the roots and later transported to the shoot, presumably remain in the roots as carboxylate salts just as they do in mature leaves (Fig. 7.8).

The synthesis of carboxylates such as oxalate from neutral precursors is not only important in the maintenance of cell pH during NO_3^- reduction, but also plays a role in the water relations of expanding cells. Expansion growth depends on the generation of turgor in the cell as a result of the osmotic pressure of the cells (Chap. 8.2). The reduction of NO_3^- and its incorporation into macromolecules potentially lowers the osmotic contribution of absorbed salts which is restored by the synthesis of carboxylates from osmotically inactive substrates such as starch (Cram, 1976; Raven and Smith, 1976). In Fig. 7.8, the expansion of young leaves of *A. spongiosa* was associated with NO_3^- reduction and oxalate synthesis, but these processes continued in mature leaves after dry weight ceased to increase. The most obviously turgid cells in *Atriplex* leaves, the bladder cells of the epidermal trichomes, contain high concentrations of oxalate when plants are grown in low Cl^- cultures (Table 7.4). The oxalate is produced during expansion of the bladder cell, and it accounts for about 50% of the cell osmotic pressure. Continued development of the bladder cells after leaf expansion has been completed may account for part of the continued increase in cation and oxalate content shown in Fig. 7.8.

Two other features of the synthesis of oxalate in response to NO_3^- reduction in *Atriplex* spp. contribute to the efficiency of nutrient utilization processes, but are difficult to evaluate. Oxalate is the most highly oxidized carboxylate and presumably its synthesis is energetically less expensive than more highly reduced carboxylates. Compared with the cost of producing highly reduced osmotica such as sucrose, the use of oxalate for this purpose must represent some saving in overall energy metabolism. Although oxalate levels in the leaves of *Atriplex* and other chenopods are unusually high, it is important to recognize that these slowly metabolized carboxylates represent only a small proportion of total carbon turnover. A second, possibly significant feature of oxalate production is the formation of insoluble salts with divalent cations. Leaf mesophyll cells of *Atriplex* spp. contain extensive and conspicuous deposits of Ca^{2+} oxalate in the cell vacuole (Osmond, 1967a). The precipitation of Ca^{2+} oxalate lowers the osmotic pressure and the carboxylate concentration of the vacuole. Further, if the capacity to absorb NO_3^- is limited by carboxylate concentration in the vacuole, or by turgor (Smith, 1973; Cram, 1973b), precipitation of Ca^{2+} oxalate would presumably increase this capacity.

7.3.3 Interactions Between NO_3^- and Cl^- During Absorption and Incorporation

The interaction between NO_3^- and Cl^- uptake processes of roots has not been investigated in *Atriplex*, nor have these interactions been examined in other plants

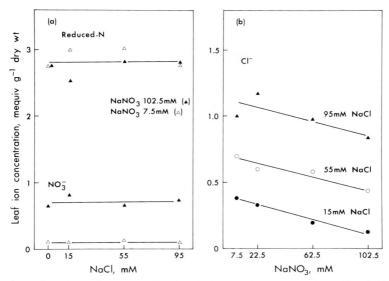

Fig. 7.10. a Effect of NaCl concentration in solution on the concentration of insoluble reduced nitrogen and NO_3^- in mature leaves of *A. spongiosa* grown at two levels of $NaNO_3$. **b** Effect of $NaNO_3$ concentration in solution on the concentration of Cl^- in mature leaves of *A. spongiosa* grown at three levels of NaCl. (Data of Osmond, 1965)

at concentrations of ions likely to be encountered by halophytes. However, the concentrations of ions in shoots of *A.spongiosa* after growth in cultures containing different concentrations of NO_3^- and Cl^- suggests that there is little interaction between these ions during uptake or reduction (Osmond, 1965). In the experiment shown in Fig. 7.10, solution Cl^- concentrations had no effect on the concentration of soluble NO_3^- or the insoluble reduced nitrogen concentration of mature leaves when the plants were grown at low or high $NaNO_3$ concentrations. Chatterton et al. (1971) found a similar lack of response of leaf N concentration to NaCl in *A.polycarpa* and Medina (1970) found it necessary to reduce NO_3^- concentrations to 1 mM in order to lower the leaf N concentration of *A.triangularis* (*patula* ssp. *hastata*) (Table 7.8).

In the experiments with *A.spongiosa,* increasing the NO_3^- concentration at constant NaCl in the culture solution increased the leaf NO_3^- and oxalate concentration and was associated with decreased leaf Cl^- concentration at each of the Cl^- concentrations used (Fig. 7.10). At a given level of NaCl in solution, a decrease in the Cl^- concentrations of leaves was correlated with increased NO_3^- and oxalate concentration, following increase in NO_3^- supply in solution. These data suggest that Cl^- uptake into leaf cells, and possibly into the plant as a whole, may respond to the vacuolar concentration of other ions such as NO_3^- and carboxylates, as observed in other tissues (Cram, 1973b; Smith, 1973).

The experiments with *A.spongiosa* were done at rather higher NO_3^- concentrations than those found in natural habitats and the potential for interactions between NO_3^- and Cl^- may have been masked. However, even in experiments with more extreme ratios of Cl^- to NO_3^- such as those of Black (1956,

Table 7.8. Effects of nitrate concentration in culture solution on expansion and nitrogen concentration of *A. triangularis* leaves. (Data of Medina, 1970)

Leaf category	Nitrate (mM)	Leaf area (cm^2)	N concentration mg g^{-1} dry wt.
Fourth pair [a]	33	30.9	54.9
	4	23.7	39.1
	1	17.1	27.5
Fifth pair [b]	33	44.5	53.4
	4	17.4	35.8
	1	11.2	26.4

[a]   10 days growth at these concentrations
[b]   14 days growth at these concentrations

Table 7.9. Nitrate content and nitrate reductase activity of *Suaeda maritima* in the Conway salt marsh. (Data of Stewart et al., 1972)

Site	Plant nitrogen status		Nitrate reductase activity	
	Protein (mg g^{-1} fr. wt.)	Nitrate (μmol g^{-1} fr. wt.)	Initial (μmol h^{-1} g^{-1} fr. wt.)	72 h after NO$_3^-$ added (μmol h^{-1} g^{-1} fr. wt.)
Lowest 1	8.0	2.7	5.6	5.7
2	8.0	0.9	—	—
3	7.5	0.8	1.6	5.8
4	7.5	0.3	0.6	5.4
Highest 5	7.2	0.1	0.1	5.6

1960) in which the ratio may have exceeded 100, there is no evidence that Cl$^-$ interfered with the uptake or incorporation of NO$_3^-$. Black observed that in *A. vesicaria* and *A. triangularis (hastata)* the cation excess remained more or less constant throughout the range of NaCl concentrations used (up to 1 M). Although growth was much reduced at high NaCl concentrations the cation excess and hence the concentration of reduced nitrogen was unchanged. The amount of NO$_3^-$ absorbed was much reduced by high salinity and the plants grew more slowly. If slow growth was due to interference in NO$_3^-$ absorption or incorporation we would have expected the N deficiency to be reflected in a lower concentration of reduced nitrogen in the plant.

Experiments with halophytes growing in salt marshes yield similar results, further indicating that Cl$^-$ does not limit NO$_3^-$ uptake or incorporation. Stewart et al. (1973) found that leaves of *A. triangularis (hastata)* from a salt marsh had among the highest nitrate reductase activities recorded in this habitat, suggesting that NO$_3^-$ was not limiting. Working with *Suaeda maritima,* these authors had previously established that nitrate reductase activity was maximal in the low marsh (nearest to the sea) and decreased in plants high in the marsh (Table 7.9). When plants in different parts of the marsh were treated with NO$_3^-$, only those high in the marsh showed an increase in nitrate reductase activity, indicating that NO$_3^-$ uptake and transport were not limiting in the low marsh. This is consistent with the presence of high NO_3^- concentration in the shoots of low marsh plants (Table 7.9). Nitrate reductase from *A. triangularis* and other halophytes has the same kinetic properties as the enzyme from nonhylophytes (Stewart et al., 1973) and inhibition due to Cl^- is much the same (Austenfeld, 1974). From the available data it seems unlikely that there is ecotypic differentiation with respect to nitrate reductase among populations of halophytes in the salt marsh (cf. Goodman and Caldwell, 1971).

Although *Suaeda* in the upper marsh contained lower NO_3^- and lower nitrate reductase activity, its protein nitrogen concentration was comparable with plants from the lower marsh (Table 7.9). Reduced growth of *Suaeda* in highly saline environments evidently does not result in nitrogen depletion, and these data thus support the inferences drawn from Black's experiments above. Because nitrogen concentration of *Suaeda* is much the same throughout the marsh, we do not share the conclusion of Stewart et al. (1972) that growth in the upper marsh was NO_3^- limited. In another marsh Jefferies (1977) found that populations from the upper marsh showed much slower growth responses to added NO_3^- than populations low in the marsh. Although NO_3^- is undoubtedly in short supply in the upper marsh, it is likely that growth is primarily limited by the effects of high salinity and low availability of water, rather than by nitrogen metabolism.

7.3.4 Efficiency of Nitrogen Utilization

Although we can piece together a reasonably adequate account of nitrogen uptake and metabolism in *Atriplex* spp. experiments to date indicate few processes which might contribute to the efficiency of NO_3^- utilization in these plants. The processes of uptake transport and incorporation seem similar to those in other plants. It may be that in halophytes the feed-back regulation of NO_3^- uptake and incorporation due to the vacuolar concentrations of NO_3^-, Cl^-, and carboxylates differs from that of glycophytes, i.e., operates with a different "set point" (Chap. 7.2.2). Synthesis of oxalate to balance NO_3^- incorporation and the subsequent precipitation of Ca^{2+} oxalate may provide a mechanism for maintaining a different set point. However, it seems equally likely that such processes, characterized by relaxation times of minutes and hours, are not the most significant processes involved in NO_3^- utilization. For example, genetic and developmental processes which determine whether NO_3^- reduction may occur more efficiently in leaves (coupled to photosynthetic energy metabolism) compared with roots (coupled to respiratory metabolism of translocated photosynthates) may be more important.

Little is known of the way different plants make use of their nitrogen resources or the way environmental factors influence their pattern of use. As described in Chap. 10.5, high temperature-adapted plants with the C_4 pathway of photosynthesis can fix CO_2 equally as rapidly as high temperature-adapted C_3 plants, but the C_4 plants achieve this with much lower concentrations of RuP_2 carboxylase. This enzyme is the major component of soluble leaf protein and it follows that the C_4 plant is, under these circumstances, able to make much more efficient use of nitrogen in leaf proteins. Broad comparisons of C_3 and C_4 species show lower leaf nitrogen concentrations in the latter (Brown, 1978) and this example simply illustrates that the efficiency of nitrogen utilization may involve interactions of processes on very different scales from those discussed above.

7.3.5 Phosphate Absorption

Of all the ions and macronutrients, the availability and absorption of phosphate present the most problems in physiology and ecology. Not only are the states and processes difficult to quantify, they are unusually difficult to integrate and to interpret in a functional sense. The success of *Atriplex* and other chenopodiaceous shrubs in arid region soils or in coastal dune sands which are unusually low in PO_4^{3-} (Fig.7.2) may imply that they are more effective in absorbing this nutrient anion. On the other hand, the success of these plants in such habitats may be related to other factors which enable them to persist and to take advantage of the limiting nutrient when it is available. For example, it is probable that in most plants, the majority of ion and nutrient absorption is restricted to periods of adequate water supply (Hsiao, 1973). In arid shrublands, it is likely that these periods also correspond to the times of maximum mineralization and release of nutrients (Chap.7.5.1). Thus following rainfall the nutrient relations of the successful species such as *Atriplex* may be similar to those of other plants. It might not be surprising, therefore, that the rate and affinity properties of PO_4^{3-} uptake in young roots of *Atriplex* spp. were not much different to those of cereal roots grown under laboratory conditions (Chap.7.2.1; Fig.7.6).

Several lines of evidence suggest it is the capacity of species such as *Atriplex* to persist and take advantage of nutrients when available, rather than the capacity to absorb ions from low nutrient, low water potential soils which may be important. The absorption of PO_4^{3-} by *Atriplex* seedlings is confined to the nonsuberized region 2–3 cm behind the tip (Cook, 1973) and when roots develop in drying soil suberization results. Cowling (1969) removed roots of *A. vesicaria* seedlings from drying soils and studied rates of PO_4^{3-} uptake by these roots from solutions. He found that the rate of PO_4^{3-} uptake was 80% inhibited in suberized roots after exposure to high levels of water stress. Only a small proportion of PO_4^{3-} absorbed by suberized roots was transported to the shoot. However, in Cowling's experiments *A. vesicaria* seedlings recovered from 14 weeks' exposure to dry soil and produced new roots which were equally as effective in PO_4^{3-} absorption as those of barley plants exposed to only 3 weeks in dry soil. Barley did not survive greater than 7 weeks' exposure to these conditions in this comparison, whereas after 60 weeks in dry soil *A. vesicaria* seedlings still produced new roots and absorbed PO_4^{3-} effectively in a few days after rewatering. From other experiments Cowling concluded that roots deep in the soil might supply water to roots in drier but nutrient-rich surface soils and permit prolonged nutrient uptake.

Although the majority of nutrient uptake takes place under the most favorable conditions, there are other properties of nutrient uptake processes which may be significant in the overall efficiency of uptake. Australian arid shrubland perennial *Atriplex* are active at any time of the year (Chap.5.1.1) and it is not surprising therefore that Cook (1973) found PO_4^{3-} uptake by *A. vesicaria* seedlings from solutions was relatively insensitive to temperature between 19° and 40 °C

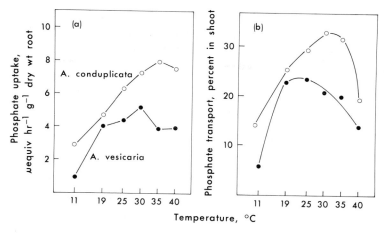

Fig. 7.11. a Effect of temperature on rate of PO_4^{3-} absorption by seedlings of *A. vesicaria* and *A. conduplicata*. **b** Effect of temperature on the percentage of PO_4^{3-} absorbed and transported to shoots of *A. vesicaria* and *A. conduplicata*. (NaH$_2$PO$_4$ solution 18 μM: data of Cook, 1973)

Fig. 7.12. Sodium deficiency in *Atriplex vesicaria*. The plants on the *left* were grown in Na$^+$-free solutions; those on the *right* were grown in solutions containing 0.01 mM Na$^+$. (Courtesy P.F. Brownell)

(Fig. 7.11). Annual species such as *A. conduplicata* are often established following summer rainfall and the higher rate and high temperature optimum for PO_4^{3-} uptake compared with *A. vesicaria* may be significant (Fig. 7.11). The difference between species in the temperature responses of ion uptake and transport presumably reflect properties of membrane lipids as well as lipid protein interactions (Carey and Berry, 1976), interactions which are discussed more fully in Chap. 9.

Cowling (1969) and Cook (1973) reported that *Atriplex* seedlings were about three times as efficient as seedlings of cereals in the transport of PO_4^{3-} from root to shoot. Figure 7.12 shows that the temperature responses of this process in *A. vesicaria* and *A. conduplicata* show a similar form to the temperature response of the uptake process. A major metabolic sink for PO_4^{3-} is photosynthesis and efficient transport from root to shoot presumably permits *Atriplex* spp. to make the most of a situation in which PO_4^{3-} uptake is limited by low concentration in the soil and a short period in which soil water relations remain favorable for nutrient uptake.

7.3.6 Micronutrient Ions and Toxicity

It is a paradox that some halophytic members of the genus *Atriplex,* capable of absorbing vast quantities of Na^+ in saline environments, should prove to have a requirement for this ion as a micronutrient (Brownell and Wood, 1957; Brownell, 1968). However, it is evident that the Na^+ requirement, which is satisfied in solution cultures at less than 0.01 mM Na^+ (Fig. 7.12), is not associated with the ability of these plants to absorb Na^+ to high concentration. Both *A. vesicaria* and *A. triangularis (hastata)* have been grown in cultures containing 1 M NaCl (Black, 1958, 1960), yet the former shows a requirement for Na^+ and the latter does not (Brownell and Crossland, 1972). It is now evident that the Na^+ requirement is found only in those species of *Atriplex* and other genera capable of C_4 photosynthesis or crassulacean acid metabolism (Brownell and Crossland, 1972, 1974). The metabolic bases for the Na^+ requirement and the metabolic consequences of Na^+ deficiency are discussed in detail in Chap. 9.

It is most improbable that the Na^+ requirement has any significant relationship to the capacity of these plants for massive Na^+ uptake. Sodium uptake in *A. vesicaria* was more or less linear with concentration up to 0.1 mM, after which there was a sudden increase in the rate of uptake (Fig. 7.13). This increase in Na^+ uptake was associated with a decrease in K^+ uptake, indicating that it probably reflects the onset of absorption processes characteristic of the high concentration ranges discussed in the next section. Whether the growth requirement for Na^+ in C_4 species of *Atriplex* is related to a lower affinity of ion uptake processes for this ion in these species or simply to a specific requirement for higher Na^+ levels for growth, is not yet clear. Preliminary experiments show, however, that labeled Na^+ is very much more mobile in species which show deficiency symptoms than in those which do not (P. F. Brownell, unpublished). Although the processes and the mechanisms of the Na^+ requirement in C_4 *Atriplex* and other C_4 species are not understood, it is improbable that Na^+ deficiency would ever be encountered under natural conditions because it is unlikely that any natural habitat would be sufficiently free of Na^+ for deficiency symptoms to be expressed.

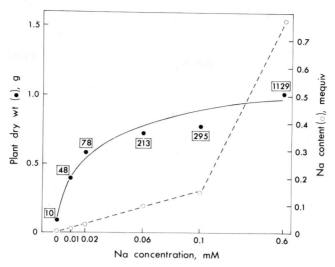

Fig. 7.13. Growth response to Na$^+$ in *A. vesicaria* seedlings (●) showing the sodium requirement of this species. The Na$^+$ content per plant is shown (○) and the Na$^+$ concentration (μEq.g^{-1} dry wt.) is shown as *boxed numbers*. (Data of Brownell, 1965)

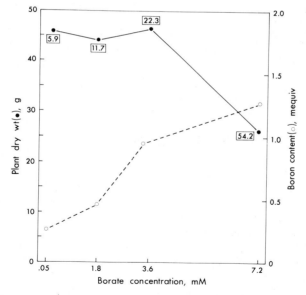

Fig. 7.14. Growth response to borate in *A. polycarpa* seedlings (●) showing the tolerance of these plants to boron. The boron content per plant is shown (○) and the concentration (μEq.g^{-1} dry wt.) is shown as *boxed numbers*. (Data of Chatterton et al., 1969, 1970)

On the other hand, *Atriplex* spp. are frequently found in arid shrubland habitats which are rich in trace element minerals such as boron (Chatterton et al., 1969, 1970) and *A. triangularis (hastata)* rapidly colonizes ash heaps which are poor in major nutrients but enriched in Mn^{2+} and Al^{3+} (Rees and Sidrak, 1956). Goodman (1973) noted that soils supporting *A. nutalli* in Utah contained about 2.5 μEq g^{-1} of boron and that this species showed no significant growth inhibition in solutions containing 36 mM boron. *A. confertifolia* (Goodman, 1973) and *A. hymenelytra* are also evidently very tolerant of boron. Chatterton et al. (1969, 1970) reported little effect of boron on growth of *A. polycarpa* in solution cultures containing up to 7.2 mM borate, a concentration about 300 times greater than the level of borate needed to satisfy trace element requirements. Figure 7.14 shows the effect of increasing borate concentration on growth and boron uptake by the most sensitive *A. polycarpa* population examined by these authors. Boron uptake was more or less linear with external concentration and levels of boron in the tissues remained far lower than those found in susceptible crop plants. Chatterton et al. (1969, 1970) proposed that the epidermal trichomes may serve to reduce the boron content of leaf mesophyll tissues and thus avoid boron toxicity.

These studies of sodium requirement and boron tolerance raise the question of whether the uptake or excretion processes involved in the ionic relations of major nutrients and ions also function in the same way for micronutrients and toxic ions. As discussed previously (Chap. 3.4.4) tolerance to heavy metal toxicity is often related to chelation in the root or shoot, rather than regulation of absorption. In the above experiments with *A. polycarpa*, roots did not increase in boron content, but the boron in the shoot could have been chelated in cell walls. Another possibility, that the metabolism of *A. polycarpa* is unusually tolerant of elevated boron concentrations, remains to be assessed. Just as Na^+ requirement seems to be associated with C_4 photosynthetic metabolism, boron tolerance may equally well be associated with specific metabolic processes.

7.4 Salinity Responses in Atriplex

Halophytes are described as plants which grow and complete their life cycle in habitats of high salinity. Although *Atriplex* spp. do not require other than trace amounts of Na^+ for normal growth, they frequently show growth stimulation in response to NaCl. That they are capable of completing their life cycle in saline habitats is shown by their distribution in saline regions throughout the world (Chap. 5.4). Most ecological accounts of halophytes begin with an elaborate classification of different halophyte types. These classifications are usually based on a minimum of data on the ionic status of the plants, such as the so-called "osmotic spectra" (Walter and Stadelmann, 1974), and are further extended to presumed processes such as the exclusion, accumulation, or excretion of ions. *Atriplex halimus*, for example, has been described as a "salt-excluding terrestro-xero-euryhalophyte" (Waisel, 1972). Although more comprehensive analyses have revealed several "physiotypes" among halophytes which merit comparative experimental study (Albert and Kinzel, 1973; Albert and Popp, 1978), many classifications only succeed in confusing the issues underlying functional ion absorption and water relationships of halophytes. Barbour (1970) has argued similarly for a reappraisal of halophyte classification. The main purpose of this section, however, is not to discuss the codification of halophytes but to outline the ion absorption processes which may be responsible for the success of *Atriplex* spp. in saline environments and to assess the significance of high salt contents to metabolic activities in these plants.

7.4.1 Salt Exclusion, Salt Accumulation, and Salinity Tolerance

One legacy of the semantic confusion which has arisen from the classification schemes for halophytes is the notion that they can be grouped into categories of salt-accumulating and salt-excluding types. This distinction is difficult to sustain, even in classic comparisons such as that of the two mangrove species *Rhizophora mucronata* and *Aegialitis annulata* (Atkinson et al., 1967). With respect to the seawater bathing medium, both exclude salt if the xylem sap concentrations are compared (17 mM and 100 mM, respectively) but both accumulate salt if the ion concentration of the leaves are compared. The ionic concentration of mature leaves of both species is slightly greater than that of seawater, showing that both species have adjusted the osmotic concentration of the cell vacuole above that of seawater and hence can maintain positive turgor in leaf cells. The species differ markedly in the way this adjustment is achieved. In *Rhizophora* the functionally significant processes involve low rates of uptake and transport to the shoot so that the salt load can be accommodated by normal growth. In *Aegialitis* functionally significant processes center on the excretion of excess salt from the leaf by salt glands which compensate for the less well-regulated ion uptake and transport processes in the roots of this plant.

Salinity tolerance is currently viewed as the interaction of ion uptake, transport, and excretion processes which permits effective osmotic adjustment by means of salt accumulation (Caldwell, 1974; Flowers, 1975; Flowers et al., 1977). In this view, successful halophytes have reduced the problem of salinity tolerance to that of maintaining cell water relations and normal biological activity in equilibrium with the low water potential of the saline substrate and of the atmosphere. This may be achieved by combinations of different processes, as indicated in the comparisons of mangroves above, but the central argument is that halophytes successfully maintain turgor by means of salt accumulation and continue to grow because in these and all other plants, dry matter increase and growth is only possible when turgor is maintained. This view of salinity tolerance may require a little modification to accommodate the possibility of toxic effects of specific inorganic ions (Greenway, 1973; Poljakoff-Mayber and Gale, 1975) and qualification to accommodate compatible solutes, but it provides a clear framework for assessing the contribution of specific ion absorption processes in salinity tolerance.

As was evident in the state descriptions of *Atriplex* spp. from different environments (Table 7.3), successful halophytes are characterized by high levels of NaCl in the shoot, particularly in halophytes from arid shrublands. Rush and Epstein (1976) compared salt-sensitive genotypes of tomato with *Lycopersicon cheesmanii,* a salt-tolerant genotype from the Galapagos Islands. The latter accumulated higher concentration of Na^+ in the shoot at all external NaCl levels like other successful halophytes. On the other hand, tolerance to moderate salinity in nonhalophytes appears to be quite distinctly associated with the prevention of salt accumulation in the shoot, and at concentrations of NaCl in excess of 50% seawater death of salt sensitive *L. esculentum* was due to a six-fold increase in leaf Na^+. Comparisons of salt-sensitive and salt-tolerant barley (Greenway, 1962) and soybeans (Abel and MacKenzie, 1964) revealed that higher yield of salt-tolerant varieties was due to their ability to maintain low levels of Cl^- in the leaves. Greenway (1962) showed that the high level of Cl^- in salt-sensitive barley was due to rapid uptake of Cl^- by the roots and rapid transport to the shoot. The relatively salt-tolerant soybean variety "Lee" does not transport large amounts of Cl^- to the leaf but instead retains NaCl in old portions of the root. In this part of the root wall ingrowths in the xylem parenchyma are found which resemble those of transfer cells in other tissues (Läuchli, 1976b; Gunning, 1977). These cells contain high Na^+/K^+ and Cl^-/K^+ ratios measured by the electron probe and are believed to function by removing NaCl from the xylem stream before it reaches the shoot. Similar xylem parenchyma development has been observed in other species and has been shown to retain Cl^- in the root. These observations explain why, in recent reviews and in the discussion which follows, emphasis is placed on those processes which regulate ion uptake to high internal concentration in cell vacuoles and on processes responsible for ion transport to the shoot.

7.4.2 Salt Absorption by Atriplex Seedlings as a Function of External Concentration

Laboratory studies have provided a wealth of state descriptions of the ionic concentration of *Atriplex* spp. following growth at different levels of NaCl. Eaton (1927) was possibly the first to conduct controlled experiments of this type and demonstrated the capacity of *A. semibaccata* to absorb high concentrations of NaCl from culture solutions. His data also show that ion absorption resulted in the maintenance of a substantial difference between the freezing point depression of the soil and plant extracts, indicating osmotic adjustment by salt accumulation (Table 7.10). Details of the osmotic adjustment processes will be discussed later (Chaps. 7.5.6 and 8.2.4) and we shall concentrate on the ion absorption properties of these plants in relation to external concentration.

Many experiments show that ion absorption from solutions of increasing NaCl concentration results in plant ionic distributions similar to those shown in Table 7.3. Greenway (1968) grew *A. nummularia* seedlings for 31 days in solutions

Table 7.10. Growth and osmotic adjustment in *A. semibaccata* at different levels of NaCl in soil culture. (Data of Eaton, 1927)

NaCl added to soil (%)	Plant growth (g)	Freezing point depression (°C)		
		Soil	Plant	Plant-soil
0	6.2, 7.9	0.34	1.37	1.03
0.05	9.9, 10.1	0.42	1.58	1.16
0.10	5.2, 8.5	0.72	1.77	1.05
0.25	10.2, 8.3	1.35	1.95	0.60

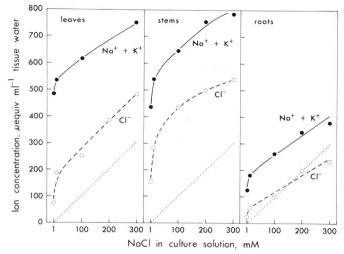

Fig. 7.15. Concentration of $Na^+ + K^+$ and Cl^- in leaves, stems, and roots of *A. nummularia* grown 31 days in culture solutions containing added NaCl. (Data of Greenway, 1968)

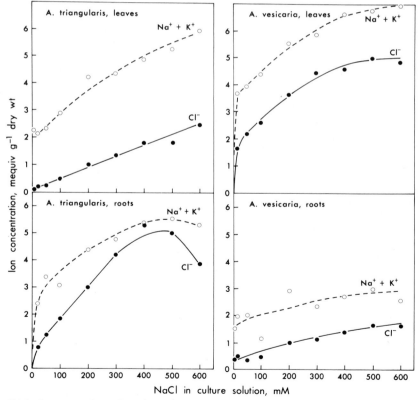

Fig. 7.16. Concentration of $Na^+ + K^+$ and Cl^- in leaves and roots of *A. triangularis* and *A. vesicaria* after 16–20 weeks grown in culture solutions containing added NaCl. (Redrawn from Black, 1956, 1960)

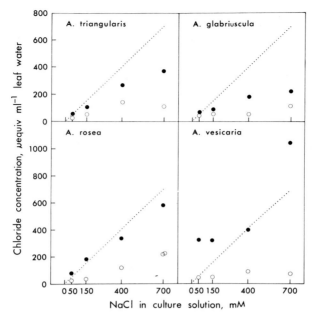

Fig. 7.17. Concentrations of Cl⁻ in roots (○) and leaves (●) of *Atriplex* spp. after 21 days growth in culture solutions containing added NaCl. (C.B. Osmond, unpublished)

containing up to 300 mM NaCl (Fig. 7.15) and found that the $Na^+ + K^+$ and Cl^- concentrations of leaves and stems were much higher than those of the roots. In the roots, Cl^- concentrations in the cell water were comparable to those of the external solution at each concentration. The difference between $Na^+ + K^+$ and Cl^- concentration shown in Fig. 7.15 was presumably largely balanced by oxalate (Chap. 7.3.3). In earlier experiments (Black, 1956, 1960) *A. triangularis* and *A. vesicaria* were grown for 16 to 20 weeks in culture solutions containing up to 600 mM NaCl. These long-term experiments resulted in similar differences between the ion concentration of roots and shoots of *A. vesicaria* (Fig. 7.16) as were observed by Greenway with *A. nummularia*.

   However, in *A. triangularis* Black (1956) noted that root Cl^- concentrations were rather higher than those in the shoot throughout the range of NaCl concentrations used and that at most concentrations, root $Na^+ + K^+$ concentrations were also higher (Fig. 7.16). This is an important difference because it suggests that *A. triangularis* may tend to retain salt in the roots rather than transport it to the shoot. It suggests that the regulation of ion uptake and transport in this species from salt marsh habitats may differ from that of halophytes from the arid shrublands.

   For this reason, the pattern of Cl^- absorption in several *Atriplex* spp. from different habitats was examined in a short-term experiment under closely comparable conditions in a greenhouse over the same 21-day period (C. B. Osmond, unpublished). Figure 7.17 shows that two salt marsh species *A. triangularis* and *A. glabriuscula* were characterized by relatively lower concentrations of Cl^- in root and leaf water at the end of this experiment. However, shoot Cl⁻ concentration was two- or threefold greater than that in the root and there was no suggestion that roots retained higher Cl⁻ concentrations than shoots, as observed by Black (Fig. 7.16). In *A. rosea* and *A. vesicaria,* shoot Cl⁻ concentrations were somewhat greater than those in the salt marsh species, consistent with the higher ion content in species from more arid saline habitats.

The discrepancy in the experiments with *A. triangularis* shown in Figs. 7.16 and 7.17 may be due to changes in ion absorption processes with plant age, or it may be due to differences in genotypes in this highly variable taxon (Chap. 4.2.3). In either event, both experiments show that processes which regulate ion uptake at high internal and external concentrations and the transport of NaCl from root to shoot play a key role in the ionic relations of *Atriplex.*

7.4.3 NaCl Uptake and Transport

The greater capacity of NaCl absorption in halophytes must ultimately be explained in terms of ion uptake processes in root cells. Several attempts to account for the greater capacity in terms of higher rates of ion uptake by the roots of halophytes have been inconclusive, simply because meaningful comparative data are not available (Greenway and Osmond, 1970; Flowers et al., 1977). Studies with the halophytes *Triglochin maritima, Plantago maritima,* and *Suaeda maritima,* reviewed by Flowers (1975) and Flowers et al. (1977), indicate that the rate of NaCl absorption by intact seedlings increases linearly with external concentration over the range 0.01 to 300 mM NaCl. These experiments, which measured uptake and transport to the shoot over a 24-h period, show much higher uptake rates for the whole plant than are observed in nonhalophytes at the maximum concentrations of electrolytes tolerated by these plants. In fact the measured rates of sustained NaCl absorption by these intact halophytes correspond to the maximum initial rates of ion uptake by excised low salt barley roots. In *Suaeda maritima* the rates of whole plant NaCl absorption, calculated on the basis of shoot dry weight, range between 120 and 320 μEq g^{-1} dry wt. day^{-1}. These compare with estimates of 240 μEq g^{-1} dry wt. day^{-1} in *A. vesicaria* (Black, 1960) and 360 μEq g^{-1} day^{-1} in *A. spongiosa* (Osmond, 1965).

These high rates of NaCl absorption suggest that ion uptake processes in the roots of halophytes differ significantly from those of nonhalophytes. Whether differences are related to the affinity properties or the maximum velocity properties of the uptake processes with respect to external concentration has yet to be determined. They result in high concentrations of salt in the vacuole of all cells in halophytes and imply that the "set point" at which regulation occurs is higher in these plants than in nonhalophytes (Osmond, 1979; see Chap. 7.2.1).

Table 7.11 shows a comparison of Cl⁻ uptake and transport in seedlings of *A. vesicaria* and *A. triangularis* treated in 1 mM NaCl. The data are based on root fresh weight and confirm that *A. vesicaria* is capable of more rapid Cl⁻ uptake than *A. triangularis* at all NaCl concentrations (Fig. 7.17). They also show that in 100 mM NaCl the rate of transport from the root to the shoot is rather greater than the rate of uptake to the vacuole of root cells. This feature of uptake

Table 7.11. Rate of chloride uptake and transport by young seedlings of *Atriplex*. (A. Läuchli, unpublished)

Species and tissue		Chloride uptake, μmol g^{-1} fr. wt. h^{-1}	
		1 mM NaCl	100 mM NaCl
A. vesicaria	Roots	0.7	7.0
	Stem and leaf	0.6	17.0
A. triangularis	Roots	0.13	4.0
	Stem and leaf	0.11	8.0

and transport presumably accounts for the lower Cl$^-$ concentration in roots, relative to shoots, of both species (Fig. 7.17) and the difference between species in these processes (Table 7.11) may also be significant. Läuchli's experiments confirm Black's (1960) observation that when *A. vesicaria* seedlings were transferred from cultures with 6 mM NaCl to 100 mM NaCl, the NaCl content of roots remained unchanged in the following 10 days, but young leaves accumulated NaCl at the rate of 240 μEq g^{-1} dry wt. day^{-1}.

Other data support two further implications of the above interpretation. If transport to the shoot is more rapid than uptake by root cells and if Cl$^-$ is largely transported from the symplast, it is possible that the symplast of the root would contain relatively low concentrations of Cl$^-$. This prediction is supported by Kramer's microprobe analysis of pericycle cells in *A. triangularis* roots in 100 mM NaCl which showed that cytoplasmic (symplasmic) Cl$^-$, relative to cations, was much lower than in the external solution. Because the rates of uptake and transport of *A. vesicaria* at 100 mM NaCl are much higher than those in *A. triangularis*, one would expect a higher turnover of Cl$^-$ in *A. vesicaria* roots. In the experiments shown in Table 7.11 A. Läuchli found that the specific radioactivity of ^{36}Cl in *A. vesicaria* roots was two- to fourfold greater than that in *A. triangularis*.

These observations on the uptake and transport of Cl$^-$ in *Atriplex* roots at high NaCl concentration suggest an "efficiency" in the transport of ions to the shoot reminiscent of that observed in comparative studies of PO$_4^{3-}$ uptake and transport at very low external concentration (Chap. 7.3.5). If this proves to be a feature of halophytes, then it may be profitable to reconsider the significance of the correlation between root cortex development and halophytism. Poljakoff-Mayber (1975) cites data of Ginzburg which indicate that roots of halophytes have only 2–5 layers of cortical cells compared with 6–14 cell layers in the cortex of nonhalophytes. A smaller sink for ions in the root may account for some of the properties of high transport rates relative to uptakes rates in halophytes.

The sustained rapid transport of NaCl from the root and accumulation to high concentration in the shoot of halophytes raises important questions as to the pathway and regulation of ion transport in halophytes. As discussed previously, transport to the shoot in glycophytes involves ion movement through the symplast and active excretion of ions across the inner plasmamembrane of xylem parenchyma cells in the stele (Pitman, 1977; see Chap. 7.2.1). This transport process is closely regulated by the shoot, so that it is largely independent of external concentration and is turned to the demands of the shoot during growth (Pitman and Cram, 1977). The situation in halophytes does not appear to be consistent with this account of ion transport in nonhalophytes.

If transport in halophytes depends on ion movement in the symplasm and active excretion into the xylem then it is possible that these processes are regulated quite differently in these plants. We have seen that the plasmamembrane of *A. triangularis* root cells (the first membrane of the symplasm) has an unusually low permeability for Na^+ (Anderson et al., 1977; see Chap. 7.2.2). Furthermore the Na^+ concentration in root cells appears to be maintained by an active Na^+ excretion process at the plasmamembrane. Together these properties suggest that the Na^+ which enters the symplast, probably by passive diffusion from external solutions of high concentration, is relatively effectively pumped from the symplasm, either into the external solution or into the stele. It is not clear, however, whether the response of this transport process in halophytes to regulatory signals from the shoot differs from that in nonhalophytes. If influx in cells of the shoots of halophytes is less sensitive to internal concentration (higher "set point") then it is probable that they can handle higher rates of salt input. That is, the generation of signals which regulate transport to the shoot would be postponed until much higher salt concentrations had been attained (Osmond, 1979).

A simpler alternative explanation of the massive salt transport to the shoot of halophytes is that at high external concentrations a greater portion of the total solute transport traverses the apoplast and effectively bypasses the symplast. There is some evidence that this occurs in nonhalophytes at higher ion concentrations. Greenway (1965) noted that at 50 mM NaCl, 40% of the Cl^- transport in barley was insensitive to metabolic uncouplers, consistent with passive transport through the apoplast rather than active transport through the symplast. Flowers (1975) commented that transport of Na^+ to the shoot of *Suaeda maritima* was less dependent on metabolism at high external concentrations. This explanation is presumably what Black (1960) had in mind when he spoke of "luxury uptake" in *A. vesicaria*. Studies of K^+, Na^+, and Ca^{2+} competition during massive salt absorption by these halophytes which conform to such a passive process are discussed below.

If apoplastic ion movement is an important feature of salt transport in halophytes, then the functional significance of the extent of suberization in the radial walls of the endodermis of halophytes and nonhalophytes needs to be assessed. Poljakoff-Mayber (1975) quotes data of Ginzburg which suggest 50%–100% suberization of this wall in halophytes compared with 27%–40% suberization in nonhalophytes, an observation which is superficially inconsistent with increased apoplasmic transport of solutes in halophytes. A further important implication of this hypothesis is that passive apoplastic transport of salt is unlikely to be responsive to hormonal signals generated when salt inputs to the shoot exceed the uptake rate of leaf cells. In this situation it seems probable that salt-excreting systems in the shoot or the potential for increasing the salt uptake by cell expansion (succulence) would assume added significance. These implications of sustained rapid salt uptake to the shoot in *Atriplex* spp. are discussed below.

7.4.4 K^+ and Na^+ Selectivity in Halophytes

Halophytes such as *A.halimus* show much lower selectivity for K^+ than nonhalophytes during uptake and transport of ions from equivalent K^+ and Na^+ solutions (Fig. 7.18). Furthermore the ratio of K^+/Na^+ decreases from root to shoot and, decreases each time ions move from the apoplast or symplast into the vacuole of another cell (Table 7.12). When *Atriplex* or other halophytes are grown in solutions containing increasing concentrations of NaCl, K^+ absorption is maintained and K^+ in the tissue is not depressed beyond what appears to be a minimal level (Fig. 7.19), even though the Na^+/K^+ ratio in solution may be greater than 100 (Black, 1960). Likewise, Na^+ uptake is maintained in solution

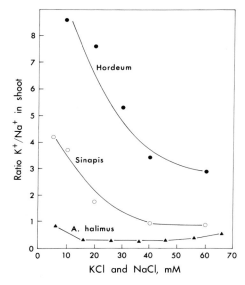

Fig. 7.18. Differences in the selectivity for K^+ and Na^+ as indicated by the K^+/Na^+ ratio in shoots of plants grown on solutions containing increasing equimolar concentrations of NaCl and KCl. (Redrawn from Pitman, 1965 and Mozafar et al., 1970)

Table 7.12. Ratio of K^+/Na^+ in tissues of *A. nummularia* seedlings grown at different NaCl concentration. (Data of Greenway 1968 and C. B. Osmond unpublished)

Tissue	NaCl concentration, mEq l^{-1}						
	1	10	50	100	200	250	300
Solution	5.0	0.5	0.1	0.05	0.025	0.020	0.016
Roots	4.0	1.5	—	0.4	0.4	—	0.3
Stems	6.0	2.8	—	0.4	0.4	—	0.2
Leaves	2.0	0.8	—	0.2	0.2	—	0.1
Leaf lamina	—	0.97	0.57	—	—	0.20	—
Leaf trichome	—	0.74	0.47	—	—	0.16	—

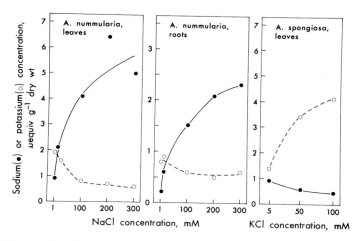

Fig. 7.19. Changes in Na$^+$ and K$^+$ concentration of *A. spongiosa* and *A. nummularia* leaves with increasing concentration of NaCl or KCl in the external solution. (Data of Osmond, 1966)

of very high KCl concentration (Osmond, 1966; Mozafar, Goodin and Oertli, 1970). Black (1960) proposed that roots of halophytes have relatively specific, low capacity uptake and transport processes for both Na$^+$ and K$^+$ and a high capacity, non-specific uptake and transport process for which both ions compete. Few advances have been made which explain these observations and which elaborate the requirement for a high K$^+$ concentration (50–100 mM) for the healthy function of the cytoplasm. This is a central assumption of most recent hypotheses and research on ionic relations of halophytes (Jennings, 1968; Jefferies, 1972; 1973; Flowers, 1975; Wyn Jones et al., 1977).

   Mechanisms for ensuring the specificity of monovalent uptake and the ratio K$^+$/Na$^+$ in nonhalophytes depend primarily on linked K$^+$ influx, Na$^+$ efflux transport at the plasmamembrane of root cells (Jeschke, 1977). In halophytes it is evident that the Na$^+$ efflux pump which maintains high cytoplasmic K$^+$ in nonhalophytes must either be bypassed to allow Na$^+$ accumulation in the vacuole, or must be augmented in some way by other processes. It is possible, for example, that the plasmamembrane transport system of halophytes has a lower K$^+$/Na$^+$ selectivity because influx and efflux are linked differently. Whether the lower passive permeability of halophytes to Na$^+$ (Chap. 7.2.2) contributes to this is not known. Lower K$^+$/Na selectivity could result in high cytoplasmic Na$^+$ concentrations which are deemed to be unhealthy. If, however, the tonoplast of halophytes had a similar K$^+$/Na$^+$ exchange system, analogous to an inside-out plasmamembrane, Na$^+$ in the cytoplasm could be transferred to the vacuole in exchange for K$^+$. Evidence that vacuolar K$^+$ in expanding cells of *Atriplex hortensis* roots was replaced by Na$^+$ when roots were bathed with 1 mM NaCl has been obtained by Jeschke and Stelter (1976). Similar results obtained with leaf slices (Osmond, 1968) are difficult to interpret because of the low Ca^{2+} status of these tissues.

This hypothetical mechanism for maintaining high cytoplasmic K^+ and allowing the accumulation of Na^+ in the cell vacuole could also account for the predominance of Na^+ transport to the shoot, if the symplast membrane adjacent to the xylem shared the same properties as the tonoplast. Alternatively, if apoplastic transport is relatively more important in halophytes exposed to high NaCl concentration, this pathway would deliver ions to the xylem more or less in proportion to their concentration in solution, and lower the ratio K^+/Na^+ in the xylem stream. Lower retranslocation is a possible further contributor to the lower K^+/Na^+ ratio in the shoot of halophytes. However, Greenway et al. (1966) found similar and quite negligible retranslocation of Na^+ from the shoots of beans and *Atriplex triangularis (hastata)*. The low permeability of *Atriplex* roots to Na^+ (Chap. 7.2.2) does not appear to be reflected in slower retranslocation of ion from leaf cells.

These and other hypotheses which account for the K^+/Na^+ ratio of halophytes, and the absorption of Na^+ to high concentrations, have yet to be extensively investigated (Flowers, Troke, and Yeo, 1977). An additional factor which should be considered is that the high-capacity salt-uptake system of halophytes is not specific for Na^+, or even for monovalent ions. In several *Atriplex* spp. the total cation $(N^+ + K^+ + Ca^{2+})$ concentration in leaves of plants grown in solutions containing $100 \, mEq \, l^{-1}$ of NaCl or KCl was not significantly greater than that of plants grown in solutions with $100 \, Eq \, l^{-1}$ of $CaCl_2$. At high $(NaCl + KCl + CaCl_2)$ concentrations, uptake and transport of Ca^{2+} was depressed relative to that in $CaCl_2$ alone, but monovalent uptake was not affected (Osmond, 1966). The preferred uptake and transport of monovalent ions over Ca^{2+} perhaps reflects the greater mobility of monovalents in the Donnan free space and further strengthens the view that the massive transport of ions to the shoot of halophytes involves relatively nonspecific passive ion movement through the apoplast.

7.4.5 Ion Uptake and Excretion in Leaves

Because leaves are in direct contact with the atmosphere they are potentially exposed to substantial fluctuations in water content which could rapidly change the ionic concentrations within the free space, cytoplasm, and cell vacuole. Because leaves are rather indirectly connected with the soil solution, control over ion input is likely to have a larger relaxation time than the change in leaf water content. The significance of these potential changes in leaf ionic status, or indeed, whether they occur, is not known. If, however, the larger salt inputs in the leaves of halophytes involve passive processes which are likely to be less susceptible to control, the potential for relatively violent changes in ion concentration in the leaf is likely to be greater. It is not surprising that leaves of halophytes often possess additional controls over leaf ion concentration. These include salt glands, which effectively unload the salt to the exterior of the leaf, and the capacity to become succulent, which effectively dilutes salt within the leaf by an increase in cell volume. The significance of both of these processes in the control of salt concentrations in *Atriplex* will be assessed.

Although little is known directly of the relationship between the rate of input to the leaf and ion uptake the vacuole of leaf cells, it is clear that leaves of many

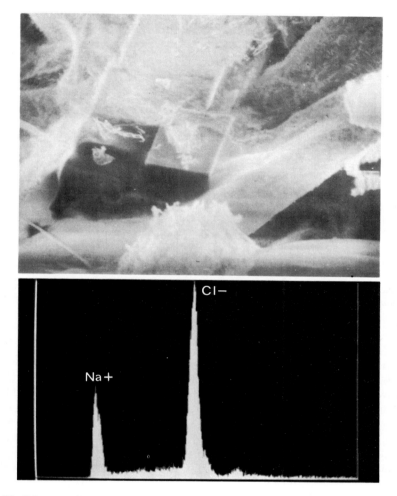

Fig. 7.20. Salt crystal among waxy debris of collapsed bladder cells on the surface of *A. hymenelytra* leaves collected in Death Valley. The nondispersive X-ray analysis of this crystal is shown with the peaks corresponding to Na^+ and Cl^-. (Mooney et al., 1974)

Atriplex spp. are equipped with a salt-excreting system which may function to regulate the salt load. The leaves of many *Atriplex* spp. are covered with epidermal trichomes which contain high concentration of salt and evidence from several laboratories has confirmed that these trichomes function as salt-excreting systems (Osmond et al., 1969a; Mozafar and Goodin, 1970; Smaoui, 1971). This process has been discussed at length in several reviews (Lüttge, 1971; Thompson, 1975; Hill and Hill, 1976).

The epidermal trichome is an unusual salt-excretion system in that ions are excreted via a gland-like stalk cell to the vacuole of the larger bladder cell. As discussed previously, the transport of Cl^- from the apoplast or symplast through

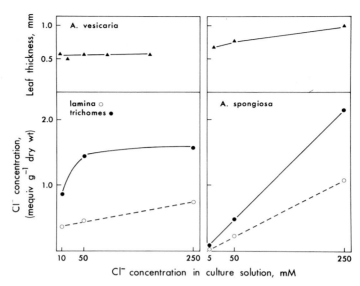

Fig. 7.21. Concentration of Cl⁻ in lamina and epidermal trichomes of *A. vesicaria* and *A. spongiosa* grown in cultures containing NaCl. Thickness of the leaf lamina is also shown (C. B. Osmond, unpublished)

the gland is an active process (Chap. 7.2.2). As the leaf matures many bladder cells collapse, leaving a highly reflective mat of wax and NaCl crystals on the leaf surface (Mooney et al., 1974; Fig. 7.20). In some *Atriplex* species more than 50% of the salt transported to the shoot may be excreted through the stalk cell of the epidermal trichomes. Although the details as to the site of the active processes, and the contribution of apoplastic and symplastic sources of NaCl are disputed, the functional significance of this process is not in doubt.

Figure 7.21 shows that the Cl⁻ concentration of the leaf lamina in *A. vesicaria* changed only slowly when culture solution concentration increases from 10 to 250 mM NaCl, whereas that in the epidermal trichomes increases markedly. In the annual species, *A. spongiosa*, this control over leaf Cl⁻ concentration is less effective (Fig. 7.21). There is no change in the thickness of the leaf lamina of *A. vesicaria* in which Cl⁻ concentration increases only about twofold, but a 40% increase in *A. spongiosa* where Cl⁻ concentration increases more than tenfold. Figure 7.22 shows that even in three *Atriplex* species with relatively few epidermal trichomes, the thickness of the leaf lamina increases as the proportion of total leaf Cl⁻ in the trichomes declines.

It is clear from these experiments that ion excretion to the bladder cell of epidermal trichomes functions to control salt concentration in the leaf lamina. When this control fails or if the species is poorly endowed with trichomes, a measure of regulation of salt concentration can be achieved by increased succulence. The latter response presumably involves a complex interaction between ion uptake, cell water relations and cell wall metabolism. It involves an elongation of hypodermal cells in *Atriplex* species with "Kranz" anatomy (Chap. 2.1.4) and of all leaf cells in

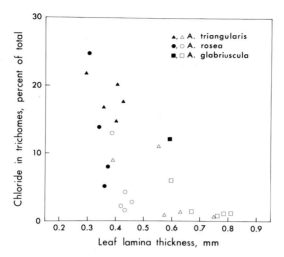

Fig. 7.22. Correlation between thickness of the leaf lamina and the percentage of total leaf Cl⁻ found in the epidermal trichomes of *Atriplex* spp.; *closed symbols*, young leaves; *open symbols*, old leaves. (C. B. Osmond, unpublished)

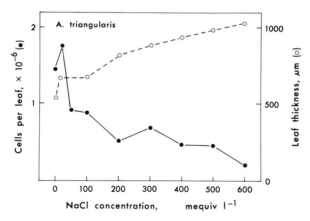

Fig. 7.23. Changes in estimated cell numbers per leaf and in leaf thickness for the oldest mature leaf on the main axis of *A. triangularis* grown in culture solutions with added NaCl. (Redrawn from Black, 1958)

species with conventional leaf anatomy (Black, 1958; C. B. Osmond, unpublished). Whether succulence is a specific response to Cl⁻ is not yet clear, but it compensates to a large extent for the depression of cell division by salinity. In *A. triangularis* *(hastata)* Black (1958) noted that the number of cells per leaf declined about six-to eightfold when NaCl concentration in culture solution was increased from $0-600\ mEq\ l^{-1}$ and that the leaf thickness doubled over this range (Fig. 7.23). About half the leaf volume was composed of spherical cells, and the other half of cylindrical cells. A doubling in cell diameter and length in these leaves represents about a six- to eightfold increase in cell volume, which compensates for the decrease in cell number.

These two processes, salt excretion and the development of succulence, probably function to regulate ion uptake by leaf cells in two very different ways. Salt excretion presumably functions to regulate the extracellular accumulation of salt in the apoplast which occurs when transport into the leaf exceeds the capacity (rate) of salt uptake to the vacuole of leaf cells. Oertli (1968) proposed that such an accumulation of salt may be responsible for salt damage in nonhalophytes. If the unregulated arrival of salt in the leaf produced an osmotic pressure in the apoplast which is near to that in the vacuole, leaf cells would presumably experience near zero turgor. Low turgor could be responsible, indirectly, for the salt damage. In terms of Oertli's hypothesis, it is clear that salt glands would allow a halophyte to maintain low extracellular (apoplastic) ion concentration by excreting salt which could not be absorbed to the vacuole of the leaf cells. There would be less likelihood that leaf cells would be exposed to zero turgor.

On the other hand, a halophyte which lacks effective salt-excreting glands, but has the capacity to respond to salinity by the development of succulence (by continued cell expansion), may be equally capable of regulating apoplastic salt concentrations. If salt uptake to the vacuole is regulated at a "set-point" by vacuolar concentration or cell turgor, expansion of cell volume and continued water uptake could prolong the approach to the "set-point".

It is arguable whether succulence is a more common or more effective means of regulating leaf ionic relationships in salt marsh halophytes, and whether salt glands are more common or more effective in arid land halophytes. Whether one regulatory mechanism is superior to the other in relation to the water status of these environments is not known. It is likely that both processes are involved in some species (Fig. 7.21) and because neither excretion nor succulence appears to have substantial effects on the metabolic capacity of halophytes, both may be regarded as mechanisms to facilitate the satisfactory compartmentation of salt and the maintenance of adequate turgor in leaf tissues.

7.4.6 Salinity, Metabolism and Growth

In the previous sections we have discussed the processes which contribute to the regulated absorption of NaCl by halophytes, and in this section we discuss evidence that ionic concentration in the cytoplasm of halophytes is low. We will show here that the in vitro metabolic processes of halophytes are no less sensitive to NaCl than those of nonhalophytes. Later we will show that in vivo processes, such as photosynthetic metabolism, are unaffected by high salt concentration in leaves of halophytes, but are much depressed in nonhalophytes. Together these observations imply that the cytoplasm of leaf cells in these halophytes is maintained at relatively low salt concentrations.

These views contrast markedly with explanations of salinity tolerance in certain bacteria and to the frequently expressed notion that the cytoplasm of halophytes is tolerant of the presence of high NaCl concentrations (Levitt, 1972; Waisel, 1972). In halophilic bacteria metabolic function is dependent on the presence of high salt concentrations in the cytoplasm (Hochachka and Somero, 1973; Lanyi, 1974). All

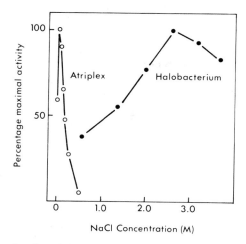

Fig. 7.24. Dependence of malate dehydrogenase activity on NaCl concentration during assay in enzyme extracts from *A. nummularia* and *Halobacterium*. (Redrawn from Larsen, 1962, and Greenway and Osmond, 1972)

structures investigated, from soluble enzymes to membrane-bound enzyme complexes, are only stable in the presence of 1 to 3 M NaCl. The salt requirement involves the stabilization of the unusually acidic cytoplasm proteins of these organisms. A combination of charge screening and long-range electrostatic interactions control the folding and volume of the cell proteins (Lanyi, 1974). In addition to these interactions with the tertiary structure, the proteins from halophilic bacteria also show interactions with NaCl at the active sites of enzymes, i.e., quite specific NaCl relationships with different proteins.

Figure 7.24 shows the NaCl dependence of the enzyme malate dehydrogenase isolated from *A. nummularia* leaves and from a halophilic bacteria. Both enzymes show stimulation and inhibition of activity due to NaCl, but the peak activity is separated by about a 100-fold difference in NaCl concentration. There is no specific experimental support for the concept that the activity of enzymes from halophytic higher plants or algae can generally be distinguished from those of nonhalophytes in terms of their response to NaCl. Several broadly based studies have now been conducted (Greenway and Osmond, 1972; Flowers, 1972a, b) and these establish that, in crude extracts, the enzymes of halophytes show similar NaCl response curves to those of nonhalophytes or those of purified enzymes from animal sources and yeast (Fig. 7.25). It is important, however, that the different kinetic properties of enzymes be considered. Because the interactions with NaCl or KCl are partially competitive, very different responses may be observed with enzymes from different sources if they differ in substrate affinity. Lower sensitivity to NaCl associated with cytoplasmic isoenzymes of malate dehydrogenase of halophytes (Cavalieri and Huang, 1977) may reflect kinetic properties. The apparently anomalous higher sensitivity to NaCl of PEP carboxylase isolated from halophytes with the C_4 pathway (such as *A. spongiosa*) when compared with that from C_3 nonhalophytes (Osmond and Greenway, 1973) may be explained

Table 7.13. Relative specific activity of enzymes in tissues of *Atriplex* spp. grown in culture solutions containing increasing levels of NaCl. Specific activity is expressed as present of control plants grown in 1 mM NaCl. (Data of Greenway and Osmond, 1972)

Species, tissue	NaCl Conc.	Relative specific activity (%)		
		Malate dehydrogenase	Glucose-6-P dehydrogenase	Isocitrate dehydrogenase
A. spongiosa, roots [a]	10	74		
	50	102	142	123
	100	67		
	250	132	44	124
	400	87	57	124
A. spongiosa, leaves [a]	10	100		
	50	97	74	100
	100	89	76	—
	250	84	80	77
	400	114	—	84
A. nummularia, leaves [b]	50	—	100	87
	250	—	142	96

[a]   12 days growth
[b]   90 days growth

in these terms. The enzyme from C_4 pathway sources has a higher K_m (Ting and Osmond, 1973a) which accounts for the higher sensitivity to NaCl and has nothing to do with salinity tolerance.

In many studies (Weimberg, 1970; Greenway and Osmond, 1972; Flowers, 1972a,b) growth of halophytes or nonhalophytes on NaCl makes very little difference to the total activity of enzymes extracted from tissues (Table 7.13). There are, however, several reports that enzyme levels do respond to salinity (Hanson-Porath and Poljakoff-Mayber, 1969; Beer et al., 1975; Huber and Sankhla, 1976). In some cases these responses are observed in tissues undergoing rapid morphological changes. These changes, which are associated with change in enzyme activity, are themselves responsive to NaCl (Osmond, 1976; Poljakoff-Mayber and Greenway, 1974). Even when the change in specific activity of an enzyme can be ascribed to NaCl alone and not to an effect of NaCl on development, the responses in halophytes are similar to those in nonhalophytes (Flowers, 1972b). Hanson-Porath and Poljakoff-Mayber (1969) and Flowers et al. (1976) have established that different isoenzymes of malate dehydrogenases may increase in activity in both halophytes and nonhalophytes. When *Suaeda* spp. were grown at high NaCl concentration isoenzymes of malate dehydrogenase showed greater stimulation with NaCl during assay than the isoenzymes from low salt plants (Flowers et al., 1976).

In view of the above it is not surprising to find very few metabolic sequences which respond in vivo in such a manner as to suggest NaCl concentration in the cytoplasm is responsible (Osmond, 1976). One much quoted response, the effect of

Table 7.14. Effect of NaCl during growth on the rate and products of 15 min dark $^{14}CO_2$ fixation in *Atriplex spongiosa* and *Phaseolus vulgaris*. (H. Greenway and C.B. Osmond, unpublished)

Species and tissue		NaCl Conc. (mM)	Rate of dark $^{14}CO_2$ fixation (%)	Percentage of total ^{14}C in carboxylic acid (%)
A. spongiosa	roots[a]	1	100	51
		400	175	51
	shoots[a]	1	100	49
		400	43	35
P. vulgaris	roots[b]	1	100	86
		200	82	56
	shoots[b]	1	100	79
		200	104	59

[a] Treated 12 days in NaCl cultures
[b] Treated 5 days in NaCl cultures

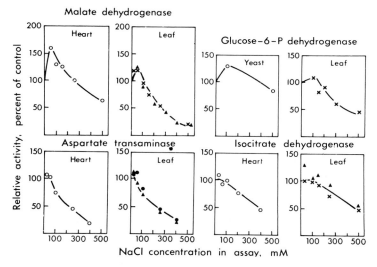

Fig. 7.25. Dependence of enzyme activity on NaCl concentration during assay in enzyme extracts from leaves of halophytes *A. spongiosa* (●), *Salicornia australis* (▲) and the nonhalophyte, *Phaseolus vulgaris* (×)

NaCl on the distribution of label between malate and aspartate during dark $^{14}CO_2$ fixation (Webb and Burley, 1965) can be observed in shoot tissues of both halophytes and nonhalophytes, but not always in roots (Table 7.14). The inhibition of dark $^{14}CO_2$ fixation may reflect the inhibition of PEP carboxylase in vivo and the increase in amino acid label with increasing NaCl concentration may reflect differences in the in vivo sensitivity of malate dehydrogenase and aspartate amino transferase (Fig. 7.25). However, the physiological significance of this response to

NaCl, beyond possible involvement in the regulation of cytoplasmic pH and ion balance discussed previously, is not clear.

In spite of extravagant claims in the literature, there is no evidence to suggest that major metabolic reaction sequences in halophytes differ in their response to NaCl from those of nonhalophytes, or that NaCl has major effects on metabolic pathways. Contentions that NaCl treatment may result in a change from the C_3 to the C_4 pathway of photosynthetic metabolism or vice versa (Huber and Sankhla, 1976) are simply not supported by the evidence. On the other hand, water stress following NaCl treatment of certain succulents with C_3 photosynthesis induces crassulacean acid metabolism in these plants (Winter and Lüttge, 1976). This change in photosynthetic pathway is primarily a response to water stress and not to the absorption of NaCl. Hall and Flowers (1973) established that the ribosomal system capable of incorporating amino acids into proteins in vitro showed similar NaCl responses whether it was isolated from nonhalophytes or halophytes. This observation, together with the fact that Cl^- has little effect on NO_3^- incorporation in vivo (Chap. 7.3.3) is consistent with the notion that halophytes maintain a relatively low Cl^- concentration in the cytoplasm.

Halophytes such as *Atriplex* show a stimulation of growth at external NaCl concentrations which are inhibitory to the growth of nonhalophytes as is evident from the early experiments of Eaton (1927) (Table 7.10) and Ashby and Beadle (1957). We have seen that growth of *Atriplex* in saline cultures is associated with substantial absorption of NaCl to the shoot, and that absorption of NaCl to similar concentrations in the shoots of nonhalophytes proves toxic. These observations, and the absence of evidence for specific metabolic responses to NaCl which distinguish halophytes from nonhalophytes, imply that the regulation of salt distribution in the shoot is the key process conferring salinity tolerance on halophytes. Flowers et al. (1977) concluded that salt concentration in the cytoplasm of leaf cells of halophytes may be less than 1/3 that of the vacuole. It is improbable, however, that plant cell membranes can retain their properties while separating compartments which differ by more than a fraction of a bar in osmotic pressure. The synthesis of osmotic solutes in the cytoplasm to compensate for the salt accumulated to the vacuole is thus a question of fundamental importance to this interpretation of salinity tolerance in halophytes (Flowers, 1975; Osmond, 1976; Flowers et al., 1977).

There is evidence that specific metabolites, the so-called compatible solutes, are involved in the regulation of the osmotic activity of the cytoplasm. These solutes are produced in response to water stress in most eukaryotic organisms and are characterized by the fact that, unlike electrolytes (Fig. 7.25) they have no effect on enzyme activity in vitro at the same osmotic concentrations. Halophytes are distinguished by their capacity to produce much higher concentrations of these solutes and in several *Atriplex* species the quaternary ammonium compound glycinebetaine is present in high concentrations (Storey and Wyn Jones, 1979).

The glycinebetaine concentration in leaves of *A. spongiosa* is directly correlated with the NaCl concentration of the leaf sap, as shown in Fig. 7.26. The slope of this curve suggests that to exert an equal osmotic pressure, the glycinebetaine would have to be retained in a volume corresponding to about 5% of that of NaCl. Storey and Wyn Jones (1979) recognized that this corresponds to approximately the

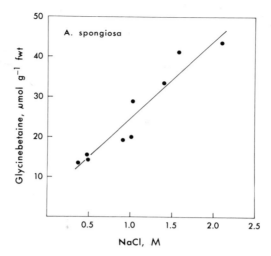

Fig. 7.26. Relationship between glycinebetaine and NaCl concentration in leaves of *A. spongiosa* seedlings grown in differing concentrations of NaCl. (Redrawn from Storey and Wyn Jones, 1979)

cytoplasmic volume of vacuolated cells in higher plants. Osmond (1976) speculated that the proline accumulated in nonhalophytes following water stress could accommodate the observed change in water potential if it was confined to the cytoplasmic volume. Retention of these compatible solutes in the cytoplasm as osmoregulatory substances provides an attractive explanation of the function of these unusual compounds. These and other aspects of osmoregulation are discussed in more detail in the next chapter, but it is clear that production of glycinebetaine is an important metabolic response to salinity in *Atriplex* and other chenopods. Wyn Jones et al. (1977) believe that the retention of adequate K^+ in the cytoplasm, and the synthesis of compatible solutes such as glycinebetaine are important processes in the maintenance of normal metabolic activity of halophytes exposed to salinity.

In summary then, the ion absorption processes in halophytes such as *Atriplex* may function to facilitate ion accumulation in the vacuole of leaf cells while maintaining low ion concentration in the cytoplasm. In this way salts do not significantly interfere with leaf metabolism but function instead to increase the osmotic pressure of the cell vacuole. The controlled accumulation of salt permits osmoregulation within the leaf cells, enabling them to maintain turgor at lower leaf water potential. Halophytes such as *Atriplex* spp. may take advantage of their salinity tolerance to maintain biological activity in leaf tissues situated between the low water potential of the saline substrate and the much lower water potential of the atmosphere. Indeed, the stimulation of growth of *A. halimus* by low salinity is only observed under conditions of atmospheric water stress (Gale et al., 1970) and, as demonstrated by Caldwell (1974), the accumulation of salt in halophytes contributes significantly to their capacity to maintain activity into the most arid period of their life cycle. These consequences of solute accumulation and water relations of halophytes are explored in more detail in the next chapter.

7.5 Ion and Nutrient Absorption Processes in the Context of Ecological Processes

In the preceding sections we have attempted an integration of ion uptake, transport, and excretion as it might relate to the growth of individual plants in soil, and to the role of some of these processes in contributing to the tolerance of plants in environments of ionic stress. Thus far we have concentrated on events relevant to the scales 10^1 to 10^5 in the continuum described in Fig. 1.1. The distribution of ions and nutrients in vegetation depends on these processes and has an impact on larger-scale ecological processes such as nutrient cycling. Because plant matter is the basic resource of the ecosystem, salt and nutrient levels may also have important effects on trophic interactions. In this final section we shall examine some of the links between ion uptake processes and these larger-scale processes characterized by relaxation times greater than 10^6 s.

7.5.1 Mineral Cycling Within the Plant Community

Mineral cycling is the overall process whereby chemical elements circulate between compartments of the ecosystem in response to biotic activities (Charley, 1978). It begins with mineral uptake by plant roots, involves the transport and excretion processes described above and is completed by the return of minerals to the soil following litter fall and mineral release. In this section we shall be primarily concerned with the latter events and it is useful to distinguish the return of nonnutrient ions (Na^+, K^+, Ca^{2+}, Mg^{2+}, Cl^-) from the nutrients (N, P, S). By and large, the former are released by leaching of litter but the nutrients are released only following decomposition of the litter which is usually facilitated by microbiological activity. Because the availability of N and P is commonly a major limitation on plant production, the mineralization of these elements in litter can be a rate-limiting process particularly in low N and P habitats such as the arid shrubland ecosystems.

Mineral cycling can be represented as a sequence of compartments, fluxes, and conductances in the ecosystem, analogous to the ionic relations of a higher plant. This representation is adequate in habitats of uniform environment or of mild and regular seasonal patterns. In other habitats such as the coastal and arid shrubland regions occupied by *Atriplex,* biological activities are primarily determined by precipitation patterns (Noy-Meir, 1973). The major fluxes of the mineral cycle in such habitats, ion uptake by roots and nutrient release from litter, are similarly initiated by rainfall, and their duration is determined by subsequent changes in soil and plant water potential. It is thus arguable whether the concept of a mineral cycle is appropriate and Charley (1978) distinguishes the mineral cycle in arid shrublands from that in other habitats on the basis that it is "pulsed in time" (see also Noy-Meir, 1973). We shall first of all consider the annual budgets for nutrient and salt cycling and then examine the key stochastic components in these processes.

In previous sections we have seen that arid shrubland species are commonly rich in the nutrient ions N and P and halophytes such as *Atriplex* spp. contain high levels of NaCl in shoots and fruits. As a consequence high concentrations of these ions are returned to the soil in leaf and fruit litter and, as we have seen, result in

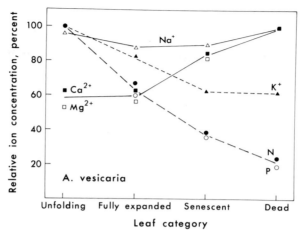

Fig. 7.27. Withdrawal of ions and nutrients from leaves of *A. vesicaria* prior to shedding. Ionic concentrations in leaves are expressed as percentage of the maximum concentration found in all leaf categories. (Redrawn from Charley, 1978)

distinct patterning of soil ionic state in horizontal and vertical dimensions (Chap. 5.8).

The nutrient recycling process in litter is to some extent short-circuited by the withdrawal before shedding of about 80% of the peak N and P content of *A. vesicaria* leaves (Fig. 7.27). The form of this storage and its mobilization for new growth has yet to be investigated, but withdrawal in this way lessens the dependence of subsequent growth on the mineralization processes in litter. Potassium is also withdrawn from older leaves but to a lesser extent. Again, retention of this ion in the plant rather than its recovery from a Na^+ dominated environment by the selective absorption processes described earlier may be favorable for new growth following precipitation. The divalent cations are presumably bound as oxalate salts which continue to accumulate throughout the growth period and hence are immobilized. Sodium concentration remains high throughout the life of leaves of these halophytes because it is the principal cationic component of the osmoregulatory system, central to maintenance of adequate plant water relations. The small increase with leaf age may reflect continued NaCl excretion to the epidermal trichomes.

The magnitude of the nutrient cycle in arid shrublands is indicated in the four case studies summarized in Tables 7.15, 7.16. In two arid region shrub communities in eastern Australia, dominated by *Atriplex vesicaria* and *Eremophila gilesii* (Charley and Cowling, 1968; Burrows, 1972), the quantity of nutrients held in the biomass is small in comparison to the total in the community but the throughput in litter is high, representing more than 1/3 of the aboveground total in *Atriplex* and 1/5 in *Eremophila* (Table 7.15). Burrows (1972) calculated that, with a decomposition time of 20 months and assuming full mineralization of the phosphorus in *Eremophila* litter, the annual return of phosphorus through the litter would account for more than half the total biomass requirement for this nutrient at

Table 7.15. Nutrient distribution and turnover in warm desert shrublands. (Data of Charley and Cowling, 1968; Burrows, 1972)

Fraction	Atriplex vesicaria		Eremophila gilseii	
	Nitrogen $g\,m^{-2}$	Phosphorus $g\,m^{-2}$	Nitrogen $g\,m^{-2}$	Phosphorus $g\,m^{-2}$
Aboveground pool				
Shoot	2.44	0.094	3.84	0.195
Litter	0.06	0.005	1.58	0.087
	2.50	0.099	5.42	0.282
Belowground pool				
Root	0.59	0.032	0.59	0.031
Soil (0–45 cm)	264.90	108.20	240.00	11.5
	265.5	108.23	240.6	111.53
Annual litter production	0.89	0.065	1.00	0.055
Annual turnover (%)[a]	36	65	18	20

[a]   Litter production/aboveground pools × 100

Table 7.16. Nitrogen distribution and turnover in cool desert shrublands. (Data of Bjerregaard, 1971)

Fraction	A. confertifolia $g\,m^{-2}$	C. lanata $g\,m^{-2}$
Aboveground pool		
Shoot	4.71	2.87
Litter	9.78	5.66
	14.49	8.53
Belowground pool		
Roots	22.4	25.9
Soil (0–45 cm)	459.5	403.4
	481.9	469.3
Annual shoot production	0.9	1.23
Annual turnover (%)[a]	8.5	9.4

[a]   Shoot production/aboveground pools × 100

the measured rates of production. Rapid litter breakdown is also implied in the *Atriplex* community, in that the litter present represented only about 12% of the total litter production in one year.

The nitrogen budget of cool desert arid shrublands differs from that of the warm deserts in that much larger litter and belowground pools have been measured (Table 7.16; Bjerregaard, 1971). If turnover is calculated by dividing the N content of new production by aboveground biomass, then lower values are obtained than in the warm desert systems. However, these estimates do not take into account the biomass or turnover of root systems. If we adopt the estimates of Caldwell and

Table 7.17. Distribution of chloride in vegetation, soil and in litter produced in 1 year by *Atriplex vesicaria* growing in virgin and degenerate gilgai communities near Broken Hill, Australia. (Data of Charley, 1959)

Fraction	Plant cover	
	100% (Virgin) $g\,m^{-2}$	47% (Degenerate) $g\,m^{-2}$
Standing leaf pool	5.8	2.8
Soil pool	3,466	113
Annual litter production		
Leaf	7.7	3.7
Fruit	1.9	0.8
	9.6	4.5
Annual chloride turnover (%)[a]	165	161

[a]   Litter/standing leaf × 100

Camp (1974) that belowground root production is three times as high as aboveground production, then the annual nitrogen turnover for these communities is 10.5% for *C. lanata* and 13.3% for *A. confertifolia*. Although these calculations are not strictly comparable with those for the warm desert systems, they do indicate that the pools of nutrients are larger and the turnover is slower in the cool desert communities. In spite of this, the cycling of nutrients is sufficient to allow these communities to be more productive than those of warm deserts. This is presumably a consequence of the larger soil reserve of N in dead roots and a longer period of adequate soil water for efficient mineralization of this organic N pool.

The magnitude of the salt cycle in *A. vesicaria* shrublands is indicated in Table 7.17. The data refer to virgin and degenerate patterned communities on the saline stony desert soils illustrated in Fig. 5.6 and provide estimates of the Cl^- turnover in the vegetation of the less saline depressions. In both the virgin and the degenerate conditions the Cl^- throughput in litter each year is 1.6 times the total Cl^- content of the vegetation. If we assume that Cl^- is retained in litter to the same extent as Na^+ (Fig. 7.27) then Cl^- turnover should correspond to about five times the N and P turnover rates because all but 20% of these nutrients are withdrawn before litter fall. Comparison of Tables 7.15 and 7.17 indicate this is so. On an equivalent basis, the annual turnover of Cl^- in *A. vesicaria* (Table 7.17) is about 20 times the annual turnover of NO_3^- and 400 times the turnover of PO_4^{3-} (Table 7.15).

Nutrient cycling in arid shrubland communities is a stochastic process dependent on the frequency and magnitude of precipitation. Charley (1972, 1978) argued that the key process in nutrient cycling is the reprocessing of the essential nutrients N and P by the soil microflora. Much of the available N and P in arid region soils tends to be cycled through plant litter, so that it is possible that the seasonal patterns of litter fall and the activity of micro-organisms involved in mineralization, rather than the uptake processes themselves, could limit the patterns of nutrient absorption and plant production. Charley (1972) reported that,

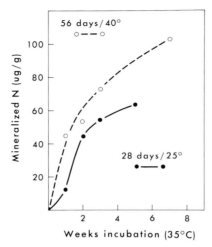

Fig. 7.28. Mineralization of nitrogen in soils from *A. vesicaria* shrublands following pretreatment of dry soils for the temperatures and times shown. (Redrawn from Charley, 1972)

like other soils, those of the *Atriplex* shrublands undergo accelerated nitrification when wetted from an air-dry state. The longer the drying and the higher the temperature (to 40 °C) the more rapid the initial rate of nitrification (Fig. 7.28). In addition, the same small precipitation events which initiate mineralization of organic N in the soil could also activate N_2-fixing systems associated with surface crust lichens and blue-green algae (Beadle and Tchan, 1955; Rogers et al., 1966). Bjerregaard (1971) observed significant increases in the N content of soil surface crusts from *A. confertifolia* communities, but not from *C. lanata* soils, when incubated in the light under controlled conditions. The increase in N content was attributed to the activity of the N_2 fixing blue-green algae *Nostoc* and *Scytonema* found associated with lichens and it was estimated that these organisms might contribute up to $3\,kg\,N\,ha^{-1}\,yr^{-1}$.

These experiments support the notion that the effective precipitation for activation of microbial activity is probably less than the effective precipitation for formation of new roots. Such small inputs of precipitation have been termed "mineralization rains" (Charley, 1972). Precipitation which is ineffective for the production of new roots, for the production of new shoots, or the germination and establishment of seedlings may nonetheless be effective in promoting the mineralization of nutrient reserves in soil and litter. Depending on whether or not more substantial precipitation falls later, these "mineralization rains" could either optimize the soil nutrient status for plant activity, or render nutrient reserves more vulnerable to erosion (Charley, 1972). A hypothetical projection of the relationships between the N pools of shoots, litter, and mineralized N is shown in Fig. 7.29. This figure portrays the way in which leaf fall in periods of low, intermittent precipitation can result in progressive loss of shoot N and increases in litter N levels. Following rainfall which is effective in initiating further microbiological activity and the mineralization of litter N but not new root growth,

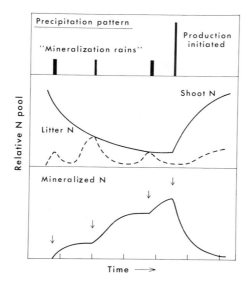

Fig. 7.29. Schematic diagram showing the response of mineralization processes to small precipitation events sufficient to initiate mineralization but inadequate to initiate growth of vegetation. Successive mineralization events deplete the litter N reserve and increase the available N pool which is later used when a rainfall event is large enough for growth to commence. (Redrawn from Charley, 1972)

the levels of litter N fall. Restoration of the pool of shoot N only occurs after a substantial rainfall event which initiates root growth and nutrient uptake which depletes the pool of mineralized N.

Although microbiological processes in litter at the soil surface may play a key role in limiting the flux of N and P back to the soil in the mineral cycle, it is evident that the stability of the plant community in an arid shrubland is very much a product of physiological processes. For established vegetation, the interaction between production of new roots and the release of nutrients from the litter is crucial to continued nutrient uptake and metabolism. For new seedlings the same reasoning applies and competition between established vegetation and the emerging seedling may be particularly important. The coordination of physiological processes at two trophic levels becomes an essential component of the larger-scale process of nutrient cycling.

7.5.2 Mineral Cycling and Stability of Arid Shrublands

Because the nutrient recycling processes are concentrated in the soil surface, the arid shrublands are particularly vulnerable to erosion and to the removal of vegetation. Charley and Cowling (1968) calculated that loss of the surface 10 cm of soil from an *A. vesicaria* shrubland corresponds to the loss of 35% of the available N and 45% of the organic matter in the rooting volume of these shrubs. Apart from its physical effects on soil stability, removal of the vegetation interrupts the nutrient cycle and renders the soil less hospitable to subsequent occupation by other species. Arid shrublands such as those dominated by *A. vesicaria* and *A. confertifolia* have been extensively exploited for the grazing of exotic herbivores. The result of over-exploitation by these herbivores is commonly the removal of the vegetation and the subsequent erosion of surface soil. As the Koonamore experiment has demonstrated, when the resulting barren areas are isolated from herbivores, regeneration processes have relaxation times of about 10^9 s (Figs. 6.18, 6.19).

Similar consequences follow if the salt cycling of halophytes is interrupted by removal of vegetation in some habitats. In the patterned *A. vesicaria* communities on gilgai soils Charley (1959) concluded that the return of salt to the depression surface in litter was more than sufficient to balance the losses due to leaching. The cycling of chloride in the dense communities of *A. vesicaria* thus maintains relatively high surface salinity and stabilizes the halophytic vegetation. In the degenerate condition, the marginal shrubs are unable to cycle sufficient salt to replace that lost by leaching and the depression soil becomes progressively less saline. Finally, the point is reached where species less tolerant to salt (such as the grass *Enneapogon*) are established. These nonhalophytes are less able to withstand arid conditions and are more susceptible to removal by grazing. Under grazing, or after a succession of drought years, the depression may quickly degenerate to a barren "scald". The significance of these small-scale processes of nutrient cycling in determining larger-scale states such as vegetation composition and pattern is also evident in the communities of the Riverine Plain. As discussed previously (Chap. 5.6.2), the redistribution of ions and the changes in soil properties consequent on the salt cycling of *A. nummularia* may, after removal of this species by heavy grazing, determine the pattern of species such as *A. vesicaria* which take its place.

7.5.3 Consequences of Plant Ion and Nutrient Absorption Processes at Higher Trophic Levels

The introduction of exotic herbivores in arid shrublands has not been without problems as far as the physiology of these animals is concerned and some of these problems can themselves be attributed to the plant physiological processes of ion and nutrient absorption. The high N status of *Atriplex* spp. and other arid shrubland species makes them desirable fodder particularly for the less fastidious herbivores such as goats (Franclet and Le Houerou, 1971). Australian studies of merino sheep-feeding preferences indicate that the perennial shrubs *Atriplex vesicaria* and *Maireana aphylla* are not very palatable. Leigh and Mulham (1966) noted that in a "good" season these shrubs comprised only 12% of the diet selected by sheep, although they represented 88% of the fodder available. Under dry summer conditions the proportion of *Atriplex* and *Maireana* in the diet increased to about 35% at a time when vegetation has the highest salt content and highest oxalate content. Increased consumption of this highly saline vegetation increases the water requirement of the animals at a time when available water is also more saline. Toxic effects due to the abnormally high NaCl intake are frequently encountered (MacFarlane, 1971). The high concentrations of soluble oxalates in leaves (Osmond, 1967), which arise as a consequence of NO_3^- uptake and incorporation during growth (Chap. 7.3.2), may be partly responsible for the unpatalability of these chenopod shrubs. Although oxalate poisoning in exotic herbivores is frequently suspected, it has been proven in relatively few cases (Jones, R.J. et al., 1970). Ellern et al. (1974) discount the likelihood that oxalate content of *Atriplex* spp. is a serious poisoning hazard to stock.

In marked contrast to these physiological stresses imposed on exotic herbivores as a consequence of ion absorption processes in *Atriplex,* Kenagy (1972) demonstrated an intriguing coupling between the salt excretion processes of *Atriplex* and the salt intake of an endemic rodent. In a study of dietry allocation in two related species of kangaroo rat (*Diprodomys*) in the Owens Valley of eastern California, Kenagy found that *D. microps* harvest leaves of *A. confertifolia* throughout the year, in contrast to *D. merriami*, which is typically granivorous. Leaves of *A. confertifolia* were selected almost to the complete exclusion of leaves of 11 other freely available species. The rat climbs into shrubs and harvests the saltbush leaves, which it then stuffs into its hair-line cheek pouches; it carries these leaves back to its burrow, where it may cache as much as 250 g of leaf material at any one time. Unlike other species of *Dipodomys* which have rounded and awl-shaped lower incisors appropriate to a primarily granivorous diet, *D. microps* possesses broad, anteriorly flattened and chisel-shaped lower incisors. The rat uses these to shave the highly saline epidermal trichomes from *Atriplex* leaves, thereby gaining access to the less saline photosynthetic parenchyma. Kenagy has demonstrated by chemical analysis of leaves and by feeding experiments that this specialized feeding behavior results in a smaller osmotic load on the animal, since the inner tissue consumed by the rat has a sodium concentration only 3% of that found in the discarded epidermal shavings.

On the basis of this incisor morphology and the associated feeding behavior of *D. microps,* Kenagy argued that phenotypes of *D. microps* with increasingly broadened and more chisel-shaped lower incisors would have been favored in selection by their ability to separate low-electrolyte, high-starch inner tissue from the outer, highly saline leaf material. Furthermore the inner leaf tissue in this C_4 plant contains the "Kranz" complex and the bulk of leaf protein. It might be expected that this selective advantage would be emphasized in those habitats where related granivorous rodents were competing for available seed resources or where the year-round source of available leaves was limited to halophytic shrubs.

7.6 Conclusions

In this review we began by considering the ionic and nutrient state of the soils in which *Atriplex* species are commonly found and then considered the ionic state of these plants on these soils. At all scales the state description provides evidence of a "craving for salt" in halophytes. Recent accounts of the processes of ion uptake, transport and secretion (at the scales of about 10^3 to 10^5 s) permit a preliminary synthesis which accounts for ionic and nutrient state of *Atriplex*. Control of uptake of ions and nutrients in cells for example may be based on a "set point" internal concentration or a corresponding "set point" in cell turgor which provides signals to regulate influx, through unknown mechanisms. We can only speculate that higher salt concentrations in halophytes imply higher "set points" for regulation of

NaCl uptake. Rapid transport of NaCl to the shoot is characteristic of many halophytes and may involve increased apoplastic transport. The presence of salt excretion systems such as the epidermal trichomes of *Atriplex* can be shown to control salt load in leaf tissues. Cell expansion and increased succulence in these plants may function in the same way to some extent.

These processes permit the orderly accumulation of salt in different tissues and cellular compartments. The metabolic machinery of plants with characteristically high salt concentrations in the cells seems to be no less sensitive to interference from NaCl than the machinery of low salt plants. This suggests that for normal function, cytoplasmic salt levels are kept low in halophytes exposed to high salt levels and that a special group of metabolites, the compatible solutes, may provide the required osmotic equivalents in the cytoplasm. The high concentration of salt in the vacuole and of other metabolites in the cytoplasm means that leaves of halophytes are able to maintain positive turgor at low leaf water potential. In turn, normal cell function, water uptake, and water transport may continue in low water potential environments.

The physiological processes of ion and nutrient absorption which result in the accumulation of high salt concentrations in individual plants leave their impression upon larger-scale processes such as the cycling of nutrients and nonnutrients in the vegetation (scales 10^7 to 10^8 s). They perhaps provide selective pressures for the speciation of kangaroo rats (scales greater than 10^{10} s). The principal fluxes in mineral cycling are determined by ion uptake by roots, reallocation in the plant before litter is shed, and the release of nutrient and nonnutrient from the litter. In arid shrublands, as in other systems, a large proportion of the nutrient capital is held in the vegetation but the cycling processes are characterized by a high turnover and by a pulsing in time, dictated by the stochastic properties of rainfall. Although we can integrate many small-scale processes to account for some large-scale processes in individual plants and plant communities at steady state, their responses when pulsed by rainfall patterns are far from clear.

The processes which ensure orderly adsorption of ions to high concentration and their accumulation in the appropriate tissues and cell compartments are of adaptive significance in two respects. Compared with plants without these processes, halophytes are able to maintain normal *performance* when confronted with electrolyte concentrations which seriously interfere with the performance of nonhalophytes. As a consequence of salt accumulation the water relations of halophytes are adjusted to enable them to *persist* at lower soil water potentials. That is, not only are plants such as *Atriplex* tolerant of high salinity, this tolerance forms the basis of improved plant water relations. In fact their "craving for salt" actually seems to slake their thirst!

Chapter 8. Water Movement and Plant Response to Water Stress

"Richard Church in one of his essays writes 'a single mature tree pumps through its veins so many hundreds of gallons of water per day. What then must be the hydraulics of the labour of a square mile of forest land?' ... A Richard Church reading this book may be disappointed to learn that the part played by trees is not much more than the provisions of an easy path through which a restless thirsty atmosphere sucks water from the moist soil."

(Briggs, 1967)

Water plays an essential role in the function and survival of plants. It serves as a reactant in numerous metabolic processes, including photosynthesis, as a medium for stabilization and functioning of biological membranes and enzymes, and for ionization and transport of metabolites. It also serves to inflate and thus maintain structural rigidity of cells and tissues. Growth is therefore necessarily associated with water uptake. In autotrophic terrestrial plants the requirement for water is at least an order of magnitude higher than in heterotrophic organisms, since in order to permit CO_2 to enter the leaf during photosynthesis the leaf must have a high permeability to gases. This inevitably results in an escape of water vapor from the wet cell surface in the leaf to the drier atmosphere, the process of transpiration. No membrane exists that would permit CO_2 to pass freely while at the same time impeding the loss of water vapor. In all higher land plants the leaf epidermis is covered by a cuticle having a very low gas permeability and gas exchange between the leaf and the atmosphere occurs almost exclusively through the stomatal openings in the leaf epidermis. The stomata function as mechanical valves that are able to regulate this gas exchange. Typically, about 99% of the total water taken up by a plant is lost by transpiration.

The actual rate of transpirational water loss depends primarily on the extent to which the stomatal pores are open (stomatal conductance) and on the difference in water vapor pressure between the interior of the leaf and the external atmosphere. The water vapor pressure inside the leaf is strongly dependent on leaf temperature and this in turn is determined by the balance between the heat gained by the leaf, mainly by absorption of solar energy, and the dissipation of heat from the leaf by several heat-transfer processes of which loss of latent heat through transpiration is an important component. This transpirational cooling of the leaf thus exerts a negative feedback effect on transpiration rate.

Table 8.1. Comparisons of relaxation times associated with plant and environmental processes which influence plant water relations. (After Passioura, 1976)

Plant process	Relaxation time (s)	Environmental process
Root, leaf growth	10^5–10^7	Depletion of soil water resources
Leaf shedding	10^5–10^6	
Leaf movement	10^4–10^5	Daily evaporative demand
Changes in hydraulic conductance	10^2–10^4	Rainfall
Changes in stomatal conductance	10^2–10^4	Cloud movement

Transpiration also provides the force that "pulls" water from the soil via the roots, stems, and leaves to the atmosphere. Water thus moves passively from the point of the highest water potential (the soil) to the point of the lowest water potential (the atmosphere). The water relations of plants are most commonly illustrated by a series of compartments connected by pathways of differing conductance. Based on a resistance analogy first suggested by Van den Honert (1948) this state description is an instantaneous view of the important processes which control the movement of water through the atmosphere-plant-soil continuum (Cowan, 1965). The significant processes in this continuum will vary from one plant to another and between environments and experiments. In this continuum, as in that described in Chap. 1, the significance of processes is most effectively evaluated by linking them with environmental or other changes of a similar relaxation time.

Table 8.1, adapted from Passioura (1976), summarizes and matches the relaxation times of plant processes and environmental processes which determine water movement and water status in plants. This table illustrates that short-term responses of water movement to changes in light, temperature, and atmospheric humidity are linked to changes in stomatal conductance and hydraulic conductance in the soil and plant. These processes are discussed in Chap. 8.1 and although they are also important as the basis of long-term changes in plant water relations, they may be insufficient to enable the plant to grow and survive during the depletion of soil water in arid habitats. For example, a sustained low stomatal conductance could well reduce carbon assimilation to the point of starvation (Chap. 9.4.1). Moreover, in hot desert habitats stomatal closure may result in excessively high leaf temperatures, causing severe heat damage and even death (Chap. 9.3.4). As shown in Table 8.1, in the long term a number of other processes are likely to be important in regulating water movement in plants. These include changes in leaf angle or leaf reflectance which modifies the radiant heat load, and changes in the ratios of water-absorbing and water-transpiring surfaces.

The interaction between these plant and environmental processes determines plant water status at any given time. If the rate of transpirational water loss exceeds the uptake of water by the roots, the water content and hence also the water potential of the leaves will gradually decrease. This desiccation has several detrimental consequences. When the leaf water potential falls to the same value as the osmotic potential of the cells the hydrostatic pressure inside the cells becomes zero, i.e., the leaf loses turgidity. Cell growth and cell division cease as does protein synthesis, and many other plant processes, including photosynthesis, are severely inhibited (Chap. 9.4.2). Unless the balance between water loss and water uptake is restored, the plant will soon die. Plants show some capacity for adjusting to desiccation. Most plants are capable of accumulating solutes, which permits plant cells to maintain positive turgor at decreasing water content (decreasing plant water potential). In some special cases, the desiccation-tolerant plants, the capacity of cells to withstand desiccation actually increases as water is lost. The role of these processes in the tolerance of plants to water deficits is discussed in Chap. 8.2.4.

Essentially, the integrated water relations processes in land plants are such as to allow the protoplasts to function at a positive pressure (turgor) of about 1 to 5 bar between the tonoplast and plasmamembrane. If these essential turgor pressure requirements are not satisfied normal cellular function is impaired (Hsiao, 1973). The protoplast functions in the presence of high concentrations of solutes and, provided these solutes are compatible (i.e., do not interfere with metabolic processes in the protoplast), osmotic pressure itself has little influence on cellular function. These pressure requirements must be maintained in the presence of a tremendous flow of water through the system from the soil to the atmosphere. Because the flow components are primarily responsible for all that follows, these will be considered first.

8.1 Water Transport

8.1.1 The States of Water

The properties of water are such that it exists in many different states in the atmosphere-plant-soil continuum. The molecular properties of water make it the most universal of solvents of many solutes. At the same time, water exists in relatively highly structured forms, similar to the crystalline structure of ice, when it is associated with colloids and other charged macromolecules such as enzymes. Changes in the properties of this "bound" water influence the reactivity of such macromolecules. At water surfaces and interfaces, water molecules are orientated so as to result in a high surface tension, a property of water which allows it to fill capillary spaces between soil particles and between cellulose microfibrils in plant cell walls. Surface tension also allows water columns to form in capillary tubes and is an important static component of water distribution in the plant.

The energy required for conversion of water from one physical state to another is another property of water important to plant life. The latent heat of vaporization of water is such that vaporization of water during transpiration from the leaf can be an important mechanism of regulating temperature of the irradiated leaf. Related to this, the high specific heat of liquid water, and the high water content of plants are effective in damping out sudden changes in temperature due to changes in radiation. These and other properties of water are treated in detail in treatises such as by Slatyer (1967), Kramer (1969), and Nobel (1970).

The most suitable parameter for specifying the state of water in any part of the atmosphere-plant-soil continuum is its chemical potential. This potential is an energy term and like other energy terms, it is relative; that is, only differences are meaningful. The reference state chosen is that of the chemical potential (μ_w^o) of pure, free water at the same temperature and atmospheric pressure as that of the system under consideration. Water potential (ψ) in any system may thus be defined as the difference in chemical potential of water ($\mu_w - \mu_w^o$) divided by the partial molar volume of water ($\overline{V}$)

$$\psi = \frac{\mu_w - \mu_w^0}{\overline{V}}. \tag{8.1}$$

Derivation of these relationships is given in the texts cited above and by Dainty (1976). Water potential is a very convenient term because it can be measured unambiguously, by vapor phase equilibration (Barrs, 1968) or by pressure chamber methods (Scholander et al., 1965). In complex systems such as soils or plants, several components of water potential may be identified but the rigorous specification of these is difficult. The convenient measurable property, water potential provides a reference point to relate plant and soil responses to changes in water status. However, it must be emphasized that changes in plant responses to water status are not usually due to changes in the total chemical potential of water, but are due rather to changes in the components of water potential (turgor pressure and osmotic pressure). The direction of water movement in plants can be deduced from gradients in water potential, detailed consideration of which is given in later sections (Chap. 8.1.4).

8.1.2 Water Potentials in the Atmosphere, Plant, and Soil

Water loss by plants occurs because the water potential of the atmosphere (ψ_a) is less than that of the plant (ψ_p). The water potential of the atmosphere is an exceedingly variable quantity. It depends on the chemical activity of water in the gas phase which is determined by the escaping tendency of water molecules from free solution. Air above a water surface is fully saturated with water vapor and the partial pressure due to the water (the saturation vapor pressure) is markedly dependent on temperature (Fig. 8.1). This partial pressure determines the chemical activity (A^w) of water in the gas phase and is readily related to relative humidity as follows

$$A^w = \frac{p_a^w}{p_o^w} = \frac{\text{relative humidity, \%}}{100}, \tag{8.2}$$

where p_a^w is the partial pressure of water vapor in the air under consideration and p_o^w the saturation vapor pressure over pure water at that temperature. The chemical potential of water as water vapor at atmospheric pressure is derived from the relation $\mu_w^0 + RT \ln A^w$, so that water potential of water vapor in the atmosphere may be calculated as follows

$$\psi_a = \frac{RT}{\overline{V}} \ln \left(\frac{\text{relative humidity, \%}}{100} \right). \tag{8.3}$$

The magnitude of the factor $RT/\overline{V}$ is such that small changes in relative humidity result in very large changes in ψ_a. For example, at 20 °C the value of

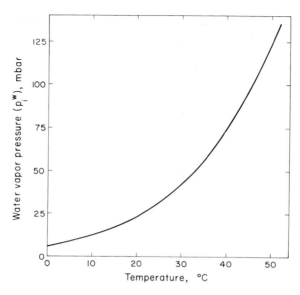

Fig. 8.1. Vapor pressure of air in equilibrium with a free water water surface as a function of temperature

ψ_a corresponding to 99% relative humidity is -14 bar; ψ_a corresponding to 95% relative humidity is -69 bar. At 28 °C, 95% relative humidity corresponds to ψ_a -83 bar, 80% to -320 bar, and 50% to -974 bar.

Plants are thus frequently exposed to water potentials in the atmosphere which range from hundreds to thousands of bars negative. No plants are known which function when plant water potential falls below about -100 bar; most plants only survive such low water potentials as seeds. However, the photosynthetic tissues of one category of plants, the desiccation-tolerant plants, may survive many months at water potentials ranging to $-3,000$ bar (Gaff, 1977; Jones et al., 1979). Most commonly, plant functions are impaired when ψ_p is reduced to below -30 bar. Thus, plant function depends on processes which allow the plant to maintain ψ_p about two orders of magnitude lower than ψ_a and to control the loss of water in the face of such a huge gradient in water potential.

Within the plant, for water to move from roots to shoots, water potential of the shoot must be lower than that of the root. In moist soils this gradient of water potential within the plant is likely to be maintained because the roots are probably in contact with a soil in which soil water potential (ψ_s) is about -1 bar and the shoot is in contact with a much drier atmosphere. The gradient in ψ_p within the plant is readily measured by the pressure bomb technique and Table 8.2 shows such measurements for the roots and shoots of *A. polycarpa*. Although ψ_s was not measured in these experiments it is clear that as the soil water content declines, the water potentials in the root and shoot decline, with the difference between the two remaining more or less constant at -10 to -14 bar.

Table 8.2. Changes in water potential of root and shoot of *A. polycarpa* in response to declining soil moisture. (Data of Sankary and Barbour, 1972)

Soil water (%)	Plant water potential (bar)		
	Root	Shoot	Shoot-root
17.5	− 2	− 16	− 14
5.9	− 17	− 31	− 13
5.4	− 31	− 41	− 10

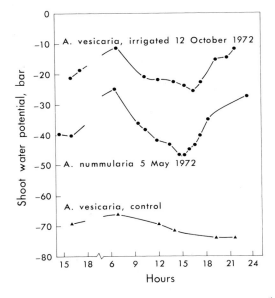

Fig. 8.2. Daily course of shoot water potential of *A. vesicaria* (irrigated and control plants) in the Riverine Plain and *A. nummularia* in the Barrier Ranges of South Eastern Australia. (C. B. Osmond, unpublished)

Plant water potential commonly varies throughout the day, because the evaporative demand changes throughout the day and is greatest about noon (Chap. 8.1.3), and plants commonly show most negative values of ψ_p in the early afternoon and recovery of ψ_p to higher values in the evening. Figure 8.2 shows the daily course of change in ψ_p for two species of *Atriplex* from Australian arid shrublands. Soil water potentials were not measured in these experiments, nor were sufficient atmospheric data recorded to construct a complete picture of water potential changes. However, it is evident that even in the well-irrigated *A. vesicaria*, there is a decrease of about 15 bar in ψ_p in the afternoon, similar to that recorded for *A. nummularia* under conditions at another site. The nonirrigated *A. vesicaria* plants were highly stressed but the extent of the afternoon depression in ψ_p could not be

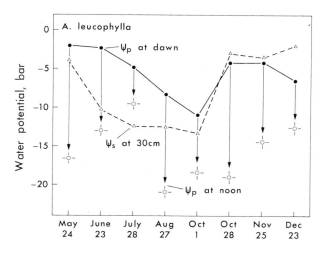

Fig. 8.3. Seasonal changes in water potential of *A.leucophylla* at dawn and at noon, and seasonal changes in soil water potential of sand dunes at Bodega Head, Northern California. (Redrawn from De Jong, 1977)

established because the pressure bomb could not measure beyond -72 bar. It was evident, however, that the highly stressed plants were still losing water and presumably still functioning to some extent. Similar results have been obtained with *A.confertifolia* in the Great Basin desert (Moore et al., 1972b).

The afternoon recovery of ψ_p in the plants shown in Fig. 8.2 implies either that water uptake increased in response to the decrease in ψ_p or that water loss decreased with increasing ψ_p. Indeed, both factors may have been important. Continued water uptake from the soil requires that ψ_p must be lower than ψ_s. If the recovery of ψ_p during the evening is sufficient to allow it to equilibrate with ψ_s, then water uptake from the soil the next day depends on the lowering of ψ_p as transpiration proceeds. Figure 8.3, from De Jong (1977), shows that *A.leucophylla* growing on sand dunes at Bodega Head usually attained dawn water potentials much higher than that prevailing in the surface sand (30 cm); that is, these plants must have access to water at high potential lower in the profile. However, at noon, ψ_p became as much as 10–15 bar more negative than ψ_s at 30 cm so that roots in the surface soil were presumably able to draw water from this layer as well. Only in winter was the surface sand at a higher water potential than the plant. Figure 8.3 shows that, throughout the year at Bodega Head, *A.leucophylla* is exposed to only mild water stress, within the range commonly encountered in crop plants.

Arid shrubland species of *Atriplex* are characterized by much lower values for ψ_p at some times of the year. Figure 8.4 shows the seasonal variation in the dawn value of ψ_p for *A. vesicaria* on the Riverine plain; the minimum values during summer months are in the vicinity of -60 bar to -80 bar. Similar values of ψ_p have been recorded in the Great Basin desert (Moore et al., 1972a) but *A.hymenelytra* growing in Death Valley does not show such marked seasonal variation in ψ_p (Mooney et al., 1977b) (Fig. 8.4).

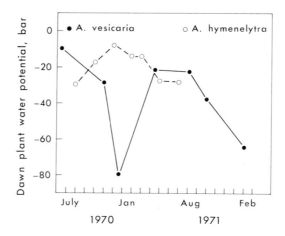

Fig. 8.4. Seasonal changes in the dawn water potential of *A. hymenelytra* in Death Valley, California and of *A. vesicaria* in the Riverine Plain, South Eastern Australia. (Redrawn from Mooney et al., 1977b; Williams, 1972)

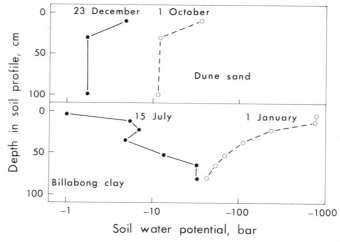

Fig. 8.5. Soil water potential with depth at different seasons for a dune sand at Bodega Head, California and clay soil at Deniliquin, Australia. Note that the abscissa has a log scale (Redrawn from De Jong, 1977; Sharma, 1976)

These differences between the values of ψ_p for *Atriplex* spp. growing in different habitats reflect differences in the water potential of associated soils. Figure 8.5 shows the variation in ψ_s with depth in the Billabong clay of the Riverine plain and in the Bodega dune sand at wet and dry seasons of the year. In the dune sand ψ_s remains at about -1 to -3 bar throughout the profile after rain. Although the surface layers of the Billabong clay show similarly high soil water potentials, ψ_s lower in the soil profile does not rise above about -30 bar. At drier times of the year the surface of the dune sand dries to between -30 and -50 bar and at 100 cm, ψ_s

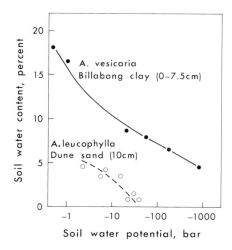

Fig. 8.6. Relationship between soil water content and soil water potential in a clay soil with
A. vesicaria and a dune sand with *A. leucophylla*. (Redrawn from Sharma, 1973b; De Jong,
1977)

increases to about -10 bar (De Jong, 1977). On the clay soil, the surface layers dry
to very low values of ψ_s, approaching -800 bar, but deep in the profile there is little
change in ψ_s. The surfaces of these clay soils thus approach water potentials similar
to those of the atmosphere, i.e., they are air-dry.

Water potential in the soil (ψ_s) has two major components, the matric
potential (ψ_m) and the osmotic pressure (π), so that $\psi_s = \psi_m - \pi$. In deep, water-
saturated soils, a pressure potential due to the height of the water column may be
added, but for relatively shallow-rooted species such as *Atriplex* and in most soils,
this component may be neglected. The relative contribution of the two major
components varies greatly with soil type. In clay soils the major portion of ψ_s is
due to matric forces which are a result of adsorption of water to colloidal materials
and to surface tension effects of water in the small pores between soil particles.
These forces are small in coarse soils such as sands, in which the principal
component is ψ_π, due to solutes dissolved in the soil solution.

Figure 8.6 shows the relationship between ψ_s and water content in the surface
layers of the clay soil and dune sands discussed above. The heavy clay soils of the
Riverine plain in Australia, which support extensive stands of *A. vesicaria,* rarely
dry below about 5% water content but because the clay particles retain water most
effectively, very low values of ψ_s (near $-1,000$ bar) may result (Sharma, 1978). The
coastal sand dunes at Bodega in California, which support *A. leucophylla* on the
drift line and foredune rarely contain more than 5% soil water in the surface. The
water potential at this water content is about -1 to -3 bar, between two and three
orders of magnitude higher than that of the Billabong clay at the same water
content. When the Bodega dune sand surface dries to about 1% soil water, ψ_s is
about -35 bar. This is almost entirely attributable to dissolved salts (De Jong,
1977).

8.1.3 Pathways for Water Movement

Having examined the chemical potential of water in different compartments of the atmosphere-plant-soil continuum, we must now consider the other essential component in the state description, the pathways which may be involved in water movement. These pathways can be described in a number of ways but it is convenient, first to characterize them briefly in structural terms, and then to examine their properties, the most important of which is their conductance.

The water transport pathway is composed primarily of connected water columns throughout the soil and plant but vapor phase transport takes place at the atmosphere-leaf interface and sometimes at the plant-to-soil interface. The phase transition at the atmosphere-leaf interface is associated with a gradient in water potential of about 10^3 bar whereas the gradient in water potential between shoot and roots is only about 10 bar (Table 8.2). Van den Honert (1948) concluded therefore that resistance to water transport in the vapor phase is greater (conductance is lower) than in the liquid phase. As pointed out by Cowan and Milthorpe (1968) this conclusion is correct but this reasoning is misleading. Although it is convenient to use the chemical potential of water in the atmosphere for comparative purposes, water vapor flux is a function of differences in vapor pressure, not of water potential. They pointed out that the vapor pressure deficit between atmosphere and leaf is commonly 6–30 mbar (and considerably higher of course, in extreme habitats such as Death Valley), whereas the vapor pressure deficit at air-water interfaces in the plant is unlikely to be more than 1 mbar. That is, the conductance of the liquid part of the water pathway in the plant must be much greater than the conductance in the vapor pathway. It follows that variation of the latter is most likely to exert the most important control on water transport. The two major components of the water transport pathway at the interface between atmosphere and leaf are the boundary layer and the stomatal conductances.

The Boundary Layer. The layer is the unstirred layer immediately above the evaporating surface which causes a build-up of humid air about the leaf and thus tends to reduce the rate of transpiration. The magnitude of this external resistance is a function of wind speed and the geometry of the evaporating surface. From Cowan and Milthorpe (1968) we can deduce that the average relationship between boundary layer conductance (C_b), the surface dimensions (b), and windspeed (μ) is

$$C_b = 0.294 \ (\mu/b)^{\frac{1}{2}} \quad . \tag{8.4}$$

Figure 8.7 shows the calculated magnitude of boundary layer conductance as a function of windspeed for a range of leaf widths corresponding to the range found in *Atriplex* spp. For the larger leaves (4–8 cm wide), such as those of *A. hortensis* it is likely that boundary layer conductance could be lower than stomatal conductance at very low windspeeds but for the small leaves (0.5 cm wide) such as those of *A. confertifolia* and most arid region species, boundary layer conductance is likely to exceed other conductances in the system.

The boundary layer conductance may be modified by leaf geometry and by leaf surface properties. In comparisons of boundary layer properties of planes and cylinders, Nobel (1974) concluded at similar windspeed, the effective boundary

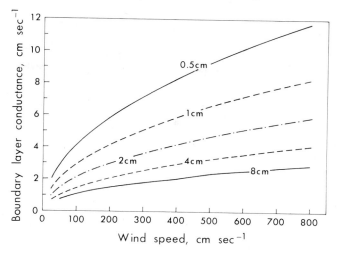

Fig. 8.7. Calculated boundary layer conductance as a function of windspeed for hypothetical leaves of different widths

layer thickness of a cylinder was only 45% greater than that of a plane surface the same width as the cylinder diameter. Leaf hairs are often thought to alter boundary layer thickness and thus reduce C_b. Ehleringer (1977) calculated the effects of pubescence on the boundary layer properties of *Encelia farinosa*. He concluded that when stomatal conductance was 1 cm s^{-1}, and windspeed was 100 cm s^{-1}, a mat of leaf hairs 0.35 mm thick would decrease boundary layer conductance only slightly and make less than 10% difference to the rate of transpiration under the equivalent of field conditions. Epidermal trichomes on the surface of *Atriplex* leaves form a dense shielding mat, particularly when they collapse in older leaves. The surface is complex, and the empirical relationships of transpiration and windspeed have yet to be evaluated with these leaves. Leaf hairs probably have greater effects on energy balance and transpiration by virtue of their reflectance properties than by their effect on the boundary layer (Chap. 8.2.2).

The Stomatal Path. The stomata constitute the second and principal component of the leaf-to-atmosphere water pathway. The stomata are effectively variable valves which control the aperture of the pores through which outward diffusion of water vapor and inward diffusion of CO_2 may take place. Figure 8.8 shows stomata of *A. littoralis* and *A. latifolia* leaves. Movements of the two stomatal guard cells regulate the size of the stomatal pore and hence the dimensions of the diffusion pathway. The processes of inflation and deflation of the stomatal guard cells are effected by changes in the turgor relations between these cells and the epidermal cells. When the solute content of the guard cells increases, water moves into them from the surrounding tissue. As a result of this inflation the guard cells become longer and bend, thereby pushing the two guard cells apart and thus forming a pore. Conversely, when the solute content of the guard cells decreases, the stomata close. The process of solute movement and metabolism involved in movements of stomatal guard cells are outside the scope of this book and have been

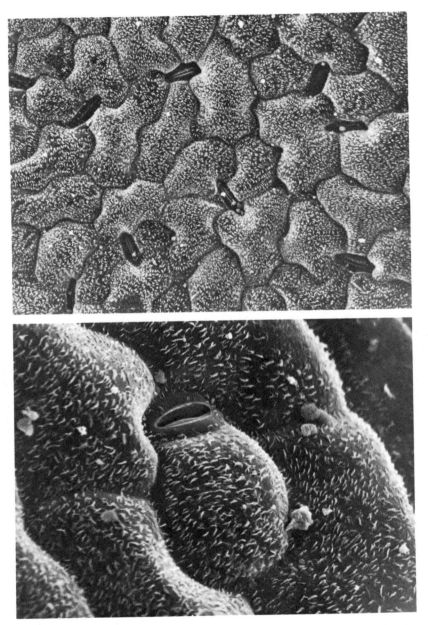

Fig. 8.8. Scanning electron micrographs of the epidermis showing stomata of *A. littoralis* (*top*, ×350) and *A. latifolia* (*bottom*, ×1,000). Plants were grown in soil in the glasshouse from seed collected in the vicinity of Helsinki, Finland. (Micrographs courtesy of R. Whitty)

recently reviewed by Raschke (1975) and Cowan (1977). Stomatal responses to environmental factors and their role in regulating the efficiency of water use are discussed in Chap. 8.2.1.

In most plants stomatal function controls the conductance of the leaf surface from between about 2 cm s^{-1} (open stomata) and about 0.02 cm s^{-1} (closed stomata). The low conductance when stomata are closed is due to the much slower movement of water through the cuticle, the relatively impermeable layer covering the remainder of the leaf surface. Cuticular conductance is usually two orders of magnitude lower than that of the stomatal pathway and can be ignored for most purposes. The properties of the cuticular lipid membrane have been reviewed recently by Schönherr (1976). In some plants, such as the succulents, the cuticle is augmented by a complex multicellular layer of water-impermeable tissue, and cuticular conductances of 0.002 cm s^{-1} have been estimated for these plants.

The Evaporation Site. The actual site of evaporation of water vapor in the leaf is a subject of much uncertainty (Cowan, 1977). Variations of stomatal geometry and the morphology of the sub-stomatal cavity in different species are substantial. It is likely that most of the evaporation takes place from the mesophyll surfaces nearest the stomatal pore and that some stomatal responses, such as that to humidity, may be influenced by the sites of evaporation in the leaf.

At the site of evaporation in the leaf, water is contained in pores of the cell wall microfibril network. These very small pores are commonly about 10–20 nm in diameter and calculations show that negative hydrostatic pressures, or tensions of about -70 to -300 bar, would be required to empty such pores. However, this water is at a higher potential than that of the atmosphere under most circumstances and so water moves from the pores to the atmosphere. Because the "bulk" water potential of the plant tissue is rather higher than that of water in these pores, water evaporated from the pores will be rapidly replaced from other regions of the tissue.

The water in the pores of mesophyll cell walls may contain solutes delivered from the root to the shoot in the transpiration stream. The concentration of these solutes will depend on the rates of solute uptake by leaf cells and on the rate of transpiration. However, it is unlikely that solute could accumulate at the site of evaporation at concentrations sufficient to interfere with the rate of water vapor exchange. The vapor pressure of a 1 M solution is only about 2% lower than that of pure water. Even in halophytes, the presence of extracellular solutes at concentrations of 1 M is likely to have more drastic effects on cellular water distribution and function than on the water vapor exchange at evaporation sites. Ashby and Beadle (1957) observed reduced water use by *Atriplex* spp. grown in saline water cultures and attributed this to a lowering of the vapor pressure of water at the site of evaporation due to the presence of salt. It is unlikely, however, that the principal evaporation sites of *Atriplex* leaves are the wet salt deposits on surfaces of dead vesicles, as suggested by these authors, and it is more probable that salinity modifies the properties of other segments of the liquid water transport pathway.

The Xylem Path. Liquid water in the pores of cell walls at the site of evaporation in the leaf is presumably in continuous contact with liquid water in the walls of other leaf cells and the water columns of the conducting elements. The liquid water columns in xylem vessels are maintained by cohesive forces due to intramolecular attraction in water molecules. The walls of xylem vessels are wettable because water is strongly held in the small pores between cellulose microfibrils, requiring pressures of several hundred bars to remove it. These strong adhesive forces of water in the cell wall are transmitted by the cohesive properties of water to the water columns of xylem vessels approximately 20 μm in diameter and impart great tensile strength to these columns. The rigidity of the xylem itself and the small size of the pores through which water enters the xylem prevent this system from collapsing. In herbaceous and short woody plants water is conducted in these continuous water columns. The situation is less clear in tall trees, in conifers with tracheids rather than vessels, and in plants such as *Atriplex* which possess complex secondary xylem structures. Cowan (1972) explored the relationships between liquid and vapor flow in plants and concluded that liquid phase conductance, expressed in the units used for vapor phase conductance, could be typically as high as 25 cm s^{-1}. This part of the water transport pathway is unlikely to control overall water movement (see however, Kaufmann, 1976b).

The Root Path. Three pathways of water movement across the root from the soil to the xylem vessels are possible. These pathways can be identified in the diagramatic representation of the root given in the previous chapter (Fig. 7.3) and correspond to water movement from one cell vacuole to another, water movement through the symplast (the connected cytoplasm), and water movement through the apoplast of the cell walls. Although the apoplast pathway may be the major pathway of water movement in leaves, the presence of the suberized and lignified walls of the endodermis in roots probably reduce water movement as effectively as does the leaf cuticle. Newman (1974, 1976) concluded that the water permeability of roots was at least an order of magnitude greater than could be accommodated by the apoplast pathway. Similarly, he concluded that movement from one cell vacuole to another requires membrane permeabilities to water at least an order of magnitude greater than those observed in higher plants. He concluded that the symplast pathway, or a combination of the cell wall pathway to the endodermis and then through the symplast, must be the major path for water movement. Calculations as to the possible water conductances of the plasmodesmata in the root endodermis (Robards and Clarkson, 1976) support this conclusion. Nevertheless, Newman (1974) concluded that "the pathway of water movement across the root and the site of the main resistance to flow... seem to me the most fundamental of all unanswered questions about the water relations of the root".

The Soil-Root Path. It could be argued, however, that the pathway of water movement from the soil to the root is more obscure. This pathway depends very much on the structure of the soil-to-root interface and the problem can be posed at several levels of complexity. The pathway can only exist if roots penetrate parts of the soil profile in which ψ_s is higher than ψ_p. The fact that plants such as *A. confertifolia* display transpiration even when ψ_p is as low as -80 bar (Moore et al., 1972a) implies that they draw upon water reserves deep in the profile. These processes are described below.

Similar observations have been made with *A. vesicaria,* in which the roots are largely confined to the surface 20 cm of soil (Jones and Hodgkinson, 1970; Sharma, 1978) which dries rapidly to less than -100 bar (Fig. 8.5); yet plants at -70 bar are capable of some recovery of water status at night (Fig. 8.2). Clearly, the penetration by a few roots into deep soil, containing water at higher potential is essential for continued water uptake. These roots are not easily recognized during excavation but must remain functional.

On a smaller scale, it is commonly held that the most active zone of water uptake corresponds to the region of root hair development (Slatyer, 1967) and it is argued that these hairs increase the surface area available for water uptake. Yet Newman (1974) calculated that, even if water flows through the high conductance symplast pathway in root hairs and cortex, water transport in the adjacent soil is likely to be 10^5 times faster than within root cells. In view of the above uncertainties and the possible role of secreted mucilaginous materials in improving the water pathway from soil to root epidermis, the role of root hairs in water uptake remains questionable. Cowan and Milthorpe (1968) drew attention to the relative geometry of plant roots and the pores between soil and particles. Even though root hairs are larger than the particles in clay soils, let alone the pores between them, the contact between the root surfaces and structurally normal water are limited, even at high values of ψ_s. It follows that some of the water absorbed by roots may involve transfer through a vapor gap. Cowan and Milthorpe calculated that this gap might be as great as that of the root radius itself without impairing water uptake. Quite clearly, the uncertainties of pathways and processes of water transport at the root-soil interface are equally as great as those of ionic relations at this interface (Chap. 7.1.2). They are discussed in more detail in recent monographs (Scott-Russell, 1977; Nye and Tinker, 1977).

8.1.4 Dynamics of Water Movement Through the Plant

At the outset of this chapter we quoted the view of Briggs (1967) that water transport from soil through plants to the atmosphere follows a path of least resistance, largely independent of the metabolically dependent processes associated with living cells. That this is so is implicit in Van den Honert's (1948) catenary hypothesis that water flow (F) depends on the conductance of the connecting pathway. The soil, plant, and atmosphere are thus conceived as a sequence of conductances in a series such that at a steady rate of transpiration

$$F = (\psi_s - \psi_r)\,C = (\psi_r - \psi_p)\,C_p = (\psi_p - \psi_a)\,(C_s + C_b), \qquad (8.5)$$

where ψ_s, ψ_r, ψ_p, and ψ_a are the water potentials in the soil, the root, the evaporating surfaces of the leaf, and the atmosphere. C, C_p, C_s, and C_b refer to the conductance of the soil, the xylem, the stomata, and the boundary layer, respectively. As stated before (Chap. 8.1.3) the expression $(\psi_p - \psi_a)$ is an inexact representation of the driving force for water vapor movement. Equation 8.5 also assumes that no active water pumps are involved in the catena, an assumption that is generally accepted. It should be pointed out, however, that the conductance in the root, involving water transport across plasmamembranes and through plasmodesmata, is sensitive to metabolism and to temperature. By the same token, the control of transpiration by C_s is also very responsive to metabolic processes in the leaf, as well as to the chemical potential of water in the leaf.

From what has been said above, it is clear that water flow will be primarily regulated by the segment of the atmosphere-plant-soil continuum across which the gradient of water potential is greatest or in which the most control of conductance can be exercised. These two factors are found together at the leaf atmosphere interface, where at the site of water loss the vapor pressure gradient is equivalent to a water potential of -10^3 bar and where C_s may vary by about two orders of magnitude. Only when ψ_s drops below ψ_r will water flow be controlled by processes of water uptake in roots.

The position of the living plant in this catena, needing to maintain its water content within relatively narrow limits for continued function, is most precarious. Water content may be regulated either by rate of water uptake or by rate of water loss. The former is largely outside the control of the individual plant, being primarily dependent on climatically determined water supply in the soil and the water extraction strategies of neighboring plants. The latter is very much within the control of the individual plant but is inextricably associated with CO_2 supply to the leaf; the dilemma of stomatal control is explored later (Chaps. 8.2.1; 9.4.2). Here it is useful to consider briefly the dynamic properties of water flow through the whole plant in response to these regulating factors and its effects on plant water status.

If the rate of water flow is altered in response to environmental or plant factors, then the effect of this on the water potential of different components in the transport pathway will depend on the water capacitance of that component and on the conductance of the pathways connecting it to other parts of the pathway. Because of the large capacitance of the shoot relative to the other components, and its proximity to the atmospheric sink for water vapor, the water potential of the shoot will decline more rapidly than that of the other components connected to it by high conductance pathways. The decline in water potential of the root will lag behind the decline in the shoot. Conversely, when transpiration slows the small capacitance of the root and the high conductance between root and soil mean that the root will equilibrate first.

The midday depression of shoot water potential shown in Fig. 8.2 presumably arises because transpiration reaches a maximum near noon and ψ_p becomes more negative [Eq. (8.5)]. The water potential recovers when transpiration declines in the afternoon. Figure 8.9 illustrates this relationship between water flow (transpiration) and plant water potential in apricot leaves throughout a day. As transpiration rate increases to a peak at noon, ψ_p declines and does not recover until transpiration slows. That this change in water flow is linked to changes in stomatal conductance is suggested by the correlations shown in Fig. 8.10. In spring, *A. hymenelytra* in Death Valley shows a daily change of about -15 bar in ψ_p, from a nightly maximum of about -7 bar and this is associated with a change in C_s from 0.05 to 0.2 cm s^{-1} (Pearcy et al., 1974). In mid-summer under stress conditions the plant regulates the diurnal change in ψ_p to about the same value (-10 bar) by substantial reductions in C_s (0.08 to 0.01 cm s^{-1}). In both cases the decrease of ψ_p in the morning is associated with an increase in C_s and the afternoon recovery is associated with a decrease in C_s. Similar responses were observed in a comparison of *A. lentiformis* in a coastal habitat and in a desert habitat (Pearcy and Harrison, 1974).

Changes in stomatal conductance enable the plant to regulate the daily change in ψ_p within a tolerable limit of about -10 to -15 bar (cf. Figs. 8.2, 8.10). As the

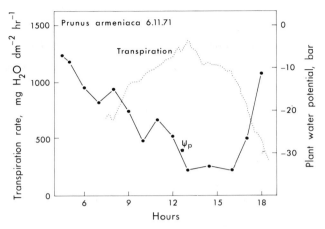

Fig. 8.9. Daily course of plant water potential in an apricot tree showing that it is minimal at the time of most rapid transpiration. (Redrawn from Schulze et al., 1972)

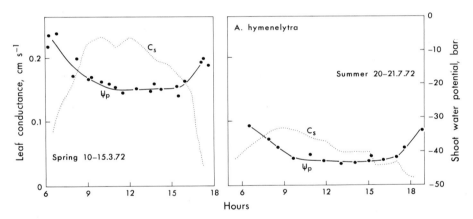

Fig. 8.10. Daily course of plant water potential and stomatal conductance in *A. hymenelytra* in winter and summer, showing reduced stomatal conductance and lower plant water potential in summer. (Redrawn from Pearcy et al., 1974)

stress imposed by decreased soil water potential and the dry atmosphere increases throughout the season, plants adjust to a lower overall water potential. This adjustment may result from changes in several of the components of plant water potential. These components, the way in which they change, and the relationship of each to metabolic and developmental processes, are considered later (Chaps. 8.1.6; 8.2.4).

8.1.5 Water Movement in Relation to Energy Balance

In the previous section we discussed the state of water in different parts of the atmosphere-plant-soil continuum and identified the "restless thirsty atmosphere" as the principal driving force for water uptake by plants. That the atmosphere acts in this way is primarily a result of fluctuations in the solar energy input. These diurnal and seasonal changes are responsible for large changes in leaf temperature and hence for changes in the partial pressure of water vapor at the evaporation sites in the leaves. Comprehensive accounts of energy and water balance at the community and individual leaf level have been provided by Slatyer (1967) and in recent monographs such as Rosenberg (1974) and Monteith (1975). The subject is considered here because *Atriplex* species most commonly grow in open, high radiation environments such as the seashore or arid shrublands, and because plant responses to the radiation environment, as well as the role of plants in dissipation of the radiation input are of considerable interest.

Of the shortwave solar energy which reaches the earth's surface, a portion is reflected and the remainder is absorbed by plants, soil, or other materials and converted into thermal energy. The reflection coefficient, termed the surface albedo, varies with angle of incidence and with the physical properties of surfaces, being highest in wet snow (about 0.8–0.9) and high on light-colored sandy soils (about 0.5). The difference between incoming and outgoing radiation, net radiation (R_n), is given by

$$R_n = R (1-r) - R_e, \qquad (8.6)$$

where R = total incident radiation, r the surface albedo, and R_e is the black body re-radiation. Net radiation falling on a plant community can be dissipated via three major components as follows

$$R_n = S + H + L, \qquad (8.7)$$

where S is the soil heat flux, H the sensible heat, and L the latent heat exchange with the atmosphere. The energy exchanges due to photosynthesis and other metabolic processes are usually smaller than the errors involved in measurement of the major components.

The majority of studies of energy balance and its impact on plant water relations at this large scale have been conducted with dense uniform stands of crop plants or regular row crops. Regularity or homogeneity is a requirement for accurate measurement in most cases and the analysis of heat and water vapor fluxes in irregular communities such as those of the arid shrubland poses substantial problems. These problems are of two types; roughness of the vegetation surface, which introduces difficulties in calculation of momentum fluxes, and advection of soil heat (S) from adjacent bare areas. In spite of these difficulties, analyses of heat and water vapor fluxes in an arid shrubland community in western New South Wales (Hasick, 1979) provide a good picture of the interrelation between energy balance, water flow and plant control of water flow.

Hasick examined the energy balance of an *A. vesicaria* shrubland at two seasons of the year; in spring when the space between individual shrubs was carpeted with a dense mat of annual *Atriplex* and other species, and in late summer when this space was unvegetated. Wind speed profiles were constructed to accommodate the roughness properties of the plant surface and soil heat flux was measured. Hasick found that in spring R_n was dissipated about equally between L and H throughout the day. As shown in Fig. 8.11, the decrease in stomatal conductance (C_s) at about 10.00 h resulted in a substantial decrease in the dissipation of latent heat. This figure

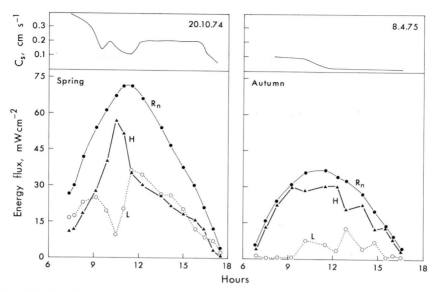

Fig. 8.11. Partitioning of net radiation (R_n) sensible heat (H) and latent heat (L) throughout the day above a shrub of *A. vesicaria* on the Riverine Plain in spring and late summer, and its relationship to stomatal conductance. (Redrawn from Hasick, 1979)

also shows that in summer, when C_s was initially much lower than in spring and decreased further throughout the day, dissipation of energy as latent heat had only a small influence on the total energy balance.

Although such studies show the way in which water movement through the plant and its control by the plant may have an impact on the energy balance of plant communities, the more important biological implications of energy balance are to be found at the level of individual leaves (Raschke, 1960; Gates, 1968). In an analogous way, leaf energy balance is determined by heat exchange due to radiation, convection, and evaporation so that total radiation absorbed by a leaf approximates

$$R_n = R\,(1-r) - R_e = H + L, \tag{8.8}$$

where R_e, the radiation emitted by the leaf, is a function of leaf temperature; H, the loss of heat by convection is a function of leaf dimensions, windspeed and leaf to air temperature difference; and L, the loss of heat by evaporation is a product of transpiration rate and latent heat of vaporization of water. Heat flux from the leaf via the petiole [analogous to S in Eq. (8.7)] is negligible. It follows from this equation that changes in leaf properties which alter the proportion of incident radiation absorbed by the leaf will influence leaf temperature and transpiration rate. Thus an increase in leaf pubescence which increases reflectance or a change in the leaf angle that decreases the absorption of radiation are effective means of lowering the heat load of leaves and may be expected to reduce both leaf temperature and transpiration rate. Adaptive aspects of these factors are discussed elsewhere (Chaps. 3.2; 8.2.2; 9.3).

8.1.6 Components of Water Potential

In the previous sections we were primarily concerned with water flow through the plant, and the plant and environmental factors controlling water flow. At different points of the water pathway in the plant, water in living vacuolated cells is in equilibrium with the potential of the flowing water which, as we have seen, decreases by about -10 to -15 bar from root to shoot. The water potential of the shoot may fluctuate by about the same amount in the course of a day and may adjust to a minimum dawn value of above -10 bar when water is readily available, or to about -80 bar when the plant is highly stressed.

It is common practice to divide plant water potential (ψ_p) into components of hydrostatic pressure (T), osmotic pressure (π), and matric potential (ψ_m) such that

$$\psi_p = T - \pi + \psi_m \qquad (8.9)$$

We have seen that in seeds and soils the pressure term is negligible (Chap. 6) and that, depending on soil structure and water content, either osmotic or matric components may predominate. Whether it is useful, or practicable, to separate osmotic and matric components of plant water potential (Briggs, 1967) is debatable (Dainty, 1976). Where, for example, in the absorption double layer associated with colloidal particles do cations cease to be a component of ψ_m and commence to contribute to π, or at what molecular weight should molecules in the cytoplasm be regarded as a solute rather than as components of the protoplasmic matrix? Such uncertainties involve scales smaller than those relevant to this discussion. For practical purposes, plant water potential may be written as

$$\psi_p = T - \pi \qquad (8.10)$$

where T is usually referred to as turgor pressure and is a positive hydrostatic pressure maintained by water movement into the osmotic compartment, the cell vacuole. Given the presently available techniques, this simplification is imperative. The biologically important component of water potential is T and this for the most part can only be measured by the sum $\psi_p + \pi$.

The conclusion that changes in T rather than changes in ψ_p and π are important is based on the following logic (Hsiao, 1973; Hsiao et al., 1976). Many plant processes change in response to a decrease of 5 to 10 bar of equilibrium ψ_p. This decrease in water potential corresponds to a reduction in the chemical activity of water by only 0.4% to 0.7%. It is difficult to ascribe large-scale biological changes to such a small overall change in the chemical activity of water. In most plants a 5 to 10 bar change in ψ_p is accompanied by about 5% change in total water content and this loss would bring about a corresponding increase in the concentration of osmotic materials (π). However, metabolic processes are not generally very responsive to such small changes in electrolytes or substrates, and thus it is difficult to ascribe the large biological changes to changes in π. When biological processes respond to a decrease in ψ_p with little change in π, it is concluded that these responses are due to the large change in T. This conclusion, that changes in turgor are principally responsible for the deleterious effects associated with decreased tissue water content, has the corollary that turgor maintenance is an important process.

Cell Wall Elasticity. The relationship between ψ_p, T, and π changes with cell volume as described by the Höfler diagram (e.g., Meidner and Sheriff, 1976). Thus in a fully turgid cell (maximum volume) $\psi = 0$, $T = \pi$, and at some smaller volume $T = 0$, so that $\psi = \pi$. Plant cells are not ideal osmometers so that for a small change in water potential, the change in volume (V) is given by

$$d\psi_p = (\varepsilon + \pi)\, dV/V \qquad\qquad (8.11)$$

where ε, the elastic modulus (bar) is a measure of cell wall rigidity (Dainty, 1976). Elastic cells have numerically small values of ε and cells with rigid cell walls have large values of ε. If tissues with elastic walls are transferred to a medium of low water potential, turgor pressure T is maintained when tissue water content and water potential decrease when water is lost, simply because the wall shrinks and the increase in π due to loss of water balances the change in ψ_p. If tissues with inelastic walls are transferred to a medium of low water potential, decreases in water content and ψ_p normally result in negative values of T because the wall cannot shrink. Plasmolysis occurs when the protoplast and vacuole pull away from the cell wall. In such cases, only a net increase in osmotic materials in the cell vacuole (due to solute uptake or synthesis via metabolism) can result in further water uptake to restore turgor.

Estimates of ε are difficult to compare because they are so dependent on turgor during measurement. However, Noy-Meir and Ginzburg (1969) showed that in *A. halimus*, $\varepsilon = 16$ bar at low turgor and 50 bar at high turgor. These estimates were compared with another xerophyte, *Ceratonia siliqua,* in which corresponding values of ε were 40 and 120 bar. These experiments imply that the more elastic, larger cells of *Atriplex* can lose more water before reaching zero turgor. However, the generation of an apparently negative turgor in the less elastic *Ceratonia* did not adversely affect the tissue water relations of this plant or preclude this plant from arid habitats. The latter observation is the antithesis of the above conclusion that maintenance of turgor is overall the most important process in maintaining biological activity of higher plant cells in the face of water deficits (Hsiao, 1973; Hsiao et al., 1976).

If we are pragmatic about the indirect and uncertain measurements of turgor, we must concede that the uncertainties surrounding apparently negative turgor are even more complex. To postulate that a particular cell organization such as small, thick-walled cells might enable plants to withstand apparently negative turgor raises rather more questions than it solves. That such cellular organization is common in some drought-tolerant species is well established, but only when direct methods for the measurement of turgor (Zimmermann, 1978) can be applied to thick-walled cells of higher plants will we be in a position to resolve these problems.

Water Potential and Water Content. It is improbable that any one component of tissue water relations is uniquely related to the capacity of plants to withstand low water potentials. Nevertheless there have been attempts to find a relationship which might be a useful indicator of drought resistance. A relationship between drought resistance and the slope of the water release curve has been explored (Weatherley and Slatyer, 1957; Connor and Tunstall, 1968). The slope of water release curve, a plot of tissue relative water content (relative to full turgor) as a function of tissue water potential, is small for leaves of drought-resistant trees such as *Acacia* and shrubs

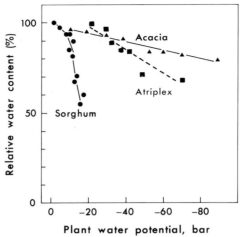

Fig. 8.12. Water release curves for leaves of species differing in drought resistance. *Acacia* is a tree which survives ψ_p − 80 bar and *Sorghum* is a grass which does not normally survive ψ_p − 20 bar. (Redrawn from Connor and Tunstall, 1968; Sankary and Barber, 1972; Jones and Turner, 1978)

like *Atriplex* but large for less tolerant species such as *Sorghum* (Fig. 8.12). The interpretation of this relationship as an indicator of drought tolerance has concentrated on water uptake mechanisms. It is held, for example, that a larger gradient of water potential can be generated for a small loss of water from plants with a shallow water release curve. This greater gradient per unit water loss would theoretically enable these species to absorb more water from a drying soil and hence increase their drought tolerance.

Such an interpretation ignores the fact that the relationships in Fig. 8.12 are the product of complex small-scale interactions among the components of cellular water relations, and that drought tolerance also involves large-scale differences in life form and rooting patterns. Water release curves are difficult to interpret because a small slope may involve a large value for π at full turgor, large capacity to accumulate solutes as ψ_p declines and a large value of ε. These measurable components are further complicated by the proportion of apoplastic water in thick- and thin-walled cells or cells of different size.

Thus there is no clear indication of why the differences in slope of Fig. 8.12 should be related to the undisputed differences in drought resistance of *Acacia* and *Sorghum*. Indeed, the water release curves of *Borya nitida* a "resurrection plant" which withstands desiccation beyond − 10³ bar is intermediate between that of *Acacia* and *Sorghum* (Gaff and Churchill, 1976). Quite clearly it is inappropriate to place emphasis on a relationship such as shown in Fig. 8.12, while ignoring differences in life forms that exist between a grass and a deep rooting shrub or tree. To do so is to ignore the truism (Fig. 1.1) that the scales of significant states and processes should be matched in space and time. Tolerance of, or adaptation to low plant water potentials in arid habitats involves a complex of responses in different water relations processes, at many scales of this continuum.

8.2 Response and Adaptation to Water Stress

Having treated the general principles and concepts of plant water relations and water transport we shall now turn to discussing the ways in which plants respond to water stress and the mechanisms they may use to maximize the efficiency of water use and to minimize the effects of drought. As pointed out in many places in this book, water is a factor of paramount importance in determining the productivity of the numerous *Atriplex* species which occupy arid lands.

There are at least three means by which plants may adjust to a limited water supply: by minimizing water loss; by increased extraction of water from the soil, and by increased ability to tolerate low tissue water potential without injury.

The principal means of minimizing water loss are by stomatal control of transpiration and by morphological changes that reduce the amount of radiation absorbed by the leaves. The only means of increased extraction of water from the soil is by increased root growth, and the principal means of tolerating low plant water potential is osmotic adjustment. Each of these means of adaptation to drought has important limitations and each results in a reduced potential for carbon gain, and these factors are evident in the discussion which follows.

8.2.1 Stomatal Regulation of Water Loss and Water Use Efficiency

The primary function of the stomata is to regulate the diffusive conductance of the leaf and, as Raschke (1976) has put it, "to them (the stomata) the plant has delegated the task of providing food while preventing thirst." Cowan (1977) and Cowan and Farquhar (1977) have extended this to suggest that stomatal conductance may vary during the course of each day so that the total loss of water is a minimum for the total amount of carbon assimilated. These authors describe stomatal behavior that conforms to this basic hypothesis as "optimal." However, they also point out that it is unlikely that such "optimal" stomatal behavior is the rule in all environments and at all stages of development. At this point we shall limit our treatment of stomatal movements to the influence of major environmental factors. The relationships between stomatal conductance, photosynthetic performance, and efficiency of water utilization are discussed in Chap. 9.4.1. We will start the present treatment by considering stomatal response to those variables that have a strong effect on the intrinsic photosynthetic rate, i.e., light, temperature, and intercellular CO_2 pressure.

The relationship between intercellular CO_2 pressure (p_i^c) and the CO_2 pressure of the ambient air (p_a^c) is given by the expression

$$p_i^c = p_a^c - \frac{P}{C_s}, \tag{8.12}$$

where P is net photosynthetic rate and C_s the stomatal conductance to CO_2 diffusion. P is in turn dependent on p_i^c. At p_i^c values encountered by the leaf in normal air ($p_a^c \approx 330$ µbar), P generally increases with increasing p_i^c. The actual relationship between P and p_i^c depends on the photosynthetic characteristics of the particular leaf and other environmental factors. For any given leaf, P tends to become increasingly dependent on p_i^c as the light intensity and temperature are increased. The above equation predicts that unless C_s also increases at the same time, an increase in P in response to increased light and temperature will result in a decreased p_i^c, thus counteracting the increased P. Efficient operation of the leaf in terms of CO_2 assimilation and water use thus demands that C_s should increase with increased light and temperature. This could be accomplished either by a direct response of the stomata to these factors, or indirectly by a stomatal response to p_i^c, so that C_s increases when p_i^c tends to decrease, or by the response of C_s to something correlated with P.

There is much evidence showing that stomatal conductance does indeed increase with increasing light intensity and that it decreases with decreasing light intensity. An example of this response is shown in Fig. 8.13 for *Atriplex triangularis*. These determinations were made at constant leaf temperature, normal ambient CO_2 pressure (p_a^c) and constant water vapor pressure deficit. It is also commonly observed that C_s tends to increase with increased leaf temperature, at least as long as light is not limiting P, and the water vapor pressure deficit remains constant. Figure 8.14 shows the response of C_s to leaf temperature in *Atriplex confertifolia*. Because of this response, p_i^c remains sufficiently high to enable photosynthesis to proceed at nearly maximum capacity over the whole temperature range in this C_4 species. These results do not permit a distinction between the possibilities that the responses of the stomata to light and temperature are direct or whether they reflect responses to changes in p_i^c, or to changes in P.

In *Atriplex triangularis* and in other plants stomata open in response to a decrease in p_i^c, even when light and temperature are kept constant. At least in certain plants, stomata open in response to decreased p_i^c even in the dark. In his review, Raschke (1975) concluded that stomatal responses to light and temperature are largely in direct result of changes in p_i^c brought about by photosynthesis. However, as pointed out by Cowan (1977) a disadvantage of stomatal control mediated by p_i^c is that it cannot cause an increase in the availability of CO_2 when the demand for it increases; since stomata open in response to a decrease in p_i^c they cannot then be responsible for an increase in p_i^c.

Recent results provide experimental support for the view that additional control mechanisms are operative. Wong et al. (1978) analyzed the relative roles of the effects of light intensity and p_i^c on stomatal conductance in *Eucalyptus pauciflora* leaves. They found that although, as expected, stomata responded to changes in p_i^c at any given light intensity, this response was much too small to have any significant role in the observed increase in stomatal conductance with increasing light intensity. The direct response of the stomata to changes in light intensity was sufficiently large to balance the effect of increased photosynthesis on p_i^c. As a result p_i^c remained substantially constant over a wide range of light intensities. We suspect that this form of response is widespread among plants. There is evidence that guard cells of certain plants contain special blue light-absorbing pigments which may be involved in a direct light control of stomatal conductance (Zeiger and Hepler, 1977). Such a control mechanism may well be important but the remarkably fine tuning between stomatal conductance and the demand for CO_2 by photosynthesis, which persists irrespective of the environmental factor that causes the potential increase in photosynthesis (Wong et al., 1979), suggests that the photosynthesizing cells are somehow able to transmit a signal to the guard cells conveying their

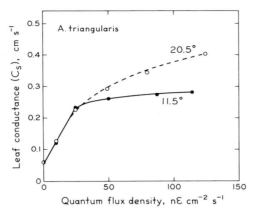

Fig. 8.13. Stomatal conductance as a function of incident quantum flux density in *A. triangularis* in its natural habitat at Bodega Head (O. Björkman, unpublished)

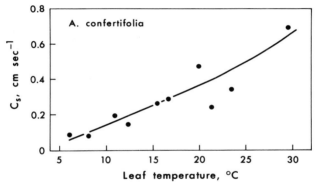

Fig. 8.14. Stomatal conductance as a function of leaf temperature in *A. confertifolia* plants raised in growth chambers at 28° day/20° night. (Redrawn from Caldwell et al., 1977a)

demand for CO_2. This hypothesis for explaining the closely parallel courses of stomatal conductance and photosynthesis is clearly in contrast with the common supposition that photosynthesis is controlled by stomatal conductance. The latter view is primarily based on the observation that when an acute water stress is imposed, stomatal conductance decreases and under these conditions may severely limit photosynthesis. It does not follow, however, that the same is true under nonstress conditions or even under chronic water stress.

It is generally recognized that stomatal conductance is affected by water status of the leaf. It appears that, in most cases, stomatal conductance (C_s) remains constant with decreasing leaf water potential (ψ_p) until a threshold value is reached, but when ψ_p is further reduced a precipitous reduction in C_s takes place. This type of response is illustrated by *Alnus rugosa* but in certain other species, such as *Simmondsia chinensis*, stomatal conductance exhibits a more or less continuous

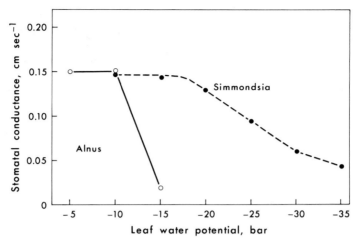

Fig. 8.15. Stomatal conductance as a function of plant water potential in leaves of *Alnus rugosa* from a streamside habitat and of *Simmondsia chinensis* from a desert habitat. (Data from Bunce, 1977)

decline over a wide range of leaf water potentials, extending from -15 to -40 bar (Fig. 8.15). In *Alnus rugosa* from mesic stream bank habitats (Bunce, 1977) the threshold value is about -10 bar. Desert species such as *Simmondsia chinensis, Atriplex hymenelytra, A. halimus* and *Larrea divaricata,* generally are able to tolerate considerably lower leaf water potentials before stomatal closure occurs, even when they have been grown with ample water supply.

Stomatal sensitivity to leaf water potential also depends on growing conditions (Chap. 9.4.2). Upland cotton had a threshold value of -16 bar when grown under ample water supply in a growth room as contrasted with less than -27 bar when grown in the field (Jordan and Ritchie, 1971). There is evidence that the threshold leaf water potential for *A. hymenelytra* is markedly lower in its native habitat in Death Valley, California, than when the plant is grown under mesic conditions. In the field, this threshold appears to move toward lower values of ψ_p between spring and fall, in concert with decreased soil water potential (Mooney and Björkman, unpublished).

There are indications that the differences between species and between plants grown under different regimes become much less pronounced when stomatal response is related to T instead of ψ_p. Turner (1974) suggests that the difference in threshold ψ_p for stomatal closure may be largely due to differences in π and thus in the ability of the plant to maintain turgor as the water potential is decreased. The nature and significance of such osmotic adjustments in response to water stress are discussed below (Chap. 8.2.4). While it is not surprising if stomata act to maintain positive turgor rather than an arbitrary water potential there is, however, no a priori reason for supposing that stomatal conductance is more closely related to leaf turgor pressure than to leaf water potential (Cowan, 1977).

Irrespective of whether stomata respond to water potential or turgor pressure, for these responses to work via a negative feedback system, the leaf must at least experience a minimum level of stress. This type of response thus primarily serves to prevent leaf desiccation rather than to conserve water. Or, as Cowan (1977) put it, "the stomata respond not to those external factors which tend to influence the rate of water loss from the leaf, but to components of the internal physiological state which are affected by the rate of water loss."

In some species stomata have an additional means of preventing water loss. Several studies show that stomata may respond *directly* to changes in the humidity of the ambient air, or rather, perhaps the water vapor pressure difference between the leaf and the ambient air ($p_i^w - p_a^w = VPD$). Such a response represents a feed-forward control instead of a feed-back control and can prevent the leaf water potential from declining. These observations have been reviewed by Farquhar (1978). Species in which stomatal responses to changes in VPD have been observed include *Helianthus annuus* and *Sesame indicum* (Camacho-B et al., 1974), *Vigna luteola* (Ludlow and Wilson, 1971), *Prunus armeniaca* (apricot), *Hammada scoparia*, and *Zygophyllum dumosum* (Schulze et al., 1972, 1973, 1974, 1975a,b). In all of these species, except apricot, the decline in stomatal conductance in response to increased VPD reduced transpiration to about one-half of the rate that would have been obtained if the stomatal conductance had remained constant. In most cases the actual transpiration rates at high VPD's were not lower than at low VPD's, however. The studies on irrigated apricot trees, *Hammada* and *Zygophyllum* in the Negev desert by Schulze, Lange, and their coworkers, are especially noteworthy. In these the decrease in stomatal conductance was so great as to result in a lower actual transpiration rate at high than at low VPD's (Schulze et al., 1972). In apricot, transpiration rate at a VPD of 30 mbar was only two thirds of the rate at 15 mbar. If no stomatal adjustment had occurred, the rate at 30 mbar obviously would have been twice that at 15 mbar. Because of the decrease in transpiration that occurred when the humidity was lowered, the water content of the leaves actually increased. Conversely, when the leaves were subjected to increased humidity, stomatal conductance and transpiration increased in spite of reduction in leaf water content.

Schulze and coworkers also observed that the sensitivity of the stomata to changes in VPD in nonirrigated apricot trees tended to increase as the leaf water potential decreased during the summer. No significant changes in this sensitivity were observed in irrigated trees. This suggests that stomatal sensitivity to changes in VPD may increase as the leaf water potential declines. Interestingly, the sensitivity of stomatal response to changes in VPD appears to decrease as the temperature increases, and stomatal conductance tends to be markedly higher at high than at low temperature (Hall et al., 1976). This counterbalancing effect of temperature on stomatal conductance is expected to be important in regulating the heat balance of the leaves. The observation that nonirrigated individuals of *Hammada scoparia* and *Zygophyllum dumosum* which are native to the Negev desert, show stomatal responses to humidity similar to those found in apricot, although they are somewhat less pronounced (Schulze et al., 1972), suggests that this kind of control may have an important role in arid habitats.

Nevertheless, stomatal control by humidity seems to be far from universal. Numerous species, including arid region plants, subjected to very high water vapor pressure differences at low soil water potential, appear to lack a stomatal response to humidity. This is also true of the three arid region *Atriplex* species that have been studied so far. Whiteman and Koller (1967) investigated the effect of humidity on the desert chenopods *Atriplex halimus, A. vesicaria,* and *Kochia indica* as well as the more mesic species *Helianthus annuus.* The plants were grown in a greenhouse as well as in the open in a botanical garden in Jerusalem and were adequately supplied with water. The *Atriplex* and *Kochia* species were subjected to water vapor pressure differences from about 10 to over 30 mbar, *H. annuus* from 7 to 28 mbar. Measurements of transpiration were conducted at 29 °–32 °C, normal and low CO_2 pressure, and under artificial as well as natural light. Transpiration increased linearly with the water vapor pressure difference (i.e., stomatal conductance remained unchanged) in all four species and under all conditions.

Barrs (1973) found no change in stomatal conductance when the vapor pressure difference was varied in maize, cotton, pepper, tomato, and sunflower *(H. annuus).* Rawson et al. (1977) obtained the same results when single attached leaves of soybean, wheat, barley, sorghum, and *H. annuus* and the arido-active species *Echinochloa crus-galli, Simmondsia chinensis,* and *Atriplex halimus* were subjected to water vapor pressure differences of 8 to 27 mbar. Stomatal conductance and photosynthesis were unaffected by humidity changes in this range. The plants were grown in a greenhouse at 27°/22 °C and 35% to 55% relative humidity with adequate water and nutrient supply. It is noteworthy that the transpiration rate of *H. annuus* was considerably higher than in the study by Camacho-B et al. (1974) and therefore a stomatal response could have been expected. However, this was not found.

It might be expected on the basis of the results obtained by Schulze et al. with apricot, *Hammada* and *Zygophyllum* in the Negev desert that certain conditions prevailing in desert habitats, such as high VPD or low plant water potentials, promote stomatal sensitivity to changes in VPD. However, this supposition is not supported by the observations of Mooney and Björkman (unpublished) in Death Valley, California, on naturally occurring plants of the desert shrubs *Atriplex hymenelytra, Larrea divaricata,* and *Tidestromia oblongifolia.* Figure 8.16 shows results obtained with *A. hymenelytra.* No changes in stomatal conductance occurred in response to changes in VPD from 15 to 50 mbar in spite of the fact that the leaf water potential was quite low (− 40 bar). Similarly, experiments conducted with irrigated as well as nonirrigated *L. divaricata* and *T. oblongifolia* plants gave no evidence for a significant stomatal response to changes in VPD in the 15 to 60 mbar range. It should be emphasized that daytime VPD below 30 mbar is uncommon during the summer in Death Valley; the VPD frequently approaches 100 mbar during the hottest months of the year in this extreme habitat. If the stomata exhibited sensitivities to VPD similar to those found by Schulze and coworkers for apricot in the Negev, they would remain closed almost continuously for many months. This they do not. In fact, growth of *T. oblongifolia* is restricted to the summer and stomatal conductance typically remains high and maximal during the hottest part of the day at a VPD of 80 mbar and above (Björkman et al., 1972b; Björkman and Mooney, unpublished). Similar stomatal behavior is indicated for the evergreen desert species, *Atriplex lentiformis* during the summer in Death Valley and at Thermal in the Salton Sea basin in southern California (Pearcy and Harrison, 1974). Nonirrigated individuals of two other evergreen Death Valley

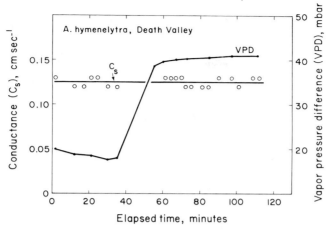

Fig. 8.16. Insensitivity of stomatal conductance of *A.hymenelytra* in Death Valley to changes in vapor pressure difference. (H.A.Mooney and O.Björkman, unpublished)

shrubs, *Atriplex hymenelytra* and *Larrea divaricata,* typically have appreciably lower stomatal conductances during the summer than in the spring but there is no evidence that stomatal conductance is controlled by the water vapor pressure deficit even when this exceeds 70 mbar and the tissue water potential falls below -40 bar (Pearcy et al., 1974; Mooney and Björkman, unpublished). Consistent with this are the observations that the tissue water potentials reach their minimum values at the time when the VPD is maximum (Björkman et al., 1972; Pearcy et al., 1974). In contrast, the water content and hence the water potential of apricot leaves in the Negev increased during periods of high VPD as a result of stomatal closure (Schulze et al., 1974).

In contrast to the desert *Atriplex* species, the cool-coastal *A.triangularis* showed a substantial stomatal response to humidity (Björkman, unpublished). Figure 8.17 shows the response of stomatal conductance to increasing vapor pressure difference in leaves of *A.triangularis* during the summer in its native habitat on a coastal bluff at Bodega Head, California. Stomatal conductance was found to fall in proportion to increases in the VPD, both at 20° and 30 °C leaf temperature, but the sensitivity was considerably smaller at the higher leaf temperature. While the observed decrease in C_s strongly counteracted the increase in transpiration due to an increased VPD, the transpiration rate did not decrease when the VPD was increased.

The effect of temperature on the dependence of C_s on VPD is important in view of the fact that, in nature, increases in VPD are largely a result of an increase in leaf temperature, and are not due to a decrease in the water vapor pressure of the ambient air. One would predict that the sensitivity of the stomata to an increased VPD due to increased leaf temperature would be smaller than that obtained from the relationship between C_s and VPD determined at a constant leaf temperature. This is confirmed by the results shown in Fig. 8.17, right. For example, the relationship between C_s and VPD measured at 20 °C (Fig. 8.17, left) shows that an increase in VPD from 8 to 19 mbar would result in a 60% decrease in C_s. However,

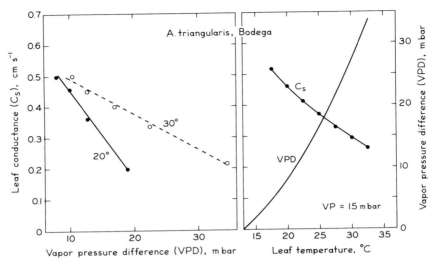

Fig. 8.17. Change in stomatal conductance of *A. triangularis* at Bodega Head in response to changes in vapor pressure difference *(left)*. The change in stomatal conductance in response to changes in leaf temperature under ambient conditions and the vapor pressure difference at different leaf temperatures is also shown *(right)*. (O. Björkman, unpublished)

an equal increase in VPD brought about by an increase in leaf temperature from $20°$ to $26\,°C$ (Fig. 8.17, right) resulted in only a 25% decrease in C_s.

The above results indicate that the strategies of plants with regard to regulation of water loss may differ between species and between habitats. Although any generalization should be treated with caution it seems that as far as xerophytic species, native to the hot floor of Death Valley (including *Atriplex hymenelytra* and *A. lentiformis*) are concerned, direct stomatal response to VPD is not a major means of regulation of water loss. Evidently the principal diurnal control of stomata in this desert habitat is by irradiance, temperature, and the water status of the leaves. In other words, the stomata open in response to increased demand for CO_2 while at the same time maintaining a high efficiency of water use (Chap. 9.4.1). Only when the loss of water threatens to cause an excessive decline in leaf water potential is this control overruled, or in other words, does prevention of thirst take precedence over prevention of hunger.

8.2.2 Changes in Leaf Inclination and Reflectance

As was pointed out in Chap. 8.1.5, changes in leaf properties which reduce the fraction of the incident radiation that is absorbed by the leaves are effective means of lowering both the leaf temperature and the transpiration rate. There are two principal ways in which this can be accomplished. One is by the alteration of the orientation of leaves so that they tend to become more parallel rather than perpendicular to the sun's rays. The other is by increasing the reflectance of the leaf surfaces by increasing leaf pubescence. Both of these kinds of change usually

Table 8.3. Calculated leaf temperature and transpiration from *A. hymenelytra* leaves of different absorptance properties at noon on a summer in Death Valley[a]. (Data of Mooney et al., 1976)

Leaf category and inclination	Absorptance	Leaf temperature (°C)	Transpiration ($\mu g \ cm^{-2} \ s^{-1}$)
Winter leaf horizontal	0.25	50	3.5
Winter leaf 70°	0.25	47	3.0
Summer leaf horizontal	0.50	45	2.7
Summer leaf 70°	0.50	43	2.5

[a] $t_a = 45\,^{\circ}C; R_n = 1.5 \ cal \ cm^{-2} \ s^{-1}; C_s = 0.05 \ cm \ s^{-1}; dewpoint = 10\,^{\circ}$

require a relatively long time ($> 10^5$ s) but certain species, e.g., Townsville stylo *(Stylosanthes humilis)* are capable of rapid leaf movements (10^1 to 10^2 s) so that the inclination of the leaves can be closely controlled throughout the day (Begg and Torssell, 1974). When water supply is ample the leaves orient so that the interception of the radiant energy is maximized, but when the plants are subject to water stress the leaves orient in a manner that minimizes the interception of radiant energy during the hours of high irradiance. Although it now appears that this ability to alter leaf orientation may be present in more species than previously thought, most plants lack this ability.

No species of *Atriplex* is known to be capable of rapid leaf movement but it is a common observation that many *Atriplex* species may change their leaf inclination seasonally. Early in the spring when the heat load is low and water supply is relatively favorable the leaves are largely horizontal and the interception of radiant energy is thus high; as the heat load and drought increase during summer the inclination of the leaves becomes steeper. Consequently, the radiation load at midday when ambient temperature is highest, is reduced. An example of this type of response is given in Table 8.3 which shows that the steep inclination of *Atriplex hymenelytra* leaves, as observed during the summer in Death Valley, significantly reduces the heat load and transpiration rate in comparison with the essentially horizontal leaves found during the winter and early spring (Mooney et al., 1977b).

The increase in the inclination of the *A. hymenelytra* leaves occurs together with an increase in their reflectance which changes from about 35% in winter to about 60% in summer. Using these changes in reflectance and substituting in the leaf energy balance equation, given in Chap. 8.1.4, under typical summer conditions in Death Valley at noon, Mooney et al. (1977b) concluded that substantial differences in leaf temperature and transpiration rate could be ascribed to reflectance alone. As shown in Table 8.3 the highly reflective leaf in the horizontal position would remain at air temperature but a winter leaf with low reflectance would be 5 °C hotter than air temperature, and would transpire 25% more rapidly. As was mentioned above, the steep inclination of the summer leaves further reduces the heat load and the transpiration rate. The combined effects of leaf inclination and reflectance changes allow the inclined, high-reflectance leaf to stay 2 °C below air temperature and to transpire 30% more slowly than the horizontal leaf with low reflectance (at a constant stomatal conductance).

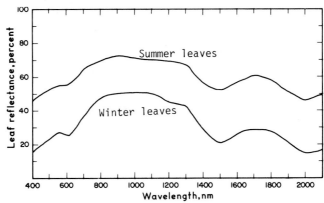

Fig. 8.18. Reflectance of *A. hymenelytra* leaves during summer and winter. (Redrawn from Mooney et al., 1977b)

High leaf reflectance is a characteristic common to most, if not all, desert species of *Atriplex*. It is also frequently found in desert species of other genera although measurements by Sinclair and Thomas (1970) showed that of all the arid shrubland species examined by them, *A. vesicaria* and *A. stipitata* had the highest reflectances. There thus appears to be little doubt that reduction of heat load and transpiration rate by means of high leaf reflectance is very common in the genus *Atriplex* and many other desert plants, as was early recognized by Billings and Morris (1951) and recently emphasized by Smith (1978). The role of leaf reflectance in the energy and carbon balance of *Encelia farinosa* and *E. californica* was discussed earlier (Chap. 3.2).

Sinclair and Thomas (1970) ascribed the high reflectance of *Atriplex* leaves to the presence of a dense mat of collapsed epidermal trichomes and salt crystals. Subsequent studies by Mooney et al. (1974, 1977b) confirmed this in *A. hymenelytra* and also provided information on the relationship between reflectance and leaf water status. As shown in Fig. 8.18, old leaves from summer shoots at a water potential of -30 bar had reflectances of about 60% whereas young leaves from winter shoots at a water potential of -10 bar had reflectances of about 35%. When old leaves were allowed to rehydrate following enclosure in reflective plastic bags, leaf reflectance decreased from 60% to 37%. When plants were grown in low salt culture solutions reflectance was 33%. This increased to 50% when the plants were grown in high salt culture. It is thus clear that the reflectance of *A. hymenelytra* leaves is directly related to leaf water content and hence to water potential. The ion absorption processes which allow these halophytes to effect osmotic adjustment, and to normalize cell water relations by salt uptake and excretion evidently also function to minimize the rate of water loss (and to avoid excessively high leaf temperatures) by reducing the interception of radiant energy during summer drought.

8.2.3 Root Extension and Water Uptake

The principal means by which plants may increase their rate of water uptake during prolonged periods of drought is by extension of the roots into previously

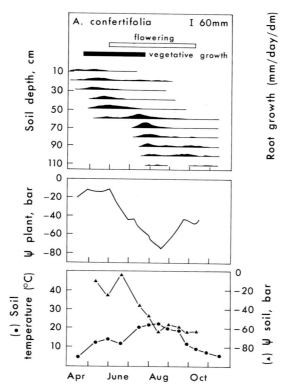

Fig. 8.19. Root growth rate at different depths beneath *A.confertifolia* throughout the growing season in the Great Basin Desert, Utah. Plant and soil water potential and soil temperatures are also shown. (Redrawn from Fernandez and Caldwell, 1975)

unexplored areas of the soil. To some extent continued exploration of the soil within the rooted zone may also increase the capacity to extract water from the drying soil. Caldwell (1976) has reviewed the role of root extension in water absorption and we shall limit our treatment of this topic to a brief discussion of a few studies based upon *Atriplex* species.

Using soil windows, Fernandez and Caldwell (1975) followed the mean daily rate of root elongation at 10 cm intervals throughout the soil profile under these shrubs throughout the growing season. Figure 8.19 shows root growth rate for *A.confertifolia,* the soil temperature and ψ_s at 40 cm depth. High in the soil profile ψ_s may be expected to be more negative than shown, except in the surface soil following summer rains, and lower in the profile ψ_s will be less negative than shown (Moore and Caldwell, 1972). Figure 8.19 also shows typical changes in ψ_p for *A.confertifolia* throughout the season but these data were taken from another set of observations at the same site (Moore and Caldwell, 1972).

The zone of maximum root elongation, and hence of water absorption, is initially confined to the surface soil but rapidly moves deeper into the profile throughout the season. By August, when ψ_p is about -70 bar and ψ_s at 40 cm is

−60 bar, root elongation is almost entirely confined to regions of the soil profile below 60 cm where higher soil water potentials are to be found. There is no root activity at all in the soil surface. Moore and Caldwell (1972) showed that as late as October, when ψ_p was −85 bar at midday, plant water potential recovered in the evening. Together these elegant data show that in the most arid season, growing roots deep in the soil profile are capable of tapping water reserves at a higher water potential than the shoot and thus function to sustain water transport and plant water status.

In another study, Hodgkinson et al. (1978) noted an increase in ψ_p of *A. confertifolia* at dawn within a few days of summer irrigation treatments which penetrated only the top 10 cm of the soil profile. Thus it seems that although these plants depend on root growth at progressively greater depths in the soil profile for the bulk of their water, root activity in the soil surface may respond to substantial summer rainfall (25 mm) and result in a temporary increase in ψ_p.

The continued growth of roots necessary for the exploration of soil water resources draws heavily on the photosynthetic assimilates of those plants at a time when photosynthesis has been drastically reduced by lowered stomatal conductance (Caldwell et al., 1977b). It is thus evident that all of the means by which the plant may control the loss as well as the uptake of water represent a complex of compromises, the net result of which is much reduced photosynthetic rate and the diversion of most of the photosynthate belowground where it is used principally to support root growth deep in the soil. The implications of these processes for productivity are explored in Chap. 10.

8.2.4 Osmotic Adjustment in Response to Low Tissue Water Potential

Although it is believed that turgor pressure is the component of plant water potential most intimately associated with plant responses to water deficits, ψ_p is commonly used as an indicator of critical plant water status for these responses. Hsiao (1973) showed that cell growth and the interrelated processes of wall synthesis and protein synthesis are inhibited following relatively small changes in ψ_p. This is not surprising because extension growth depends on the generation of adequate turgor as the driving force and if plant water status is such that turgor is near zero, growth will be reduced. Following the reduction in turgor, inducible enzymes such as nitrate reductase, which are intimately associated with growth processes, also decrease in activity. It is possible that these metabolic changes and the reduced demand for metabolites in wall and protein synthesis may allow the diversion of some carbon into solutes which can be utilized for purposes of osmotic adjustment. However, calculations suggest that the quantity of solutes required for osmotic adjustment is small and is not a substantial drain on biosyntheses. Osmotic adjustment is thus an important means of adjusting plant cell water relations at low water potential.

Continued physiological activity in plants at low water potential under natural conditions depends on their capacity for osmotic adjustment. This simple statement lies at the heart of a major long-term misunderstanding in the water relations literature which hinges on two different interpretations of the same data. Over many years Walter and his colleagues (Walter and Stadelmann, 1974) have

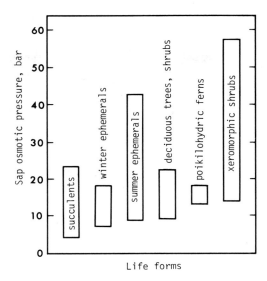

Fig. 8.20. Range of values for osmotic pressure of sap expressed from plants in the Arizona Desert. (Redrawn from Walter and Stadelmann, 1974)

correlated the osmotic pressure of sap from different life forms with the ability of these life forms to survive in different habitats. Figure 8.20 shows, for example, that winter ephemerals and summer deciduous species of the Arizona desert display a lower range of values for π than summer ephemerals or xeromorphic shrubs. High tissue ionic contents are indeed a feature of arid shrubland vegetation as was pointed out in Chap. 7. Walter and his colleagues concluded that these latter plants were more resistant to drought because they survived *in spite* of the higher π. That is, survival was related to protoplasmic tolerance of high π. On the other hand, in keeping with the overall view of the importance of turgor in plant responses to water stress, Hsiao et al. (1976) point out that the summer ephemerals and xeromorphic perennials survive the period of water stress *precisely because* the capacity to maintain higher π enables them to maintain positive turgor at lower ψ_p. That is, these successful species presumably have the capacity to adjust osmotically.

In nonhalophytes accumulation of common metabolites such as sugars and carboxylic acids serve to increase osmotic pressure by about -5 to -10 bar and are major components of the limited osmotic adjustment possible in these plants (Jones et al., 1979). In some, the accumulation of inorganic ions may also be important. Osmotic adjustment by means of salt uptake in response to salinity treatments was indicated in the early experiments of Eaton (1927) with wheat and *A. semibaccata*. He observed that both species were able to maintain sap osmotic pressures, measured as freezing point depression, in excess of the osmotic pressure of the salt-treated sand in which the plants were grown (see Table 7.10). Ashby and Beadle (1957) noted a similar relationship in *A. nummularia* seedlings grown in saline water cultures (Fig. 8.21) and the phenomenon was independently confirmed in crop plants by Bernstein (1961) and Slatyer (1961).

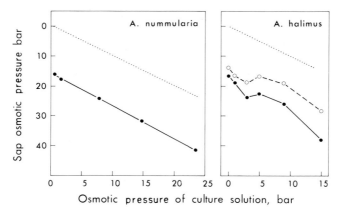

Fig. 8.21. Osmotic pressure of sap expressed from *A. nummularia* and *A. halimus* before (●) and after (○) removal of epidermal trichomes. Plants were grown in culture solutions containing increasing concentrations of NaCl. Dotted line refers to equivalent osmotic pressure in solution and sap. (Redrawn from Ashby and Beadle, 1957; Gale et al., 1970)

    An unspecified part of the change in π in *Atriplex* species could be due to salt excreted to the leaf surface in the epidermal trichomes which have subsequently collapsed. This salt would not be in equilibrium with that in the cells and may lead to spuriously high values for π. In later experiments, Gale et al. (1970) took the precaution of washing leaves of *A. halimus* before the sap was expressed. As shown in Fig. 8.21, soluble salt on the leaf surface may account for about 7 bar of the measured leaf π at high external salt concentration. The water potential of intact bladder cells on the leaf surface is not known, but in the absence of evidence of a low conductance between the bladder cell and bulk leaf, it is presumably similar. If the bladder cells accumulate salt to the high concentrations estimated (Chap. 7.5.5) these cells must be subject to osmotic pressure up to 550 bar and turgor pressures of the same magnitude (Mozafar and Goodin, 1970). It is not surprising in these circumstances that the bladder cells burst and release their contents to the leaf surface. The ion absorption processes which continue despite such pressure, deserve further attention.

    In spite of the complications that the deposition of salt on the leaf surface of *Atriplex* spp. introduces to estimates of π in leaf sap, it is likely that osmotic adjustment due to salt uptake in these halophytes takes place under natural conditions. Sharma et al. (1972) noted, for example, that Na^+ and Cl^- concentrations (leaf dry weight basis) increased in leaves of *A. vesicaria* and *A. nummularia* throughout spring and summer and decreased again during winter. Increase in salt content was correlated with decreasing soil osmotic and matric potential and decreasing relative leaf water content. A more complete account of osmotic adjustment can be deduced from the data of Moore et al. (1972a) for *A. confertifolia* from the Great Basin Desert. Figure 8.22 shows measured changes in ψ_p, π and leaf water content throughout the season, as well as the sap chloride concentration reported by these authors, and the chloride concentration on a dry weight basis, calculated from sap chloride concentration and water content.

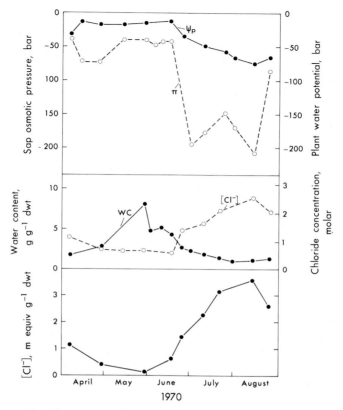

Fig. 8.22. Changes in salt concentration, water content, sap osmotic pressure and plant water potential of *A.confertifolia* growing in the Great Basin Desert, Utah. (Redrawn from Moore et al., 1972a)

Together these data provide a relatively complete account of the water relations of *A.confertifolia* in this habitat. It is likely, for example, that the leaves of the shoots of these plants remained at positive turgor throughout the season because π was always greater than ψ_p. During the season of most rapid growth and minimum soil water stress (April–late June) ψ_p was a minimum (-11 to -14 bar) and leaf chloride concentration remained constant. Leaf water content increased markedly to a peak in late May when the tissue was most succulent but then declined steadily throughout the remainder of the period. Associated with this was a dramatic increase in chloride uptake (dry weight basis) and chloride concentration in the leaf sap which was reflected in a large decrease in π. Chloride uptake was balanced by sodium uptake (Moore et al., 1972a) but the proportion of salt excreted to the leaf epidermis is not known. The excreted salt presumably accounts for the probable large overestimate of π. The data in Fig. 8.22 provide strong evidence for osmotic adjustment in *A.confertifolia* by increased salt uptake, resulting in increased osmotic pressure and presumably turgor maintenance as cell water content and water potential decline. Whether the decline in cell water content was a product of incompletely effective stomatal control of water loss or of a decrease in conductance of the root-to-shoot transport pathway, cannot be determined from the available data.

In the previous chapter it was pointed out that because the metabolism of halophytes is not particularly insensitive to electrolytes, osmotic adjustment by means of electrolyte accumulation in the cytoplasm may have undesirable metabolic consequences. It was proposed that the so-called "compatible solutes", including quaternary ammonium compounds such as glycine-betaine, may be particularly important to osmotic adjustment in the cytoplasm.

Other plants have developed mechanisms of resistance to drought periods which apparently do not depend on osmotic adjustment. The succulent plants capable of crassulacean acid metabolism resist drought by maintaining high tissue water potential and almost completely eliminate the exchange of external CO_2 and H_2O. The desiccation-tolerant plants, on the other hand, display a remarkable degree of protoplasmic tolerance to extremely low water potentials (Jones et al., 1979). However, neither the succulents surviving at high tissue water potential nor the desiccation-tolerant plants, surviving at extraordinarily low tissue water potentials, are capable of net photosynthetic production during the stress period. It seems that osmotic adjustment is the principal process which makes it possible for terrestrial plants to adjust their cellular water relations at the same time as they adjust water flow, thereby maintaining turgor and metabolic capacity, as well as access to external CO_2. These plants can then continue to function throughout the stress period. Some, such as *A. confertifolia*, maintain only modest levels of production at low water potentials in arid environments. Others, such as *Tidestromia*, by virtue of better water supply display less osmotic adjustment and maintain maximum productivity at the time of peak evaporative demand.

The integration of different components of plant water relations in the control of growth is well illustrated in the interactions of salinity and atmospheric humidity during the growth of *A. halimus*. Like most *Atriplex* species and many halophytes, *A. halimus* usually shows more rapid growth in the presence of modest concentrations of NaCl. Gale et al. (1970) noted, however, that this growth stimulation only occurred when plants were treated with NaCl in a dry atmosphere. In a humid atmosphere all NaCl treatments were inhibitory to growth. These authors implied that in the dry atmosphere the accumulation of NaCl contributed to an improvement in plant water relations. Subsequent experiments by Kaplan and Gale (1972) demonstrated that treatment with NaCl modified both flow and pressure components of plant water relations. The effect of growing plants in -10 bar NaCl solution, in a dry atmosphere, on the stomatal conductance and mesophyll conductance to water vapor movement as well as on transpiration rate and photosynthesis is shown in Table 8.4. Growth in NaCl produced substantial decrease in all flow-related parameters and the decreases in C_s and C_m were similar. As a result, transpiration rate was approximately one-half that of the control plants under the same conditions. As can be expected, the rate of photosynthesis was reduced also but this reduction was only about 20%. The reason for the small reduction in photosynthesis compared with transpiration is possibly related to the fact that photosynthesis in this C_4 species was not directly proportional to stomatal conductance (or to intercellular CO_2 pressure). However, calculations of

Table 8.4. Changes in flow-related parameters of plant water relations and in photosynthesis of *A. halimus* grown in control and NaCl (-10 bar) cultures in a dry atmosphere. (Data of Kaplan and Gale, 1972)

Parameter	Control	$+$NaCl
Stomatal conductance (cm s^{-1})	0.24	0.09
Intrinsic conductance (cm s^{-1})	0.29	0.07
Root conductance (mg g^{-1} root min^{-1})	8.4	5.3
Transpiration (mg dm^{-2} h^{-1})	590	289
Photosynthesis (mg dm^{-2} h^{-1})	10.7	8.5

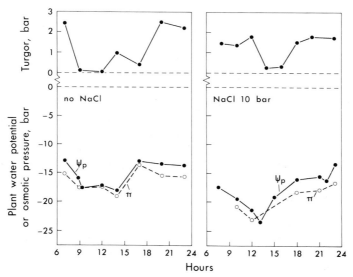

Fig. 8.23. Daily course of plant water potential, sap osmotic pressure, and turgor pressure in leaves of *A. halimus* grown in culture solution and culture solution with NaCl equivalent to 10 bar. (Redrawn from Kaplan and Gale, 1972; note that osmotic pressure has a positive sign)

intercellular CO_2 concentration from these data give impossible values, suggesting there may have been errors in the estimation of the water pathway conductance. The relationship between photosynthetic rate and stomatal conductance is discussed in Chap. 9.

Reduced water loss in these plants treated with NaCl in a dry atmosphere was associated with improved shoot water relations as shown in Fig. 8.23. The shoot water potential in plants grown in -10 bar NaCl declined to lower levels at midday than it did in the control plants. However, the salt-treated plants also showed higher osmotic pressure and as a result appeared to be able to maintain positive turgor throughout the day. In the absence of salt, plants spent 3 h at zero turgor in the morning and turgor was less than 1 bar for most of the photosynthetic period.

This presumably reduced the period of active photosynthesis or interfered with the utilization of photosynthates for growth. Although growth in NaCl was associated with slight reduction in photosynthesis due to decreased C_s, this was evidently compensated by the capacity to retain turgor. Other aspects of the interactions between CO_2 and H_2O exchange of stomata in relation to plant water status are discussed in the next chapter.

8.2.5 Conclusions

Each of the plant responses to water stress discussed above provides a marginal benefit to plants in arid habitats. Together they may contribute to the maintenance of cellular water relations and the maintenance of turgor pressure within tolerable limits when plants are exposed to low water potential environments. Although these marginal benefits undoubtedly contribute to the survival of plants in arid habitats, they only do so at the expense of net carbon gain.

The cost of the different plant responses to water stress discussed above may be evaluated as follows. Control of water flow by stomatal closure inevitably reduces the conductance of the CO_2 diffusion pathway and reduces CO_2 assimilation. Changes in leaf inclination and reflectance reduce the interception of light and hence the energy available for photosynthesis. Increased root growth in response to water stress depends on the diversion of photosynthates from leaf production to root production and maintenance. It can thus be considered in terms of reduced development of photosynthetic surface area and lowered photosynthetic potential. Maintenance of cell turgor by means of osmotic adjustment may be the least expensive of these responses. However, the energetic costs of salt accumulation and of the diversion of metabolites for osmotic adjustment are not known.

Although all of the above plant responses to water stress may be costed in terms of foregone photosynthesis, it should not be concluded that satisfactory plant water relations in arid habitats can only be maintained with low primary productivity. It is true that many plants adapted to arid habitats are not highly productive and this is particularly so in the case of species which remain active at relatively low water potentials. It is also true of succulent plants which maintain high water potential by stomatal closure and root abscission and then maintain the status quo by nocturnal CO_2 assimilation via crassulacean acid metabolism. However, some species such as *Tidestromia oblongifolia* are only active in the hottest and most arid period of the year. By virtue of its effective root system, by virtue of the high thermal stability of membrane systems, and by virtue of the C_4 photosynthetic pathway, *T. oblongifolia* is able to maintain extraordinarily high rates of productivity in the extremely arid habitat of Death Valley. This species can exploit the available water to maintain high tissue water potential and can thus take advantage of the high temperature coefficient of metabolism, thus permitting high productivity at high temperature. Adaptations of plant water relations to water-limited habitats may be many and varied and, at least for some plants, it is clear that the desert is not a place of hunger, in spite of its extraordinarily "restless thirsty atmosphere."

Chapter 9. Photosynthesis

"Although all life depends directly or indirectly on photosynthesis the chemical nature of this remarkable energy-converting process is not at all well understood ... The reason for our ignorance about this extraordinary natural process is not so much due to lack of scientific effort as to the inherent complexity of the system ... A frequently raised question is: How long will it take to solve the problem of photosynthesis? To answer this question we must be clear as to what we mean by solving a problem ... most scientific questions have found their answers again and again – but each time within the limits of contemporary knowledge and vocabulary."

(French, 1967)

For most of its life span, from the emergence of the cotyledons and depletion of the seed resources until the completion of its life cycle, the plant derives all of its carbon and energy from photosynthesis alone. Consequently, the growth of a plant in terms of dry matter production is directly proportional to the difference between the amount of photosynthetic activity that it has been able to achieve and the losses it has incurred as a result of respiratory activity during the same time period. This implies that in any given environment plant productivity is to a large extent governed by the efficiency with which the photosynthetic machinery is able to operate under the prevailing conditions and its ability to cope with environmental stresses. However, while this is undoubtedly true, the relationship between plant productivity and photosynthetic efficiency and performance is very complex. In part, this is so because the concepts of photosynthetic efficiency and performance are not easily defined and the criteria used to determine what constitutes photosynthetic efficiency must necessarily depend on the environment under consideration.

In this chapter we will attempt to review available information on the environmental and biological control of photosynthesis in higher plants, particularly in relation to the success and performance of *Atriplex* species in natural habitats. Wherever possible we will make an effort to analyze phenomena of photosynthetic adaptation in terms of underlying physiological and molecular mechanisms.

As in other chapters, we will start with a brief outline of the small-scale states and processes which underlie photosynthetic mechanisms and will then proceed to discuss the responses of photosynthesis to each of the major environmental variables. Special consideration will be given to the phenomenon of photosynthetic acclimation, i.e., the long-term phenotypic adjustment of photosynthetic characteristics to environmental changes, and to photosynthetic acclimation potential, i.e., the ability of a given species or genotype to so adjust. In the next chapter we shall attempt to relate these responses of photosynthesis to primary productivity of plants in different environments.

9.1 The Photosynthetic Process

Photosynthesis may be defined as the conversion of light energy to the chemical energy contained in the plant constituents. In green plants the principal reactants are CO_2, the ultimate electron acceptor, and H_2O, the ultimate electron donor. Electrons are removed from water, yielding O_2 as the product, and donated to CO_2 which is reduced to the level of carbohydrate, $(CH_2O)_n$. While CO_2 is by far the most important substrate, a portion of the energy generated in photosynthesis is used for the reduction of certain other inorganic compounds, notably NO_3^- and SO_4^{2-}. The products of these latter reduction processes are essential in the biosynthesis of proteins and other cell constituents.

9.1.1 Photochemistry

The overall photosynthetic process is a continuum of photochemical and biochemical events which are interdependent and generally considered in a linear sequence. Kamen (1963) conceptualized the primary processes of photosynthesis in a sequence scaled on the basis of the relaxation times of the component reactions which range from 10^{-18} to 10^1 s. This treatment of photosynthesis was seminal to the development of the space–time scale described in Fig. 1.1, although the processes dealt with by Kamen have much smaller relaxation times than those dealt with in detail here.

The photochemical and biochemical reactions of photosynthesis in higher plants take place in discrete organelles, the chloroplasts. The photoacts, with associated electron transport and phosphorylating machineries, occur on the internal chloroplast membrane system called lamellae or thylakoids. This lamella membrane system is surrounded by a matrix, the stroma, in which the CO_2 fixation and reduction processes take place. Recent comprehensive accounts of the structure and of the primary processes in photosynthesis are contained in Trebst and Avron (1977) and in Hall et al. (1978).

The location of components of the photosynthetic electron transport pathway in the chloroplast has been intensively studied in recent years and is most elegantly illustrated by the particles distributed on the surfaces exposed when chloroplasts are freeze-fractured for observation in the electron microscope. Figure 9.1 shows that particles are embedded in chloroplast membranes. Analysis of the distribution and size classes of these particles in the different fracture faces in mutant and developing chloroplast systems has led to the interpretation shown in the schematic diagram. The thylakoid membranes are differentiated into stacked and unstacked regions, known as the grana lamellae and stroma lamellae respectively. The stacking of thylakoid membranes will be discussed in more detail later (Chap. 9.2.1); for our present purposes it is sufficient to note that the two fracture

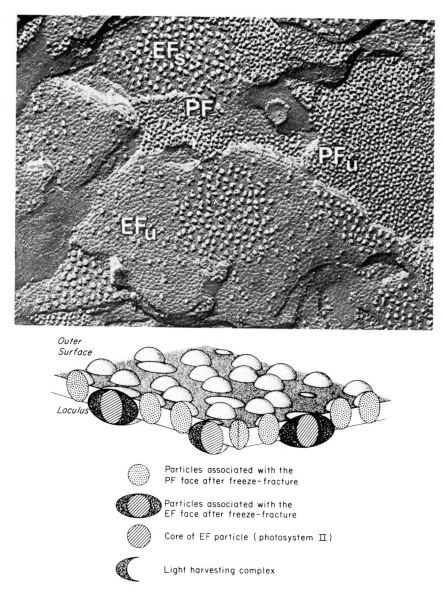

Fig. 9.1. Freeze-fracture electron micrograph of chloroplast membranes *(top)* and model illustrating the relationships of the particulate subunits of this membrane *(bottom)*. The prefix P or E is used to indicate outer or inner membrane leaflet fracture faces which are differentiated into stacked (PF$_s$, EF$_s$) and unstacked (PF$_u$, EF$_u$) regions. Particles on the EF face have been identified as photosystem II "core" units surrounded by aggregates of the light-harvesting chlorophyll a + b complex; particles on the PF face are considered to be photosystem I (containing P-700, cytochrome complexes and the integral membrane portion of the coupling factor). The particles seen in this state description are about 10^{-3} μm in diameter, which is outside the scale of Fig. 1.1. (*Nerium oleander* chloroplast membranes × 125,000; micrograph courtesy P. Armond; model from Armond et al., 1977)

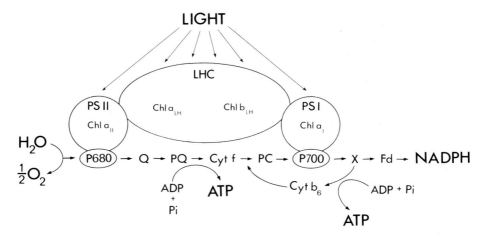

Fig. 9.2. Diagram of the light-harvesting and electron transport system of photosynthesis. Light quanta are absorbed by: Chl a $_{II}$ (chlorophyll a closely associated with photosystem II); Chl a$_{LH}$ and Chl b$_{LH}$ (chlorophylls bound in the light-harvesting pigment–protein complex, LHC). The light energy is then transferred from these chlorophyll molecules to P-680 (the reaction center chlorophyll of photosystem II) and to P-700 (the reaction center chlorophyll of photosystem I). Electron transport from H_2O to $NADP^+$ involves the following intermediates: Q, the primary electron acceptor from photosystem II; PQ, plastoquinone; $Cyt\,f$, cytochrome f; PC, plastocyanin; X, the primary electron acceptor for photosystem I, and Fd, ferredoxin. Products of this process are O_2, NADPH, and ATP. ATP is formed concomitantly with noncyclic electron flow ($H_2O{\rightarrow}NADP^+$). (Diagram by courtesy of P. Armond)

faces PF and EF, shown in Fig. 9.1, are differentiated into stacked (PF$_s$, EF$_s$), and unstacked membrane regions (PF$_u$ and EF$_u$). The particles on the PF fracture face are believed to contain components of the photosystem I complex and those on the EF fracture face to contain the components of the photosystem II complex and the light-harvesting chlorophyll complex (Armond et al., 1977). The functional relationships between these components are shown in Fig. 9.2. At this scale it is not helpful to further separate structure and function (states and processes) and these will be integrated in the following simple treatment.

Photosynthesis begins with the absorption of light quanta by the light-harvesting chlorophyll (a + b) pigment–protein complex (denoted LHC in Fig. 9.2) and chlorophyll a$_I$ and a$_{II}$ which are closely associated with photosystems I and II, respectively. As a result the energy level of these "antenna" chlorophyll molecules is raised; they are said to be in an excited state. The energy is then transferred from these chlorophyll molecules to the reaction centers of photosystem II (P-680) and photosystem I (P-700), at which oxidizing and reducing entities are formed. Each complex, consisting of several hundred antenna chlorophyll molecules together with one P-680 and one P-700 molecule is often referred to as a photosynthetic unit (PSU). The more antenna chlorophyll molecules per unit, the larger the photosynthetic unit is said to be. The first reductants formed upon excitation of photosystem II and I are denoted Q and X, respectively. These are the primary photochemical events. They are very fast, having half-times in the order of 10^{-12} s. The rate at which these photochemical reactions can proceed is solely dependent on the intensity and quality of the light. It is independent of other environmental factors such as temperature, as long as the integrity of the system is intact.

In contrast to the primary photoreactions, photosynthetic electron transport and photophosphorylation are only indirectly dependent on the quality and intensity of the light and they are strongly temperature-dependent. The primary oxidants and reductants produced by the photochemical events drive the transport of electrons from H$_2$O to NADP via a number of electron carriers, shown in Fig. 9.2. These include plastoquinone (PQ), cytochrome f (cyt f), plastocyanin (PC), and ferredoxin (Fd). The final electron acceptor is NADP. For each O$_2$ molecule evolved by oxidation of water (four electrons), two molecules of NADP are reduced to NADPH$_2$.

The photochemical reactions generate a higher chemical potential than needed to support the transfer of electrons from H$_2$O to NADP. This difference in chemical energy can be used to convert ADP to ATP in the process called photophosphorylation. Current data indicate that there are two "sites" of ATP formation. One of these "sites" is located in the electron transport chain between the two photochemical reactions and shown in Fig. 9.2. Another "site" of ATP formation is thought to be located close to photoact I, involving a cyclic flow of electrons mediated by cytochrome b$_6$ (cyt b$_6$). The overall ratio of ATP to NADPH formation depends not only on the number of phosphorylation "sites", but also on the efficiency with which photochemical energy is converted to ATP formation.

As discussed in the following section, at least 3.2 ATP and 2 NADPH$_2$ are required for the fixation and reduction of 1 molecule of CO$_2$ to sucrose or starch. The requirement is substantially greater in air of normal O$_2$ and CO$_2$ pressures and in C$_4$ photosynthesis is at least 5.2 ATP and 2 NADPH$_2$, i.e., the ATP:NADPH$_2$ ratio is 2.6. Measurements of the stoichimetry of electron transport and noncyclic photophosphorylation in isolated chloroplasts have yielded values rarely exceeding 1.3 ATP:NADPH$_2$. Unless the efficiency of noncyclic photophosphorylation in vivo greatly exceeds that found in isolated chloroplasts, this process is unable to meet the overall ATP requirement. Presumably, the additional ATP is generated in cyclic photophosphorylation, using photochemically reduced NADPH$_2$. These photophosphorylation processes would take place at the expense of an increased light requirement (decreased quantum yield) for the overall process of photosynthesis.

9.1.2 CO$_2$ Fixation and Metabolism

Three major biochemical variants of photosynthetic carbon metabolism are known in terrestrial plants: C$_3$ photosynthesis, so termed because the first product of CO$_2$ fixation is 3-phosphoglycerate (3-PGA), a compound having three carbon atoms; C$_4$ photosynthesis, so termed because the first product of CO$_2$ fixation is oxaloacetic acid (OAA), a compound having four carbon atoms; and crassulacean acid metabolism (CAM) so termed because this type of carbon assimilation was

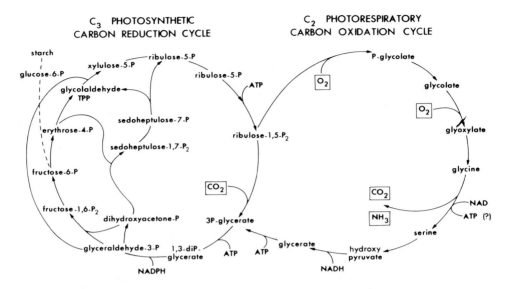

Fig. 9.3. The interlinked photosynthetic carbon reduction (PCR) cycle and photorespiratory carbon oxidation (PCO) cycle in leaves of C₃ plants. (Redrawn from Lorimer et al., 1978)

first discovered in succulent members of the family Crassulaceae. In plants capable of CAM the first products of CO_2 fixation may be either 3-phosphoglyceric acid or oxaloacetic acid, depending on prevailing plant or environmental factors (Osmond, 1978; Kluge and Ting, 1978). CAM is not known to occur in *Atriplex* or any other member of *Chenopodiaceae* but, as mentioned in Chap. 2, about 40% of the *Atriplex* species investigated so far are C_3 plants and the remaining 60% are C_4 plants.

The C_3 *Pathway*. In the C_3 photosynthetic pathway CO_2 is fixed and subsequently reduced to the level of carbohydrate in a cyclic process, which is here termed the photosynthetic carbon reduction (PCR) cycle and also commonly known as the Calvin-Benson cycle or the reductive pentosephosphate cycle (Bassham and Calvin, 1962). The PCR cycle is the only pathway capable of reducing CO_2 to carbohydrate and simultaneously regenerating the CO_2 acceptor. It is common to all green plants, whether C_3, C_4, or CAM, and the basis of autotrophic metabolism. As depicted schematically in the left portion of Fig. 9.3, CO_2 diffusing into leaf from the external air is fixed directly by ribulose-1,5-bisphosphate (RuP₂) carboxylase, the fundamental carboxylation enzyme of all photoautotrophs. In the presence of high CO_2 pressures or low O_2 pressures essentially all of the carbon flows through the PCR cycle. However, there is now strong evidence that in normal air (325 μ bar or 0.032% CO_2 and 210 mbar or 21% O_2), a substantial fraction of the carbon flows through the photorespiratory carbon oxidation (PCO) cycle shown in the right part of Fig. 9.3 (Lorimer et al., 1978).

The PCO cycle is initiated by the oxygenase activity of RuP₂ carboxylase, which results in the synthesis of P-glycolate, and the two cycles are linked by conversion of

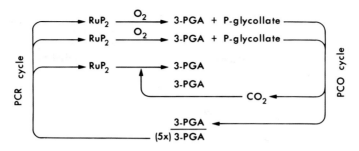

Fig. 9.4. The balanced photosynthetic carbon reduction (PCA) and photorespiratory (PCO) cycles at the CO$_2$ compensation point in leaves of C$_3$ plants. (From Lorimer et al., 1978)

P-glycolate to 3-PGA. Whereas the enzymes of the PCR cycle are entirely contained in the chloroplast stroma, the enzymes of the PCO cycle are partitioned between the chloroplast and two other organelles, the peroxisomes and mitochondria. The principal site of CO$_2$ release, termed photorespiration, is believed to be the mitochondrion and it is possible that this step may be coupled to electron transport and ATP synthesis under some circumstances (Woo and Osmond, 1976). The release of CO$_2$ in the PCO cycle is accompanied by the stoichiometric release of NH$_3$, and in the green cells of C$_3$ plants this NH$_3$ is rapidly reassimilated (Woo et al., 1978). The intracellular localization of component reactions and the flow of carbon through the PCR and PCO cycles have been reviewed by Schnarrenberger and Fock (1976) and Lorimer et al. (1978).

Many responses of net CO$_2$ exchange in plants with C$_3$ pathway photosynthesis can be explained in terms of the in vitro kinetic properties of RuP$_2$ carboxylase and the stoichiometry of the integrated PCR and PCO cycles. One particularly important consequence of the linked cycles is that as the CO$_2$ partial pressure in air is lowered, the rate of net CO$_2$ exchange will approach zero at the CO$_2$ compensation point, when CO$_2$ uptake in the PCR cycle is exactly balanced by CO$_2$ output in the PCO cycle. The reactions of Fig. 9.3 can be simplified to those shown in Fig. 9.4 which describes the balanced carbon flow at the CO$_2$ compensation point. Studies of ^{18}O$_2$ fixation in vivo provide strong support for the stoichiometry shown in Fig. 9.4 (Berry et al., 1978). When stomata close in the light the CO$_2$ concentration inside the leaf will fall to the CO$_2$ compensation point which will vary with temperature and O$_2$ concentration. That is, the cycles of carbon metabolism will continue to turn over rapidly and act as a sink for the products of photochemistry. The possible significance of these reactions under such conditions is discussed later (Chap. 9.2.2).

Many models of photosynthesis have been developed which are based on the competitive interactions of CO$_2$ and O$_2$ with RuP$_2$ carboxylase-oxygenase (Laing et al., 1974; Peisker, 1974; Hall and Björkman, 1975). Figure 9.4 shows that when this enzyme reacts with one mol of O$_2$, 0.5 mol of CO$_2$ is subsequently released via the PCO cycle. Net photosynthetis (P) is then given by Eq. (9.1)

$$P = v_c - 0.5\,v_o, \tag{9.1}$$

where v_c is the rate of carboxylation and v_o the rate of oxygenation. These rates are determined by the following kinetic relationships

$$v_c = \cfrac{V_c}{1 + \cfrac{K_c}{p_i^c}\left(1 + \cfrac{K_o}{p_i^o}\right)} \tag{9.2}$$

$$v_o = \cfrac{V_o}{1 + \cfrac{K_o}{p_i^o}\left(1 + \cfrac{K_c}{p_i^c}\right)}. \tag{9.3}$$

Where V_c, V_o are maximum velocities of the carboxylase and oxygenase activities; K_c, K_o are the Michaelis-Menten constants for CO_2 and O_2 and where p_i^c and p_i^o are the intercellular partial pressures of CO_2 and O_2. This simple model requires further modification to take into account the high concentration of RuP_2 carboxylase-oxygenase in vivo which determines the relationships with RuP_2, to accommodate the activation of both catalytic activities by CO_2, and to accommodate the temperature dependencies of kinetic properties. These considerations are outside the scope of our treatment but are fully developed in the models of Farquhar (1979a) and Farquhar et al. (1979). These analytical models are particularly useful in exploring the molecular basis of photosynthetic responses to temperature, light intensity, and atmospheric composition.

The C_4 Pathway. The C_4 pathway of photosynthesis is a complex biochemical and physiological accessory to the PCR cycle in which C_4 dicarboxylic acids are both primary products and intermediates of carbon assimilation, serving to transfer carbon to the PCR cycle. The primary carboxylation involves phosphoenolpyruvate (PEP) carboxylase, producing oxaloacetate (OAA) which is converted by malate dehydrogenase or aspartate transaminase to malate (MAL) or aspartate (ASP). Kinetic studies show that these C_4 acids are subsequently decarboxylated and carbon released is refixed via RuP_2 carboxylase. The residual 3-carbon compound, usually pyruvate (PYR), is reconverted to PEP, the substrate for the initial CO_2 fixation processes (Björkman, 1973; Hatch and Osmond, 1976).

The first reports of the C_4 pathway in dicotyledons were based on experiments with *Atriplex spongiosa* (Osmond, 1967b) and as a result of experiments with *A. rosea,* it was first recognized that the intracellular compartmentation of enzymes was the functional basis of the pathway (Björkman and Gauhl, 1969).

The separation of the two carboxylase enzymes of the C_4 pathway, and the enzymes involved in the regeneration of their substrates in adjacent cell layers has now been firmly established by several methods, and by the partial and complementary metabolic properties of isolated mesophyll and bundle-sheath cells (Björkman, 1973; Hatch and Osmond, 1976). Figure 9.5 shows a schematic localization of these reactions in one subtype of C_4 plants, the NAD malic enzyme type which includes C_4 *Atriplex* spp. This state description implies that an essential process for the functioning of the C_4 pathway is the movement of aspartate and alanine (ALA) between adjacent cell layers. Calculations suggest that the concentrations of these metabolites and the likely gradients in concentration are

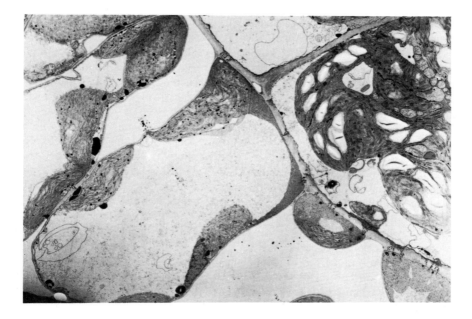

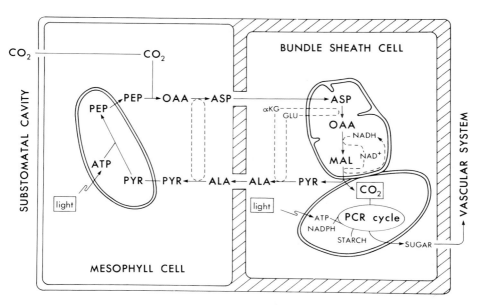

Fig. 9.5. Electron micrograph of leaf transection of the C$_4$ plant *Atriplex rosea* showing mesophyll cell *(left)* and bundle-sheath cell *(right)*. (From Boynton et al., 1970). The *lower* schematic diagram shows the locations of component processes of C$_4$ photosynthesis in *A. rosea*. The primary fixation of CO$_2$ occurs in the mesophyll cells and CO$_2$ is then transported to the bundle-sheath cells as the C-4 carboxyl of aspartate. This is decarboxylated in the mitochondria of the bundle-sheath cells and the CO$_2$ is refixed by the bundle-sheath cell chloroplasts. (Modified and redrawn from Hatch and Osmond, 1976)

sufficient to support the required fluxes of C_4 acids via the simple diffusion of these
metabolites in the symplasm through plasmodesmata connecting mesophyll and
bundle-sheath cells (Hatch and Osmond, 1976). This metabolite exchange involves
pathways between cells which are similar to those involved in ion movement
through the root cortex (Chap. 7.1.2). The metabolite exchange of C_4
photosynthesis is one of the most rapid symplastic transport systems known in
plants (Osmond and Smith, 1976).

The intracellular compartmentation of the component processes in C_4
photosynthesis is at least as complex as that discussed for the PCO cycle. They
involve the coordinated operation of chloroplasts in two adjacent cell layers, as well
as leaf mitochondria. In all C_4 plants, the ATP- and $NADPH_2$-requiring steps of
PEP synthesis and malate production take place in mesophyll cell chloroplasts, but
the carboxylation of PEP occurs in the cytoplasm (Fig. 9.5). *Atriplex* spp. belong to
a group of C_4 plants in which aspartate seems to be the principal C_4 acid formed in
mesophyll cells. The release of CO_2 from aspartate in bundle-sheath cells is rather
complex, and involves decarboxylation in the bundle-sheath cell mitochondria
(Fig. 9.5). Spatial separation of the enzymes responsible for C_4 acid synthesis and
degradation in different cells enables the C_4 pathway to function as a CO_2-
concentrating mechanism. The high activity of PEP carboxylase in mesophyll cells
can maintain low CO_2 concentration and relatively high malate or aspartate
concentration in these cells. At the same time, the high activity of decarboxylases
and lower activity of RuP_2 carboxylase in bundle-sheath cells maintain a high CO_2
concentration and low C_4 acid concentration in these cells. The high internal CO_2
concentration in C_4 plants has been measured directly (Hatch, 1971). A model of C_4
pathway photosynthesis, based on the sub-model for C_3 pathway photosynthesis,
but incorporating a resistance analogy which limits back-diffusion of CO_2 from the
bundle sheath to mesophyll, has been developed by Berry and Farquhar (1978).

The CO_2 concentrating function of the C_4 pathway is significant in another way.
The concentration of CO_2 in bundle-sheath cells is only possible if this
compartment is relatively "CO_2-tight"; i.e., if most of the carbon which enters as the
C-4 carboxyl of malic or aspartic acid is assimilated in these cells (Hatch and
Osmond, 1976). Because practically all the CO_2 fixed by PEP carboxylase is
subsequently refixed by RuP_2 carboxylase in a closed compartment, there is little
probability of discrimination among the isotopes of carbon during the latter step in
C_4 photosynthesis. In an open system, such as prevails in C_3 plants, and in
experiments with purified RuP_2 carboxylase in vitro, CO_2 fixation by this enzyme
shows discrimination against ^{13}C. This results in a $\delta^{13}C$ value, relative to a limestone
standard, of about $-30^0/_{00}$ (Lerman, 1975; Benedict, 1978). In C_4 plants, and with
purified PEP carboxylase in vitro, CO_2 fixation shows relatively little discrimination against ^{13}C. This results in a $\delta^{13}C$ value of about $-11^0/_{00}$ relative to the
limestone standard, which is not very different from the $\delta^{13}C$ value of normal air
($-7‰$). The less negative $\delta^{13}C$ value of C_4 plants is thus a product of two processes
which are the essence of the C_4 pathway: the primary assimilation of CO_2 via PEP
carboxylase and the subsequent conversion to reduced compounds via RuP_2
carboxylase in a relatively "CO_2-tight" space which precludes expression of the
discrimination properties normally associated with this enzyme. It will be recalled
that $\delta^{13}C$ values were examined to determine if hybrids of C_3 and C_4 *Atriplex* species

were capable of integrated C_4 metabolism (Chap. 4.4). Although several hybrids displayed anatomical and enzyme properties similar to those of the C_4 present, in no case examined did CO_2 assimilation result in $\delta^{13}C$ values comparable to *A.rosea* (Fig. 4.17). We concluded that these plants were functionally C_3 plants and a similar conclusion must be applied to all so-called "intermediary" forms of C_3 and C_4 plants identified up to the present.

9.2 Photosynthetic Responses to Light

Light, the driving force of photosynthesis, shows great variation from one habitat to another and may also show considerable seasonal variation within a given habitat. In extreme cases such as dense forests, the total number of quanta reaching the understorey vegetation in the course of a day may be 100 to 200 times smaller than that falling on the top of the canopy. Considerable variation in light intensity, or more correctly, quantum flux density (i.e., the number of quanta incident on a unit area per unit time), also exists within the canopy of a given plant stand and, of course, also occurs during the course of the day.

The great spatial and temporal variation in quantum flux density imposes a potential problem since on theoretical grounds one would expect that the photosynthetic characteristics yielding optimum performance at high quantum flux densities would be very different from those giving optimum performance at low quantum flux densities. For example, in dim light a high efficiency of quantum absorption by the light-harvesting pigments and an efficient conversion of excitation energy into chemical energy are of primary importance. As the quantum flux density increases, so does the need for a higher capacity of subsequent component "dark" reactions of photosynthesis. These include the photosynthetic electron transport chain, the carbon fixation process, and the diffusive transport of CO_2 via the stomatal path to the CO_2 fixation sites in the chloroplasts. To achieve this high capacity, an increased quantitative investment in electron carriers, enzymes, and other cellular components is presumably required. A balance point eventually would be reached at which the cost of this investment would outweigh the benefit. This balance point would obviously be strongly dependent on the light environment.

It has long been known that plants possess the ability to alter this balance in direct response to the availability of light, and this phenomenon has been termed light acclimation. It is also well known that the range of quantum flux densities in which different plants are able to respond in an adaptive manner is related to the light environment prevailing in their native habitats. Plants restricted to densely shaded habitats (obligate shade plants) are capable of photosynthetic light acclimation over a quantum flux density range, extending from extremely weak to moderate light, while in plants occupying exposed, sunny habitats (sun species) this range extends from moderately weak to very high light. As mentioned in a previous part of this book (Chap. 3.4.1) such genetically based differences in photosynthetic light acclimation potential also exist at the intraspecific levels. For example, "sun ecotypes" of *Solidago virgaurea* and *Solanum dulcamara* differ from "shade ecotypes" of the same species in a similar manner as do "sun species" and "shade species". Two recent reviews (Björkman, 1973; Boardman, 1977) present detailed treatments of photosynthetic adaptation to light; the latter review specifically deals with this topic.

9.2.1 Photosynthetic Light Acclimation in Atriplex triangularis

No *Atriplex* species is known to inhabit such low light habitats as the floors of dense forests, and all *Atriplex* regardless of whether they are C_3 or C_4 plants can probably be regarded as "sun plants". Nevertheless, many *Atriplex* species experience considerable temporal and spatial variation in the light climate. For example, *A. triangularis* may occupy bright, sunny beaches; it also occurs in marshes dominated by tall grasses or sedges such as *Spartina, Phragmites,* and *Scirpus.* In addition, substantial seasonal variation in quantum flux density occurs in many habitats occupied by this species. Capacity for photosynthetic light acclimation therefore can be expected to play an important part in the success of *A. triangularis.*

Björkman et al. (1972a) studied the effects of daily quantum flux density on the photosynthetic characteristics of *A. triangularis* and some of their results will be presented here in some detail. Seedlings of this species were grown in the Canberra Phytotron under three different light regimes: 5300, 3630, and 530 µeinstein cm^{-2} day^{-1} (einstein = mol of quanta). In the following, these will be referred to as high, intermediate, and low. The high regime is approximately the same as the average daily quantum flux density in the native seashore habitat at Pescadero, California during the peak for vegetative growth of this plant. The low regime may be considered as representing the minimum sustained daily quantum flux density experienced by this species; it is near the lower limit at which the experimental plants were able to grow. For comparison it might be mentioned that this low light regime is still some 20 times higher than that endured by extreme shade rainforest shade plants such as *Alocasia macrorrhiza* (Björkman and Ludlow, 1972). Other growing conditions for *A. triangularis* were identical. Daytime and nighttime temperatures were 25° and 20 °C, respectively; relative humidity was 70% and the plants were amply supplied with water and nutrients. All measurements were conducted on fully exposed, horizontally oriented leaves which had reached 75% to 90% of full expansion.

The light dependence of photosynthesis of single attached leaves from plants grown under the three light regimes is shown in Fig. 9.6. A detailed view of the initial portions of the same curves are given in Fig. 9.7. All rates are expressed on the basis of unit leaf area. It is clear that the net rate of photosynthesis at low quantum flux densities is highest in the low-light-grown leaves and lowest in the high-light-grown leaves, and that the quantum flux density required for light compensation increased with increased light during leaf development. These differences are wholly attributable to differences in the dark respiratory activity of the leaves for the photosynthetic quantum efficiencies are virtually the same. (The quantum yields on the basis of incident light (Φ_i) are given by the initial slopes of the light-dependence curves in Fig. 9.7. Thus, the increased overall efficiency in carbon gain (on a leaf area basis) at low quantum flux densities that occurs when the light during growth is decreased is wholly attributable to a decreased maintenance cost.

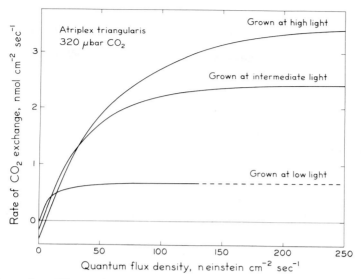

Fig. 9.6. Rate of net CO_2 uptake as a function of incident quantum flux density at 330 μbar CO_2 and actual stomatal conductances for *A. triangularis* leaves grown under three different light regimes. (From Björkman et al., 1972a)

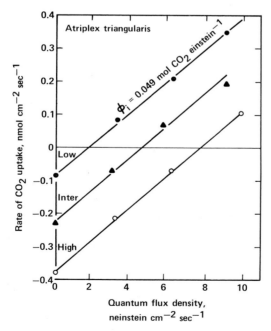

Fig. 9.7. Linear parts of the photosynthesis versus light dependence curves for *A. triangularis*, grown under three different light regimes. The slope of these lines (Φ_i) gives the quantum yield of photosynthesis. (O. Björkman, unpublished)

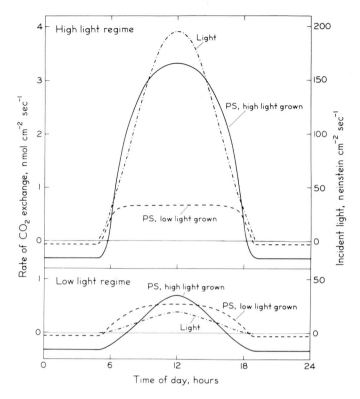

Fig. 9.8. Daily course of net CO_2 exchange of *A. triangularis* under a high light regime *(top)* and a low light regime *(bottom)*. The graphs show the actual responses of plants grown under the same light regimes as well as the simulated response of a low-light-grown plant in the high light regime and that of a high-light-grown plant in the low light regime. (O. Björkman, unpublished)

The capacity to respond favorably to increased quantum flux densities is very limited in the low-light-grown plants (Fig. 9.6). The rate of photosynthesis saturates already at quite low quantum flux densities and most of the light is wasted at high quantum flux densities. (In fact, sustained light levels in excess of 140 neinstein $cm^{-2} s^{-1}$ cause an inhibition of photosynthesis. This photoinhibitory effect will be discussed in Chap. 9.2.2.). The capacity to utilize bright light is much higher in the plants grown in the intermediate light regime and reaches a maximum in the plants grown in the high light regime. No photoinhibitory effect was observed in these plants at quantum flux densities up to 220 neinstein $cm^{-2} s^{-1}$.

Daily Carbon Balance. The importance of photosynthetic light acclimation to the success of the plant is illustrated in Fig. 9.8 and Table 9.1. The upper part of Fig. 9.8 shows the simulated course of quantum flux density during a clear day in a natural, open *A. triangularis* habitat. Also shown are the daily course of CO_2 exchange for the leaves of plants grown under the high light regime as well as the *predicted* course for plants grown under the low light regime. It is assumed that the

Table 9.1. Daily net photosynthetic productivity of *A. triangularis* leaves under two contrasting light regimes. (Data from Björkman et al., 1972a)

	Light = 5,300 μein-stein cm^{-2} day^{-1}			Light = 530 ein-stein cm^{-2} day^{-1}		
	A Actual High light- grown	B Predicted Low light- grown	Ratio, A/B	C Actual Low light- grown	D Predicted High light- grown	Ratio, C/D
(1) Leaf area basis (μmol CO_2 cm^{-2} day^{-1})						
Daytime net CO_2 uptake	117.5	30.5	3.85	17.89	11.85	1.51
Nighttime CO_2 release	11.2	2.2	5.1	2.25	11.25	0.20
Daily net CO_2 gain	106.3	28.3	3.76	15.64	0.65	24.1
(2) Leaf dry weight basis (μmol CO_2 g^{-1} day^{-1})						
Daytime net CO_2 uptake	2,641	1,725	1.53	1,011	266	3.80
Nighttime CO_2 release	253	127	1.99	127	253	0.50
Daily net CO_2 gain	2,388	1,598	1.50	884	13	68

photosynthetic CO_2 exchange characteristics of the low-light-grown plant remained unchanged. The lower part of Fig.9.8 shows the converse situation, i.e., the daily course of CO_2 exchange on a leaf area basis in a low light environment of a plant grown under this regime, and the *predicted* course for a plant grown in the high light, assuming that its CO_2 exchange characteristics would remain constant.

The total daily gains and losses obtained by integration of these curves are summarized in the upper part of Table 9.1. Under the high light regime, the high-light-grown plant is capable of almost four times as much daily net carbon gain as that predicted for a low-light-grown plant on a leaf area (or ground area) basis. This is entirely attributable to the increased capacity of the high-light-grown plants to make use of high quantum flux densities. The low-light-grown plant is capable of a substantial daily net carbon gain under the low light regime. By contrast, the daily net carbon gain by the leaves of the high-light-grown plant approaches zero, primarily because of the high respiratory activity of the leaves. Estimates show that the carbon balance of the whole plant would become negative because of the respiratory activity by roots and stems.

So far we have only considered the carbon balance of the plants in terms of photosynthetic carbon gain and respiratory maintenance costs on the basis of leaf area. We have not considered the difference in the costs involved for producing the leaves. Such costs are difficult to assess in absolute terms; however, an approximate estimate of the relative costs can be obtained if it is assumed that they are proportional to the specific leaf weight (dry organic matter per unit leaf area). This assumption is reasonably valid in comparisons of sun and shade leaves of the same

Atriplex triangularis

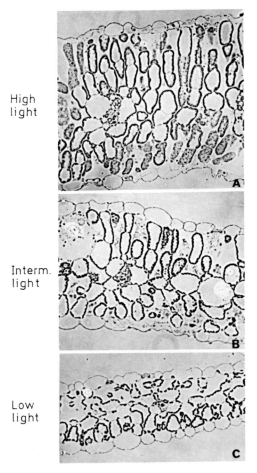

High
light

Interm.
light

Low
light

Fig. 9.9. Phase contrast light micrographs of leaf sections from *A. triangularis* plants, grown
under three light regimes. (From Björkman et al., 1972a)

genotype; it becomes much more questionable in comparisons of unrelated species
having strikingly different ratios of photosynthetic to nonphotosynthetic leaf
tissue. As shown in Fig. 9.9, there are large differences in leaf thickness between the
leaves grown under the three light intensity regimes. Such modifications have been
known for at least a century and are well documented for a multitude of species. The
low-light-grown leaves of *A. triangularis* had, on the average, only three to four
cells, the intermediate-light-grown leaves, five to six, and the high-light-grown
leaves six to seven cells across each section. The proportion of columnar palisade
cells also increases with increasing light. The number of chloroplasts across the leaf
section in the low-light-grown leaves was about one-half of that in the high-light-
grown leaves, as was also the number of cells. These anatomical differences are

Table 9.2. Photosynthetic activities and composition of the photosynthetic apparatus in leaves of *Atriplex triangularis*, grown under three different regimes. The numbers in parenthesis indicate the relative values compared with the low-light-grown plants. (Data from Björkman, unpublished)

Characteristic	High light-grown		Intermediate light-grown		Low light-grown	
Photosynthetic rate in normal air, nmol cm^{-2} s^{-1} [a]	13.33	(4.85)	9.58	(3.48)	2.75	(1.0)
Leaf specific weight, mg day wt. cm^{-2}	4.45	(2.51)	3.54	(1.99)	1.77	(1.0)
Leaf soluble protein, µg cm^{-2}	703	(2.40)	608	(2.07)	293	(1.0)
Cytochrome f content, n mol cm^{-2}	0.172	(2.53)	0.112	(1.65)	0.068	(1.0)
Cytochrome b_6 content, n mol cm^{-2}	0.253	(1.85)	0.227	(1.66)	0.136	(1.0)
Cytochrome b_{559} content, n mol cm^{-2}	0.241	(1.76)	0.218	(1.60)	0.136	(1.0)
Chlorophyll (a + b), n mol cm^{-2}	56.9	(1.21)	56.8	(1.21)	46.7	(1.0)
Leaf absorptance (400–700 nm)	0.83	(1.04)	0.83	(1.04)	0.80	(1.0)
P_{700} content, n mol cm^{-2}	0.132	(1.32)	0.131	(1.31)	0.100	(1.0)
Q content, n mol cm^{-2}	1.2	(1.2)	1.2	(1.2)	1.0	(1.0)
Quantum yield of PS II activity, rel. units cm^{-2} [a]	1.2	(1.2)	1.2	(1.2)	1.2	(1.0)
Ratio, mol chlorophyll (a + b) per mol P_{700}	430	(0.93)	433	(0.93)	468	(1.0)
PSI-driven electron transport, nEq. cm^{-2} s^{-1} [b]	20.8	(2.9)	14.5	(2.0)	7.3	(1.0)
PSII-driven electron transport, nEq. cm^{-2} s^{-1} [b]	19.3	(3.8)	11.7	(2.3)	5.1	(1.0)
RuP$_2$ carboxylase activity, nEq. cm^{-2} s^{-1}	26.0	(4.17)	19.24	(3.08)	6.24	(1.0)

[a] Measured at rate-limiting quantum flux densities
[b] Measured at high quantum flux density; 4 n equivalents correspond to 1 n mol CO_2 or O_2

reflected in the average specific leaf weight which was 17.7, 35.3, and 44.5 mg dry matter cm^{-2} leaf area, for the low, intermediate, and high-light-grown leaves, respectively. Leaf protein content increased in similar proportions (Table 9.2). Thus, specific leaf weight probably serves as satisfactory measure of the relative cost of producing the leaves.

The daily photosynthetic carbon gains of the high- and low-light-grown leaves on the basis of leaf dry matter content are shown in the lower part of Table 9.1, left column. After correction for carbon allocation to nonleaf organs and respiratory activity of these organs, the daily net CO_2 values shown in this table should be proportional to the relative growth rate (R_w) of the plants. Under the high light regime both the daytime and the daily net carbon gain values are substantially higher for the high-light-grown plants than those predicted for the low-light-grown plants. It is thus clear that the benefit received in terms of photosynthetic gain outweighs the additional cost involved in producing a high-capacity photosynthetic machinery. It may be mentioned that in reality the gains would probably be considerably greater than indicated by these estimates, since the failure to increase the photosynthetic capacity is likely to result in damage of the photosynthetic machinery induced by high light (Chap. 9.2.2). Other considerations are, of course, the additional space and the support structures required for a full display of the

Table 9.3. Stomatal conductance, stomatal frequency, and photosynthetic capacity in leaves of *Atriplex triangularis* grown under three different light regimes. The numbers in parenthesis indicate the relative values in comparison with the low-light-grown plants. (Data from Björkman et al., 1972a)

Characteristic	High light-grown	Intermediate light-grown	Low light-grown
Stomatal conductance, $cm^{-2} s^{-1}$	1.02 (2.91)	0.82 (2.34)	0.35 (1.0)
Stomatal frequency, number mm^2	407 (2.18)	335 (1.80)	186 (1.0)
Photosynthetic rate in normal air, $nmol\ cm^{-2} s^{-1}$			
a) Actual stomatal conductances	3.33 (4.85)	2.40 (3.48)	0.69 (1.0)
b) Infinite stomatal conductances	3.96 (5.36)	2.77 (3.74)	0.74 (1.0)
Photosynthetic rate in low O_2 and saturating CO_2, $nmol\ cm^{-2} s^{-1}$	5.27 (5.48)	3.66 (3.80)	0.96 (1.0)

much larger area of thinner leaves in exposed environments, as well as possible effects on the water relations of the plant.

As shown in the right lower column of Table 9.1, the ability for photosynthetic acclimation to low light environments is vital. Failure of the plant to produce leaves of the shade type would have led to a near-zero daily carbon balance for the leaves and a negative carbon balance for the plant as a whole.

Capacities of Component Steps of Photosynthesis. It is evident that photosynthetic light acclimation in *Atriplex triangularis,* and many other species, involves changes in the amount of photosynthetic tissue per leaf area. The thickness of the leaf parenchyma (Fig. 9.9), the specific leaf weight, and the leaf soluble protein all increase with increasing light during growth. As shown in Table 9.2, the capacities of several component steps of photosynthesis also increase in a similar manner when expressed on a leaf area basis. The capacity for electron transport, driven by photosystem (PSI), and the contents of the photosynthetic electron carriers, cytochrome f, b_6, and b_{559} (high potential), increase in approximately the same proportions as specific leaf weight and protein content. However, light acclimation is not simply a matter of adjusting the number of cells across the leaf section. Important changes in the relative capacities of component reactions and in the composition of the photosynthetic apparatus occur as well (Tables 9.2 and 9.3). One striking feature is that the amount of light-harvesting chlorophyll does not increase in proportion to specific leaf weight, leaf protein content, or the number of chloroplast-containing cell layers. In fact, the chlorophyll per unit leaf area remains virtually constant and so does leaf absorptance. Thus, the fraction of the incident light that is absorbed by the leaves remains high, irrespective of the growth light regime. Similarly, only small changes occur in the amount per unit leaf area of P_{700}, the reaction center of photosystem I. Neither do any significant changes take place in the amount of the primary electron acceptor of photosystem II (denoted Q in Table 9.2 and the scheme shown in Fig. 9.1). Consequently, there are only small differences in the ratio of total chlorophyll to reaction centers between the low-light-grown and high-light-grown leaves.

It thus appears that the capacity per unit leaf area of those steps which determine the efficiency of quantum absorption and conversion in primary photochemical steps remains nearly as high in the low-light-grown as in the high-light-grown leaves, despite the fact that the former have many fewer chloroplasts or less photosynthetic machinery per unit leaf area. This is in agreement with the constancy of the quantum yield at low quantum flux densities of photosystem activity by isolated chloroplasts (Table 9.2), as well as the constancy of the quantum yield of CO_2 fixation by intact leaves (Fig. 9.7).

By contrast, the capacity of those component steps which determine the photosynthetic capacity at light saturation, show considerable increases as the light during growth is increased. As already mentioned, light-saturated photosystem I activity and the amounts of associated electron carriers (per unit leaf area of chlorophyll) show increases that are approximately proportional to the increase in specific weight. Similar changes in these and other electron carriers [ferredoxin (Fd), plastoquinone (PQ), α-tocopherol quinone] were found by Grahl and Wild (1972) for *Sinapis alba,* grown under different light regimes. It is noteworthy that at least in *A. triangularis,* the increase in the light-saturated capacity of photosynthesis by intact leaves, on a leaf area basis, that took place when the plants were grown at high as compared with low light, was greater than the increase in specific leaf weight. This was also observed for certain component photosynthetic reactions, notably the light-saturated activity of photosystem II-driven electron transport and the activity of RuP_2 carboxylase (Table 9.2). Similar changes in RuP_2 carboxylase activity have also been reported for a number of other species (Boardman, 1977). It seems probable that this increase in enzyme activity is caused by an increased content of Fraction-I protein. (This protein is considered to be identical to RuP_2 carboxylase).

The large changes that occur in the ratio of chloroplast components that govern the efficiency of quantum absorption and conversion to those that determine the capacity of photosynthesis at high quantum flux densities are also reflected in the chloroplast structure. Figure 9.10 shows electron micrographs of chloroplasts from high and low-light-grown leaves of *Atriplex triangularis*. Similar changes are also known to occur in many other species (e.g., Ballantine and Forde, 1970; Skene, 1974). The most prominent differences are that chloroplasts from leaves grown at low light are larger and have a greater number of thylakoids in their grana stacks. The amount of chlorophyll per chloroplast volume is much greater than in chloroplasts from high-light-grown leaves. This serves to counterbalance the effects of a decreased number of cell layers, thereby maintaining a high leaf absorptance. Conversely, chloroplasts from high-light-grown leaves appear to have a greater volume of stroma per chloroplast; this is consistent with their greater activity of RuP_2 carboxylase and content of Fraction-I protein (and perhaps also of other Calvin-cycle enzymes).

Diffusive Transfer of CO_2. There is little doubt that the light-induced changes in both the overall size in the photosynthetic apparatus and the concomitant changes in the relative capacities of its constituent processes, discussed above, are prerequisite to photosynthetic light acclimation. However, it is evident that the much larger intrinsic capacity for photosynthesis at high quantum flux densities of high-light-grown leaves might not be fully expressed unless a concomitant increase

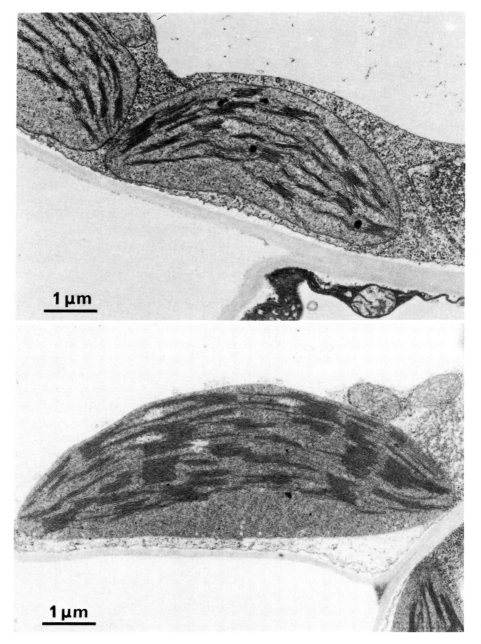

Fig. 9.10. Electron micrographs of chloroplasts from *A. triangularis* leaves, grown under a high light regime *(top)* and a low light regime *(bottom)*. (From Björkman et al., 1972a)

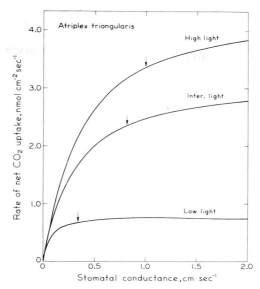

Fig. 9.11. Rate of light-saturated CO_2 uptake as a function of stomatal conductance, calculated from the experimentally determined relationship between CO_2 uptake and intercellular CO_2 pressure. *Arrows* indicate the actual stomatal conductances of the different leaves at light saturation and 320 µbar ambient CO_2 pressure. (From Björkman et al., 1972a)

in the capacity of the diffusive transport of CO_2 takes place also. Two diffusive CO_2 transport components can be distinguished: (1) the transfer of CO_2 between the external leaf surface via the stomata to the intercellular spaces, and (2) the transfer of CO_2 within the leaf. The effect of the first component can be assessed by simultaneously measuring the stomatal conductance to water vapor exchange and the CO_2 dependence of photosynthesis. From such measurements the relationship between photosynthesis and stomatal conductance can be readily obtained (cf. Fig. 9.29). The results obtained with *A. triangularis* are shown in Fig. 9.11. The actual stomatal conductance values, indicated by arrows, were determined under saturating light in normal air (21% O_2, 320 µbar CO_2). In each case the leaves so adjusted their stomatal conductance that it imposed only a minor effect on photosynthesis. A further increase in stomatal conductance would have only a small effect on photosynthesis, but it would, of course, result in an increased water loss (in direct porportion to the stomatal conductance). Clearly, the low light-saturated rate of photosynthesis of low-light-grown leaves cannot be attributed to their lower stomatal conductance; the differences in photosynthetic capacity between growth regimes would be even greater if the stomatal conductance had been infinite (Table 9.3). However, if increased light during growth had not resulted in an increased stomatal conductance, then the light-saturated photosynthetic rates of the intermediate and high-light-grown plants would have risen to about only two-thirds of the rates actually achieved.

Adjustment of stomatal conductance is therefore a significant component in photosynthetic light acclimation in *A. triangularis* as well as in other species

(Ludlow and Wilson, 1971; Charles-Edwards and Ludwig, 1975). In *A. triangularis*, the differences in stomatal conductance are probably a direct result of differences in stomatal frequency (Table 9.3). On a leaf area basis, the high-light-grown leaves had three times as many stomata on the upper epidermis and nearly twice as many on the lower epidermis as the low-light-grown leaves. No significant differences were detected in the stomatal length or guard cell size.

The quantitative importance of physical barriers to the diffusive transfer of CO_2 between the intercellular spaces and the photosynthesizing cells cannot be determined by direct measurements. Calculations show that, in general, the transfer resistance in the gaseous phase inside the leaf is likely to be small in comparison with that imposed by the stomata. The internal transfer resistance is directly related to the exposed cell wall area of the chloroplast-containing cells in the leaf. Fortunately, as the number (or size) of the photosynthetic cells contained under a given leaf surface area increases, so does the internal cell wall surface area. As a result, the ratio of internal cell wall surface area to leaf area automatically increases with increasing leaf thickness (Turrell, 1936; Nobel et al., 1975). The conductance to CO_2 transfer therefore increases in proportion to the increased demand for CO_2 supply to the cells.

A note of caution may be warranted in interpreting the effects of CO_2 transfer limitations in photosynthesis. The correlation that exists between light-saturated rate, photosynthetic rate, and conductance to CO_2 transfer on the basis of unit leaf area does not necessarily mean that these conductances, singly or taken together, are themselves the limiting factors in photosynthesis. At least for *A. triangularis*, and probably for many other species, the size and the composition of the photosynthetic apparatus are undoubtedly the primary determinants in photosynthetic light acclimation. This is further supported by the observation that the differences in photosynthetic capacity between the plants, grown under the three light regimes, are approximately the same regardless of whether the comparisons are made at limiting or saturating CO_2 concentrations (Table 9.3).

9.2.2 Photoinhibition

It is well known that attempts to grow obligate shade plants in strong light result in damage to the photosynthetic apparatus. These detrimental effects include disturbances of the chloroplast structure, chlorophyll bleaching, and severe inhibition of photosynthesis. Such effects can also be seen in leaves of sun plants which have developed in weak light and are then abruptly transferred to strong light. It may be recalled that low-light-grown leaves of *Atriplex triangularis* showed a time-dependent decline in photosynthetic rate at high quantum flux densities. New leaves developing under the high light regime show no such detrimental effects and even mature leaves may gradually recover and attain the photosynthetic characteristics of sun leaves. This acclimation capacity is very limited in shade plants and if the light is sufficiently strong the detrimental effects show a progressive increase with time (Björkman and Holmgren, 1963; Gauhl, 1969, 1970, 1976).

Although chlorophyll bleaching occurs at more advanced stages of photoinhibition by strong light, this is not the primary effect or the cause of the

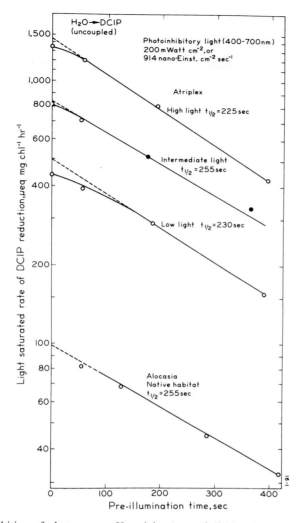

Fig. 9.12. Inhibition of photosystem II activity (rate of dichloroindophenol reduction) as a function of time of preillumination of isolated chloroplasts from different sources to photoinhibitory light. (From Björkman et al., 1972)

inhibition of photosynthesis. Following sudden exposure of shade-grown leaves to bright light, a substantial decrease in the quantum yield of photosynthesis can be observed as early as one or several hours before any change in chlorophyll content can be detected (Björkman and Holmgren, 1963; Björkman, 1966a; Gauhl, 1976). Similar results were also obtained when isolated chloroplasts from spinach (Jones and Kok, 1966a,b), *Atriplex triangularis* and the shade plant *Alocasia macrorrhiza* (Björkman et al., 1972a) were exposed to bright light (see Fig. 9.12). Jones and Kok concluded that such photoinhibition is caused by an inactivation of the photosynthetic reaction centers by transfer of excitation energy from the light-harvesting pigment molecules in excess of that which can be used in photosynthesis. An important question is, which factor, or combination of factors, is responsible for the much higher sensitivity to photoinhibition in obligate shade plants, and low-light-grown leaves of sun plants in comparison with high-light-grown leaves of sun plants. One possibility is that the intrinsic sensitivity of the pigment-reaction center complex is somehow greater, or that there is a much larger number of light-harvesting pigment molecules per reaction center in the shade leaves so that, at any given quantum flux density, each reaction center would receive more excitation energy. These proposed mechanisms are not supported by experimental evidence, however.

Björkman et al. (1972a) compared the sensitivity of isolated chloroplasts from *Atriplex triangularis,* grown under the three different light regimes, and from the shade species *Alocasia macrorrhiza,* grown in the extremely weak light of its native rainforest habitat. The chloroplasts were illuminated for different time periods with a xenon arc source in the absence of an electron acceptor. The activities of photosystems I and II were subsequently measured in the presence of appropriate electron donors and acceptors. No loss of chlorophyll could be detected. An example of the kinetics of photoinactivation of photosystem II activity is shown in Fig. 9.12. Identical first-order slopes were obtained for photosystem II activity with the *Atriplex* chloroplasts, irrespective of growth light regime, and also with the *Alocasia* chloroplasts. The sensitivity to photoinhibition of photosystem I activity was also the same for the different plants, although the decrease in activity was not first-order. These results indicate that there are no appreciable differences in the intrinsic photolability of the chloroplasts between high- and low-light-grown leaves of *Atriplex* or between this species and the shade species *Alocasia.* Thus, another explanation must be sought for the different tolerances observed with intact leaves.

A simple and attractive explanation is that resistance to photoinhibition is related to the photosynthetic capacity at high quantum flux densities. In a high-light-grown leaf the capacities for electron transport and carbon fixation are high relative to the light-harvesting capacity. Thus, a large fraction of the absorbed light is used in photosynthesis even at high quantum flux densities. In a low-light-grown plant (or in an obligate shade plants) this ratio is considerably lower. Consequently, at any given high quantum flux density, a much smaller fraction of the absorbed light can be utilized for photosynthesis, thus resulting in a great excess of excitation energy. In other words, the low capacity for electron transport or carbon fixation would cause the reaction centers to be closed for a higher proportion of time, thus increasing the probability of their photoinactivation. In sun plants such as *Atriplex triangularis,* light acclimation enables the plant to adjust the ratio of electron transport and carboxylation capacity to light-harvesting capacity. As we have seen, this adjustment enables the plant to make much more effective use of the available light. If the above hypothesis is correct, then the resistance to photoinhibition would change as a direct consequence of this adjustment. In obligate shade plants the limited genetic capacity for such an adjustment would thus result in both a low capacity for photosynthesis at high quantum flux densities and a low resistance to photoinhibition.

Osmond and Björkman (1972) suggested the possibility that photorespiration may serve to alleviate photoinhibition in high light environments. This would

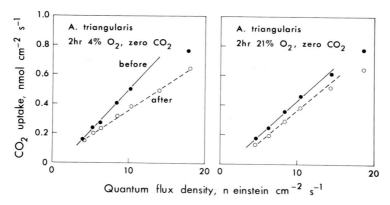

Fig. 9.13. Photoinhibition of photosynthesis in *A. triangularis* leaves as a result of exposure for 2 h to strong light in a CO_2-free atmosphere, containing 4% O_2 and 96% N_2. Photoinhibition is indicated by the decreased slope of the light dependence curve *(left)*. When leaves were illuminated in the same way but in the presence of 21% O_2, there is no change in this slope, indicating that internal production of CO_2 via the PCO cycle is sufficient to prevent photoinhibition. (C.B. Osmond, unpublished)

become especially important when the supply of CO_2 to the leaf is strongly reduced because of stomatal closure. Under such conditions the photochemical traps would evidently remain closed, rendering them very vulnerable to photoinactivation unless an alternative terminal electron acceptor can be used in place of CO_2. According to Osmond and Björkman, the photorespiratory cycle (Chap. 9.1.2) provides a mechanism that could permit O_2 to serve this function. The production of phosphoglycolate by the reaction of RuP_2 with O_2 (catalyzed by RuP_2 oxygenase/carboxylase) and the subsequent photorespiratory metabolism of phosphoglycolate may enable the carbon cycle to turn over without net gain of carbon while the consumption of photochemically generated energy per turn of the cycle remains high. This would cause the photochemical traps to remain open for a greater proportion of the time, thus decreasing the probability of photoinactivation.

Evidence that the internal generation of CO_2 in photorespiration and the turnover of carbon through the linked PCR and PCO cycles at the compensation point protects against photoinhibition in leaves exposed to CO_2-free air has been obtained with several species of C_3 plants. If leaves are exposed to high irradiance for 1–3 h in CO_2-free N_2 with 1% O_2, conditions which minimize the oxygenation of RuP_2 and prevent photorespiration, their capacity for CO_2-saturated, light-saturated photosynthesis is reduced by about 75% (Powles and Osmond, 1978). The apparent quantum requirement for CO_2 fixation (Chap. 9.2.3) is also substantially inhibited. Figure 9.13 shows that the apparent quantum yield in leaves of *A. triangularis* was halved following 2-h irradiation at 200 neinstein $cm^{-2} s^{-1}$ in an atmosphere of 4% O_2 and no CO_2. However, no photoinhibition was observed in similar experiments conducted in the presence of 21% O_2.

Photoinhibition following illumination in low O_2 seems to be qualitatively similar to that observed when shade leaves are transferred to high radiance

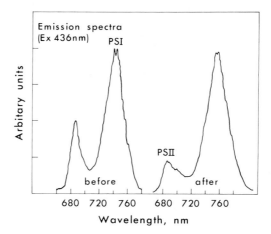

Fig. 9.14. Fluorescence emission spectra at 77 K of chloroplasts isolated from *Phaseolus vulgaris* leaves before *(left)* and after *(right)* exposure of the leaves to strong light for 3 h in a CO_2-free atmosphere, containing 1% O_2 and 97% N_2. The peaks at approximately 685 and 740 nm are thought to correspond to fluorescence emission by photosystem II and photosystem I, respectively. Loss of fluorescence emission at 685 nm *(left)* is indicative of a loss of integrity in the association between the light-harvesting chlorophyll and the reaction center of photosystem II (cf. Fig. 9.2). (Graph by courtesy of S.W. Thorne)

environments under normal CO_2 and O_2 pressures. Preliminary studies of the chloroplasts isolated from photoinhibited leaves of *Phaseolus* after illumination under low CO_2 and O_2 pressures show that the activity of photosystem II-driven electron transport was about 50% inhibited. Photosystem I-driven electron transport was less reduced following photoinhibition (Powles and Osmond, 1979). That photosystem II is particularly sensitive to inactivation under these conditions is further indicated by the fluorescence emission spectra measured at liquid N_2 temperature. Following photoinhibition the fluorescence emission at 680 nm, characteristic of photosystem II, was drastically reduced (Fig. 9.14).

These responses suggest that the oxygenation of RuP_2 and the subsequent internal generation of CO_2 permit the orderly dissipation of photochemical energy in the linked PCR and PCO cycles of illuminated leaves when external CO_2 supply is limiting. Thus, although the presence of 21% O_2 in the atmosphere limits the efficiency of CO_2 fixation in C_3 plants, it may provide a "fail safe" mechanism which protects the photosynthetic apparatus when the supply of CO_2, the natural electron acceptor, is cut off following stomatal closure in response to water stress (Chaps. 8.2.1; 9.4.1).

9.2.3 Light Utilization by C_3 and C_4 Atriplex spp.

The responses to light regime, discussed in the preceding section, are probably valid for other *Atriplex* species as well as for higher plants in general. There are also no particular reasons to expect that the mode of light acclimation in C_4 species would be substantially different than in C_3 species. This is consistent with available

Table 9.4. Quantum yield for CO_2 uptake (mol CO_2/absorbed einstein) of different C_3 and C_4 species measured at a leaf temperature of 30 °C and an atmospheric CO_2 pressure of 325 µbar. (Data of Ehleringer and Björkman, 1977)

Photosynthetic pathway	O_2 partial pressure	
	20 mbar	210 mbar
C_3 species		
Atriplex glabriuscula	0.072	0.051
Atriplex heterosperma	0.073	0.053
Atriplex hortensis	0.073	0.055
Atriplex triangularis	0.073	0.051
Encelia californica	0.074	0.052
Encelia farinosa	0.074	0.052
Plantago lanceolata	0.074	0.052
Mean and standard deviation	0.0733 ± 0.0008	0.0524 ± 0.0014
C_4 species		
Atriplex argentea	0.052	0.052
Atriplex rosea	0.054	0.052
Atriplex sabulosa	0.054	0.054
Atriplex serenana	0.055	0.054
Tidestromia oblongifolia	0.054	0.054
Mean and standard deviation	0.0538 ± 0.0011	0.534 ± 0.0009

experimental results (Louwerse and Van der Zweerde, 1977). However, because of the increased efficiency in the utilization of limiting CO_2 pressures, conferred by C_4 photosynthesis, and the higher intrinsic energy cost of this pathway, it may be expected that C_3 and C_4 plants would exhibit marked differences in photosynthetic light response characteristics. As will be shown in this section, experimental results are in accordance with these expectations.

Quantum Yield of Photosynthesis. At low quantum flux densities the quantum yield (ϕ) is constant and maximal. Ideally, this maximum yield should be independent of external conditions such as temperature or CO_2 pressure and, if expressed on the basis of absorbed quanta, should reflect the efficiency of energy conversion in the photoacts alone. All plants having the same energy requirements in terms of ATP and $NADPH_2$ needed for the fixation of one CO_2 and the evolution of one O_2 molecule should therefore have the same quantum yield (ϕ_a) if expressed on the basis of the light absorbed by the light-harvesting pigments. In C_3 photosynthesis a theoretical minimum of eight to ten absorbed quanta is considered to be needed; however, in practice the quantum requirement (ϕ^{-1}) is higher. Several factors contribute to this: (1) components other than the photosynthetic pigments account for some of light absorption by a leaf, (2) light of shorter wavelengths, especially blue light, cannot be used as efficiently as red light; this is partly attributed to a less efficient energy transfer of light absorbed by carotenoid pigments than of that absorbed by chlorophyll, (3) conditions under which the quantum yield measurements are made are not completely ideal. These factors raise the quantum requirement to about 12 quanta per CO_2 (i.e., $\phi_a = 0.08$) for a normal

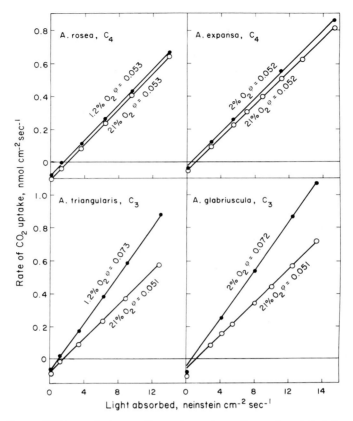

Fig. 9.15. Rate of CO_2 uptake in two C_3 species and two C_4 species of *Atriplex* as a function of absorbed quantum flux density in 21% and 2% O_2. Leaf temperature was 30° and CO_2 pressure was 325 μbar. The *slopes of the lines* give the quantum yield. (Redrawn from Björkman et al., 1970 and Ehleringer and Björkman, 1977)

green leaf in light of the spectral composition of natural daylight. This is what is commonly measured under high CO_2 or low O_2 partial pressure. However, in air of normal O_2 and CO_2, ϕ_a is even lower (Björkman, 1966b, 1971a). Figure 9.16 shows the ϕ_a values for the C_3 species *Atriplex glabriuscula* and *A. triangularis* and the C_4 species *A. rosea* and *A. expansa*, determined at a CO_2 partial pressure of 325 μbar, 21% and 2% oxygen, and 30 °C leaf temperature. Similar values have also been obtained for many other plants; some of these values are listed in Table 9.4.

In C_3 species of *Atriplex* and other genera, the quantum yield at low O_2 is independent of CO_2 pressure over a wide range (Björkman, 1971a; Ehleringer and Björkman, 1977). It is also independent of temperature and approaches the expected value of about $\phi_a = 0.08$ mol CO_2 einstein^{-1} (Björkman, 1971; Ehleringer and Björkman, 1977). This independence of ϕ_a on temperature and CO_2 is consistent with theoretical considerations, namely that the efficiency of energy transfer from the light-harvesting pigments to the photosystem reaction centers and

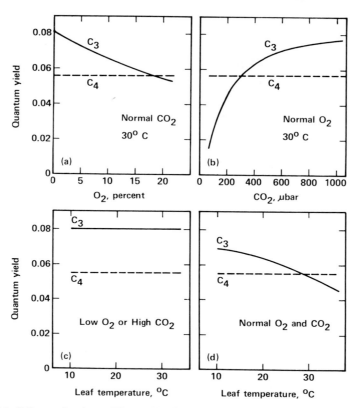

Fig. 9.16. Effects of oxygen, CO_2, and leaf temperature on the quantum yield of CO_2 uptake for C_3 and C_4 species. The indicated relationships are based on many different studies, e.g., Björkman (1966b, 1971a,b) and Ehleringer and Björkman (1977)

the conversion of this energy to primary photoproducts should be the sole determinant of ϕ_a. It is also assumed, however, that the energetic requirement for CO_2 fixation remains constant. In C_3 plants this requirement is met only under conditions where the O_2 inhibition of net photosynthesis is absent, i.e., at low O_2 or high CO_2 concentrations (Fig. 9.15, lower left). In normal air, O_2 inhibition is pronounced both at limiting and saturating quantum flux densities. O_2 and CO_2 act in a competitive manner so that high CO_2 abolishes the inhibitory effect of O_2 (Fig. 9.15, upper parts). At 325 μbar and 30 °C, 21% O_2 causes a reduction in quantum yield of about 30% (Fig. 9.15, upper left). This inhibition decreases at lower temperatures and increases at high temperatures (Fig. 9.16, lower right). These interactions between quantum yield and temperature, CO_2, and O_2 partial pressure are largely the product of the kinetic properties of RuP_2 carboxylase-oxygenase (Chap. 9.1.2) and have been accurately modeled on these bases (Berry and Farquhar, 1978).

In C_4 plants, including all C_4 *Atriplex* investigated to date, the quantum yield is not inhibited by 21% O_2 even at low CO_2 pressures (Fig. 9.16, upper right). In accordance with this, ϕ_a is also independent of temperature (Fig. 9.16, lower part). The probable reason for this lack of an O_2 inhibition of photosynthesis is the operation of a CO_2-concentrating mechanism in C_4 plants so that the effective CO_2 pressure at the site of CO_2 fixation in the PCR cells is sufficiently high to completely suppress the inhibitory effect of 21% O_2 (Chap. 9.1.3). However, the intrinsic quantum yield for C_4 plants is approximately 30% lower than in C_3 plants. This is in accordance with a higher intrinsic energy cost for the operation of C_4 photosynthesis. As shown earlier (Chap. 9.1.3), C_4 photosynthesis theoretically requires 5 ATP and 2 $NADPH_2$ per CO_2 fixed while C_3 photosynthesis requires 3 ATP and 2 $NADPH_2$. The intrinsic quantum yields for C_3 and C_4 plants are here considered to be 0.080 and 0.056 mol CO_2 einstein^{-1}, respectively. These values are based on numerous determinations on higher plants. The absolute values may vary somewhat depending on the method of measurement, but should fall within 10% of the above values.

In a primitive atmosphere of high CO_2 or low O_2 partial pressure, or both, the quantum yield of C_3 plants would thus be substantially superior to that of C_4 plants, irrespective of temperature. However, under the present atmospheric composition (21% O_2, 330 µbar CO_2), the quantum yield of C_3 plants exceeds that of C_4 plants only at low temperatures and the situation is reversed at high temperatures. This is so because the quantum yield of C_4 plants remains independent of environmental conditions having a constant value of $\phi_a = 0.056$, whereas in C_3 plants, the quantum yield becomes temperature-dependent, varying from $\phi_a \approx 0.070$ at 10 °C to $\phi_a \approx 0.048$ at 35 °C. This temperature dependence of ϕ_a in C_3 plants is evidently caused by the inhibitory effect of O_2. This inhibition almost precisely offsets the additional ATP requirement of the C_4 pathway at 25° to 30 °C, resulting in nearly identical ϕ_a values in this temperature range. The comparative efficiencies of C_3 and C_4 plants in utilizing low quantum flux densities thus depend on the temperature regime.

The comparative photosynthetic responses of C_3 and C_4 plants to increased quantum flux densities beyond the light-limited range are strongly temperature-dependent and will be discussed later (Chap. 9.3.1).

Light Utilization in Relation to Leaf Absorptance. As we have seen, the efficiency of utilization of low quantum flux densities, in terms of absorbed light, shows little or no variation among plants with the same pathway of CO_2 fixation. Any variation in quantum yield in terms of incident light (Φ_i) can probably be attributed to differences in absorptance (α). Glabrous leaves with a normal chlorophyll content per unit leaf area (40 to 60 µg Chl cm^{-2}) have very similar absorptance values ($\alpha = 0.80$ to 0.85). Absorptance values substantially lower than 0.75 are almost exclusively attributable to a high reflectance of the leaf surface, usually because of the presence of a pubescent layer on the upper epidermis. Extreme examples of such pubescence were given in Chap. 3.1.1 for *Encelia farinosa*. Many *Atriplex* species, especially those occupying hot arid environments, have leaves which are covered by bladder-like trichomes. Examples of such *Atriplex* leaves were illustrated in Figs. 2.6 and 7.20, and the function of epidermal trichomes in salt excretion was discussed (Chap. 7.5.4).

In many desert *Atriplex* a considerable seasonal shift in the reflective properties of the leaves can be observed and this is undoubtedly related to the development and hydration state of the leaf trichomes. The study of Mooney et al. (1977b) provides an example of such seasonal changes in the optical properties of *A. hymenelytra* leaves. New but fully expanded leaves of this species formed during the mild winter in its native habitat on the floor of Death Valley, California, reflect about 31% of the incident light; leaf transmittance is only about 1% giving an absorptance (α) of 0.68. At this time the leaves have a high water content (about 83%) and the trichomes are filled with liquid. As the season progresses, the heat load and aridity increase and the already existing, as well as the newly developing leaves, increase their reflectances, reaching a maximum of about 58% in the early autumn. This is equivalent to an absorptance of only 0.41. At this time the water content of the leaves has decreased to about 50% and the trichomes have dried out and collapsed. Salts, mostly NaCl, previously contained within the trichomes, crystallize on the leaf surface as these glands collapse (Fig. 7.20), resulting in an increased reflectance throughout the spectrum (Fig. 8.9). Mooney et al. provided evidence indicating that this increased reflectance is an important adaptive feature; it minimizes the heat load on the leaves, resulting in markedly lower leaf temperatures and increased water use efficiency during the extremely hot and dry summer and autumn months. During this time stomatal conductance is strongly reduced in this species and photosynthesis thus remains light-saturated for most of the day in this high light environment in spite of the high leaf reflectance. Under these conditions, the decrease in quantum yield (ϕ_i) due to low leaf absorptance is far outweighed by the benefits gained by a lowered leaf temperature and reduced water loss. An additional advantage of low leaf absorptance could be that it helps in protecting the photosynthetic machinery from photoinhibitory effects which may otherwise be expected to take place in this intense light environment when the overall rate of photosynthesis is strongly reduced as a result of a low stomatal conductance. As expected, the quantum yield (ϕ_i) at low quantum flux densities of *A. hymenelytra* leaves was approximately proportional to the reduction in leaf absorptance, indicating that ϕ_a remained constant throughout the season. Thus, under certain conditions, high efficiency of light utilization at low quantum flux densities is evidently sacrificed in order to meet other requirements that are imposed by the environment. Regulation of the reflective properties of the leaves is a simple and effective means to achieve the proper balance.

9.3 Photosynthetic Responses to Temperature

Although heat is neither a driving force nor a reactant in photosynthesis, this process, like all other metabolic processes, is strongly influenced by temperature. With the exception of the primary photochemical reactions, which are independent of temperature as long as the functional integrity of the chloroplast membranes is maintained, the rates of all component photosynthetic reactions are strongly temperature-dependent. The potential increase in reaction rate that takes place when the temperature is increased depends on the activation energy of the particular component reaction. In general, a 10 °C temperature rise causes a 1.5- to 2.5-fold increase in the potential reaction rate. The actual rise in the overall rate of photosynthesis is strongly dependent on other interacting factors such as light and

CO_2 availability. The effect of temperature is smallest in weak light and is maximal in strong light and saturating CO_2. Conversely, the light dependence of photosynthesis is smallest at low temperature and maximal at high temperatures and saturating CO_2.

While high temperature potentially is beneficial to photosynthesis by, allowing the temperature-dependent reactions to proceed at high rates, a point is reached at which a further temperature rise results in a decline in the photosynthetic rate. There are several causes for this decline. One is related to the increase with temperature of the inhibitory effect of O_2 under conditions of limiting CO_2. This applies only to C_3 plants since no inhibition of photosynthesis by atmospheric O_2 concentrations occurs in C_4 plants. The implications of the O_2 inhibition on the temperature dependence of the quantum yield of photosynthesis were treated in Chap. 9.2.3. The temperature dependence of photosynthesis at high quantum flux densities is also affected by the O_2 inhibition. The O_2 effect is especially pronounced at low intercellular space CO_2 pressures and increases with temperature. This tends to decrease the optimum temperature for net photosynthesis in C_3 plants. High CO_2 pressure eliminates the O_2 inhibition effect as well as any limitation to photosynthesis caused by diffusive transfer of CO_2. In general, the optimum temperature of photosynthesis thus is considerably higher under saturating CO_2 than it is in normal air. When the temperature is increased beyond the optimum for CO_2-saturated photosynthesis, a time-dependent and largely irreversible decline in photosynthetic rate is observed, indicating a high-temperature inactivation of vital components of the photosynthetic machinery. The temperature at which such thermal inactivation sets in is strongly dependent on the species and growth temperature. It is obviously of paramount importance that this critical temperature exceeds the maximum temperature encountered by the plant in its native habitat.

The habitats occupied by different *Atriplex* species show great differences in temperature regime. Cold arctic coastal habitats and hot, low deserts probably represent the extremes for the genus (Chap. 5). For example, members of the *A. triangularis* complex occur on the Arctic Ocean coasts of Norway and the Soviet Union where daytime air temperatures during the growing season are generally below 10 °C. In contrast, species such as *A. lentiformis, A. hymenelytra,* and *A. polycarpa* occur on the floor of Death Valley, California, one of the hottest places on earth, where summer temperatures generally reach 46 °C and not infrequently exceed 50 °C. However, temperature also shows considerable seasonal variation in a given habitat. The thermal load imposed on a desert plant which is active in the winter is not necessarily greater than that imposed on a summer-active plant in the Arctic.

When considering photosynthetic adaptation to temperature it should also be realized that it is the *leaf* temperature, prevailing during that period of the day when light intensity is relatively high, which is important. The temperature that a photosynthesizing leaf will assume depends primarily on the radiant energy absorbed by the leaf, the convective transfer of sensible heat between the leaf and the surrounding air, the loss of latent heat from the leaf by transpiration, and reradiation of heat from the leaf surfaces (Chaps. 8.1.5; 8.2.2). The temperature of a sunlit leaf may thus deviate considerably from that of the air. In general, the leaf tends to be warmer than the air at low air temperature whereas it tends to be cooler at very high air temperatures. The transpiration rate may have a strong influence on leaf temperature. A rapidly transpiring leaf may be significantly cooler than the surrounding air whereas a leaf exposed to full sunlight and transpiring at a low rate may be as much as 5° to 10 °C warmer than the air. Wind speed also affects leaf temperature, primarily by increasing the transfer of sensible and latent heat between the leaf and the air.

In addition to its direct and profound effect on photosynthetic metabolism, temperature strongly affects the efficiency of water use in photosynthesis because of the strong temperature dependence of the saturation vapor pressure of the water inside the leaf (Fig. 8.1).

9.3.1 Photosynthetic Characteristics of Cold-Adapted and Heat-Adapted C_3 and C_4 Plants

Interactions Between Light and Temperature. In Chap. 9.2.1 we discussed the various leaf factors that underlie the photosynthetic capacity of C_3 plants at high quantum flux densities. A major factor in determining this capacity is the amount of photosynthetic apparatus (electron acceptors, photosynthetic enzymes, etc.) present per unit area of leaf. Where the intrinsic capacity is high, stomatal conductance also becomes an important factor. The photosynthetic capacity at high quantum flux densities of a given leaf is also strongly dependent on external factors, especially CO_2 pressure and temperature.

In comparative studies of C_3 and C_4 plants aimed at assessing the influence of the C_4 photosynthetic pathway on the photosynthetic capacity at high quantum flux densities it is therefore highly desirable that these factors are as similar as possible in the plants being compared. It is further desirable that the plants be comparable from an ecological viewpoint. For example, comparisons between a C_3 shade plant and a C_4 sun plant would obviously introduce an undesirable bias. As will be shown in the following section, similar considerations apply to comparisons involving cold-adapted C_3 plants and heat-adapted C_4 plants.

Figure 9.17 shows the light dependence of photosynthesis for a pair of cold-adapted C_3 and C_4 species, *Atriplex glabriuscula* and *A. sabulosa*. Both of these species are native to cool seashore habitats in northern Europe and northeastern North America where they often are found growing together. These plants were grown at a daytime temperature of 16 °C and the same high light regime. Leaf absorptances, chlorophyll contents, specific leaf weights, and leaf protein contents, as well as stomatal conductances, were similar in the two species. The light dependence of photosynthesis was determined at 16 °C and air of normal O_2 and CO_2 pressures.

On the whole, the light responses of the species are remarkably similar but two differences can be distinguished. The first is the somewhat greater efficiency of the C_3 *Atriplex* in the utilization of low quantum flux densities; this is in accordance with our previous conclusion that the quantum yield of C_3 photosynthesis exceeds that of C_4 photosynthesis at low temperatures. The second is the somewhat higher capacity for light-saturated photosynthesis of the C_4 *Atriplex*. This is in accordance with the expected higher efficiency of C_4 photosynthesis in utilizing CO_2 of low concentrations. That this efficiency is indeed higher in the C_4 species is shown in

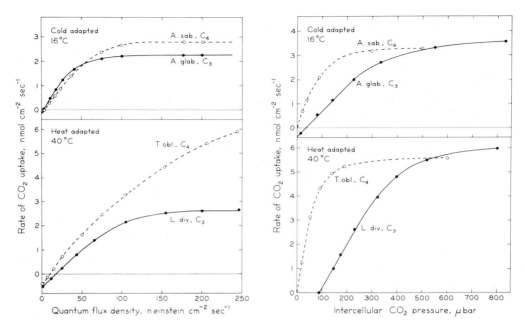

Fig. 9.17. Rate of CO_2 uptake as a function of quantum flux density *(left)* and intercellular CO_2 pressure *(right)* in cold-adapted *(top)* and heat-adapted *(bottom)* C_3 and C_4 species. The light dependence curves were determined at 330 µbar ambient CO_2 pressure. In each case the temperature of measurement was the same as the daytime growth temperature

Fig. 9.17. The calculated daily carbon gain does not differ appreciably between these two species at a temperature of 16 °C since the higher capacity of the C_4 species for photosynthesis at high quantum flux densities is approximately counterbalanced by its lower efficiency at low quantum flux densities. This is also consistent with the observation that the relative growth rates are quite similar in the two species under this temperature regime (Chap. 10). At even lower temperatures a point is reached at which photosynthesis also becomes CO_2-saturated in the C_3 *Atriplex;* the rates of photosynthesis at high quantum flux densities then become equal in the two species while the quantum yield of the C_3 plant becomes even more superior to the C_4 plant (not shown). As a result the daily carbon gain by the C_3 plant will exceed that of the C_4 plant.

At higher temperature this relationship is reversed. Because of the higher efficiency of utilization of low CO_2 pressures, the photosynthetic capacity at high quantum flux densities can be considerably expanded with increasing temperature in the C_4 species. The quantum yield of the C_4 *Atriplex* remains constant but declines in the C_3 *Atriplex*. Thus the relative advantage of C_4 photosynthesis in terms of daily carbon gain increases with temperature. However, for reasons unrelated to presence or absence of C_4 photosynthesis neither of these cold-adapted *Atriplex* species is able to respond favorably to an increase in temperature above 30 °C.

To illustrate the influence of C_4 photosynthesis on the photosynthetic light dependence at high temperature regimes, similar to those prevailing in hot desert habitats, we will use another pair of C_3 and C_4 species, *Larrea divaricata* and *Tidestromia oblongifolia*. These plants are native to the floor of Death Valley, California, and both are active during the hot season although the C_3 species, *L. divaricata*, also remains active during the cooler part of the year. Both species possess exceptionally high thermal stabilities of their photosynthetic machineries (Chap. 9.3.4). So far no C_3 *Atriplex* species with an equally high thermal stability has been found, but certain hot desert C_4 *Atriplex*, such as *A. lentiformis*, approach *T. oblongifolia* in this respect.

The light-dependence curves determined at 40 °C for the two heat-adapted species are shown in Fig. 9.17. Both species were grown at a daytime temperature of 40 °C, and the same high light regime. Again, the leaf absorptances, chlorophyll contents, and stomatal conductances were similar in the two species, but in this case the specific leaf weights and leaf protein contents were somewhat higher in the C_3 species. Nevertheless, their photosynthetic capacities at CO_2 saturation were almost identical (Fig. 9.17) so they should well serve the purpose of our comparisons. The difference in utilization of low CO_2 pressures is comparable to that generally found between other C_3 and C_4 species. It is evident from Fig. 9.17 that the photosynthetic rate of the C_3 plant is much inferior to that of the C_4 plant at all quantum flux densities. As expected, the quantum yield of the C_3 plant is much lower than in the C_4 plant at this high temperature. The light-saturated capacity of photosynthesis of the C_4 plant is so high that light continues to exert a partially limiting influence even at quantum flux densities comparable to full noon sunlight (approximately 200 neinstein cm^{-2} s^{-1}). This is in accordance with the extremely high efficiency of utilization of low CO_2 concentration in this C_4 species (Fig. 9.17). In spite of the very high rate of CO_2 uptake, the intercellular CO_2 pressure remains essentially saturating over the full range of quantum flux densities. In the C_3 plant the CO_2 pressure strongly limits photosynthesis; this effect is especially pronounced at high quantum flux densities.

The examples given above and many other studies clearly demonstrate that C_4 photosynthesis confers the potential of an increased ability to fix carbon at high quantum flux densities. This effect may be small or nonexistent at low temperature and paramount at high temperature. C_4 photosynthesis also affects the quantum yield at low quantum flux densities; at low temperature the extra energetic cost in terms of ATP requirement exceeds the benefit derived from the presence of this pathway whereas at high temperature the opposite is found.

Temperature Dependence at High Quantum Flux Densities and CO_2 Pressures. The interaction between the light dependence of photosynthesis and temperature in C_3 and C_4 plants was discussed in detail in the previous section. A few conclusions regarding the effect of temperature emerged. The rate of light-limited photosynthesis is essentially independent of temperature in C_4 plants, but in C_3 plants this rate decreases with increasing temperature. This was attributed to the presence of an inhibitory effect of O_2 in C_3 but not in C_4 plants. At high quantum flux densities photosynthesis is strongly temperature-dependent in both C_3 and C_4 plants, but C_4 photosynthesis confers a particular advantage at higher temperatures. This was attributed to the higher efficiency of utilization of low CO_2

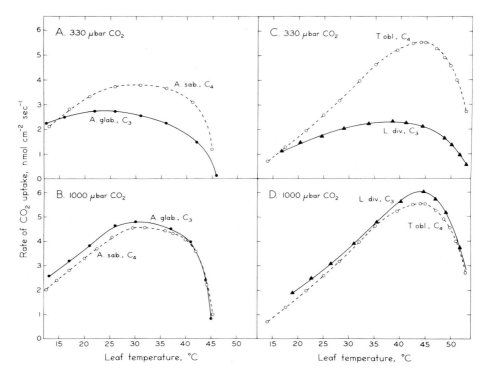

Fig. 9.18. Temperature dependence of net CO_2 uptake at 330 μbar CO_2 *(top)* and 1,000 μbar *(bottom)* of cold-adapted *Atriplex sabulosa* (C_4) and *A.glabriuscula* (C_3) and of heat-adapted *Tidestromia oblongifolia* (C_4) and *Larrea divaricata* (C_3). The two *Atriplex* species were grown at 16°; the other two species were grown at 45° daytime temperature. Based on data of Björkman et al. (1975), Mooney et al. (1978), and O. Björkman (unpublished data)

pressures in C_4 plants. The higher efficiency in all probability is due to the C_4 pathway serving as a CO_2-concentrating mechanism, thereby increasing the effective CO_2 pressure at the site of CO_2 fixation by the Calvin cycle and thus also suppressing the inhibitory effect of 21% O_2.

If these conclusions are correct then we should expect that the difference in temperature dependence between our paired C_3 and C_4 species would largely disappear if the external CO_2 pressure was raised to the point where photosynthesis in both types of plant is CO_2-saturated. As shown in Fig.9.18, this is indeed what is observed. At a CO_2 pressure of 330 μbar the photosynthetic rates of the two cold-adapted *Atriplex* are about the same below 16 °C (Fig.9.18a). As the temperature is raised, the photosynthetic rate of C_4 *A.sabulosa* shows a marked increase reaching a maximum at about 34 °C. In C_3 *A.glabriuscula* the increase in photosynthesis in response to increasing temperature in the low range is much smaller and the rate reaches a maximum at about 25 °C. Increasing the CO_2 pressure to 1,000 μbar (Fig.9.18b) results in a much more pronounced temperature dependence in the 12° to 30 °C range in *A.glabriuscula* while the change in *A.sabulosa* is much smaller; the temperature-dependence curves for the two species thus become virtually identical over the full temperature range with a temperature optimum at about 30 °C.

Similar effects of increasing the CO_2 pressure were obtained with the heat-adapted species even though the temperature dependencies of photosynthesis of these plants are strikingly different from the cold-adapted species. Both heat-adapted species have similar low rates at low temperatures and their temperature optima occur at very high temperatures (Fig. 9.18 c). The C_4 plant is able to respond very favorably to increasing temperature up to the optimum which occurs at 45 °C. In the C_3 plant the rate ceases to increase with temperatures above 35 °C and the optimum is broader. At 45 °C the rate of the C_3 plant is only about 4% of that of the C_4 plant. Increasing the CO_2 pressure to 1,000 µbar (Fig. 9.18 d) causes no change in the rate or temperature dependencies of photosynthesis in the C_4 species, confirming that normal atmosphere CO_2 pressure is sufficiently high to saturate photosynthesis in this plant. In contrast, the photosynthetic rate of the C_3 plant is strongly enhanced at all temperatures above 20 °C, especially at higher temperatures, reaching a three-fold increase at 45 °C. As a result, the rates and temperature dependencies of photosynthesis of the C_3 and C_4 species become virtually indistinguishable at the high CO_2 pressure.

Thus there is little doubt that the superior performance of the C_4 species at high temperature in normal air is caused by the operation of a CO_2-concentrating mechanism. It is also evident that the potential advantage conferred by C_4 photosynthesis can only be fully expressed if certain other prerequisite factors are also present. One such factor obviously is a sufficiently high thermal stability of the photosynthetic apparatus. The temperature-dependence curves, determined at high CO_2 concentrations, indicate that the cold-adapted C_3 and C_4 species possess similarly low thermal stabilities while the two heat-adapted C_3 and C_4 species possess similarly high stabilities. Thermal stability thus appears to be independent of C_4 photosynthesis per se.

That C_4 plants are often tolerant of high temperatures is therefore likely to be a result of their evolutionary history. It seems probable that C_4 photosynthesis evolved (polyphyletically) under conditions where the selective advantage of this CO_2-concentrating mechanism would be especially great, i.e., in environments with high insolation, high temperatures, and limited water supply. The C_3 progenitors of these C_4 plants were therefore likely to possess a number of adaptive characteristics, including a high thermal stability, that enabled them to occupy these environments. C_4 plants now growing in these types of environment are likely to have retained these adaptive features, whereas those that have succeeded in colonizing very different habitats, such as the cool oceanic species *A. sabulosa*, may have lost them and acquired other adaptive features instead. It is likely that many features such as rooting habit, stomatal behavior, and other response patterns to the water and temperature regimes must be altered in order for a plant to succeed in a new contrasting environment because the characteristics required for efficient performance in one type of environment may be mutually exclusive to those required for efficient performance in the other. For example, high temperature tolerance of the photosynthetic machinery may well preclude a high photosynthetic capacity at low temperature and vice versa. This would be the situation if high thermal stability of chloroplast membranes or proteins requires structural or compositional alterations that are incompatible with their efficient functioning at low temperature.

9.3.2 Environmentally Induced Acclimation to Temperature

In the preceding section comparisons of photosynthetic characteristics were made between "cold-adapted" and "heat-adapted" plants. The term cold-adapted was used to designate plants that are native to and were grown in a low temperature environment; heat-adapted was used to designate plants that are native to and were grown in a high temperature environment. It is well known that the growth-temperature regime may have pronounced influence on the photosynthetic characteristics of a given plant. In certain instances the changes in photosynthesis characteristics brought about by a change in the temperature regime result in a superior photosynthetic performance under the new regime. This phenomenon is often called photosynthetic temperature acclimation. The term should be reserved for changes that confer an advantage and not be used for any change in photosynthetic response that occurs when the growth regime is altered. Frequently, such changes may rather indicate a failure to tolerate the new conditions than an adaptive change.

The potential for temperature acclimation may vary considerably between different species in a manner reflecting the temperature regimes of their native habitats. In plants from environments exhibiting only a moderate temperature variation during the period of vegetative growth, the potential for photosynthetic acclimation appears to be relatively limited. Figure 9.19 illustrates the effects of the growth temperature on the temperature dependence of photosynthesis in plants representing three different habitat types. *Atriplex glabriuscula* and *A. sabulosa* are restricted to cool oceanic habitats in which the temperature variation during the growing season is relatively small. *T. oblongifolia* is a winter dormant plant whose growth in its native Death Valley habitat is restricted to three or four extremely hot summer months. *Atriplex hymenelytra, A. lentiformis,* and *Larrea divaricata* occur in this same habitat but these species are evergreen and must cope with a great seasonal change in temperature. The mean daily maximum temperature in Death Valley ranges from 18 °C in December and January to 46 °C in July and August and summer temperatures frequently reach or exceed 50 °C. The highest temperature ever recorded in Death Valley is 56 °C and the lowest is -9.4 °C.

As shown in the upper part of Fig. 9.19, coastal *A. glabriuscula* (C_3) and *A. sabulosa* (C_4) are incapable of photosynthetic acclimation to a 40 °C daytime temperature growth regime. Although an upward shift in the optimum temperatures for photosynthesis does take place when the plants are grown at 40 °C, the absolute photosynthetic rates at this temperature are not higher than in the plants grown under the cool regime. Instead, the rates are reduced to about one-half of the values found in the 16 °C grown plants. Similar results were obtained with a coastal ecotype of *A. lentiformis* (Chap. 3.4.2, Fig. 3.5). Growing this plant under a 43 °C regime resulted in a 50% reduction of photosynthesis at this temperature in comparison with 23 °C grown plants (Pearcy, 1977). Although these coastal species are capable of some degree of photosynthetic acclimation within the 15° to 30 °C range (Björkman, unpublished), it is evident that the acclimation potential is very limited.

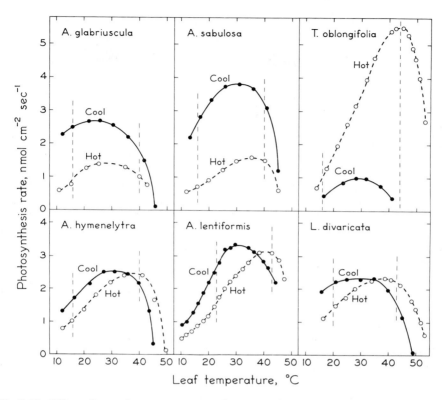

Fig. 9.19. Effect of growth temperature on the rate and temperature dependence of light-saturated net CO_2 uptake for a number of C_3 and C_4 species, native to habitats with contrasting thermal regimes. The "hot" growth regimes were 40 °C for *A.glabriuscula*, *A.sabulosa*, and *A.hymenelytra*, 43 °C for *A.lentiformis* and 45 °C for *T.oblongifolia* and *L.divaricata*. The "cool" growth regimes were 23 °C for *A. lentiformis*, 20 °C for *L.divaricata* and 16 °C for the other species. (Data for *A.glabriuscula*, *A.sabulosa*, *A.hymenelytra*, and *T.oblongifolia* from Björkman et al., 1977; and O.Björkman, unpublished; *A.lentiformis* from Pearcy, 1977; and *L.divaricata* from Mooney et al., 1978)

The response of *Tidestromia oblongifolia* is in sharp contrast with those of the coastal species. This thermophilic species is capable of extremely high photosynthetic rates at high temperatures when grown in a hot regime, but the rate declines steeply when the temperature is reduced to below 40 °C (Fig. 9.19 upper right). Growing this species at 30° to 35 °C rather than 40° to 45 °C results in a downward shift of the optimum temperature by several degrees with little or no effect on the absolute photosynthetic rate at the respective optima (not shown), indicating that this species is capable of some temperature acclimation. However, lowering the growth temperature to below 20 °C causes a drastic reduction in photosynthetic capacity at all temperatures (Fig. 9.19 upper right) and growth comes to an almost complete halt. The severely depressed photosynthetic capacity of such low-temperature-grown plants is evident under a wide range of measurement conditions including any combination of quantum flux densities, CO_2

pressures and temperatures. The low photosynthetic activity is not caused by stomatal closure or other factors affecting the diffusive transport of CO_2.

Although the temperature range for efficient photosynthetic performance of *T.oblongifolia* is in sharp contrast with that of the coastal species, both kinds of plants evidently possess a very limited potential for photosynthetic temperature acclimation. A considerably greater potential is found in the evergreen desert species that are subjected to a very large seasonal variation in temperature. Field studies of photosynthesis conducted on *Atriplex hymenelytra* (Pearcy et al., 1974), *A.lentiformis* (Pearcy and Harrison, 1974), and *Larrea divaricata* (Mooney et al., 1978) show that these species remain photosynthetically active throughout the year. A seasonal shift takes place in the photosynthetic temperature-dependence characteristics so that the photosynthetic capacity at low temperatures is increased during the cool season and the thermal stability of photosynthesis is increased during the hot season, permitting the plants to remain photosynthetically active at temperatures that in the absence of such acclimation would result in a severe inhibition of photosynthesis. This seasonal shift in the temperature dependence is especially pronounced in *A.lentiformis* and *L.divaricata*. While the optimum temperature for photosynthesis exhibits considerable seasonal changes there is little change in the absolute capacity for photosynthesis at this temperature as long as the water supplied to the plants is sufficient to avoid severe water stress and stomatal closure.

Changes in the temperature-dependence characteristics of photosynthesis similar to those occurring seasonally in the field can be effected by growing the plants under different temperature regimes while other environmental conditions are kept constant (Fig.9.19, lower part). In all three species the temperature dependence of light-saturated photosynthesis shifted with the temperature regime under which the plants were grown in a similar manner as the seasonal shifts observed in the field. The plants grown under the cool regimes exhibit considerably higher rates at low temperatures than do the plants grown in the hot regime and vice versa. Photosynthesis measurements made on *L.divaricata* and *A.hymenelytra*, grown under an intermediate temperature regime, also showed that the temperature dependences of these plants were intermediate between those grown under the cool and the hot regimes (Mooney et al., 1978; Björkman, unpublished data). The absolute photosynthetic rates at the respective optimum temperatures were similar irrespective of growth temperature regime in each of the three species. Analyses were made of a number of factors that potentially can contribute to the observed changes in the temperature dependence of net photosynthesis in normal air. These include stomatal conductance and nonstomatal limitations of CO_2 diffusive transport, dark respiration rates, and in the C_3 species, oxygen inhibition of photosynthesis. None of these factors, separately or taken together, can account for the observed acclimation responses, indicating that they are primarily the result of changes in intrinsic characteristics of the photosynthetic apparatus at the cellular or subcellular levels.

Another species which is capable of photosynthetic acclimation over a very wide temperature range is *Nerium oleander*, a C_3 perennial shrub native to the arid regions of South West Asia and North Africa (Björkman et al., 1978). This species is extensively grown in a wide range of climates in California, from the cool coast to

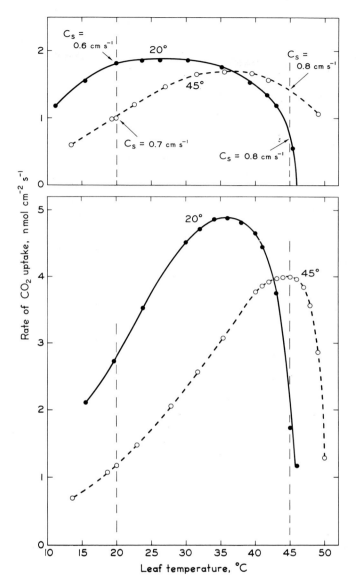

Fig. 9.20. Temperature dependence of light-saturated CO_2 uptake for *Nerium oleander*, grown at two contrasting temperature regimes ($20°$ day/$15°C$ night and $45°$ day/$32°C$ night). Measurements were made in normal air ($330\,\mu bar$ CO_2, 21% O_2) *(top)* and at high CO_2 pressure (750 to $850\,\mu bar$ CO_2, 20% O_2) *(bottom)*. Also shown are the stomatal conductances (C_s) measured at the two daytime growth temperatures. These temperatures are indicated by the *broken vertical lines*. (Data from Björkman et al., 1978)

the floor of Death Valley where it remains active throughout the year. As shown in Fig. 9.20 (top) the photosynthetic capacity in normal air, measured at 20 °C is twice as high in plants grown at 20 °C as in those grown at 45 °C. Conversely, the photosynthetic rate measured at 45 °C is twice as high in plants grown at 45 °C as in those grown at 20 °C. Stomatal conductance measurements showed that the differences in photosynthetic rates between the 20 °C-grown and the 45 °C-grown plants, measured either at high or low temperatures, cannot be attributed to stomatal factors. Figure 9.20 (bottom), which shows the photosynthetic temperature dependence of the same plants measured at saturating CO_2 pressures, clearly demonstrates that growth at low temperature increases the intrinsic photosynthetic capacity of the leaves. The possible factors underlying this increased photosynthetic capacity are discussed in the next section (Chap. 9.3.3).

The superior photosynthetic performance of 45 °C-grown plants at high temperature is primarily due to an increased thermal stability of the photosynthetic machinery. The temperature optimum and the threshold temperature above which CO_2-saturated photosynthesis exhibits a steep decline are shifted upward by about 6 °C compared with the 20 °C-grown plants, and the temperature at which an irreversible heat inactivation begins is shifted upward to a similar extent. The nature of this increase in thermal stability is discussed in Chap. 9.3.4. One can distinguish between two separate effects of growth temperature on the photosynthetic characteristics: an increased capacity for photosynthesis at low, rate-limiting temperatures in plants grown under a cool regime and an increased thermal stability of key components of the photosynthetic apparatus in plants grown under a hot regime.

It is especially noteworthy that even fully expanded, mature leaves are capable of full acclimation following transfer of *N. oleander* plants from a cool to a hot growth regime and vice versa (Fig. 9.21). The time required for a complete reversal of photosynthetic characteristics is less than two weeks. Recent measurements by J. Berry further indicate that *N. oleander* plants, initially grown at 20 °C may acquire

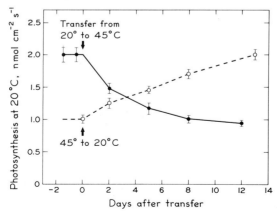

Fig. 9.21. Time course of change in light-saturated photosynthetic capacity, measured at 20 °C, following transfer of 45°- and 20 °C-grown *Nerium oleander* plants to the opposite growth temperature regimes. (Data from Björkman et al., 1979)

the same high thermal stability as plants grown at 45 °C, as early as one to two days after transfer to a 45 °C growth regime. Since seasonal changes in temperature regime in any natural habitat are much more gradual than those imposed in these experiments, it appears highly probable that leaves developed in the cool season would be capable of effective photosynthetic function in the hot season and vice versa.

There is little doubt that the ability of photosynthetic acclimation over a very wide temperature range plays an important part in the photosynthetic performance in habitats with large seasonal temperature variation. Interestingly, however, the species possessing this ability do not appear to be as efficient at low temperatures as the cool-coastal species, nor are they as efficient at high temperatures as the highly thermophilic *T.oblongifolia*. The high thermal stability and extraordinary photosynthetic performance at high temperatures of the latter species seem to preclude the ability to function at low temperatures which permits an excellent performance in the coastal plants. Whether the mechanisms required for an unusually efficient functioning at one extreme preclude an efficient functioning at the other extreme is unknown.

9.3.3 Leaf Factors Determining Photosynthetic Capacity at Low Temperatures

Chilling Tolerance in Relation to C_4 Photosynthesis. Inability to operate efficiently at low temperatures is a feature common to many C_4 species from tropical or warm arid environments. Several tropical C_4 grasses are unable to photosynthesize below 10 °C (Ludlow and Wilson, 1971), and many show chlorosis and impaired photosynthesis even after only brief exposure to temperatures below 10° to 12 °C (McWilliam and Naylor, 1967; Shneyour et al., 1973; Taylor and Rowley, 1971). When treated at low temperature these chilling-sensitive C_4 plants show substantial changes in the kinetics of CO_2 assimilation via the C_4 pathway and in severe cases the changes indicate a total disruption of the operation of the C_4 pathway. However, these changes are also accompanied by an abnormal chloroplast structure and disruption of photochemical activity by the chloroplasts, presumably resulting in an impaired $NADPH_2$-generating capacity (Brooking and Taylor, 1973; Slack et al., 1974; Taylor and Craig, 1971).

These observations have led to the proposal that the C_4 pathway is intrinsically incapable of efficient functioning at low temperatures. However, this cannot be generally valid since, as we have seen, several C_4 species such as *A.sabulosa,* *A.hymenelytra,* and *A.lentiformis* perform well at low temperatures. This is also true of *A.confertifolia,* a C_4 species native to cold-temperate deserts of western North America and *A.vesicaria,* an evergreen species of arid and semi-arid parts of the Australian continent (Caldwell et al., 1977a). Both of these plants grew well at a daytime temperature of 8 °C and no disruption of the metabolic sequence of C_4 photosynthesis could be detected at low temperatures in these plants, irrespective of whether the plants had been grown at cold or warm temperatures. Only relatively minor kinetic adjustments during low temperature photosynthesis took place. Neither did low growth temperatures result in any impairment of chlorophyll synthesis or disorganization of chloroplast structure. As shown in Fig.9.22, the

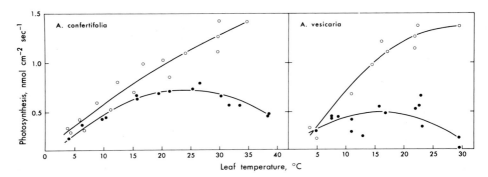

Fig. 9.22. Temperature dependence of light-saturated CO_2 uptake in *A.confertifolia* and *A.vesicaria*, grown at 8° day/6 °C night (●—●) and 28° day/20 °C night (○—○) regimes. (Redrawn from Caldwell et al., 1977a)

photosynthetic capacity of intact leaves at low temperature was not significantly reduced by growing the plants under a 8°/6 °C regime although there is no evidence that any photosynthetic acclimation to low temperature occurred. The plants grown under the warm regime had as high photosynthetic rates at low temperatures as those grown under the cold regime and they possessed much higher rates at high temperatures. (It seems probable that also these species are capable of photosynthetic acclimation over a certain temperature range but the cold regime used in these experiments may well be below this range).

It thus appears that the chilling sensitivity observed in many warm-habitat C_4 plants is not due to any intrinsic property of the C_4 pathway per se. In this connection, it is noteworthy that also many tropical C_3 species are chilling-sensitive. This does not exclude the possibility that in chilling-sensitive plants components of the C_4 pathway, just as those of the C_3 pathway, may be incapable of proper functioning at low temperature. It is noteworthy that those C_4 species that have been found to be chilling-sensitive are of the "NADP malic enzyme" type, while the chilling-tolerant C_4 species appear to belong to the "NAD malic enzyme" type (Chap. 9.1.3). It is not known whether this correlation reflects a causal relationship between chilling sensitivity and the "NADP malic enzyme" type of C_4 photosynthesis or whether it is just incidental. The abnormal kinetics of CO_2 assimilation via the C_4 pathway, observed at low temperatures in certain chilling-sensitive plants, could well be an indirect result of a breakdown in chloroplast membrane functions, and thus unrelated to the C_4 pathway of CO_2 fixation.

Chilling Tolerance in Relation to Chloroplast Membrane Lipids. There is evidence that chilling tolerance of both C_3 and C_4 plants is related to the properties of the lipid-protein matrices of the chloroplast as well as the mitochondrial membranes. As the temperature is changed, a point is reached at which a change in the physical phase of these membrane lipids occurs. The membrane lipids are considered to be in a liquid-crystalline (functional) state above this phase transition temperature but start to change to a solid (nonfunctional) state at this temperature. Drastic changes in various physiological activities have been observed in the vicinity of these transition temperatures. For example, abrupt changes in the activation energy of respiration of plant tissues and isolated mitochondria have been found to occur in many chilling-sensitive higher plants of tropical and subtropical origin in the 10° to 14 °C range and these changes were coincident with phase transition changes in the mitochondrial membranes. No such changes have been observed at temperatures above the freezing point in chilling-resistant plants (Lyons, 1973; Raison et al., 1971). More recently, similar effects have been observed with chloroplasts (Raison, 1974; Shneyour et al., 1973; Murata et al., 1975; Murata and Fork, 1977). For example, the Arrhenius activation energy for light-driven NADP reduction by chloroplasts from the chilling-sensitive species bean (C_3), tomato (C_3), and maize (C_4), shows an abrupt increase as the temperature is decreased to below 11° to 14 °C. Changes in the activation energy for the motion of spin label infused into the chloroplast membranes occurred at these same temperatures (Raison, 1974). This indicates that the chilling sensitivity of photosynthetic activity is caused by a phase-shift in the chloroplast membrane lipids. Again, no breaks in the Arrhenius plots of either photosynthetic activity or spin-label motion were observed with chloroplasts from chilling-resistant plants such as pea and lettuce (Raison, 1974).

As mentioned in Chap. 3.4.2, the temperature at which the lipid phase change and the resulting change in the activation energy of membrane-bound enzymes occurs can differ within a single species. The grass *Themeda australis* is found mainly in tropical regions of Australasia but possesses races which extend into temperate climates. Chloroplasts isolated from this grass from tropical New Guinea and temperate Tasmania exhibited phase transition temperatures of 14° and 7 °C, respectively (Raison, 1974).

All of these results suggest that physiological tolerance to low but nonfreezing temperatures is at least in part caused by an alteration in the lipoproteins of the chloroplast and mitochondrial membranes. In general, the transition temperature can be expected to depend on the lipid species and the chain length and saturation of the fatty acid constituents. One would predict that the phase transition temperature of the lipoprotein membrane would increase with increasing saturation of the lipids. This prediction is in agreement with experimental evidence obtained with blue-green algae. In one such alga, *Anacystis nidulans,* the degree of saturation of the photosynthetic membrane lipids is much higher than in green algae and higher plants and it is strongly dependent on growth temperature (Holton et al., 1964). As the growth temperature is increased, the degree of saturation of the membrane lipids increases and so does the phase-transition temperature of these membranes as determined by spin-label techniques and by in vivo chlorophyll a fluorescence kinetics measurement. Moreover, the temperature at which photosynthetic O_2 evolution, electron transport, and other indices of photosynthetic function show abrupt changes in activation energy shifts in concert with the phase transition temperature, being approximately 13° to 16 °C below the respective growth temperatures (Murata and Fork, 1975; Murata et al., 1975). Similar results have been obtained with the extreme thermophilic blue-green alga *Synechococcus lividus* (Fork et al., 1979). These results indicate that a causal relationship exists in these algae between the degree of lipid unsaturation and phase transition temperature on the one hand and photosynthetic function on the other.

Increase in growth temperature also causes an increased degree of saturation of the chloroplast membrane lipids in higher plants, including *Atriplex lentiformis* (Pearcy, 1977), *A.glabriuscula, A.sabulosa,* and *T.oblongifolia* (Björkman et al., 1976). However, these changes are relatively small in comparison with those observed in the blue-green algae mentioned above. Moreover, although *T.oblongifolia* has more saturated chloroplast membrane lipids than *A.sabulosa* and *A.glabriuscula,* these differences are not very large and on the whole the content of polyunsaturated fatty acids, especially linolenic acid, is quite high in all higher plants investigated to date, irrespective of growth temperature and chilling sensitivity. It appears unlikely that the differences in the degree of saturation of the bulk chloroplast lipids, a major portion of which consists of mono- and digalactosyl diglycerides, can account for the observed differences in the phase-transition temperatures of the chloroplast membranes and chilling tolerance among higher plants. It is more likely that changes in the saturation of certain charged lipid components such as phosphatidyl glycerol play a specific role in the lipid-protein associations within the membrane and may thus be responsible for the differences.

In the *Atriplex* species investigated no phase-transitions have been observed in the 0° to 20 °C range, nor are there any other indications of chilling sensitivity, irrespective of growth temperature. The decreased photosynthetic capacity at low temperatures observed when *A.lentiformis* and *A.hymenelytra* are grown under warm regimes must therefore be attributed to other factors. The same argument may be made for the poor photosynthetic performance at low temperatures of *T.oblongifolia*. The phase-transition temperature in the chloroplast lipids of this species, as determined by spin-label motion, was below 5 °C (Raison and Berry, unpublished). Changes in the activation energy of a number of activities related to chloroplast membrane function occurred at about 5 °C even though the plants were grown at very high temperatures (Murata and Fork, 1977). These temperatures are much lower than the minimum temperature required to sustain growth and photosynthetic competence in this species.

Other Leaf Factors Affecting Photosynthesis at Cool Temperatures. Very little is known about the mechanisms responsible for the increased photosynthetic capacity that occurs when plants with high temperature acclimation potential, such as the desert species *Atriplex lentiformis, A.hymenelytra,* and *Larrea divaricata* are grown under a cool as compared with a warm regime. However, a number of potential causes can be eliminated. It is clear that this response cannot be attributed to stomatal conductance or other aspects of CO_2 diffusion; neither is it related to photorespiratory activity (Björkman et al., 1975; Mooney et al., 1978; Pearcy, 1977). The results discussed in the preceding paragraphs also argue against a role of a phase transition in the membrane lipids in explaining the lower capacity of photosynthesis at low temperature in high-temperature-grown plants.

It is obvious that an increase in the amount of any rate-limiting catalyst would result in an increased photosynthetic capacity at low temperature. Where photosynthetic capacity is considered on a leaf area basis, this increase could be achieved by simply increasing the overall amount of photosynthetic machinery per leaf area. However, experimental results indicate that this does not take place in the species under consideration.

The amounts of chlorophyll, total protein, and dry matter per unit leaf area did not show any marked differences between *A.sabulosa* and *T.oblongifolia* leaves in spite of the large differences in the photosynthetic capacities of the two species (O. Björkman, unpublished). Neither did different growth temperature regimes significantly affect the gross leaf composition, thickness, or anatomy in *Larrea divaricata* (Mooney et al., 1978) and *Nerium oleander* (Björkman et al., 1978). This indicates that the increased photosynthetic capacity is caused either by a quantitative change in specific rate-limiting catalysts or by a qualitative change in the kinetic properties of these catalysts, resulting in an enhanced catalytic activity at low temperature.

There is very little information as to which catalytic step of photosynthesis exerts the primary limitation to the overall rate of the process at low temperatures in any plant. RuP_2 carboxylase is one of many possibilities. Interestingly, as shown by Pearcy (1977) the activity of this enzyme (measured at a constant temperature) was about twice as high in low-temperature-grown *A. lentiformis* plants as in high-temperature-grown plants of the same species. The photosynthetic capacity at low rate-limiting temperature also was twice as high in the low-temperature-grown plants. In contrast, growth temperature had no significant effect on the activity of PEP carboxylase (Pearcy, 1977). Similar results have also been obtained with *T. oblongifolia,* grown at daytime temperatures of 25°, 35°, and 45 °C (Björkman et al., 1976). The photosynthetic capacity at low, rate-limiting temperatures was correlated with the RuP_2 carboxylase activity, but no such correlation was found with PEP carboxylase; the activity of this enzyme was rather higher in the high-temperature-grown plants (Björkman, unpublished).

A comparison of the activities of a number of enzymes of photosynthetic carbon metabolism between *Nerium oleander* plants, grown under a 20° and a 45 °C regime, is shown in Table 9.5. The CO_2-saturated photosynthetic capacity of intact leaves,

Table 9.5. Comparison of photosynthetic capacity at 20 °C of intact leaves, photosynthetic electron transport and photophosphorylation capacities of isolated chloroplasts, and activities of enzymes of photosynthetic carbon metabolism in leaf extracts of *Nerium oleander*, grown under two contrasting thermal regimes. (Data of Björkman et al., 1978)

Process	μmol g fresh wt.$^{-1}$ min^{-1}		Ratio A/B
	(A) Grown at 20 °C	(B) Grown at 45 °C	
Photosynthetic CO_2 uptake at 20 °C (saturating CO_2 pressure and light intensity)	3.88	1.33	2.92
Electron Transport ($H_2O \rightarrow MV$) at 20 °C	6.42	4.69	1.37
Photophosphorylation (noncyclic) at 20 °C	3.11	2.79	1.11
RuP_2 Carboxylase activity at 20 °C	4.91	3.29	1.49
$FruP_2$ Phosphatase	3.64	1.40	2.46
Ru5P Kinase	93.3	91.0	1.14
3-Phosphoglycerate Kinase	104.0	91.0	1.14
NADP Glyceraldehyde-3P Dehydrogenase	43.0	43.0	1.02
$FruP_2$ Aldolase	25.7	21.9	1.17
Phosphoglucomutase	9.8	8.3	1.18
Phosphohexose Isomerase	10.7	12.4	0.86
NADP Malate Dehydrogenase	2.07	2.55	0.81
Adenylate Kinase	8.9	7.9	1.13

measured at 20 °C, and electron transport and photophosphorylation in chloroplasts isolated from these leaves, are also shown. Growth at low temperature did not result in a general increase in electron transport and photophosphorylation capacities, or in the general level of enzymes of photosynthetic carbon metabolism. However, two Calvin cycle enzymes, RuP_2 carboxylase and fructose biphosphate ($FruP_2$) phosphatase, both of which have been implicated in the literature as potentially limiting steps of photosynthesis under certain conditions, exhibited considerably higher activities in the 20 °C-grown than in the 45 °C-grown plants. Calculations show that the higher RuP_2 carboxylase activity of the 20 °C-grown plants may to some extent contribute to their higher photosynthetic capacity. However, the difference in the level of this enzyme is too small to account for the difference in photosynthetic capacity at 20° between the 20 °C-grown and the 45 °C-grown *N.oleander* plants. $FruP_2$ phosphatase is the only enzyme of those examined which is affected by growth temperature to a similar degree as photosynthetic capacity. Moreover, in transfer experiments of 45 °C-grown plants to a 20 °C-growth regime (cf. Fig.9.21) the time course for the increase in photosynthetic capacity closely followed that for the increase in $FruP_2$ phosphatase activity. Parallel decreases in these two factors also occurred following transfer of 20 °C-grown plants to a 45 °C-growth regime, except that the initial decline in photosynthetic capacity was not accompanied by a decrease in $FruP_2$ phosphatase activity. This initial decline in photosynthesis is probably a result of a direct thermal inhibition after the abrupt increase in temperature. Thus, the experimental data strongly implicate $Fru P_2$ phosphatase as a component which may well be largely responsible for the superior photosynthetic capacity at low temperatures of low-temperature-grown *N. oleander* leaves. Whether similar differences in the level of this enzyme may also explain the difference observed between such species as *Atriplex sabulosa* and *Tidestromia oblongifolia* is not yet known.

It is probable that the differences in $FruP_2$ phosphatase and RuP_2 carboxylase activities between the plants grown under different temperature regimes are caused by a change in the amount of protein rather than by a change in their catalytic activities. The enzyme assays were carried out at saturating substrate concentrations, so differences in the affinity of the enzymes for their substrates could not have been responsible for the differences in activity. Also, the specific activity of purified RuP_2 carboxylase from *T.oblongifolia* is not significantly affected by the growth temperature and it is also the same as that of the *A.glabriuscula* enzyme (Björkman et al., 1976). Moreover, measurements of the activation energy for RuP_2 carboxylase from a range C_3 and C_4 species, native to contrasting thermal habitats and grown under a wide range of temperature regimes, show little variation in this kinetic property (Björkman and Pearcy, 1971).

It is at least conceivable that the temperature responses of some catalytic properties of photosynthetic enzymes may change during temperature acclimation. This possibility is suggested by studies of ectothermic animals in which temperature acclimation is accompanied by substantial changes in the temperature response of enzymic properties. In rainbow trout, for example, the temperature response profile for K_m of acetylcholine esterase and pyruvate kinase shifts to give minimum values at temperatures corresponding to the growth temperature of 2° and 18 °C

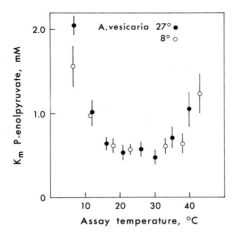

Fig. 9.23. Temperature dependence of K_m for phosphoenolpyruvate (PEP) of PEP carboxylase, extracted from leaves of *A. vesicaria*, grown at 28° day/20 °C night (●—●) and 8° day/6 °C night (○—○) regimes. (Data from C.B. Osmond, unpublished)

(Hochachka and Somero, 1973). However, no correspondingly large changes of the temperature response of K_m for plant enzymes during growth at different temperatures have been obtained.

When *A. vesicaria* was grown at 8° or 27 °C, the temperature response profile for the K_m (PEP) or PEP carboxylase remained unchanged (Fig. 9.23). In other studies (J. Teeri, unpublished) the temperature response profile of other enzymes also remained unchanged when plants were grown at different temperatures. Although available data are limited, there is little evidence that temperature acclimation of photosynthesis involves changes in the temperature response profiles of enzyme kinetic properties.

A particularly noteworthy discovery is that in certain heat-sensitive plants, such as winter rye, even moderately high temperatures may induce a deficiency of chloroplast ribosomes while other ribosomes are more heat-stable. There is much evidence that assembly of the large subunits of RuP_2 carboxylase takes place on the 70S chloroplast ribosomes. The synthesis of this enzyme was found to be strongly retarded at high growth temperatures (above 30 °C) that caused a deficiency of these ribosomes in developing rye seedlings (Feierabend and Schrader-Reichhardt, 1976). Presumably, the synthesis of other components of the photosynthetic machinery that are assembled on these ribosomes would be affected also, although in rye, chlorophyll and RuP_2 carboxylase are the only chloroplast constituents found to be affected so far. Ten other soluble enzymes which are exclusively or partially localized in chloroplasts were not affected, neither were microbody and mitochondrial enzymes affected.

It is conceivable that failure to acclimate photosynthetically above (and perhaps also below) a given temperature range is in part related to the temperature sensitivity of chloroplast ribosome formation. One can also speculate that growth temperature may regulate the synthesis of RuP_2 carboxylase and certain other chloroplast constituents via its effect on these ribosomes even within the temperature range of photosynthetic acclimation. However, no direct experimental evidence for or against this hypothesis is available in any higher plant species.

9.3.4 High Temperature Stability of Photosynthesis

Many factors must be taken into account in analyzing the cause of differences in photosynthetic performance that exist between cold-adapted C_3 and C_4 species on the one hand and heat-adapted C_3 and C_4 species on the other. Differences in the ability to photosynthesize at high temperatures under normal atmospheric CO_2 pressure could well be caused by differences in the temperature response of the stomata. However, detailed studies of stomatal responses in the plants, whose photosynthetic temperature dependence are shown in Fig. 9.19, clearly demonstrate that the differences in photosynthetic performance between cold-adapted and heat-adapted plants at high temperatures cannot even in part be ascribed to differences in stomatal conductance (Björkman et al., 1975; Mooney et al., 1978; Pearcy, 1977).

At saturating CO_2 pressures, photosynthesis of course is entirely independent of stomatal conductance or any other aspect of diffusive CO_2 transport. Examples of temperature response curves of cold-adapted and heat-adapted plants, determined at high CO_2 pressure, were shown in Fig. 9.18. The much superior performances of the heat-adapted C_3 and C_4 species in comparison with the cold-adapted C_3 and C_4 species must therefore reflect differences in the heat stability of the leaves at the cellular level. This does not necessarily mean, however, that this difference is attributable to a higher thermal stability of the photosynthetic machinery per se. High temperature damage to other membrane systems could have secondary effects on photosynthesis. For example, damage to the plasmamembrane would dramatically change the intracellular environment, presumably resulting in an inactivation of photosynthesis and other aspects of metabolism.

Cellular Integrity. Experiments elucidating the relative heat sensitivity of photosynthesis, respiration, and ion leakage from the leaves (a measure of damage to the plasmamembrane) in the cold-adapted and heat-adapted C_4 species *A. sabulosa* and *T. oblongifolia* are illustrated in Fig. 9.24. In these studies illuminated leaves were subjected to a series of high temperature treatments, each lasting for 10 min. Following each treatment the rates of photosynthesis (measured at high light and CO_2), dark respiration, or ion leakage, were determined at a constant, noninhibitory temperature. Large differences between the species are seen in the temperature at which thermal damage sets in according to all three criteria. However, in each species photosynthesis is much more heat-sensitive than respiration and ion leakage can only be detected after exposure to temperatures that cause complete inhibition of both photosynthesis and respiration. Of special interest in this connection is the finding of Krause and Santarius (1975) that the outer chloroplast envelope membrane is considerably more resistant to heat damage than the photosynthetic (thylakoid) membranes. These results indicate that in neither the cold-adapted nor the heat-adapted species is the thermal inhibition of photosynthesis caused by a general breakdown of cellular metabolism, inferring that the difference in high temperature sensitivity between these plants reflects

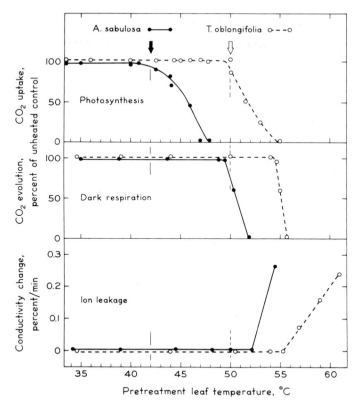

Fig. 9.24. Comparison of the heat sensitivity in vivo of light- and CO_2-saturated photosynthetic rate, dark respiration, and ion leakage in leaves of *A.sabulosa* and *Tidestromia oblongifolia*. "Control rates", of photosynthesis were measured at a standard noninhibitory temperature; the attached, illuminated leaves were subsequently treated for 15 min at the temperature indicated on the abscissa and then quickly returned to the standard noninhibitory temperature and the photosynthetic rate measured again. (Data from Björkman and Badger, 1977). The procedure for the respiration measurements was similar to that used for photosynthesis, except that the leaves were kept in the dark. (Data from O. Björkman, unpublished). Ion leakage was measured by the increase in the electrical conductivity of a bathing solution in which leaf slices were submerged. (Data from O. Björkman, unpublished)

differences in the thermal stability of components of the photosynthetic machinery itself.

Chloroplast Membrane Reactions. Further studies provide positive evidence in support of the above inference and also point to the component steps of photosynthesis responsible for this difference. Figure 9.25 (upper left) shows the inhibition of the quantum yield of photosynthesis by intact leaves following pretreatment to high temperatures in similar manner as in the studies of inactivation of light-saturated photosynthesis. The quantum yields of photosystem I and photosystem II activities in chloroplast preparations, isolated from similarly pretreated leaves, are shown in the upper and lower right

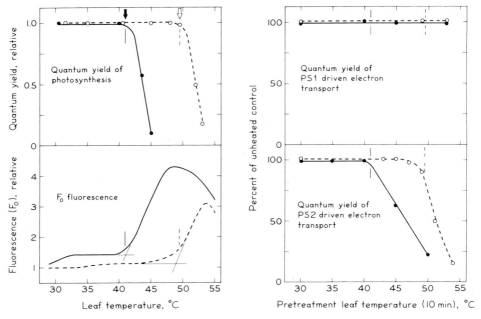

Fig. 9.25. Effect of leaf temperature on the quantum yield for CO_2 uptake by intact leaves of *A.sabulosa* (●-●) and *T.oblongifolia* (○-○), and of pretreating illuminated detached leaves for 10 min at different temperatures on the quantum yields for photosystem I- and photosystem II- driven electron transport by chloroplasts isolated from these leaves. (Redrawn from Björkman et al., 1976). Also shown is the effect of leaf temperature on the fluorescence yield of detached leaves. (Redrawn from Schreiber and Berry, 1977)

respectively. Evidently, the quantum yield for photosystem I activity shows no detectable inhibition over the temperature range investigated in either the cold- or the heat-adapted species, indicating that this photosystem possesses a high thermal stability in both species and is not responsible for the inhibition of the quantum yield of overall photosynthesis. In contrast, inactivation of photosynthesis and the quantum yield of photosystem II activity sets in at very similar temperatures in each of the species. This strongly suggests that one or several components of photosystem II, or of components closely associated with this photosystem, possess a low thermal stability and may thus be a primary cause of the heat inactivation of whole leaf photosynthesis. The thermal stability of these components is evidently much greater in the heat-adapted species.

Measurements of the variable yield of chlorophyll fluorescence provides an additional approach indicating the functional state of photosystem II in vivo. Extensive studies of high temperature inhibition of photosynthesis in leaf discs utilizing fluorescence kinetics measurements have been made on cold-adapted *A.sabulosa*, *A.glabriuscula* and heat-adapted *T.oblongifolia* and *L.divaricata* (Schreiber and Berry, 1977; Armond et al., 1978). Similar studies have also been made on *A.lentiformis* which, like the two preceding species, was native to Death Valley (Pearcy et al., 1977). An example of the temperature dependence of the F_o

state fluorescence, emitted by leaf discs of *A. sabulosa* and *T. oblongifolia*, is depicted in Fig. 9.25, lower left. At temperatures that are noninhibitory to photosynthesis the fluorescence yield is very low and essentially constant with temperature, but when the temperature reaches the point at which inhibition of photosynthesis occurs there is a sudden rise in fluorescence yield. This rise is interpreted to mean that energy transfer from the light-harvesting chlorophyll to the reaction center of photosystem II is impaired. Results obtained with cold-adapted *A. glabriuscula* were very similar to those shown for *A. sabulosa*. In heat-adapted *A. lentiformis* the rise in fluorescence yield occurred at a few degrees lower temperature than in similarly heat-adapted *T. oblongifolia*. Inhibition of the quantum yield for photosynthesis of intact leaves and photosystem II activity of chloroplasts, isolated from heat-treated leaves of *A. lentiformis*, occurred at approximately the same temperature as that which caused an abrupt rise in fluorescence yield. As in the other species examined, photosystem I activity exhibited much higher heat stability.

Several lines of evidence indicate that heat inactivation of photosystem II activity is also accompanied by an inhibition of ATP production. Apparently, this inhibition is at least in part attributable to an uncoupling of noncyclic photophosphorylation (Björkman et al., 1976; Pearcy et al., 1977). Experiments in which the isolated chloroplasts rather than the leaves were subject to heat treatment also indicate that inactivation of photosystem II activity and uncoupling of noncyclic photophosphorylation take place at about the same temperature (Emmett and Walker, 1973; Mukohata, 1973). However, interpretations of results obtained in experiments in which isolated chloroplasts were heated, must be made with caution, especially when these chloroplasts exhibit much greater heat sensitivity than they do when heated in situ. There is evidence that the composition of the suspending medium may substantially modify the apparent thermal stability of the chloroplasts, especially where the integrity of the chloroplast envelope has been lost in the isolation procedure (Krause and Santarius, 1975; Oku and Tomita, 1971; Santarius, 1973). The heat stability of such chloroplasts is generally much lower than that of chloroplasts with intact envelopes or of intact leaves. Similar considerations apply to studies of high temperature inactivation of enzymes; in many cases the thermal stability of enzymes is much lower in isolation than in situ, presumably because the ionic environment in the cell is more favorable than that provided by the isolation medium.

There is evidence that the heat stability of chloroplast membrane reactions as well as whole leaf photosynthesis increases with increased growth temperature, at least in species which are capable of photosynthetic acclimation to high temperature. Desert clones of *A. lentiformis*, grown at 43 °C, exhibited considerably greater heat stability than did the same clones grown at 23 °C. Heat treatment of leaves from 23 °C grown plants at 46 °C resulted in a marked reduction in the quantum yield of CO_2 fixation in intact leaves of 23 °C grown plants as well in photosystem II activity of chloroplast isolated from them. In contrast, similar heat treatments of leaves from 43 °C-grown plants caused no inactivation of these functions. Other in vivo estimates of photosystem II functioning indicated a similar difference in the thermal stability between plants grown at 23° and 43 °C (Pearcy, 1977). Similar increases in the thermal stability of photosystem II-driven electron transport with increasing growth temperature, as well as whole leaf photosynthesis, are also evident in *Larrea divaricata*, which, like desert clones of *A. lentiformis*, is capable of photosynthetic acclimation over a wide temperature range (Armond et al., 1978). In this species the heat stability of excitation energy transfer between the light-harvesting pigments and the photosystem II reaction centers is markedly enhanced by high growth temperature. This and other observations suggest that acclimation to high temperature involves changes of the thylakoid membranes which render the pigment system embedded in this membrane less heat-sensitive. The particular components that are damaged or the structural changes that take place in the thylakoid membranes as a result of heat damage are unknown.

One may speculate that the much greater thermal stability of the thylakoid membranes of heat-adapted plants could be caused by an increased hydrophobicity of the membrane proteins or by an increased saturation or other compositional changes in the membrane lipids. To our knowledge, no comparative studies of chloroplast membrane proteins in relation to heat stability have been reported. The saturation of the membrane lipids does indeed increase when species such as *A. lentiformis, A. sabulosa, A. glabriuscula,* and *T. oblongifolia* are grown at high temperatures (Pearcy, 1978; J. Berry, personal communication).

However, as mentioned in Chap. 9.3.3, the content of polyunsaturated chloroplast lipids, especially the galactolipids, is quite high even in heat-tolerant, high-temperature-grown plants. Also, increased saturation or chain length of the constituent fatty acids is not necessarily correlated with higher thermal stability of

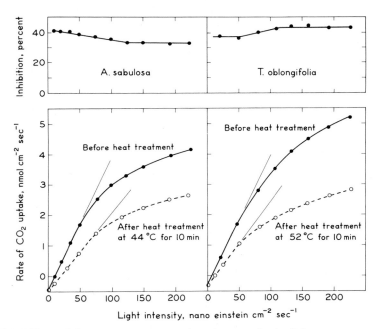

Fig. 9.26. Effect of heat treatment on the photosynthetic light response curves for *A. sabulosa* and *T. oblongifolia*. Inhibition by heat treatment is expressed as the percentage reduction in the sum of net photosynthesis and dark respiration. (From Björkman and Badger, 1977)

whole leaf photosynthesis or thylakoid membrane function. For example, when grown at a high temperature, which is optimal to photosynthesis and growth for *T.oblongifolia* but above optimum for *A.lentiformis,* the latter species tends to have more saturated chloroplast lipids than the former. Moreover, no significant differences were detected between the lipid compositions of coastal and desert clones of *A.lentiformis* at either high or low growth temperatures in spite of the fact that the desert clones have considerably greater heat tolerance (Pearcy, 1978). It seems doubtful, therefore, that differences in the thermal stability of photosynthesis can be attributed to differences in saturation of the bulk chloroplast lipids alone.

Enzymes of Carbon Assimilation Processes. As illustrated in Fig. 9.26, heat treatment of leaves from the two heat-adapted and cold-adapted species at any given inhibitory temperature affects the quantum yield and the light-saturated rate of photosynthesis in a similar manner. The simplest explanation for these results

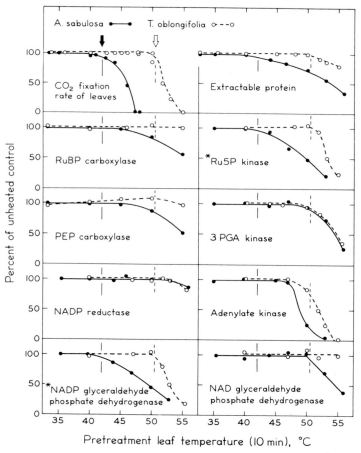

Fig. 9.27. Responses of light-saturated photosynthesis, extractable protein, and indicated enzyme activities to pretreatment of illuminated *A.sabulosa (solid lines)* and *T.oblongifolia (broken lines)* leaves to different temperatures. (From Björkman and Badger, 1977)

would be that the heat damage is limited to a single site in the thylakoid membranes which controls both of these aspects of photosynthesis. However, the same results could also be obtained if separate sites or components, governing the capacity of photosynthesis in its light-limited and in its light-saturated state, possess similar thermal stabilities. Indeed, it seems reasonable to expect that the ability of the photosynthetic apparatus to tolerate unusually high temperatures might necessitate an increased thermal stability of more than a single component.

Recent studies of the heat stabilities in situ of a number of photosynthetic enzymes contained in the chloroplast stroma and in the cytoplasm indicate that several of these enzymes are inactivated at similar temperatures as photosynthesis (Björkman and Badger, 1977). Heat treatment at temperatures just sufficiently high to cause significant inhibition of photosynthesis also resulted in a significant loss of the amount of protein that can be extracted from the leaves, indicating that one or several soluble proteins undergo heat denaturation and aggregation so that they become insoluble (Fig.9.27, top right). Apparently, soluble photosynthetic enzymes account for at least part of this loss.

The functions of the enzymes investigated in these C_4 species are depicted in Fig.9.28 and some of the results are illustrated in Fig.9.27, lower part. To ensure that enzymes requiring light for their activation were fully activated, experiments were made on intact leaves, illuminated with bright light both before and during the

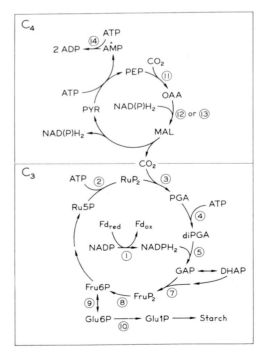

Fig. 9.28. Diagram depicting carbon fixation pathway in C_4 plants. *Numbers* indicate the enzyme steps referred to in the text

heat treatment. The presence in the extraction buffer of dithiothreitol prevented dark deactivation of the light-activated enzymes following their extraction from the leaves, which was made immediately upon termination of the heat treatment.

Several enzymes such as NADP reductase (1), phosphoglucomutase (9), phosphohexose isomerase (10), and NAD malate dehydrogenase (12) possessed very high heat stabilities and little or no difference between the species was detected. The heat stabilities of certain other enzymes were considerably greater in the heat-adapted than in the cold-adapted species. These differences may well be the result of adaptive changes in these proteins but for several of these enzymes the heat stabilities are too high to explain the heat inhibition of whole leaf photosynthesis in the same plant. This group of enzymes includes carboxylase (3), phosphopyruvate carboxylase (11), NAD glyceraldehyde 3-phosphate dehydrogenase (6) and fructose bisphosphate aldolase (7). In a third group of enzymes the heat stabilities approach that of photosynthesis of the heat-adapted species; this is especially evident with adenylate kinase (14) and NADP malate dehydrogenase (13), the latter being a light-activated enzyme. However, in the cold-adapted species the heat stabilities of these enzymes appear too high to account for inhibition of photosynthesis; another enzyme falling in this group is 3-phosphoglycerate kinase (4). For two enzymes, NADP glyceraldehyde-phosphate dehydrogenase (5) and ribulose phosphate kinase (2) heat inactivation occurs near the same temperatures that cause inhibition of photosynthesis in each of these species. These enzymes may thus be in part responsible for the heat inactivation observed for whole leaf photosynthesis in both species. In *T. oblongifolia* this may also be true for adenylate kinase and NADP malate dehydrogenase. Since NADP glyceraldehydephosphate dehydrogenase, ribulosephosphate kinase, and NADP malate dehydrogenase, require photochemically generated reducing power for their activation, further studies are needed to ascertain whether the observed high temperature inactivation of these enzymes is attributable to a low intrinsic heat stability of these enzyme proteins. The possibility that it is an indirect effect of heat inactivation of the light activating system cannot be completely ruled out.

Further studies are also needed to determine the extent to which, if any, the heat stabilities of photosynthetic enzymes in vivo as well as in vitro are affected by growth temperature and hence are subject to acclimation. The influence of temperature on the biosynthesis of these enzymes in heat-tolerant and heat-resistant plants is another area of investigation which must be pursued in order to resolve the mechanisms that underlie photosynthetic adaptation to temperature.

9.4 Photosynthesis in Relation to Water Use and Water Stress

9.4.1 Photosynthetic Water Use Efficiency

As mentioned previously (Chap. 8.4), the rate of transpirational water loss (T) is a direct function of the stomatal conductance to gaseous diffusion (C_s) and to the difference between the water vapor pressure inside the leaf (p_i^w) and that of the ambient air (p_a^w). Hence for any leaf

$$T = C_s \cdot (p_i^w - p_a^w). \tag{9.4}$$

The water vapor pressure inside the leaf (p_i^w) is equal to the saturation vapor pressure of water at the leaf temperature. The temperature dependence of p_i^w is specifically pronounced at high temperatures where a further rise in leaf temperature causes a disproportionately large increase in water vapor pressure and consequently also in transpiration rate. At any given stomatal conductance and ambient water vapor pressure, increased leaf temperature will thus cause a large increase in transpiration rate (Fig. 8.1).

The effect of a change in stomatal conductance on the rate of photosynthesis depends on two interacting factors: the extent to which the intercellular CO_2 pressure (p_i^c) is affected and the dependence of photosynthesis on this pressure. The relationship between p_i^c and photosynthesis (P) is given by the expression:

$$p_i^c = p_a^c - \frac{P}{C_s} \cdot 1.56 \tag{9.5}$$

where p_a^c is the ambient CO_2 pressure (and thus essentially constant), and C_s is the stomatal conductance to diffusive transport of water vapor. Since the diffusivity of water vapor is 1.56 times that of CO_2, C_s is divided by this factor to give the conductance to CO_2 diffusion.

The relationship between photosynthesis (P) and intercellular CO_2 pressure (p_i^c) is determined by the intrinsic photosynthetic characteristics of the leaf. The effect of stomatal conductance on photosynthesis in normal air will thus depend on these characteristics as well as on temperature and light intensity. If the relationship between photosynthesis and intercellular CO_2 pressure is experimentally determined it is then possible to calculate the relationship between photosynthesis and stomatal conductance. Examples of this are given in Fig. 9.29. The transpiration rate for any stomatal conductance, temperature, and external water vapor pressure can be readily calculated without any experimental measurements. We can then also obtain the photosynthetic water use efficiency, P/T, i.e., the amount of CO_2 fixed per amount of water expended, for any stomatal conductance and environmental condition without actually varying the stomatal conductance experimentally. In comparing the water use efficiencies of different species it is convenient to use an expression in which the influence of external factors on transpiration is eliminated. For this purpose we will use the term intrinsic water use efficiency which is given by P/C_s, i.e., the rate of photosynthesis obtained for any given stomatal conductance.

High P/C_s values are of course of utmost importance to primary productivity in any environment where the amount of water available to the plant is limited. As is evident from the preceding paragraphs this property becomes increasingly critical as the dryness of the air and the leaf temperature increase. The selective advantage of the ability to maintain high P/C_s values therefore should be particularly great in plants that occupy arid habitats where the temperature is high during the growing season.

In any plant P is essentially linearly related to C_s at low conductance values and the water use efficiency (P/T) is maximal and nearly constant. At higher conductances, P increases less than linearly with C_s; hence, high rates of photosynthesis can only be achieved at the expense of a reduced water use efficiency. Figure 9.30 (curve a) describes a generalized relationship between water

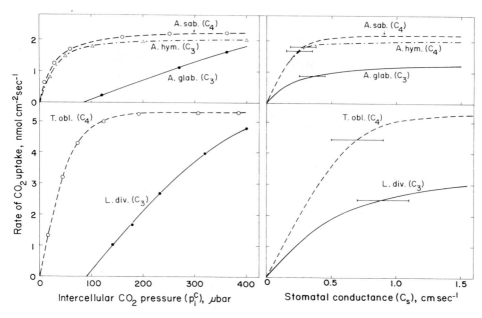

Fig. 9.29. Photosynthetic CO_2 uptake as a function of intercellular CO_2 pressure *(left)* and stomatal conductance *(right)* for C_3 and C_4 species having different photosynthetic capacities at CO_2 saturation. *Horizontal bars* indicate the range of actual stomatal conductances, measured at an ambient CO_2 pressure of $330\,\mu bar$. Plants were grown and photosynthesis measurements were made at $40\,°C$ (*A.glabriuscula, A.sabulosa,* and *A.hymenelytra*) or $45\,°C$ (*L.divaricata* and *T.oblongifolia*). (Based on data from Björkman et al., 1975; Mooney et al., 1978; and O. Björkman and H.A. Mooney, unpublished)

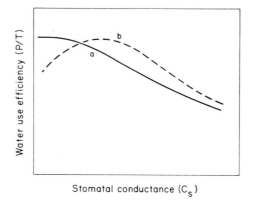

Fig. 9.30. Generalized relationship between photosynthetic water use efficiency and stomatal conductance (see text)

use efficiency and stomatal conductance, assuming that the leaf temperature remains constant as the conductance changes. It should be noted, however, that in reality the water use efficiency may not fall as much with increasing C_s as is indicated by this curve. Because of the increased transpirational cooling of the leaf that takes place when C_s increases, the water vapor pressure in the leaf (p_i^w) will also fall, partially counteracting the increase in T that would occur as a result of an increase in C_s. This change in leaf temperature may of course also influence the water use efficiency through its effect on photosynthesis. At low temperatures the additional transpirational cooling of the leaf may result in a decrease in P_i thus offsetting the decrease in T. In contrast, at high temperatures the effects of a lowered leaf temperature work together in enhancing the water use efficiency by increasing P and decreasing T. Under these conditions the dependence of water use efficiency may be better described by curve b in Fig. 9.30. Water use efficiency would then reach a maximum at some intermediate rather than a very low conductance value. The actual deviation of curve b from curve a is obviously strongly influenced by the air temperature and the photosynthetic temperature dependence characteristics of the species under consideration. At air temperatures near those which would result in thermal damage to the photosynthetic machinery transpirational cooling of the leaf can be expected to have overriding importance. It is noteworthy in this connection that many plants, including the species in Fig. 9.29, appear to lose control of stomatal conductance at such high temperatures, i.e., the stomata open fully.

The above considerations of photosynthetic water use efficiency in relation to stomatal conductance apply to the instantaneous performance of single leaves. In considering the water use efficiency in terms of the carbon balance of the whole plant stand over a 24-h period, several other factors must be taken into account. One such factor is that increased transpiration by one leaf may well increase the water vapor pressure around adjacent leaves, thus tending to decrease their transpiration rates. This effect would tend to result in an upward shift of the stomatal conductance value which provides the maximal P/T ratio for the whole plant in comparison with the single leaf. Another factor is that the respiratory losses (R) by the whole plant during the night and by nonleaf materials during the day is largely independent of the daytime stomatal conductance. Thus, the conductance value which would yield the optimum daily net carbon gain (P-R) per amount of water transpired during the day (T) would be considerably greater than the conductance value which would give the optimum water use efficiency in terms of daytime photosynthetic rate per amount of water simultaneously transpired by a single leaf.

We may thus conclude that in reality, maximal water use efficiency is not attained when the stomata are almost closed (as is indicated by curve a in Fig. 9.30), but will occur at a considerably higher stomatal conductance value. This value will be strongly dependent on a number of factors, particularly the CO_2 and temperature dependence of photosynthesis, the thermal load imposed on the plant, the ratio of photosynthetic to respiratory activity, and the density and aerodynamic characteristics of the plant stand. Cowan and Farquhar (1977) have proposed that plants may well be capable of a rather close control of the stomatal conductance permitting them to achieve maximum overall water use efficiency under the range of conditions normally encountered in their respective natural environments. These authors have also designed a mathematical model which takes into account some of the interacting factors mentioned above. Because of the complexity of factors that must be considered in any realistic evaluation of photosynthetic water use efficiency it has not yet been possible quantitatively to verify these predictions in actual field experiments. However, laboratory studies as well as the very limited field observations that are available on studies of stomatal responses in relation to photosynthetic water use efficiency appear to be in general accordance with the proposal that plants adjust their stomata to the prevailing conditions in a manner that enables them to achieve a high water use efficiency while at the same time maintain a relatively high photosynthetic rate (Cowan and Farquhar, 1977; also see following paragraphs).

The relationships between photosynthesis and stomatal conductance for certain C_3 and C_4 species are given in Fig. 9.29. The C_3 species *Atriplex glabriuscula* and the two C_4 species, *A. sabulosa* and *A. hymenelytra* represent plants with similarly low CO_2 saturated photosynthetic rates (2.6, 2.2, and 2.0 nmol CO_2 cm^{-2} s^{-1} respectively). In contrast, the C_3 species *Larrea divaricata* and the C_4 species *Tidestromia oblongifolia* represents plants with very high and approximately equal CO_2 saturated rates (5.9 and 5.6 nmol CO_2 cm^{-2} s^{-1}, respectively). In each of these comparisons, the C_4 species possess a much higher efficiency of utilizing low intercellular CO_2 pressures than the C_3 species (Fig. 9.29). As a result, for any given stomatal conductance (and hence transpiration rate) the C_4 species has a higher photosynthetic rate than the C_3 species. Conversely, for any given rate of CO_2 uptake the C_4 species can keep its stomata more closed than the C_3 species. In either case, the C_4 plant would have a much superior photosynthetic water use efficiency.

This does not necessarily mean that C_4 plants must inevitably have a superior water use efficiency under all conceivable conditions. It is certainly possible to envisage situations where the opposite result would be obtained. For example, this would be the case if *A. hymenelytra*, a C_4 species with low photosynthetic capacity, were to operate at a high conductance of 1.0 cm s^{-1} while *L. divaricata*, a C_3 species with high photosynthetic capacity, were operating at a stomatal conductance below 1.0 cm s^{-1}. However, this situation is contrived and is probably a very infrequent occurrence, if, indeed, it ever occurs. A stomatal conductance of 1.0 cm s^{-1} is far higher than needed to provide maximum photosynthetic rates in *A. hymenelytra*. To our knowledge, stomatal conductance values exceeding those required to saturate photosynthesis have not been recorded for any species either in the laboratory or in the field. Commonly measured maximum C_s values for *A. hymenelytra* under conditions of adequate water supply range from 0.15 to 0.35 cm s^{-1}. The approximate ranges of stomatal conductance values for the species in Fig. 9.29 are indicated by horizontal bars. At the stomatal conductance values normally encountered in these plants, the C_4 species have considerably superior photosynthetic water use efficiencies.

It is noteworthy that in each of these species stomatal conductance appears to be so adjusted that a relatively high photosynthetic rate and high water use efficiency can be maintained at the same time, thus lending support to the proposal of Cowan and Farquhar, mentioned in a previous paragraph. In species with high photosynthetic capacity (e.g., *T. oblongifolia*) a much higher stomatal conductance is obviously needed to achieve this adjustment than in species with low photosynthetic capacity (e.g., *A. hymenelytra*). Thus, although the generalization frequently has been made in the literature that C_3 plants have higher stomatal conductance than C_4 plants, it is evident that this need not be the case. However, the ratio of photosynthesis rate to stomatal conductance is, indeed, higher in C_4 than in C_3 plants. This is undoubtedly a direct result of the C_4 pathway serving as an internal CO_2-concentrating mechanism.

On the average, C_4 species have P/C_s ratios about 2.5 times those of C_3 species when compared under similar environmental conditions (Björkman et al., 1970; Downes, 1970; Ludlow and Wilson, 1972; Rawson et al., 1977; Slatyer, 1970). *Atriplex* species are no exception to this general rule. For example, *A.rosea*, a C_4 herbaceous annual, had twice the P/C_s ratio of *A.triangularis* (C_3) at 26 °C when grown together in growth chambers under ample supply of water and nutrients (Björkman et al., 1970). In this case the C_s values were high (approx. 1.0 cm s^{-1}) in both species. Rawson et al. (1977) found that *A.halimus*, a C_4 perennial desert shrub, had a P/C_s ratio similar to those of the C_4 grasses, *Sorghum bicolor* and *Echinochloa crus-galli* (barnyard grass); the P/C_s ratios of these C_4 species were about 2.4 times those of the C_3 species, *Triticum aestivum* (wheat), *Glycine max* (soybean), and *Helianthus annuus* (sunflower). The P/C_s ratios remained constant over a wide range of water vapor pressure deficits in all of these species. The data of Björkman et al. (1975) provide evidence that the coastal C_4 annual, *A.sabulosa*, and the C_4 desert shrub, *A.hymenelytra*, possess similar P/C_s ratios, being 2 to 3 times higher than that of the coastal C_3 annual, *A.glabriuscula*. This difference was present both when the plants were grown and compared at 16 °C as well as at 40 °C. Other observations show that the C_4 annuals *A.expansa* and *A.serenana* have P/C_s ratios two to three times those of the C_3 annuals, *A.hortensis* and *A.heterosperma* (Björkman, unpublished). Slatyer (1970) compared the water use efficiencies of the C_4 annual, *A.spongiosa*, with the C_3 annual, *A.triangularis* at different developmental stages and leaf ages (Fig.9.31). Although considerable changes occurred in photosynthetic rate and stomatal conductance with developmental stage, the P/C_s ratio was consistently 2.5 to 3 times higher in the C_4 as compared with the C_3 species at any given point in time.

These and other studies show that the P/C_s ratio is remarkably constant among plants having the same CO_2 fixation pathway, irrespective of their taxonomy, morphology, and ecological origin, at least as long as the plants are not subject to severe stress. C_4 grasses and other C_4 monocots have P/C_s ratios very similar to those of C_4 dicots. Moreover, the P/C_s ratios of C_4 desert shrubs, grown with adequate water supply, do not appear to show any marked differences from those of C_4 plants from mesic habitats, and comparisons between C_3 xerophytes and mesophytes have yielded similar results. One report suggested that the C_3 desert shrub, *Simmondsia chinensis* may have a P/C_s ratio that is intermediate between those generally found in C_3 and C_4 plants (Rawson et al., 1977). However, the stomatal conductance value of this species was very low in these determinations; a detailed study of the water relations of *S.chinensis* (Collatz, 1977) gave higher C_s values and the P/C_s ratio of unstressed plants was very similar to those found in other C_3 species.

Most of the P/C_s ratios reported above were obtained at high quantum flux densities. The absolute values for the ratio obtained if P and C_s are

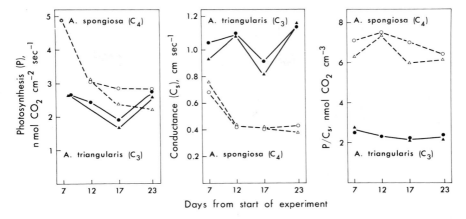

Fig. 9.31. Photosynthetic CO_2 uptake *(left)*, stomatal conductance *(center)* and water use efficiency *(right)* at different developmental stages in *A. triangularis* (C_3) and *A. spongiosa* (C_4). (Redrawn from Slatyer, 1970)

integrated over an entire day may well be somewhat different. However, a close regulation of C_s by the plant as both quantum flux density and P vary, tends to maintain the P/C_s ratio relatively constant through most of the day (and growing season). The difference between C_3 and C_4 plants thus remains approximately the same irrespective of whether instantaneous or daily P/C_s ratios are considered. Moreover, the relative differences in the P/C_s ratio, determined at the leaf level, are reflected largely unchanged in the photosynthetic water use efficiency at the plant stand level. Ludlow and Wilson (1972) found that the difference in P/C_s ratio between leaves of C_4 pasture grasses and C_3 legumes was expressed in a comparable difference in water use efficiency by the individual plants as well as whole swards; it was also reflected in a similar difference in the water use efficiency of dry matter production over the growing season.

That the higher intrinsic photosynthetic water use efficiency, conferred by the C_4 photosynthetic pathway, is generally expressed in an increased water use efficiency for growth has been documented in numerous comparative studies with many different C_3 and C_4 species. These growth studies which also include *Atriplex* species are discussed in Chap. 10. Thus, there is no evidence indicating that this aspect of the superiority of C_4 photosynthesis at the single leaf level is continuously eroded as one proceeds through the whole plant level and eventually to the level of plant productivity in the field (Gifford, 1974). However, this should not be taken to mean that the higher intrinsic water use efficiency, conferred by C_4 photosynthesis, will always result in a greater ratio of annual carbon gain to water loss. The actual amount of carbon fixed per amount of water transpired is obviously strongly dependent on the water vapor pressure deficit ($p_i^w - p_a^w$) during the growing season. As was shown in Fig. 8.1, this deficit increases exponentially with temperature. Consequently, the P/T ratio is likely to be much lower during the warm than during the cool season even if the P/C_s ratio remains constant. This prediction is in close agreement with experimental determinations of the water use efficiency for dry matter production. Studies conducted under semi-arid field conditions in Israel

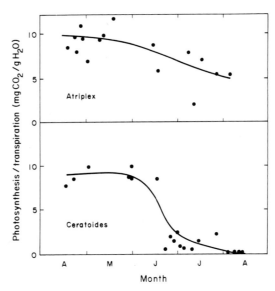

Fig. 9.32. Seasonal course of photosynthetic water use efficiency (P/T) for *A.confertifolia* (C₄) and *Ceratoides lanata* (C₃) in a cold-desert area in Utah. (Redrawn from Caldwell et al., 1977b)

showed that measured variation in these characteristics in response to seasonal changes in water vapor pressure deficit agreed closely with predictions based on gas exchange theory (Van Keulen, 1975). In their recent review, Fischer and Turner (1978) state that careful studies of six wheat cultivars, planted at monthly intervals in a mediterranean climate in south-eastern Australia demonstrate that water use efficiency was inversely related to the evaporative demand. In winter when the evaporative demand is at a minimum the estimated water use efficiency for dry matter production was 3.4 times as high as during the summer when the evaporative demand is at a maximum. Thus, it would not be surprising to find that a C_3 plant, growing in a cool-temperate environment, or restricting its growth to the cooler part of the year (as many warm-desert annuals do) would exhibit the same or perhaps even a greater actual water use efficiency than a C_4 plant whose main activity occurs during the hot season.

This emphasizes the importance of taking differences in seasonal activity into consideration when comparing actual water use efficiencies of different species. Caldwell et al. (1977b) compared the daily rates of CO_2 fixation and transpiration during different times of the growing season in the C_4 species, *A.confertifolia* and the C_3 species, *Ceratoides lanata,* growing together either in mixed communities or in mono-specific stands in close proximity in a cold desert area in north-western Utah. These workers concluded that, contrary to their expectations, they observed no marked differences in the total annual carbon fixation in terms of photosynthetic water use between the two species. As shown in Fig.9.32, the daily P/T rates of the two species did not exhibit any marked differences during the period of maximum photosynthetic rates which occurred during the cool spring when soil moisture was still reasonably high. During the warm dry summer months the C_4 *Atriplex* did indeed have considerably superior daily P/T ratios. However, this was evidently counterbalanced by the fact that the C_3 species completed a greater proportion of its annual productivity earlier in the season when the temperatures were low (Chap. 10.6.1). The estimated annual average water use efficiencies for growth were 4.3 and 2.9 mg dry wt. $g^{-1} H_2O$ in *A. confertifolia* and *C. lanata*, respectively.

9.4.2 Photosynthesis in Relation to Water Stress

The higher intrinsic water use efficiency conferred by C_4 photosynthesis is obviously an extremely important feature in arid habitats since it enables the plant to produce more dry matter before its water supply is depleted. There is, however, no theoretical reason to expect, nor is there any experimental evidence to indicate, that C_4 photosynthesis confers an increased tolerance to low water potential once this has developed. It is probable that a strong correlation exists between the presence of the C_4 photosynthetic pathway and drought tolerance both among *Atriplex* species and among all plants. However, as discussed in Chap. 8.2, there is a wide range of drought tolerance both among C_3 and among C_4 plants. All available evidence indicates that drought tolerance is related to the environment to which the plants have adapted and any correlation with photosynthetic or taxonomic type is likely to be indirect (cf. Ludlow, 1976).

The reasons for the unusually great ability of arid region species of *Atriplex* and other genera to survive and even remain metabolically active at very low leaf water potentials were discussed earlier (Chap. 8.2). It was concluded that a high capacity for osmoregulation probably plays a key role in drought tolerance. It was also mentioned that water stress affects virtually every aspect of the physiological and metabolic activities of the plant. Here we shall discuss the effect of water stress on photosynthesis.

Stomatal and Nonstomatal Inhibition of Photosynthesis by Water Stress. It is well known that as leaf water potential is decreased a level is reached at which stomatal conductance decreases (Chap. 8). Since the stomata act as regulators for gas exchange it follows that stomatal closure must also affect photosynthesis and the rates of photosynthesis and transpiration thus change in concert as the leaf water potential decreases. This has generally been interpreted to mean that the effect of drought stress on photosynthesis can be explained on the basis of stomatal closure alone, and in many cases this is supported by experimental evidence. Changes in stomatal conductance were shown fully to account for the effects of mild water stress on photosynthesis in C_3 and C_4 crop plants such as cotton (Troughton, 1969; Troughton and Slatyer, 1969), tomato (Duniway, 1971), maize (Boyer, 1970), soybean (Boyer, 1970; Mederski et al., 1975); kidney bean (Moldau, 1973); mung bean (Mederski et al., 1975), and wheat and millet (Slatyer, 1973).

On the other hand, effects of water stress that could only be partly accounted for by stomatal responses have also been reported. Boyer (1971) found that in sunflower, nonstomatal inhibitions of photosynthesis occurred at approximately the same leaf water potential as stomatal closure. Concurrent stomatal and nonstomatal reductions of photosynthesis have also been observed in crop plants such as maize (Heichel and Musgrave, 1970), wheat (Lawlor, 1976), kidney bean (O'Toole et al., 1977), and tobacco (Redshaw and Meidner, 1972). Doley and Trivett (1974) found that both stomatal and nonstomatal reductions of photosynthesis occurred when the xerophytic grass, *Astrebla lappacea,* was subjected to water stress, although stomatal conductance was more sensitive than nonstomatal components. Similarly, Ludlow and Ng (1976) found that in *Panicum maximum* stomatal conductance started to decrease when the leaf water potential fell below -6 to -8 bar, whereas nonstomatal inhibition occurred only at leaf water potentials below -10 bar. Collatz et al. (1976) working with the effect of water stress on the xeromorphic shrub, *Eucalyptus socialis,* found that reduction of stomatal conductance as well as of nonstomatal aspects of photosynthesis took place when leaf water potential was reduced to below -18 to -20 bar. The major part of the reduction of photosynthesis could be accounted for by stomatal effects, at least at the milder stress levels (-20 to -30 bar). By comparison, in the desert xerophytic shrub, *Simmondsia chinensis,* stomatal and nonstomatal factors showed very similar sensitivities to decreasing water potential (Collatz, 1977). Both were essentially unaffected until the leaf water potential had fallen to approximately -22 bar; further decreases in the water potential resulted in parallel reductions in stomatal conductance and nonstomatal factors. In this species the greater fraction of the overall inhibition of photosynthesis was apparently attributable to nonstomatal factors. Bunce (1977) determined the effect of low leaf water potentials on photosynthesis of intact leaves of 12 species of woody plants, native to habitats ranging from streamsides to deserts. All plants were initially grown in a controlled environment with ample water supply. Water stress was imposed by termination of watering. Pronounced differences in the sensitivity of photosynthesis to low water potentials were evident; streamside species such as *Alnus oblongifolia* and *A. rugosa* were the most sensitive, while desert plants such as *Vauquelinia californica, Simmondsia chinensis,* and *Larrea divaricata* were the least sensitive. In each of the 12 species, the calculated intrinsic efficiency of CO_2 utilization (C_i, the reciprocal of "mesophyll resistance" according to most authors) started to decline at the same water potential as the stomatal conductance.

A further analysis of the effect of experimentally induced water stress on photosynthesis in *Larrea divaricata* is provided by the studies of Mooney et al. (1977a). The response of photosynthesis at a high quantum flux density to CO_2 pressure was measured on the same plants before and after exposure to water stress. Since stomatal conductance and the calculated CO_2 pressure in the intercellular spaces were continuously monitored, any reduction in intercellular CO_2 pressure due to partial stomatal closure could be compensated for by increasing the external CO_2 pressure. Water stress was imposed on initially well-watered plants either by withholding water for several days or by rapid cooling of the roots. The latter method resulted in reductions of leaf water potential within several minutes. Either method of imposing water stress caused an inhibition of photosynthesis when the leaf water potential reached -20 to -23 bar and the inhibition became increasingly severe as the water potential was further reduced (Table 9.6). As also shown in this table, inhibition of photosynthesis was accompanied by reduction in stomatal conductance, but this could only in part account for the reduction in

Table 9.6. Effect of water stress on stomatal conductance and photosynthesis in leaves of *L. divaricata*[a]. (Data of Mooney et al., 1977)

Water potential, bar	Stomatal conductance, (to water vapor) cm s^{-1}	Photosynthesis, nmol cm^{-2} s^{-1}		Percentage of total inhibition caused by:	
		At actual conductance	At constant conductance[b]	Stomatal closure	Nonstomatal components
−10	0.80	2.32	2.32	0	0
−27	0.55	1.49	1.70	25	75
−30	0.44	1.14	1.38	20	80
−36	0.18	0.65	0.88	19	81

[a] Measurements were made in normal air (325 µbar CO_2), 30 °C leaf temperature and a light intensity of 180 neinstein cm^{-2} s^{-1}
[b] Calculated rates of photosynthesis that should have been obtained if the stomatal conductance had remained constant at 0.8 cm s^{-1}, i.e. the value for the unstressed plant. These calculations were based on the relationships between photosynthesis and intercellular CO_2 pressure shown in Fig. 9.33 and Eq. (9.5)

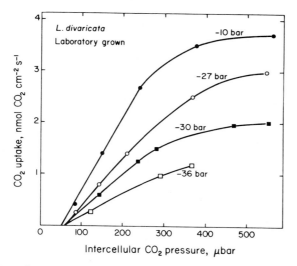

Fig. 9.33. Effect of water stress on photosynthesis in leaves of *L. divaricata* at different intercellular CO_2 partial pressures. Plants, originally from the floor of Death Valley, were grown under a 35° day/25 °C night regime at the laboratory with ample water and nutrient supply. Water was then withheld and the xylem water potential (ψ) monitored until the desired value was obtained. For unstressed plants $\psi = -10$ bar. Photosynthesis measurements were made at a light intensity of 180 n einstein cm^{-2} s^{-1}, a leaf temperature of 30 °C, and an O_2 concentration of 21%. (From Mooney et al., 1977a)

photosynthesis. Figure 9.33, shows that decreasing water potential caused an inhibition of photosynthesis at any given intercellular CO_2 pressure. Using the relationships between photosynthesis and intercellular pressure shown in this figure it is possible to calculate the rate of photosynthesis at normal ambient CO_2 pressure which would have been obtained if the stomatal conductance of the water-stressed plants had remained at the value of unstressed plants. These data (Table 9.6, right columns) indicate that 20% to 25% of the reduction in photosynthesis can be accounted for by stomatal closure. Laboratory studies on the C_4 desert species, *Tidestromia oblongifolia* and *Atriplex hymenelytra*, yielded results similar to those obtained for *L. divaricata* although the water potentials at which inhibition occurred differed somewhat between the species (Mooney, personal communication).

In *A. hymenelytra* marked inhibition of nonstomatal factors is evident at a leaf water potential of -20 bar and below (data not shown).

There is an inherent weakness in most studies attempting to distinguish between stomatal and nonstomatal components of water-stress-induced inhibition of photosynthesis on the basis of gas exchange measurements at limiting CO_2 pressures. The relationship between photosynthesis and intercellular CO_2 pressure is derived from simultaneous measurements of photosynthesis and stomatal conductance and relatively small measurement errors can result in large errors in intercellular CO_2 pressure. This difficulty is especially pronounced at low, limiting CO_2 pressures and low stomatal conductance values. The consequences of this problem are much less important under conditions where photosynthesis is no longer strongly dependent on CO_2. The results depicted in Fig. 9.33 show that water stress has a pronounced inhibitory effect on photosynthesis even at CO_2 saturation. Moreover, Mohanty and Boyer (1976) reported that in sunflower the quantum yield of photosynthesis at rate-limiting quantum flux densities was substantially reduced when the leaves were subjected to even moderate reductions in water potential. Similarly, Mooney et al. (1977a) found that the quantum yield of *L. divaricata* leaves was inhibited to approximately the same degree as the light saturated photosynthetic rate (Fig. 9.34). The inhibitions of the quantum yield and of photosynthesis at high CO_2 pressures in *L. divaricata* can in no part be attributed to stomatal effects. The inhibition of photosynthesis in its light-limited and in its CO_2-saturated state (at high quantum flux densities) thus appears to be similar to that observed at limiting CO_2 pressures. This provides corroborative evidence that nonstomatal aspects of photosynthesis are indeed affected by water stress even under conditions of normal air and high quantum flux densities.

There seems to be little doubt that reduction in stomatal conductance is a primary response to water stress, especially in short-term experiments. This does not, however, exclude the possibility that where the intrinsic capacity of photosynthesis has been reduced by water stress, stomatal adjustment may occur as a result of this lowered capacity. Adjustment of the stomatal conductance to match the change in intrinsic photosynthetic capacity would be essential in order to maintain a reasonably high water use efficiency (cf. Fig. 9.29).

Effects of Water Stress on Chloroplast Function. Very little is known about the mechanism of inhibition of the photosynthetic machinery by water stress or the relative sensitivities of the various component reactions. Nevertheless, there is sufficient evidence to conclude that photosynthetic electron transport and photophosphorylation are inhibited by desiccation. Nir and Poljakoff-Mayber (1967) found that chloroplasts isolated from flaccid *Beta vulgaris* leaves that had lost 29% and 61% of their water had 32% and 85% lower photosynthetic electron transport activity than chloroplasts isolated from unstressed leaves. The capacity

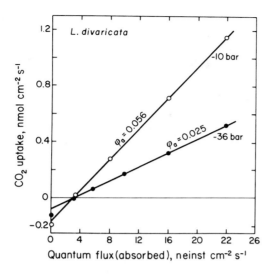

Fig. 9.34. Effect of water stress on the quantum yield of CO_2 uptake in *L. divaricata*, grown with ample water supply. Plants were grown and measurements made as given in Fig. 9.33, except that the CO_2 partial pressure was kept constant at 320–340 µbar and light was the experimental variable. (From Mooney et al., 1977a)

for cyclic photophosphorylation was reduced to similar extents. Mild stress (incipient wilting of the leaves) had no effect. Santarius (1967) working with intact leaf cells of *Beta vulgaris* and *Spinacea oleracea* (spinach), found no decrease in light-dependent ATP formation until 40% to 50% of the leaf water content was lost and NADP reduction did not decline until the water loss exceeded 50% to 60%.

Fry (1970) found that severe stress of cotton leaves resulted in reduced electron transport activity of chloroplasts subsequently isolated from the leaves. Chloroplasts from wilted leaves ($\psi_w = -28$ bar) had about 50% of the activity of those from fully turgid leaves ($\psi_w = -4$ bar). Electron transport activity of chloroplasts from *Pisum sativum* (pea) and *Helianthus annuus* (sunflower) leaves was already inhibited when leaf water potentials fell below -12 and -8 bar, respectively, and the inhibition was proportional to leaf water potential below these levels (Boyer and Bowen, 1970). Carbon fixation by isolated spinach chloroplasts was reduced when these were exposed to reduced osmotic potentials (Plaut, 1971); a decrease in CO_2 fixation rate of isolated chloroplasts was also found when leaves were desiccated in air prior to chloroplast isolation (Plaut and Bravdo, 1973).

While the above studies show that photosynthetic activities at the subcellular level are clearly inhibited by water stress, it is very difficult to evaluate the significance of these findings to photosynthetic behavior at the whole leaf level. For one thing, it seems possible that in most instances overall photosynthetic activity may well have ceased before the water stress was sufficiently severe to cause a marked inhibition of the capacity of electron transport and photophosphorylation. Nevertheless, the extensive studies by Boyer and coworkers on *Helianthus annuus* provide strong evidence that at least in this species inhibition of chloroplast reactions is an important factor in the inhibition of whole leaf photosynthesis by water stress. At high quantum flux densities, the reduction in photosystem-II-driven electron transport capacity paralleled the reduction in the rate of CO_2 fixation by intact leaves when the leaf water potential was reduced below -8 bar; stomatal conductance showed an almost identical response (Boyer and Bowen, 1970). Later studies (Keck and Boyer, 1974) showed that whole-chain electron transport was appreciably inhibited at leaf water potentials of -9 bar and continued to decline in activity as the leaf water potential decreased. Photophosphorylation capacity (both cyclic and noncyclic) was unaffected at -10 to -11 bar but then started to decline, reaching total inhibition at -17 bar. This decline in photophosphorylation was coincident with uncoupling of electron transport. These results indicate that at high quantum flux densities low leaf water potential affects photosynthesis both by inhibiting electron transport and photophosphorylation and by reducing stomatal conductance.

As was mentioned in a previous paragraph, water stress not only affects light-saturated photosynthesis but the quantum yield at rate-limiting quantum flux densities is inhibited as well. Mohanty and Boyer (1976) found that reducing the water potential of *H. annuus* leaves from -4 to -15.3 bar resulted in a decrease in the quantum yield of photosynthesis (at $2\% \ O_2$) from 0.079 to 0.020 mol CO_2/einstein. The quantum yield for photosystem-II-driven electron transport (for 4 equivalents) by chloroplasts isolated from leaves with a water potential of -3.6 bar also was 0.079; but when the water potential of the same leaf tissue had fallen to -14.8 bar the quantum yield of the chloroplasts decreased to 0.028. Thus, water stress affected the efficiency of light conversion of whole leaves and isolated chloroplasts in a very similar manner. Stomatal conductance is unlikely to exert an appreciable influence on the quantum yield. The close similarity in the response of whole leaf photosynthesis and chloroplast activity at rate-limiting quantum flux densities to water stress in sunflower indicates that inhibition of primary photochemical events rather than stomatal closure is responsible for the reduction in the quantum yield of photosynthesis at low leaf water potentials. The results obtained with *Larrea divaricata* (Fig. 9.34) and preliminary observations on C_3 and C_4 species of *Atriplex* (Björkman, unpublished) suggest that reduction in the efficiency of light conversion at low leaf water potentials may occur in many higher plant species. Alberte and Thornber (1977) suggest that the content of the light-harvesting chlorophyll a + b protein complex (cf. Fig. 9.1) may be a rather specific target for water stress. When maize was subjected to drought stress a substantial loss of chlorophyll occurred. Most of this loss occurred in the mesophyll cells; the chlorophyll content of the bundle-sheath cells was affected to a lesser degree. All of the loss of chlorophyll could be accounted for by a reduction in the lamellar content of the light-harvesting chlorophyll a + b protein complex. These workers conclude that since this complex is a major intrinsic component of the thylakoid membranes, it is likely that the losses in this component may lead both to perturbation of the structural organization of the membrane, a reduction in the efficiency of photosynthetic electron transport and, hence, a reduction of whole leaf photosynthesis. However, in many studies of the effect of water stress on photosynthesis no significant loss of chlorophyll was observed, yet severe nonstomatal inhibition occurred. For example, in the studies of Mohanty and

Boyer (1976), the quantum yields of whole leaf photosynthesis and photosystem-II-driven electron transport were inhibited by 74% and 64%, respectively, when sunflower leaves were subjected to a water potential of -14.8 to -15.3 bar. The inhibition of photosynthesis was even more severe in sunflower leaves at -20.2 bar and in soybean leaves at -25.5 bar. Yet, in none of these cases could any loss in chlorophyll be detected, and the absorption spectra were identical for chloroplasts that had been exposed to high and low water potentials. The absorption spectra of the intact leaves were identical as well. While the inhibition of the quantum yield for electron transport in all probability was caused by a fundamental alteration in the ability of the thylakoid membrane to convert absorbed radiation to chemical energy, this alteration obviously was not due to a loss of light-harvesting chlorophyll.

Effects of Water Stress on Chloroplast Structure. The postulate that the inhibition of photosynthetic electron transport by water stress is caused by structural changes in the thylakoid membranes at first sight appears to be supported by electron microscopic studies. Da Silva et al. (1974) reported that in water-stressed cotton leaves chloroplast thylakoid structure was profoundly affected; in many instances the characteristic fine structural organization was lost. Chloroplast ribosomes were among the first structures to disappear. Alieva et al. (1971) and Kurkova and Motorina (1974) reported that water stress resulted in losses in structural integrity of the chloroplasts, such as unstacking and vesiculation of the thylakoid membranes. Giles et al. (1974) observed that in maize leaves subjected to a water potential of -19 bar swelling of the outer chloroplast membranes occurred in most of the mesophyll cells; about 25% of these cells had broken tonoplasts and the cells became filled with chloroplast debris. The bundle-sheath cells showed no gross ultrastructural damage. *Sorghum bicolor* leaves were much more resistant to water stress than maize leaves (Giles et al., 1976). Swelling of the outer chloroplast membrane, disarrangement of the stoma lamellae and tonoplast fragmentation were not apparent in the mesophyll cells until the leaf water potential had fallen to -37 bar. As in maize the bundle-sheath cells of *S. bicolor* seemed more resistant than the mesophyll cells. It is noteworthy that a similar difference in resistance to water stress between mesophyll and bundle-sheath cells has also been observed in *Atriplex halimus* (Blumenthal-Goldsmith and Poljakoff-Mayber, 1968). In plants grown in salinized nutrient solutions, swelling of the thylakoid membranes was barely evident at low NaCl concentration in the growth medium (-3 to -7 bar) but became pronounced at medium concentrations (-11 to -19 bar) and the chloroplast structure was completely disrupted at high concentrations (-19 to -23 bar). Similar differences were observed between *A. halimus* plants growing in nonsaline and highly saline natural habitats. The first signs of changes in the chloroplast structure occurred in the mesophyll cells; only when these were extensively damaged did the structure of the bundle-sheath cells become affected.

One problem in structural studies of desiccated tissue is that the use of standard aqueous fixatives may result in rehydration of the tissue during the fixation. Unless the osmolarity of the fixation medium is adjusted to match the water potential of the leaf tissue the medium would become increasingly hypotonic as the leaf water potential declines. No such osmotic support was provided in the structural studies mentioned above. Any artifacts caused by rehydration can thus be expected to become worse as desiccation becomes more severe. In a recent study, Fellows and Boyer (1976) found that if sunflower leaves having a water potential of -5.3 bar were fixed in a fixative having a water potential of -6.5 bar, chloroplast structure appeared perfectly normal and the thylakoid lamellae were tightly stacked. If the same fixative was used with leaves having a water potential of -21.9 bar the thylakoid membranes appeared to be only loosely stacked, but if sucrose was added to the fixative to provide osmotic support for the desiccated tissue chloroplast structure again appeared normal and nearly all of the thylakoid lamellae were tightly stacked. Indeed, the thylakoid membranes retained much structural detail even in air-dried leaves having a water potential of -123 bar. However, breakage of the tonoplasts and plasmalemmae became evident at leaf water potentials below -26 bar even though the fixation medium was osmotically supported. Fellows and Boyer conclude that inhibition of photosynthetic electron transport in water-stressed sunflower leaves is not caused by a loss of integrity of the chloroplast membranes, at least at leaf water potentials above -26 bar. They propose that much more subtle changes in the *conformation* of the lamellar membranes are responsible. Their results also suggest the possibility that the observed reduction in chloroplast activity at low water potentials may somehow be related to cytoplasmic alterations.

Mechanism of Nonstomatal Photosynthetic Inhibition by Water Stress. In Chap. 8 we discussed the physical and chemical mechanisms that may underlie the various effects of water stress on growth processes. It was concluded that although leaf water potential is a very useful indicator of the water status of the plant, it is improbable that reductions in the activity of water, as represented by the leaf water potential, is in itself the immediate cause of the physiological effects of water stress. Of the various parameters of water status, turgor pressure is the only component of water potential which has been shown to be directly linked to cellular processes. There appears to be little doubt that changes in turgor pressure play a critical role both in cell growth and stomatal movement. It would seem that such changes in pressure across the cell wall might also affect photosynthesis and other metabolic events as a result of changes in the spatial relations of cellular components, e.g., by causing conformational alterations in membrane-bound electron carriers or enzymes. On the basis of their work on sunflower leaves, Boyer and Potter (1973) concluded, however, that turgor had little effect on nonstomatal inhibition of photosynthesis. While turgor became zero in sunflower leaves at a leaf water potential of -10 bar, light-limited photosynthesis and chloroplast activities were mainly affected at leaf water potentials in the -10 to -19 bar range. These authors also concluded that decreases in the Gibbs free energy of water due to decreased osmotic potentials during desiccation of sunflower leaves are much too small to account for the major effects of leaf desiccation on electron transport (Potter and Boyer, 1973). At present, no mechanism is known that provides a satisfactory explanation of the observed effects of moderate water stress on photosynthesis at the chloroplast level.

Species Differences and Acclimation to Water Stress. As was mentioned in Chap. 8 large differences have been observed between species from arid and mesic habitats in the water potential at which stomatal closure and inhibition of cell growth occur. Much less is known about interspecific differences with regard to inhibition of nonstomatal aspects of photosynthesis. It is, however, clear that such differences do exist. For example, while nonstomatal inhibition sets in at about − 10 bar in sunflower leaves, no effects were observed until the water potential dropped below − 20 to − 23 bar in the desert shrubs *Eucalyptus socialis* (Collatz et al., 1976), *Simmondsia chinensis* (Collatz, 1977), and *Larrea divaricata* (Mooney et al., 1977a). As previously mentioned, Bunce (1977) found a wide range of sensitivities among species whose natural habitats ranged from wet to arid habitats. These results were all obtained on plants that had been grown under ample water supply prior to the drought experiment.

It has long been known that many plants possess an increased resistance to drought when they have been subjected to water stress during development (Levitt, 1972). The sensitivity of the stomata to low leaf water potentials is markedly decreased by such preconditioning but there is only limited information on the extent to which the sensitivity of nonstomatal aspects of photosynthesis is decreased also. Collatz et al. (1976) found no significant effect of drought hardening of *E.socialis* plants in the laboratory on subsequent photosynthetic performance although the hardening treatment resulted in an increased xeromorphy of the leaves. Ludlow and Ng (1976) also did not find any marked influence of such hardening treatments on the photosynthetic drought sensitivity of *Panicum maximum* in their laboratory studies. However, no inhibition of photosynthesis was evident in *P.maximum* plants growing in the field even when their water potentials had fallen to considerably lower levels than those which caused severe inhibition of photosynthesis in potted plants. The potted plants showed inhibition when grown either in growth chambers or outdoors, and whether or not they had been subjected to drying cycles. Ludlow and Ng suggest that the difference in response between field-grown and potted plants is associated with the much larger soil volume available to the former, resulting in a much slower decline of leaf potential during a drying cycle. This slower drying would allow enough time for osmotic adjustment of existing leaves and/or development of new, more drought-tolerant leaves.

Dramatic differences in the sensitivity of photosynthesis to low leaf water potentials are evident between unstressed, laboratory-grown plants of *Larrea divaricata* and plants of the same species growing in their natural habitat on the floor of Death Valley (Mooney et al., 1977a). Figure 9.35 compares the response of the quantum yield of photosynthesis in laboratory grown plants to an experimentally lowered water potential with the quantum yield of field plants in situ whose water potentials naturally ranged from − 20 to − 49 bar. (No attempts were made to change the water potentials of the field plants experimentally.) The quantum yields of the field plants remained the same irrespective of the water potential even at − 49 bar. These results demonstrate that this desert plant is capable of a remarkable degree of acclimation to water stress that includes the ability of the photosynthetic apparatus to operate at leaf water potentials that in the absence of such acclimation would result in severe or total inhibition of photosynthetic activity. It seems likely that most plants, irrespective of ecological origin, possess the capacity for at least some degree of acclimation, but that this acclimation potential is especially high in arido-active and in halophytic plants. Osmotic adjustment is probably an essential component of this acclimation.

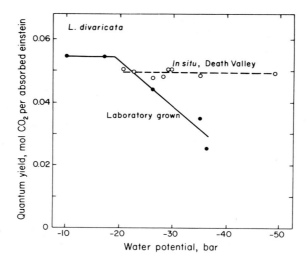

Fig. 9.35. Quantum yield as a function of water potential in *L.divaricata,* grown under ample water supply at the laboratory, and in situ on the floor of Death Valley. Measurements were made as indicated in Fig.9.34. (From Mooney et al., 1977)

Indeed, field measurements in Death Valley during different seasons of *L. divaricata* and *Atriplex hymenelytra* show that osmotic adjustment permits these plants to remain at positive turgor at water potentials at least as low as -45 bar (Bennert and Mooney, 1979). Whether or not acclimation to low water potentials involves any specific changes at the chloroplast level is unknown, nor is there any information on the question whether or not species differences in tolerance to water stress involve any differences in the intrinsic properties of the chloroplasts.

Water Use Efficiency at Low Leaf Water Potentials. The above considerations of the effects of low leaf water potential on stomatal and nonstomatal aspects of photosynthesis together with the phenomenon of drought acclimation raise the question how photosynthesis and water use efficiency are affected by water stress under natural conditions. It is obvious that the information gained in laboratory studies so far is insufficient to allow any realistic quantitative predictions of how photosynthesis and water use efficiency will respond to seasonal changes in water status of the plants in their native habitats. On theoretical grounds, one would expect that the effect of water stress on photosynthetic water use efficiency will be strongly dependent on the relative response of stomatal and nonstomatal factors. If the reduction of photosynthesis due to nonstomatal factors is large in comparison with the reduction caused by decreased stomatal conductance, then the water use efficiency should decline as the water stress increases. Conversely, if stomatal conductance is reduced but the intrinsic efficiency of photosynthesis is little affected by water stress, then the water use efficiency should tend to increase. Actual measurements of the responses of photosynthetic water use efficiency to seasonal changes in plant water status in the field are notably scarce, especially as far as wild plants are concerned. The results from field studies in Death Valley on the evergreen drought-tolerant shrubs *Larrea divaricata* and *Atriplex hymenelytra* give some insight into the relationship between photosynthetic performance and water stress in an extremely arid habitat and we shall therefore discuss the results of these studies in some detail.

Table 9.7 shows the response of light-saturated photosynthesis, stomatal conductance, and intrinsic water use efficiency of naturally occurring, unwatered *L. divaricata* plants in Death Valley, as the plant water potentials decrease owing to seasonally increased evaporative demand and decreased soil moisture. (For simplicity all values shown in Table 9.7 refer to a constant leaf temperature of 30 °C.) The daytime leaf water potential fell from -26 bar in late January to -49 bar in early October; the predawn water potential showed similar seasonal changes. During the same time stomatal conductance declined from 1.2 to 0.35 cm s^{-1}, i.e., a 71% reduction, while photosynthesis fell from 2.12 to 1.45, only a 32% decrease. As a result, the intrinsic water use efficiency (P/C_s) more than doubled, from 1.8 to 4.1 nmol CO_2 cm^{-3}. Determinations of the CO_2 response curves showed that the photosynthetic rate at any given intercellular CO_2 pressure remained essentially constant throughout the year, demonstrating that nonstomatal inhibition of photosynthesis did not occur.

The lower actual photosynthetic rates at low leaf water potentials can thus be fully attributed to reductions in stomatal conductance. The observed increase in the intrinsic water use efficiency with increasing water stress is in accordance with the prediction that this should occur if stomata are solely responsible for the reduced photosynthetic rate.

Field studies on *Atriplex hymenelytra,* also conducted in its native Death Valley habitat (Pearcy et al., 1974), show that the intrinsic photosynthetic water use efficiency increased with increased water stress in this species as well, even though in this case the photosynthetic capacity was considerably reduced. Photosynthetic gas exchange characteristics and environmental parameters determined on two occasions, March 15, 1971 and July 20, 1972 are summarized in Table 9.8. On both of these days skies were cloudless and water potentials, leaf temperatures, photosynthetic rates, and stomatal conductances were representative for these times of the year. On March 15, plant water potentials were moderate, ranging from -12 bar in the morning to -22 bar at noon. Maximum photosynthetic rates and leaf conductances were reached at about 9 h and remained essentially constant until 15 h. Leaf temperatures were generally between 30° and 35 °C during the period when photosynthetic rates were highest. The intrinsic water use efficiency (P/C_s) showed only a small variation during the day, ranging from 5.1 at 8 h to 6.8 nmol CO_2 cm^{-3} at 16 h. The P/C_s value integrated over the entire daytime period was 5.3 nmol CO_2 cm^{-3}. This is about 2.5 times as high as for the C_3 species *Larrea divaricata* under similar environmental conditions.

On July 15, both early morning and mid-day leaf water potentials were much lower than in March, ranging from -36 bar at 8 h to a minimum of -43 bar during mid-day. The daily course of water potential closely followed that of leaf temperature and hence the water vapor pressure deficit. Photosynthesis and stomatal conductance reached their maximum values at about 9 h when the leaf temperature was 35° to 37 °C and gradually declined later in the day as leaf temperature rose to a maximum of 46 °C. Determinations of the temperature dependence of photosynthesis could be in part attributed to temperature effects

Table 9.7. Seasonal course of water potential, photosynthesis and stomatal conductance of *L. divaricata* in Death Valley.[a] (Data of Mooney et al., 1977)

Date	Water potential (bar)		Photosynthesis, P (nmol cm^{-2} s^{-1}) at actual conductance	Stomatal conductance, C_s (cm s^{-1})	Ratio P/C_s (nmol cm^{-3})	Photosynthesis[b] (nmol cm^{-2} s^{-1}) at constant conductance
	Predawn	At time of measurement				
January 23, 1976	−21	−26	2.12	1.20	1.77	1.97
March 23, 1976	−22	−29	1.95	1.01	1.93	1.90
May 17, 1975	−26	−31	1.67	0.40	4.18	2.44
October 4, 1975	−34	−49	1.45	0.35	4.14	2.21
September 27, 1975 (Irrigated plant)	−18	−34	2.30	0.90	2.56	2.30

a   Measurements were made in normal air (325 μbar CO_2), 30 °C leaf temperature and a light intensity of 180 neinstein cm^{-2} s^{-1} on attached twigs of naturally occurring, unwatered plants (except where indicated otherwise)
b   Calculated rate at a constant stomatal conductance of 0.9 cm s^{-1} (see Table 9.6[b])

Table 9.8. Environmental parameters, photosynthetic rate, stomatal conductance and photosynthetic water use efficiency of *Atriplex hymenelytra* shrubs in Death Valley, California, during the spring and summer. (Data of Pearcy et al., 1974)

Momentary values	March 15			July 20		
	8 h	12 h	16 h	8 h	12 h	16 h
Leaf temperature, °C	19	35	35	37	45	46
Leaf water potential, bars	−12	−22	−20	−36	−43	−42
Photosynthesis, nmol cm^{-2} s^{-1}	0.87	1.23	1.16	0.63	0.63	0.33
Conductance (C_s), cm s^{-1}	0.17	0.23	0.17	0.09	0.07	0.04
P/C_s, nmol CO_2 cm^{-2} cm^{-1}	5.12	5.36	6.82	7.03	8.75	8.50
Daily totals	7 to 17 h			6 to 18 h		
Daily photosynthesis, μmol cm^{-2}	37.27			23.41		
Daily P/C_s, nmol cm^{-2} cm^{-1}	5.32			7.82		
Daily P/T, mmol CO_2 (mol H_2O)$^{-1}$	4.20			3.80		

alone. In spite of this decline the intrinsic water use efficiency rose during the day from 7 nmol CO_2 cm^{-3} in the morning to 8.8 nmol CO_2 cm^{-3} in the afternoon. This rise was undoubtedly a result of stomatal adjustment. The P/C_s value integrated for the whole day was 7.8 nmol CO_2 cm^{-3}, almost 50% greater than in March.

The photosynthetic rate was considerably lower in the summer than in the spring. This is in part due to the high supraoptimal summer leaf temperatures but the difference is substantial even if the comparison is made at a constant temperature. At 33 °C leaf temperature the maximum photosynthetic rates were 1.32 and 0.68 nmol CO_2 cm^{-2} s^{-1}, for March and July, respectively. The corresponding stomatal conductances were 0.18 and 0.074 cm s^{-1}, giving P/C_a ratios of 7.3 in March and 9.2 in July. Subsequent field studies on *A.hymenelytra* in Death Valley (Mooney and Björkman, unpublished) clearly shows that the intrinsic capacity of photosynthesis is considerably lower late in the season when the water potential is low than early in the season when it is relatively high. This is in contrast with *L.divaricata,* whose intrinsic capacity of photosynthesis remained largely unchanged. Nevertheless, because the stomatal conductance changes in concert with the intrinsic photosynthetic capacity in *A.hymenelytra* there is no decline in the P/C_s ratio as the water stress increases. As we have seen, the P/C_s ratio instead increases.

This greater intrinsic water use efficiency counterbalances the effect of the increased water vapor pressure deficits that develops because of high leaf temperatures. As a result, the actual water use efficiency (P/T) shows only a slight decline between March and July (Table 9.8, bottom line). The results obtained by Caldwell et al. (1977b) with *A.confertifolia,* growing in a much colder desert in Utah, indicate that a similar compensation occurs in this species (cf. Fig. 9.32).

We can thus conclude that at least as far as the arido-active species *A.hymenelytra* and *L.divaricata* are concerned, the intrinsic photosynthetic water use efficiency tends to rise rather than decline as the water stress increases in the natural habitats. The results further show that the difference in photosynthetic water use efficiency between these C_3 and C_4 plants is present not only at high but also at low leaf water potentials. The water use efficiency of *A.hymenelytra* was two to three times that of *L.divaricata* down to water potentials at least as low as − 46 bar.

9.5 Photosynthesis in Relation to Salinity and Mineral Nutrition

The interactions between nutrient ions (N and P) and nonnutrient ions (Na$^+$, K$^+$, Cl$^-$) and metabolism were discussed in Chap. 7 and the same categories of interaction may be considered in relation to photosynthesis. Ions may be involved as metabolic intermediates (NH_4^+, PO_4^{3-}), as nonspecific effectors of metabolic processes (Na$^+$, Cl$^-$) or as essential micronutrients (Na$^+$). In Chaps. 7 and 8 it was concluded that the salinity tolerance of halophytes was largely based on the orderly accumulation and compartmentation of salt in the shoot which was in turn responsible for osmoregulation and the maintenance of satisfactory water relations in the tissue. The common processes which underlie photosynthetic responses to water stress and salinity have been recently reviewed by Winter (1978) and for our purposes, represent a convenient link between this section and the preceding one.

9.5.1 Response to Salinity

Photosynthesis of salt-sensitive plants is depressed by the presence of NaCl (Nieman, 1962) and Gale et al. (1967) distinguished between the effects of NaCl on stomatal and nonstomatal components of photosynthesis. *Allium cepa* was unable to adjust osmotically so that treatments in NaCl (5.4 bar) resulted in loss of turgor, stomatal closure, and inhibition of photosynthesis. Osmotic adjustment occurred in *Phaseolus* and *Gossypium* (treated with 3.4 and 4.5 bar NaCl, respectively) but photosynthesis was reduced. In *Phaseolus* the initial depression of photosynthesis was due to decreased stomatal conductance but after several days treatment with NaCl, the slope of the light response curve for photosynthesis in air was depressed, showing that NaCl treatment also interfered with non-stomatal components of photosynthesis. This is indicated by a decrease in C_i (the reciprocal of "mesophyll resistance"; c.f. p. 356). In *Gossypium*, most of the inhibition of net photosynthesis was associated with nonstomatal factors, as confirmed by Longstreth and Nobel (1978). Downton (1977) concluded that the depression of photosynthesis in grapevines in response to increased leaf NaCl content was principally associated with nonstomatal factors. Although stomatal and nonstomatal factors changed markedly with increasing salinity in *Vitis*,

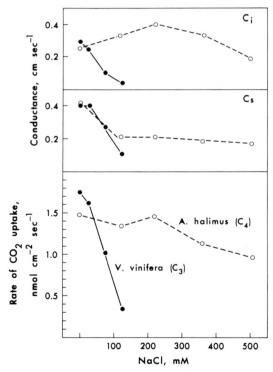

Fig. 9.36. Changes in nonstomatal (C_i) and stomatal (C_s) components of photosynthesis, and photosynthetic rate in leaves of salt-sensitive *Vitis vinifera* and the halophyte, *A. halimus,* following growth in saline culture solutions. (Redrawn from Downton, 1977; and Gale et al., 1970)

the change in nonstomatal components (C_i) was greater than change in stomatal conductance (C_s) (Fig. 9.36). In contrast to this, photosynthesis in the halophyte *Atriplex halimus* is much less sensitive to NaCl, declining by about 30% over the range of 0 to 500 mM NaCl (Gale et al., 1970). In this halophyte, C_s declined to about half its control value in response to low salinity but remained practically unchanged thereafter (Fig. 9.36). On the other hand, C_i increased slightly in moderate salinity treatments.

The above studies suggest that stomatal control of photosynthesis predominates when salt-sensitive plants are unable to adjust leaf water relations by osmotic adjustment following exposure to NaCl. Even if the salt-sensitive species are able to adjust osmotically, the absorbed salt interferes with biochemical processes and the inhibition of photosynthesis is then reflected in changes in C_i. Evidently the salt absorbed by salt-sensitive plants is not properly compartmented in the cytoplasm or interferes with cytoplasmic water relations.

In halophytes on the other hand, both stomatal and nonstomatal components of photosynthesis show relatively little response to high levels of salinity. This generalization is sustained by studies of marsh halophytes such as *Spartina* spp. (Mallott et al., 1975; Longstreth and Strain, 1977) and by comparisons of C_3 and C_4 *Atriplex* or other species of the sand dunes (Fig. 9.37) in addition to *Atriplex halimus* (Fig. 9.36). In halophytes, the relative insensitivity of stomatal and nonstomatal

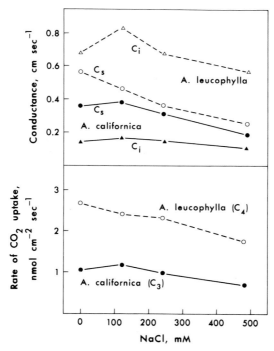

Fig. 9.37. Changes in nonstomatal (C_i) and stomatal (C_s) components of photosynthesis, and photosynthetic rate in leaves of *A. californica* (C_3) and *A. leucophylla* (C_4) from coastal sand dune habitats, following growth in saline culture solutions. (Redrawn from de Jong, 1977)

components of photosynthesis to high levels of salt in culture solutions presumably reflects the adequacy of osmotic adjustment and intracellular compartmentation of salt described previously (Chap. 7.4.5).

Salinity has a marked effect on the intrinsic water use efficiency (P/C_s) of halophytes. In the leaf CO_2 and H_2O exchange experiments shown in Figs. 9.36 and 9.37, this ratio increased from 4.6 to 7.6 in *A. leucophylla* (C_4) and from 3.8 to 6.8 in *A. halimus* (C_4) but only from 3.0 to 3.6 in *A. californica* (C_3). The improved water use efficiency measured in these short-term experiments is retained when halophytes are grown in long-term experiments at different levels of NaCl. Eaton (1927) noted that the water use efficiency for dry matter production in *A. semibaccata* (C_4) increased from 2.8 to 4.1 mg dry wt g^{-1} H_2O when soil NaCl level was raised from zero to 0.25% w/w. This improved water use was associated with a positive growth response to added salt. In salt-sensitive wheat (C_3) the same level of NaCl also produced an increase in water use efficiency from 1.8 to 2.8 mg dry wt g^{-1} H_2O but this was accompanied by a 76% decrease in dry weight of the plants grown in 0.25% NaCl. Obviously an improvement in water use efficiency following osmoregulation is only beneficial if salinity does not interfere with non-stomatal components of photosynthesis.

Calculations of C_s are usually based on the assumption that the water vapor pressure at the sites of evaporation on the wet cell walls surrounding the leaf intercellular spaces (p_i^w) is equal to the saturation vapor pressure of pure water at the leaf temperature (Chap. 9.4.1). The presence of solutes at these sites causes a reduction in p_i^w but this effect is small and p_i^w for a 1M NaCl solution is only about 2% lower than p_i^w for pure water (p. 263). Hence, direct effects of salinity on transpiration by lowering p_i^w may be considered negligible. In spite of this, Kaplan and Gale (1972) reported that when *A. halimus* was grown in salinized culture solutions ($\psi = -10$ bar), transpiration was reduced to a greater degree than could be accounted for by the reduction in C_s and these authors concluded that "... for plants grown in saline media, only the presence of a factor, such as r_m [= mesophyll resistance to water loss], could explain the considerable decrease of overall leaf resistance to water loss, while there was only a small reduction in photosynthesis (which could be ascribed to changes in r_s = [stomatal resistance = $1/C_s$])".

Although the possibility cannot be ruled out that under some extreme conditions p_i^w may be significantly lower than obtained from the saturation vapor pressure of water, the conclusion reached by Kaplan and Gale (1972) appears to be an artifact of the methods used calculate r_m and to difficulties in obtaining sufficiently accurate measurements of C_s. Undoubtedly, changes in the cuticle development and in suberization at different parts of the leaf water transport path may occur in response to salinity and several authors conclude that changes in leaf architecture may be involved in salinity-dependent changes in photosynthetic characteristics (Longstreth and Strain, 1977; Longstreth and Nobel, 1978). However, the most likely explanation for the improved water use efficiency in salt-treated halophytes is that stomata adjust to give a higher P/C_s ratio (see Section 9.4.1). This is evident in the data of Kaplan and Gale (1972) and implies that photosynthesis in halophytes may adjust to function at a lower intercellular CO_2 pressure in response to salinity. Thus if transpiration is regarded as a cost and carbon fixation as a benefit, at low stomatal conductance the marginal cost is less.

By closing stomata slightly in response to salinity (Figs. 9.36, 9.37) halophytes operate at a greater marginal benefit, lower marginal cost in the presence of salt. The other essential feature of this account is that C_i for CO_2 should remain unchanged or increase in response to salt, as shown in Figs. 9.36 and 9.37. Presumably halophytes are able to maintain adequate cellular compartmentation so that biochemical processes are not subject to electrolyte or water stress in leaves of plants growing in saline habitats. The available data thus suggest a simple, plausible account of the basis for the photosynthetic response functions of halophytes and salt-sensitive species which can be readily integrated with our earlier treatments of the ionic and water relations of these plants.

9.5.2 Response to Mineral Nutrition

A wide range of photosynthetic responses to mineral nutrition have been reported in the literature (Nátr, 1975) but in few instances have these been explored in sufficient detail to permit evaluation of the component processes. The nutrient treatments applied a range from mild limitation through to nutrient deficiency and it is frequently difficult to distinguish primary responses from secondary effects following overall breakdown in tissue functions. In some cases, such as K^+ deficiency, a reasonably direct sequence of events can be traced. Peaslee and Moss (1968) noted, for example, in the early stages of K^+ deficiency in maize the reduction in photosynthesis was almost entirely attributable to decreased C_s. This is not surprising in view of the particular role for K^+ in turgor generation and stomatal response (Raschke, 1975).

Peaslee and Moss observed that at later stages of K^+ deficiency nonstomatal components were increasingly responsible for further reductions in photosynthesis. Terry and Ulrich (1973) found that in sugar beet C_i was more affected by K^+ withdrawal than was C_s, possibly because this halophyte has a less specific K^+ requirement for stomatal opening. These authors have studied the effect of a wide range of nutrient responses on photosynthesis in sugar beet. When photosynthesis was inhibited by deficiency of P, S, or Mg^{2+}, the inhibition was almost entirely due to a decline in C_i.

The best-documented responses of photosynthesis to nutrition are those involving nitrogen. In C_3 *Beta vulgaris* (Nevins and Loomis, 1970) and C_4 *Zea mays* (Ryle and Hesketh, 1969) depression of photosynthesis in nitrogen-limited plants was attributed almost exclusively to decreased C_i. In *Zea mays* C_i decreased from 0.66 to 0.06 with time after transfer to low nitrogen culture, whereas C_s only ranged from 0.25 to 0.17. Medina (1970, 1971) investigated the relationship between nitrogen nutrition and photosynthetic processes in *A. triangularis*. He confirmed the above observations and obtained a rather complete account of the process underlying the increase in intrinsic conductance of leaves grown in low nitrogen cultures.

The photosynthetic properties of leaves of *A. triangularis* grown in 33 mM and 1 mM NO_3^- cultures are summarized in Table 9.9. Medina (1971) found that changes in leaf nitrogen content were almost exactly mirrored in the soluble protein

Table 9.9. Concentration of nitrogen, chlorophyll, soluble protein, and Fraction-1 protein, RuP_2 carboxylase activity and photosynthetic properties in leaves of *A. triangularis* grown at high light intensity with two levels of nitrate in culture solution. (Data of Medina, 1980)

Leaf property		Nitrate concentration in culture (mM)	
		33 mM	1 mM
Nitrogen concentration	(mg g^{-1} dry wt.)	58.2	23.8
Chlorophyll	(mg g^{-1} fresh wt.)	1.2	0.8
Soluble protein	(mg g^{-1} fresh wt.)	18.1	10.5
Fraction-1 protein	(mg mg^{-1} sol protein)	0.44	0.16
RuP_2 carboxylase activity	(nmol cm^{-2} s^{-1})	4.43	1.13
CO_2 fixation rate[a]	(nmol cm^{-2} s^{-1})	1.76	0.82
Stomatal conductance[a]	(cm s^{-1})	0.60	0.70
Nonstomatal conductance[a]	(cm s^{-1})	0.19	0.11

[a]   Based on measurements in 21% O_2 with intercellular CO_2 pressure adjusted to 200 μbar

content. Fraction-1 protein (RuP_2 carboxylase) comprised 44% of the soluble leaf protein in the high nitrogen plants, but only 16% in the low nitrogen plants. The activity of this enzyme was substantially lower in leaves of the low nitrogen plants, presumably as a result of lowered enzyme level rather than lower catalytic activity. The low activity of this enzyme in leaves of low nitrogen plants was associated with much reduced photosynthetic rate. Indeed, in a series of observations with leaves of widely ranging nitrogen contents, RuP_2 carboxylase activity and CO_2 fixation rate were highly correlated (Medina, 1970). Table 9.8 also shows that in *A. triangularis* leaves C_i was lower than C_s and although C_s remained virtually unchanged in leaves of low nitrogen-grown plants, C_i declined significantly. These data suggest that changes in the level of RuP_2 carboxylase in leaves of nitrogen-deficient plants may be responsible for the lowering of C_i and for the reduced CO_2 fixation rate.

This conclusion was supported by further studies of other processes which could conceivably contribute to change in nonstomatal factors reflected in C_i. Medina found for example, that at dawn, leaves of low nitrogen-grown *A. triangularis* contained about 20 times the starch content of the high nitrogen-grown plants. In electron micrographs chloroplasts of low nitrogen plants are distended by large starch grains (Fig. 9.38) and Medina suggested that the accumulation of starch in leaves could interfere with the photochemical activities of chloroplasts. This possibility was discounted on two grounds. As shown in Fig. 9.39, the initial slopes of the curves for assimilation as a function of light intensity are nearly identical in both low nitrogen and high nitrogen leaves, suggesting that the quantum requirement for CO_2 fixation is similar in both. In addition, Medina found that Hill reaction activity of isolated chloroplasts from leaves of high and low nitrogen plants was similar. He concluded that the decrease in C_i in low nitrogen leaves could be largely attributed to reduction in RuP_2 carboxylase activity. In other experiments Neales et al. (1971) found large differences in the photosynthesis of bean leaves following selective defoliation which could not be ascribed to changes in

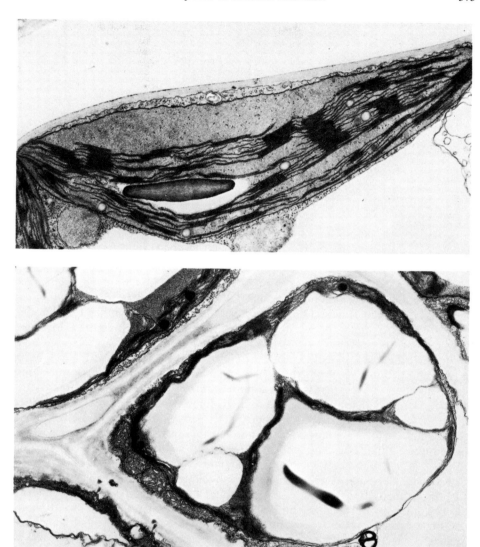

Fig. 9.38. Electron micrographs of chloroplasts from *A.triangularis* grown under ample nitrogen supply *(top)* and severe nitrogen deficiency *(bottom)*. The latter chloroplasts are distended by prodigous starch grains. (From Medina, 1970)

stomatal conductance or in quantum requirements. Like Medina, they found a good correlation between RuP_2 carboxylase activity and photosynthetic rate.

Although these data show a strong correlation between photosynthesis, C_i, and the activity of RuP_2 carboxylase, other data in other species suggest that C_s and photosynthesis are closely linked over a range of leaf nitrogen levels, and presumably also RuP_2 carboxylase levels. In *Oryza* and in *Phaseolus* stomatal conductance and photosynthetic rate are correlated regardless of leaf age or

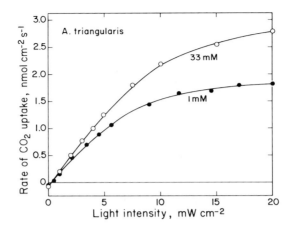

Fig. 9.39. Light-dependence curve of photosynthetic CO_2 uptake for *A. triangularis* grown in 33 mM and 1 mM NO_3^- cultures. (Redrawn from Medina, 1970)

nitrogen content (Yoshida and Coronel, 1976; Davis and McCree, 1978). During aging of *Phaseolus* leaves stomatal conductance and photosynthesis declined in such a way as to maintain constant intercellular CO_2 partial pressure (Davis and McCree, 1978).

In *Zea* the level of RuP_2 carboxylase also varies markedly with the level of nitrogen nutrition. S. C. Wong (unpublished) monitored photosynthesis and C_s in a large number of *Zea* plants with different nitrogen contents and found that stomatal conductance adjusted so that intercellular CO_2 partial pressure remained more or less constant, despite changes in level of RuP_2 carboxylase. Stomatal conductance in air was adjusted so that any further increase in conductance would have only a small effect on photosynthesis, although it would of course have resulted in a much larger increase in water loss. That is, in the terminology of Cowan and Farquhar (1977), stomatal conductance adjusts to the level of RuP_2 carboxylase activity so that CO_2 gain is optimal with respect to water loss. In terms of relationship between assimilation (P) and transpiration (T), stomatal conductance is adjusted to an optimal value of $\partial P/\partial T$. The marginal benefits of increased stomatal conductance in terms of increased assimilation are outweighed by the costs of increased transpiration (Farquhar, 1979a).

In *A. triangularis* intrinsic water use efficiency (expressed in terms of P/C_s as outlined above) increased with increasing leaf nitrogen content and RuP_2 carboxylase activity. In the low nitrogen leaves studied by Medina (1971) this ratio was 1.2 and in high nitrogen leaves of the same species the ratio was 2.9. That is, in normal air the leaves of low nitrogen plants have a higher intercellular CO_2 concentration than those of the high nitrogen plants during steady-state photosynthesis. Increased water use efficiency in response to nitrogen nutrition was clearly not due to reduction in stomatal conductance, as in the salinity responses described above (Chap. 9.5.1). Rather, the ratio P/C_s increased as a result of higher photosynthetic capacity of the leaves, as a result of higher RuP_2 carboxylase activity. Again, the changes in water use efficiency in response to nitrogen nutrition

Table 9.10. Soluble protein and RuP_2 carboxylase levels in leaves of C_3 and C_4 species. (Data of Medina, 1970; Björkman et al., 1976)

Species	Fraction-1 Protein (% total soluble protein)	Soluble Protein (mg g^{-1} dry wt.)
C_3 species		
A. triangularis	44	24.2[a]
A. glabriuscula	41–46	130
C_4 species		
A. sabulosa	19–21	117
T. oblongifolia	4– 8	64

[a]   mg g^{-1} fresh weight

in short-term laboratory studies may be extended to long-term growth trials in the field. Trumble (1932) grew *A.semibaccata* in soil with the addition of 0.03% $NaNO_3$, and in other experiments phosphate and NaCl were added to the $NaNO_3$ supplement. Water use efficiency in the control plants was 2.8 mg dry wt. g^{-1} H_2O. In plants receiving $NaNO_3$, water use efficiency increased to 3.5 mg dry wt. g^{-1} H_2O.

Another important aspect of nitrogen nutrition and photosynthesis is the evident high nitrogen use efficiency of C_4 plants. As discussed previously (Chap.9.1.2), the integrated physiological and biochemical processes of C_4 photosynthesis represent a mechanism for concentrating of CO_2 in bundle-sheath cells. This CO_2-concentrating mechanism effectively abolishes the inhibition of CO_2 fixation due to O_2 and optimizes the activity of RuP_2 carboxylase (Björkman, 1971b). Because this enzyme is able to function at CO_2 concentrations near its V_{max} in C_4 plants, rather than at concentrations in the vicinity of its K_m as in C_3 plants, it follows that equal or higher rates of photosynthesis can be sustained by lower enzyme concentrations in C_4 plants. This speculation has recently been evaluated by Björkman et al. (1976) and by Brown (1978). Table 9.10 shows that Fraction-1 protein (RuP_2 carboxylase) comprises about 41%–56% of soluble leaf protein in C_3 *Atriplex* species grown with adequate nitrogen nutrition at moderate day temperature. In C_4 *A.sabulosa* grown under the same conditions only 19%–21% of soluble leaf protein was associated with this enzyme and in C_4 *Tidestromia oblongifolia* grown at day temperatures of 45 °C, RuP_2 carboxylase comprised only 4% to 8% of soluble leaf protein. In air at the respective temperature optima, these data from Fig.9.18, show that 108 mg RuP_2 carboxylase protein per g^{-1} dry wt. of leaf in *A.glabriuscula* would be required to achieve an equivalent rate of photosynthesis as achieved by 4 mg RuP_2 carboxylase protein g^{-1} dry wt. in *T. oblongifolia*.

Quite clearly the high nitrogen use efficiency of C_4 plants, particularly in those growing in high temperature environments, may compensate for the low nitrogen status of many arid shrubland habitats. The high water use efficiency of C_4 photosynthesis, particularly in saline conditions, and the high nitrogen use efficiency of this process together represent complex mechanisms to maximize photosynthetic CO_2 assimilation in water- and nutrient-limited environments. Whether this high potential for productivity of C_4 species is realized, and the ways in which assimilated carbon is deployed in these plants in natural habitats, is addressed in the next chapter.

9.5.3 Photosynthesis and Micronutrients

Deficiencies of many nutrient ions may depress photosynthesis and the response of photosynthesis to micronutrient deficiencies is no exception. In general, however, it is difficult to identify specific cause and effect relationships. Two interesting exceptions which deserve further study are those involving Zn^{2+} and Na^+. Zinc is tightly bound to carbonic anhydrase and Zn^{2+} deficiency in plants is usually associated with substantial depression of carbonic anhydrase activity (Wood and Sibly, 1952). Carbonic anhydrase is associated with chloroplasts and it may play a role in facilitating the diffusion of CO_2 during photosynthesis (Everson, 1970). Randall and Bouma (1973) grew spinach plants under severe Zn^{2+} stress so that carbonic anhydrase activity was reduced to below 10% of the controls. Although they observed a small reduction in photosynthetic rate in air, the slope of the CO_2 response curve for photosynthesis was unchanged. That is, C_i in spinach leaves did not decrease following Zn^{2+} deficiency. Within the deficiency range tested, there was little evidence of a relationship between carbonic anhydrase activity and C_i.

The essentiality of Na^+ as a micronutrient was established in the halophyte *Atriplex vesicaria* (Brownell and Wood, 1957). As discussed previously (Chap. 7.3.6) and shown in Table 9.11, subsequent experiments indicated that the requirement for Na^+ as a micronutrient was confined almost exclusively to plants with the C_4 dicarboxylic pathway of photosynthesis (Brownell and Crossland, 1972). *Atriplex hortensis* is the only C_3 plant to have shown a slightly significant ($p = 0.05$) response to this micronutrient ion. This intriguing correlation has yet to be explained, but it is quite clear that the whole apparatus of the C_4 photosynthetic pathway remains functional in these plants under deficiency conditions. The carbon

Table 9.11. Relationship between photosynthetic pathway and requirement of Na^+ as an essential micronutrient, indicated by the significance of the response of growth to the presence of 0.1 mM Na^+ in the nutrient solution. (Data of Brownell and Crossland, 1972)

C_3 plants	Growth response[a]	C_4 plants	Growth response[a]
Atriplex glabriuscula	N.S.	*Atriplex inflata*	0.001
A. hortensis	0.05	*A. leptocarpa*	0.01
Beta vulgaris	N.S.	*A. lindleyi*	0.05
Chenopodium capilatum	N.S.	*A. nummularia*	0.001
Maireana pyramidata	N.S.	*A. paludosa*	0.05
Poa pratensis	N.S.	*A. quinii*	0.05
		A. semibaccata	0.01
		A. spongiosa	0.001
		A. vesicaria	0,001
		Amaranthus tricolor	0.001
		Echinochloa utilis	0.001

[a] N.S. = not significant at p 0.001

Table 9.12. δ^{13} C values of C_4 plants grown in the presence and absence of Na^+. (Data of Boag and Brownell, 1979)

Species	δ^{13} C Value ($^0/_{00}$)	
	+Na	-Na
Atriplex spongiosa	-11.7	-10.2, -11.2
Chloris barbata	-12.4, -11.8	-13.5, -12.0
Kochia childsii	-13.3, -13.3	-13.3, -14.6

isotope discrimination ratio (expressed as δ^{13}C value) and the CO_2 compensation point in 21% O_2 which are indicators of integrated C_4 metabolism, remained unchanged in Na^+ deficient C_4 plants (Table 9.12).

There is no evidence to support the notion (Shomer-Ilan and Waisel, 1973) that an increased proportion of net CO_2 assimilation passes directly through RuP_2 carboxylase in leaves of Na^+-deficient C_4 plants. However, larger pools of C_3 compounds have been found in Na^+-deficient plants (Boag and Brownell, 1979) which may imply that Na^+ deficiency interferes with the balanced recycling of these compounds. The interference of Na^+ deficiency with the synthesis, activation or activity of pyruvate, Pi, dikinase would be one way in which the deficiency could be associated almost exclusively with the C_4 photosynthetic pathway, consistent with these observations. Such speculation requires further analysis and we should recall that, though of academic interest, Na^+ deficiency is unlikely ever, to occur in natural habitats and is unlikely therefore to be of physiological significance in an ecological context.

9.6 Conclusions

The account of photosynthetic processes "within the limits of contemporary knowledge and vocabulary" given in this chapter permits an evaluation of photosynthetic responses and adaptations to single factors of the physical environment, occurring over the time scale of 10^0 to 10^6 s. There are, however, at least two very important aspects of photosynthetic response and adaptation that we have only been able to touch upon. One concerns the interactive effects of major environmental factors, such as, for example, the combined effects of excessively high temperature and water stress. Another is that in the longer term, feedback interactions occur between photosynthesis and other growth processes, such as those determining the utilization of the products of photosynthesis and the allocation of these products to different tissues and organs. Changes on photosynthate utilization and allocation may in turn affect the formation of new leaves and their photosynthetic capacity, the reproductive output of the plant, root growth for exploration of new soil volume for water and nutrients, et cetera. Such interactions may be either negative, leading to a progressive depression of plant growth, or positive, leading to a relative improvement in the response of photosynthesis and growth to the imposed stress.

To understand adaptation, to account for the success of plant species in natural environments, it is imperative that we have at least some information on the various kinds of interactions mentioned above. In the next chapter we shall attempt to approach the difficult problem of relating photosynthetic performance, plant growth and productivity, to the success of plants in different environments.

Chapter 10. Productivity and Environment

"While recognizing that uncertainty as to the exact long-term goals of evolution exists, we suggest that maximization of carbon gain, which in water-limited environments amounts to maximizing the efficiency with which water is acquired and 'traded' for carbon, remains central to other goals, such as stress resistance or survival, which may also have to be satisfied."

(Fischer and Turner, 1978)

There is little doubt that the survival of an autotrophic plant is based on its ability to maintain a positive carbon balance even under severe stress conditions, and also that its reproductive potential is dependent on its ability to gain carbon and energy by photosynthesis. These relationships are complex and it is not necessarily true that selection pressure has always favored maximization of productivity as such. Survival obviously depends on a combination of a number of characteristics, only a few of which may be fully exploited. Nevertheless, we share the view of Fischer and Turner (1978) that maximization of photosynthetic efficiency in terms of a limited resource (such as water) remains central to maximization of survival. But we wish to again re-emphasize the importance of considering characteristics of possible adaptive significance in the proper time and space perspective. As Eckardt (1975) has pointed out ... "activities of living beings seem to be directed towards the attainment of goals that lie in the future, [but] their behaviour just reflects the fact that they have evolved under changing conditions which favoured great variability." Having discussed the responses and adaptations of the photosynthetic characteristics we shall now examine the difficult subject of the relationship between productivity and the success and survival of plants in natural environments.

10.1 Photosynthesis and Photosynthate Partitioning in Relation to Primary Productivity

It is obvious that the productivity of a plant over a given time span is directly proportional to the integrated net carbon gain (photosynthesis minus respiration) over the same period of time. However, it is equally obvious that the photosynthetic capacity measured for single leaves under a certain combination of light, temperature, CO_2, and humidity does not necessarily reflect the productive performance of the plant stand under natural conditions where the environmental factors, especially light, vary both temporally and spatially. One pre-requisite in predicting plant productivity on the basis of photosynthetic capacity thus is that the responses of photosynthesis of the different leaves in the canopy to both short-term and long-term changes in the major environmental parameters are known.

Another set of factors that must be taken into account is the cost involved in producing and maintaining the photosynthetic machinery itself. For example, it may be of little use for a plant to double the photosynthetic capacity of its leaves on some arbitrary basis such as leaf area, if the cost of making such a leaf is more than twice as high in terms of cellular components (enzymes, chloroplast membranes, cell walls, etc.). The plant might have been at least as well off if it had produced two of the less costly leaves unless, of course, a reduction in the total leaf area of the canopy confers certain advantages which are unrelated to photosynthetic capacity of the leaf as such. There is thus no intrinsic reason why a plant possessing leaves with a higher maximum photosynthetic capacity per leaf area would have a higher productivity unless the gain received exceeds the additional cost in energy, carbon, nitrogen, and other constituents it took to produce the leaf. The point at which the costs of producing leaves with increased photosynthetic capacity start to outweigh the gains derived is likely to be strongly dependent on the environment, especially light intensity, but also dependent on the temperature and water relations. The photosynthetic acclimation of *A. triangularis* to different light regimes, discussed at length in Chap. 9.2, provides a good example of these relationships. Yet other component considerations of the cost versus benefit relationship are leaf longevity and the costs involved in maintaining effective photosynthetic functioning of mature leaves. Unfortunately, very little research has been done on these important problems in any plant and the present lack of information precludes any quantitative assessment of these factors (Penning de Vries, 1975; McCree, 1976).

Photosynthate Allocation. In addition to its dependence on the net rate and efficiency by which photosynthesizing leaves are able to acquire energy and carbon, productivity is also strongly influenced by the fate of the fixed carbon. Obviously, the proportion of the carbon that the plant allocates to the production of new photosynthetic tissues greatly affects the growth rate. For example, Monsi (1968) calculated that two plants having the same photosynthetic characteristics, maintenance cost, and respiratory losses, but with different ratios of allocation to photosynthetic versus nonphotosynthetic tissues, would have very different productivities. A plant which allocates 70% of the carbon to photosynthetic tissues (primary leaves) and 30% to support and root system, would produce 15 times as much dry matter in three growth cycles as one which allocates 30% to the leaves and 70% to nonphotosynthetic parts. Conversely, of two plants both allocating 70% to photosynthetic tissues, the one with 2.5 higher photosynthetic rate (on a dry weight basis) would produce 50 times as much dry matter. The effect of different environmental factors on the allocation of carbon in *Atriplex* spp. will be discussed in later sections.

Growth Analysis. Richards (1969) and Evans (1972) have provided reviews of the use of growth analysis methods for meaningful description of plant growth and the application of these methods to an assessment of productivity. Nicholls and Calder

(1973) have commented on the selection of appropriate techniques on the basis of their studies on *Atriplex* species.

The classical growth analysis approach is based on the determination of major growth parameters over a given harvest interval on the assumption that dry matter accumulation of the whole plant obeys the laws of compound interest over the time of the harvest interval. For short time intervals between harvests the relative growth rate, R_w, is obtained from the relationship

$$R_w = (\log_e W_2 - \log_e W_1)/t \qquad\qquad (10.1)$$

where W_2 and W_1 are the whole plant dry weights at two consecutive harvests and t is the time interval between harvests.

For this relationship to be strictly valid several conditions must be satisfied. First, a constant fraction of the carbon gain must be continuously reinvested in new photosynthetic tissue having the same photosynthetic activity per unit dry matter as the existing tissue and all photosynthetic tissue must remain fully active. Second, the leaf display in the canopy must remain constant so that the light interception by the leaves is unaltered. Third, the ratio of leaf area to leaf dry weight must remain constant. These conditions may not be met in many cases, especially where the harvest intervals are long or during times of rapid ontogenetic changes. However, these difficulties are not a serious problem in the context of our present purpose and will not be further considered here.

The growth parameters used in this chapter are:

Parameter	Symbol	Unit
Net assimilation rate	E_A	g plant dry wt. gained dm^{-2} leaf area day^{-1} (or $week^{-1}$)
Relative growth rate	R_W	g plant dry wt. gained g^{-1} plant dry wt. day^{-1} (or $week^{-1}$)
Leaf area ratio	F_A	dm^2 leaf area g^{-1} plant dry wt.
Leaf weight ratio	F_W	g leaf dry wt. g^{-1} plant dry wt.
Specific leaf area	S_A	dm^2 leaf area g^{-1} leaf dry wt.

E_A is obtained from measured values of R_W and the mean leaf area ratio F_A during the harvest interval so that $E_A = R_W/F_A$.
The relationship between F_A, F_W and S_A is given by $F_W = F_A/S_A$.

10.2 Productivity in Relation to Light

In the previous chapter we concluded that photosynthesis is dramatically affected by the amount of available light over a very wide range. It was demonstrated that plants from open habitats possess the ability of acclimating to changing light conditions so that the leaves are capable of an efficient use of the available light. Under favorable temperature, water status, and nutrient supply, the integrated daily photosynthesis of many plants continues to increase with the integrated daily quantum flux density to at least the maximum insolation levels found on the earth's surface. Because partial light saturation of the photosynthetic

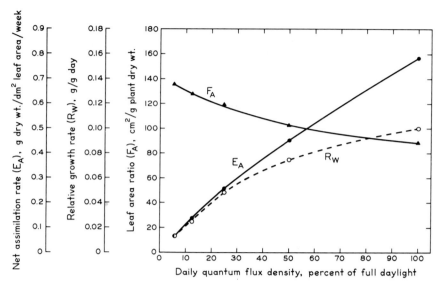

Fig. 10.1. Effect of mean daily radiation receipt on net assimilation rate (E_A), relative growth rate (R_W) and leaf area ratio (F_A) in sunflower plants grown in a garden at Oxford, England. The mean daily radiation (100% of daylight) is estimated to about 325 cal cm^{-2}. (Redrawn in modified form from Blackman and Black, 1959)

machinery occurs in the fully sunlit leaves of the canopy at times of peak insolation, the integrated daily photosynthetic activity of the whole plant canopy is not directly proportional to the daily quantum flux density. The extent of departure from a linear relationship is species-dependent; it tends to be smallest in C$_4$ plants from high light environments and greatest in plants which are native to densely shaded habitats. It also depends on other environmental factors. Generally, light saturation effects are smallest at high (but not excessively high) temperatures, and under conditions that permit the stomata to remain fully open. In short, the most efficient utilization of light should occur where environmental stresses are absent.

Much of the information on the effect of light level on the growth and productivity of higher plants has been obtained using classical growth analysis methods. Space permits only a brief treatment of this topic and the reader is referred to the reviews of Blackman (1961, 1968), Kuriowa et al. (1964), Singh and Gopal (1973), and Whitehead (1973).

These growth analyses confirm the semi-quantitative predictions that can be made on the basis of our knowledge of the response of photosynthesis to light. This is illustrated in Fig. 10.1 (redrawn from Blackman and Black, 1959), which shows the effect of shading on the vegetative growth of sunflower plants grown at Oxford, England. The light level incident on potted seedlings in a garden was reduced in steps from 100% to 5% of full daylight by employing a series of metal screens. Plant samples were harvested at the beginning and end of the experiments, which usually lasted for three weeks. Leaf blade area was determined for each plant and leaves, stems and root were dried separately and weighed.

It is evident from Fig. 10.1 that any reduction in light level reduced the net assimilation rate (E_A). E_A is strongly dependent on light level over the full range although as might be expected the relationship is curvilinear. The "relative growth rate" (R_W) is also strongly dependent on light level over the same range but the departure from a linear relationship is more pronounced. Since R_W is the product of E_A and F_A it is evident that the decline in F_A that takes place when the light level is increased in part accounts for the difference in light dependence between R_W and E_A. However, as shown by Warren-Wilson (1967) for sunflower and other species, the increase in F_A with decreased radiation is due to an increase in the leaf area/leaf weight ratio (S_A) rather than an increase in the leaf weight/plant weight ratio (F_W). It should be noted that change in S_A in response to light level is associated with changes in the photosynthetic capacity per unit leaf area which alter the photosynthetic light saturation characteristics of the leaves in an adaptive manner (Chap. 9.2.1). This should be kept in mind in the following discussion of compensating changes in F_A.

It seems to be a common observation that increased light level causes an increased allocation to root growth, although published quantative data are relatively scarce. Whitehead (1973) reported that in *Filipendula ulmaria,* decreases in light level caused a proportional increase in the partioning of photosynthate to leaf growth, and the extra leaf proportion was almost entirely at the expense of root growth. Similarly, shading of *Chamaenerion angustifolium* reduced the root fraction to less than one-half that of unshaded plants. Considerable decreases in the root fraction as the light level was decreased were also observed with the *Atriplex triangularis* plants, used in the photosynthetic light acclimation experiments described in Chap. 9.2.1 (Björkman, unpublished). However, it would appear that in "obligate heliophytes", such as *Datura stramonium* (C_3) and *Amaranthus spinosus* (C_4), there is little change in photosynthate allocation to the roots (Singh and Gopal, 1973).

In these heliophytes and in nearly all of the open-habitat species investigated by Blackman and coworkers (e.g., Blackman and Wilson, 1951 a,b; Blackman and Black, 1959), a reduction in daily radiation reduced E_A. From these data it can be concluded that during the season of active growth the quantity of light reaching the individual plants was limiting the rate of photosynthesis. In obligate shade plants full solar radiation is known to inhibit photosynthesis (Chap. 9.2.2) and hence also E_A and R_W. Full sunlight is also known to inhibit E_A and R_W in pinto beans (Lockhart, 1963), *Cassia tora* (Singh, 1969), *Anagallis arvensis* (Singh and Gopal, 1973), *Stellaria media* (Sinha, 1968), and *Impatiens parviflora* (Hughes and Evans, 1962).

There is a wide species (and intraspecies) variation in the response of R_W to light level, even among plants for which a reduction in light level below full sunlight causes a reduction E_A. The sunflower data (Fig. 10.1) showed that the reduction in E_A with decreased light level was partially compensated for by a concomitant increase in F_A. The degree of such compensatory changes in F_A varies considerably with species and genotype. For many of the species studied by Blackman and coworkers, the fall in E_A as a result of shading was greater than the corresponding fall in F_A and it was deduced that the optimal level for R_W is above that of normal daylight. In *Amaranthus spinosus* and *Datura stramonium* there was no compensation at all and R_W fell proportionately or even more than proportionately when the light level was reduced (Singh and Gopal, 1973) and Wassink (1969) found a linear decrease in total dry matter production at reduced light levels. In contrast, in shade-tolerant plants such as *Geum urbanum* (Blackman and Wilson, 1951 a) and *Filipendula ulmaria* and *Iris pseudocarpus* (Whitehead, 1973) the compensatory change in F_A was such that complete compensation for the reduction in E_A was obtained over a wide range of light levels. In *G. urbanum*, the increase in F_A more than counterbalanced the fall in E_A so that maximum R_W occurred at about 40% of full daylight.

In Chap. 9.2 we concluded that the ability of the photosynthetic apparatus to adjust its light response characteristics to changes in the light level is of great importance to productivity in natural habitats which are subject to large temporal changes in the light climate, such as those caused by natural shading. It would appear that the ability to alter the partitioning of photosynthate to leaves versus nonphotosynthetic plant parts is an additional important characteristic which enables the plant to cope with changes in the light environment and it may not be suprising to find that species that are able to compete in such environments all seem to possess this characteristic. However, as emphasized by Whitehead (1973), compensation for a lowered photosynthetic rate by an increased allocation to leaf growth is not without cost. There is evidence that in many species increased partitioning to leaf growth is at the expense of a decreased reproductive capacity. This is illustrated by the response of *Chamaenerion angustifolium*, which showed a perfect compensation so that over a longer period of time the same rate of dry matter production was obtained at 40%, 70%, and 100% daylight (Whitehead and Myerscough, 1962). An increase in the leaf fraction exactly balanced the reduction in E_A over this range of light levels. However, the 40% daylight plants flowered four weeks later than the 100% daylight plants. Moreover, fewer flowers, fruits, and seeds were produced under the lower light levels. The increased relative cost of investment into more photosynthetic machinery was closely correlated with the decrease in the production of seed capsules.

As was mentioned in an earlier paragraph, increased allocation to leaf growth may also be at the expense of root growth. In the *C. angustifolium* plants just re-ferred to, the root system of shaded plants can be less than half that of unshaded plants and, as mentioned earlier, increases in the leaf fraction were almost entirely at the expense of root growth when *F. ulmaria* plants were grown at reduced light levels. Such reduced root growth may not have any significant disadvantages in shaded habitats where nutrient levels are adequate and water relations are favorable, but is likely to have serious consequences where this is not the case, and especially in situations where the light level suddenly increases. Whitehead (1973) commented that when some branches are removed from a tree under which *C. angustifolium* plants are growing in a wood, thereby exposing the plants to full sunlight, permanent wilting occurred in a few hours even if the soil was saturated with water, indicating that the root system was inadequate to support the increased transpiration rate.

It is apparent that in most, if not all, species which occupy open habitats, both E_A and R_W are limited by light even at full daylight irradiances as long as the nutrient and water relations are favorable. Short-term measurements of the interactions between light and temperature on photosynthesis (Chap. 9.3.1) indicate that the light limitation of photosynthetic rate even at high light levels would increase with temperature. This predicts that increased daytime temperature would increase light limitation of E_A and R_W. There is some experimental evidence from growth studies which point in this direction (e.g., Blackman, 1956; Hiroi and Monsi, 1963) but direct attempts to determine such relationships are few even in crop species in spite of the importance of climate in agricultural crop production.

Using a combination of multiple regression analysis of field data determined during different seasons and the results obtained on the same species in controlled-environment studies, Warren-Wilson (1967) sought to determine the effects of radiation and temperature on E_A and R_W in rape, sunflower, and maize. The field experiments were conducted on widely spaced plants in nutrient cultures in the semi-arid climate of the Riverine Plain at Deniliquin, New South Wales, where the daily solar radiation and mean monthly temperature vary from a minimum of about 240 cal cm^{-2} day^{-1} and 9 °C for June and July to a maximum of 700 cal cm^{-2} day and 23 °C in December and January. For all three species the two variables accounted for over 90% of the seasonal variation in E_A and R_W in the multiple regression analysis. As shown in Fig. 10.2, E_A increased linearly with radiation in all species. In rape, a cool season crop, multiple regression showed no effect of temperature on E_A. In sunflower, a warm season crop, there was a positive interaction between the effects of temperature and light level, whereas in maize, a chilling-sensitive crop, increase in temperature not only increased the level of E_A but also steepened the dependence of E_A on radiation. These field responses are in agreement with the results obtained in controlled-environment studies which showed that E_A for rape was little affected by temperature in the range under consideration, while E_A for sunflower was considerably affected, and E_A for maize was strongly affected. R_W increased with both radiation and temperature in the field experiments, in general following the response of E_A although, especially in rape in cool weather, temperature may be a more potent determinant of R_W than radiation.

Using the relationship between E_A and R_W on the one hand, and radiation receipt on the other, together with monthly weather records, Warren-Wilson (1967) calculated the seasonal course of growth for the three species and compared these with observed data. As shown for sunflower in Fig. 10.3, the calculated and observed data follow roughly the same pattern. Similar results were obtained with rape and maize.

As was shown in Fig. 10.2, maximum values of E_A and R_W for sunflower occurred when high mean temperatures (~ 25 °C) were combined with high radiation (~ 700 cal cm^{-2} day^{-1}). Such combinations of high mean temperature and high radiation are maintained for longer periods of time only in arid, lower latitude regions of the world which are characterized by clear skies. The peak values of R_W for sunflower obtained under these conditions were about twice the maximum reported by Blackman and Black (1959) for sunflower grown at Oxford, England, and approach the levels regarded maximal for plants in natural climates. Surprisingly, in the Riverine Plain experiments the regression of E_A on radiation was linear up to the highest level of radiation in all three species. Firstly, one might expect that there would be some degree of light saturation of photosynthesis at the

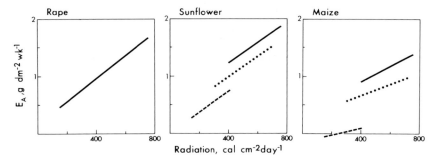

Fig. 10.2. Relationship between net assimilation rate and daily radiation receipt, at selected temperatures, for rape, sunflower, and maize plants calculated from multiple regressions. (Redrawn from Warren-Wilson, 1967). Rape grown at 10 °C, for other species, 10° (- - - -), 16° (● ● ●), 22 °C (——)

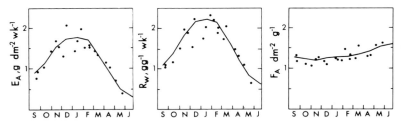

Fig. 10.3. Seasonal courses of net assimilation rate (E_A), relative growth rate (R_W) and leaf area ratio (F_A) for sunflower plants at Deniliquin, New South Wales. *Lines* are based on multiple regressions and average weather records; *points* are actual determinations. (Redrawn from Warren-Wilson, 1967)

highest radiation levels. Secondly, it seems likely that the combination of high irradiance and the very high water vapor deficits of this arid climate would result in an unfavorable leaf water status even in plants grown in nutrient solution. Both of these factors would tend to cause E_A to level off at the highest irradiances, but evidently this was not the case.

Growth analysis experiments with *Atriplex* species, specifically designed to evaluate the effect of light level on E_A and R_W are not available. However, Jones, R. et al. (1970), determined the production rates of young plants of *A. nummularia* and *A. vesicaria* which are dominant components of the climax vegetation of the Riverine Plain, under conditions similar to those used by Warren-Wilson (1967). The two *Atriplex* species, together with *Danthonia caespitosa,* an important species in the disclimax grasslands developed from *Atriplex* communities in parts of the Riverine Plain, and sunflower were included for comparison. The results of some of these experiments are shown in Table 10.1. It is evident that the two *Atriplex* species are capable of production rates equal to or even exceeding those of sunflower. The highest R_W was obtained for *A. nummularia* (0.22 g g^{-1} day^{-1}) and the highest E_A value for *A. vesicaria* (0.31 g dm^{-2} day^{-1}). The very high R_W and E_A values obtained for sunflower by Warren-Wilson (1967) were not achieved in these experiments and

Table 10.1. Mean relative growth rate (R_W) and mean net assimilation rate (E_A) for seedlings of four species grown under two conditions of daily radiation and mean temperature in nutrient culture in the open, at Deniliquin, New South Wales. The daily radiation receipts and mean temperature were $572\,cal\,cm^{-2}\,day^{-1}$, $23\,°C$ and $693\,cal\,cm^{-2}\,day^{-1}$, $27.8\,°C$ for experiments 1 and 2, respectively. (After Jones, R. et al., 1970)

	R_W (g g^{-1} day^{-1})		E_A (g dm^{-2} day^{-1})	
	Exp. 1	Exp. 2	Exp. 1	Exp. 2
Atriplex vesicaria	0.16	0.22	0.25	0.31
Atriplex nummularia	0.11	0.17	0.13	0.24
Danthonia caespitosa	0.11	0.11	0.15	0.16
Helianthus annuus	0.13	0.19	0.15	0.24

the performance of sunflower was inferior to that of the *Atriplex* species when all three species were grown together under similar conditions. This and other studies (Chap. 10.4.2) suggest that these *Atriplex* species are capable of productivities as high as agronomic species under favorable conditions, and that they are capable of efficient utilization of the high levels of solar radiation received in this semi-arid environment when water supply is unlimited.

10.3 Productivity in Relation to Temperature

In the preceding section it became evident that there is a considerable degree of interaction between light level and temperature on productivity. In general, the steepness of the response of E_A and R_W to light level tends to increase with increased temperature. In this section we shall focus on temperature as the main variable in determining plant productivity of *Atriplex* species in environments where the light level is high. We shall begin by examining overall responses of growth in contrasting natural environments, and then proceed to an analysis of growth in controlled-temperature environments. Finally, we shall consider the effects of air and soil temperatures on the partitioning of photosynthate between the roots and the shoot and the resulting influence of this partitioning on productivity.

10.3.1 Performance in Thermally Contrasting Habitats

Björkman et al. (1974a, b) reported on their comparative growth experiments on plants native to habitats with contrasting thermal regimes. These studies were conducted in the climatically contrasting transplant gardens at Bodega Head and Death Valley, California, and in the controlled growth facilities as the Carnegie laboratory. The experimental plants included several cool-coastal *Atriplex* species such as the annual C_3 species *A. triangularis* (from the Northern California coast), *A. glabriuscula* (from the Atlantic coast of England), *A. heterosperma* (from the San Joaquin Valley of California); the C_4 annual species *A. sabulosa* (from the Atlantic coast of Northern France), and *A. rosea* (from the San Joaquin Valley of California)

Table 10.2. Summer performance under ample water supply in the Death Valley and the Bodega Head transplant gardens. (Modified after Björkman et al., 1974 b)

Unable to survive	Death Valley Garden Slow growth	Very rapid growth
A. glabriuscula C_3 A. triangularis C_3 A. sabulosa C_4 A. rosea C_4	A. hymenelytra C_4 A. lentiformis C_4	T. oblongifolia C_4

Unable to survive	Bodega Head Garden Slow growth	Rapid growth
T. oblongifolia C_4	A. hymenelytra C_4 A. lentiformis C_4	A. glabriuscula C_3 A. sabulosa C_4 A. heterosperma C_4 A. rosea C_4

and the evergreen C_4 species *A.hymenelytra* and *A.lentiformis* (both from Death Valley) and *A.lentiformis* ssp. *breweri* from the coast of Southern California. *Tidestromia oblongifolia,* a summer-active but winter-dormant perennial C_4 species, native to Death Valley, was also included in these comparative studies.

Performance in the Death Valley Garden. On the basis of their performance under well-watered conditions in the Death Valley garden these plants may be divided into three categories: (1) those unable to survive the hottest period of the year (June 1 to October 1), (2) those which survived the summer and retained their leaves and whose net growth occurred during cooler times of the year, and (3) those with their main productivity occurring during the period June 1–October 1. The results are summarized in Table 10.2.

All of the annual *Atriplex* species, whether C_3 or C_4, fall in the first category. The coastal annuals died before June 1, before they were able to produce seed, and the annuals from inland habitats were dead before mid-July.

Although incapable of surviving the Death Valley summer, most of the *A.rosea* seedlings very rapidly completed their life cycle in the period April 1–May 1, producing very small, stunted plants less than 15 cm tall, with only a few leaves and a few mature seeds. This contrasts with the behavior of this species in its native Eurasian habitats and in the California habitats where it is widely naturalized. In the latter habitats this species usually begins to grow in early May and produces a prodigious seed crop in September when it commonly reaches a height and a diameter exceeding 1 m.

The two evergreen C_4 perennial *Atriplex* species fall in the second category. In *A.hymenelytra,* growth was very active until early June when it slowed down and net growth apparently ceased to resume again in late September. Most leaves were retained and some new leaves developed during the summer although marked changes took place in leaf size, reflectance, water content and water potential

(Chaps. 8.2.2; 9.2.3; 9.3.2). The *A. lentiformis* plants of Death Valley origin showed a behavior similar to that of *A. hymenelytra* but active growth continued longer into the summer. Interestingly, the performance of both of these taxa in the irrigated Death Valley garden closely resembled that of naturally occurring unwatered plants outside the garden, indicating that the seasonal activity of these plants in their native habitats is to a large extent controlled by temperature and not by water supply alone.

The coastal form of *A. lentiformis* was the only *Atriplex* species which survived the summer in Death Valley. However, net growth ceased earlier in the summer, resumed earlier, and the amount of dry matter production during the period May 1–November 1 was significantly lower in the coastal than in the desert form of *A. lentiformis*. Conversely, this is consistent with the findings of Pearcy and co-workers that ecotypic differentiation with regard to photosynthetic characteristics is present with this taxon (Chaps. 3.4.2; 9.3.2).

T. oblongifolia is the only species of the many species tested which falls in the third category. In Death Valley, little growth occurs until May but during June growth is extremely rapid. At peak growth rate, doubling of the dry matter occurs in about three days ($R_w = 0.26$) and very high productivity is maintained throughout July and August when temperatures often approach or even exceed 50 °C. Flowering takes place in September and an abundant seed crop is produced in late October. In November the plants usually die back, retaining only a few basal leaves until the following May. Although a true perennial, this species resembles an annual plant both in terms of its rapid vegetative growth and prodigious reproductive capacity. Surprisingly, it exhibits little or no growth during February–April, the main growing season for the vast majority of other species that are native to the desert floor, both annual and perennial plants. As was found with *A. hymenelytra* and the Death Valley form of *A. lentiformis,* the seasonal activity of watered *T. oblongifolia* was very similar to those of natural, unwatered stands.

Performance in the Bodega Head Garden. All eight of the *Atriplex* taxa that were transplanted into the coastal garden in April 1973 were able to grow and survived until mid-November or until they had produced mature seed. The only species which completely failed to grow is *T. oblongifolia.* Seedlings of this species which were repeatedly transplanted into the Bodega Head Garden during spring and summer slowly lost weight and all of them died within two months, indicating that this thermophilic species is unable to cope with the prevailing cool temperatures in this coastal environment.

Although they were all able to grow, there was a great diversity in performance among the *Atriplex* species (Table 10.3). The highest relative growth rates (R_w) measured during the period of what appeared to be most rapid vegetative growth for each species were exhibited by the coastal annuals, *A. glabriuscula* (C_3), *A. triangularis* (C_3), and *A. sabulosa* (C_4). Interestingly the growth rate of *A. sabulosa* closely approached those of the two coastal C_3 species despite the fact that it was supporting rapid reproductive growth during the period of the R_w determinations while the other species remained vegetative during this period. The *A. sabulosa* plants had essentially completed their life cycles and had produced an abundant crop of ripe seed in late July, while the two other coastal annuals remained completely vegetative until mid August and both vegetative and reproductive

Table 10.3. Daily growth rates, final dry weight and reproductive behavior of *Atriplex* species in the Bodega Head transplant garden during 1973. (Data from Björkman et al., 1974b and M. Nobs and O. Björkman, unpublished)

Species	Relative growth rate determinations		Final harvest data		Reproductive behavior	
	$R_w g g^{-1}$ day^{-1}	Period for R_w determination	Date of harvest	Dry wt. g plant^{-1}	Date of first flowers	Date of first ripe seed
A. glabriuscula	0.083	June 1–July 10	Oct. 17	2,333	Aug. 6	Sept. 12
A. triangularis	0.085	June 20–July 30	Oct. 17	2,489	Aug. 10	Sept. 16
A. sabulosa	0.080	July 1–July 30	Aug. 6	370	Jun. 29	July 24
A. heterosperma	0.073	July 1–Aug. 10	Oct. 17	2,300	Sept. 17	Nov. 3
A. rosea	0.063	June 20–July 30	Oct. 17	1,400	Aug. 20	Oct. 20
A. lentiformis ssp. breweri	<0.03	July 1–Aug. 10	Nov. 19	60	Sept. 17	No seed
A. lentiformis	<0.03	July 1–Aug. 10	Nov. 19	31	No flowers	No seed
A. hymenelytra	<0.03	July 1–Aug. 10	Nov. 19	50	Sept. 17	No seed
T. oblongifolia	0	July 1–Aug. 10	Oct. 17	0.5	No flowers	No seed

growth continued until about October 1. As a result, the final dry matter yields of these plants greatly exceeded that of *A. sabulosa*.

The two annual species from the San Joaquin Valley of California, *A. heterosperma* (C_3) and *A. rosea* (C_4) had somewhat lower R_w than the coastal annuals. Of these, *A. heterosperma* produced the highest final dry matter yield (comparable to those of *A. glabriuscula* and *A. triangularis*), but the onset of flowering was very late and only a small fraction of the seed reached maturity before the plants died in mid-November. *A. rosea* flowered earlier than *A. heterosperma* and successfully completed its life cycle but produced less total dry matter.

The perennial C_4 shrubs had very much lower R_w and produced much less final dry matter than the annual species in spite of the fact that they remained essentially vegetative throughout the duration of the experiments, and none produced ripe seed, although *A. hymenelytra* and *A. lentiformis* ssp. *breweri* produced some flowers late in the season. The Death Valley form of *A. lentiformis* barely survived and completely failed to flower. Although *A. lentiformis* ssp. *breweri* performed better than *A. lentiformis* from Death Valley its performance in the Bodega Garden was poor. It should be noted, however, that this coastal subspecies originated from the coast of southern California which has a considerably warmer climate than Bodega Head.

It is evident from these reciprocal transplant experiments that striking differences exist in the performance among the different species in a manner reflecting the climates of the respective native habitats. It is especially noteworthy that in these experiments the relative performance of the different species in the cool Bodega Garden in all cases was inversely related to their relative performance during the hot summer in the Death Valley Garden. The possible factors that may underlie this differential behavior will be discussed below.

Table 10.4. Effect of temperature on dry matter yield, daily growth rate, allocation and daily growth rate on a leaf weight basis, for species from cool and from hot environments. (Data from Björkman et al., 1974b)

Species	Yield in 22 days, (final/initial wt.)		Daily growth rate g(g total dry wt.)$^{-1}$		Dry matter distribution						Daily growth rate g(g leaf wt.)$^{-1}$	
					Leaves		Roots		Other			
	Cool[a]	Hot[a]	Cool	Hot	Cool	Hot	Cool	Hot	Cool	Hot	Cool	Hot
A. glabriuscula	2.45	< 1	0.15	—	0.55	—	0.22	—	0.23	—	0.26	0
A. sabulosa	17.3	< 1	0.14	—	0.53	—	0.12	—	0.35	—	0.26	0
A. hymenelytra	14.8	8.1	0.13	0.09	0.65	0.60	0.20	0.22	0.15	0.18	0.19	0
T. oblongifolia	< 2.7	88.7	<0.04	0.22	0.67	0.43	0.22	0.25	0.11	0.32	<0.05	0.50

[a] Cool and Hot regimes were 16° day/11 °C night, and 45° day/31 °C night, respectively

10.3.2 Response of Growth to Temperature Under Controlled Conditions

The above field studies strongly suggest but do not prove that the contrasting responses exhibited by the different species in the coastal and the hot desert environments are in large part attributable to differences in the response to temperature as such. Nevertheless, in addition to temperature, other important environmental variables, notably air humidity and edaphic and biotic factors, also show important differences between the two types of habitat. In this section we shall review the results of comparative growth experiments of *Atriplex* species under controlled temperature regimes in which all other environmental factors were kept constant. This includes the air humidity and the supply of water and nutrients, which were maintained at high and nonlimiting levels. Other environmental stresses that may differ between the coastal and the desert habitats such as excessive wind speeds, salt spray, and predators were completely eliminated. These experiments further permit us to evaluate the role of photosynthate allocation in the response of growth to temperature.

Table 10.4 summarizes the results obtained in such controlled-growth experiments with two of the coastal and two of the desert species used in the field studies reviewed in the previous section. The temperature regimes of 16° day/11 °C night and 45° day/31 °C night simulate the summer conditions of the cool coastal and the hot desert habitats, respectively. The overall responses of growth closely resemble those obtained in the two transplant gardens. The two coastal species, *A.glabriuscula* (C_3) and *A.sabulosa* (C_4) both produced high yields in the cool regime but were unable to grow and survive in the hot regime. Conversely, the thermophilic desert species *T.oblongifolia* (C_4) produced very high yields in the hot regime but performed very poorly under the cool regime. The desert evergreen *A. hymenelytra* (C_4) was able to grow in both the cool and the hot regimes but its yield was lower than those of the two coastal species under the cool regime and very much lower than that of *T. oblongifolia* under the hot regime. Other growth experiments under a series of different temperatures (cf. Fig. 10.4) showed that the growth rates of the coastal species under the cool regime were as high as 70% of the rate at optimum temperature (25°–27 °C day). The hot regime was very close to optimum for *T.oblongifolia* (46 °C day) but too hot for optimum growth of *A.hymenelytra* (30°–35 °C day).

That the temperature regime of the native habitats is reflected in temperature response of growth among a number of *Atriplex* species is further shown in Table 10.5 which compares the ratio of the yield under a 36° day/31 °C night regime to that obtained by the same species under a 15° day/10 °C night regime. In the species originating from cool temperature climates the yields under the cool regime are only about one half of those under the hot regime. This is in sharp contrast to the response of the *Atriplex* species from warm interior habitats. In the latter species the yields obtained under the warm regime are several times as high as those obtained under the cool regime. It should be noted that all of the species in this group are C_4 plants, while with the notable exception of *A.sabulosa* the cool-climate species are C_3 plants. The performance of *A.sabulosa* shows that C_4 species are not necessarily inferior to C_3 plants at low temperatures. *A.sabulosa* is capable of relative growth rates similar to those of the C_3 cool-climate *Atriplex* species at low temperatures,

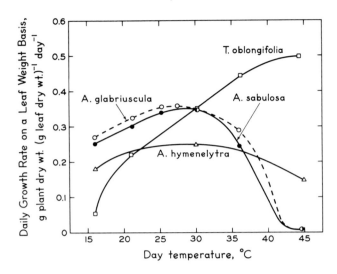

Fig. 10.4. Temperature dependence of daily growth rate on a leaf dry weight basis (R_W/F_W) for *Atriplex glabriuscula*, *A. sabulosa*, *A. hymenelytra*, and *Tidestromia oblongifolia*. (Based on data of Björkman et al., 1974b and unpublished data of H.A. Mooney and O. Björkman)

Table 10.5. Response of growth to low and to high temperatures in C_3 and C_4 species of *Atriplex* from different habitat types. Response is expressed as the ratio of dry weight yields in 22 days of plants grown at 36° day/31 °C night, to those grown at 15° day/10 °C night. Data for the thermophilic desert species *Tidestromia oblongifolia* are also included

Species	Plant type	Yield at 36°/31 °C / Yield at 15°/10 °C
Coastal Origin		
A. glabriuscula	C_3 Annual	0.40[a]
A. triangularis	C_3 Annual	0.68[a]
A. hortensis	C_3 Annual	0.35[a]
A. sabulosa	C_4 Annual	0.26[a], 0.75[a]
Interior Origin		
A. spongiosa	C_4 Annual	3.04[a], 3.74[c]
A. rosea	C_4 Annual	3.73[a]
A. expansa	C_4 Annual	5.31[a]
A. lindleyi	C_4 Annual	3.14
A. hymenelytra	C_4 Evergreen	3.04[b]
T. oblongifolia	C_4 Perennial (winter dormant)	>25.0[b]

[a]   Unpublished experiments of O. Björkman, conducted in the Canberra Phytotron during December, 1971–January, 1972
[b]   Data from Björkman et al. (1974b) on plants grown during the summer of 1974 at Stanford, California
[c]   Calculated from data of Nicholls (1972) on plants grown in the Canberra Phytotron during February–March, 1971

but it lacks the superior performance at high temperatures exhibited by the warm-climate *Atriplex* species and especially the thermophilic *T.oblongifolia*.

As was mentioned earlier in this chapter, the fraction of the photosynthetically fixed carbon which is allocated to the production of new leaves versus nonphotosynthetic tissue has a profound influence on relative growth rate and yield. Consequently, the response of overall growth to temperature may in large part be the result of temperature effects on carbon allocation. As will be shown below, temperature, especially root temperature, is indeed often found to exert a strong influence on the distribution of dry matter between the shoot and the root. Nevertheless, the different responses of overall growth to cool and hot temperature regime exhibited by the species listed in Table 10.4 cannot be attributed to differences in allocation patterns. As is also shown in this table, the ratio of total dry matter allocated to the leaves under the cool regime is rather smaller in the two coastal species which exhibit higher relative growth rates and total yields than in the two desert species. As a result, daily growth rates in the cool regime expressed on a basis of *leaf* dry weight (or leaf area) tend to show even greater differences between the coastal and the desert species than do growth rates expressed on a *total* dry weight basis. Also, as shown in Table 10.4, growth of *T.oblongifolia* under a hot regime caused a smaller fraction to be allocated to the leaves than when it was grown under a cool regime, primarily because of a larger allocation to stems. Similar effects of temperature on allocation patterns were obtained with *A. glabriuscula*, *A. sabulosa*, and *A. hymenelytra*. As a result, the differences in the temperature dependence of daily growth rate among the species becomes further accentuated when expressed on a basis of *leaf* dry matter (Fig. 10.4).

It follows from these studies that the contrasting interspecific response of growth to temperature observed in these experiments must primarily be a result of intrinsic differences in the temperature dependence of primary growth processes, notably photosynthesis. Positive evidence supporting this conclusion is provided by the temperature-related photosynthetic characteristics of the same species (Chap.9.3). The similarities and differences among the various species in their response of growth to temperature are in close agreement with the similarities and differences among the species in photosynthetic characteristics. This is especially evident in the ability to perform at high temperatures, which in large part must be attributable to an increased high-temperature stability of the photosynthetic apparatus itself. The results of these growth experiments are also consistent with the conclusion that C_4 photosynthesis potentially confers an advantage at high temperatures by enabling photosynthesis to proceed closer to CO_2 saturation.

For reasons discussed at length in Chap.9.3, CO_2 limitation of photosynthesis becomes increasingly important as the temperature increases. However, as exemplified by the performance of *A.sabulosa,* whose photosynthetic machinery does not possess an unusually high-temperature stability, C_4 plants are not necessarily superior to C_3 plants at high temperatures. Conversely, the performance of *A.sabulosa* also shows that the presence of C_4 photosynthesis need not impose a disadvantage in cool environments. *A.sabulosa* is capable of reaching relative growth rates matching those of coastal C_3 species of *Atriplex* species under cool temperatures, a performance in full agreement with that predicted on the basis of photosynthetic characteristics (Chap.9.3).

10.3.3 Effect of Temperature on Photosynthate Partitioning

An especially important aspect of allocation is the fraction distributed to the roots. One would expect that the root/shoot ratio optimal to overall growth is reached when the root development is just sufficient to provide an adequate supply of water and nutrients to the growing shoot. Any further increase in the allocation to the roots would result in decreased growth due to a reduction in the amount of photosynthesizing tissue. Temperature strongly and directly affects both the rate of photosynthesis and the rate by which the roots are able to take up water and nutrients. Complex interactive effects of temperature on the root/shoot ratio can be expected since the temperature responses of photosynthesis and root activity are not the same and also since root and shoot temperatures are often markedly different, especially in nature. Thus, one can expect that the balance between root and shoot which is optimum to growth will be strongly dependent on both root and shoot temperatures as well as on species.

Perhaps the most striking interspecific differences are those between the perennials *Atriplex confertifolia* from the cool deserts of the Great Basin and *Atriplex vesicaria* of the warm deserts in South Eastern Australia. Analysis of biomass distribution in the field shows that the root/shoot ratio of *A. confertifolia* is about 7 (Caldwell and Camp, 1974) and of *A. vesicaria* about 0.3 (Jones and Hodgkinson, 1970). The remarkable diversion of photosynthate for growth below ground in *A. confertifolia* is discussed later (Chap. 10.6.1). It is evidently a genotypic property of these cool desert shrubs, but the effect of temperature on assimilate partitioning in these plants has not been investigated.

In general, it is found that as temperatures (of root and shoot together) are raised from low to moderate the fraction allocated to the root decreases, while at the same time the shoot as well as total yield increases. This type of response has been found in a number of *Atriplex* species grown under controlled conditions in which shoot and root temperatures have been varied simultaneously. Nicholls (1972), in an investigation of temperature effects on growth and allocation in seedlings of the two annual C_4 species, *A. spongiosa* and *A. lindleyi,* and the two perennial C_4 species, *A. vesicaria* and *A. nummularia,* from the interior of Southern Australia, found that in all four species the most important influence of temperature was a marked reduction in the allocation to the roots. The root/shoot ratios gradually fell by almost one half as the temperature was raised from 15° day/10 °C night to as high as 36° day/31 °C night. Similar trends were observed with the *Atriplex* species investigated by Björkman et al. (1974b). However, in species from cool-coastal origin the root/shoot ratio fell to minimum at moderate temperatures and then increased again as the temperature was further increased.

This type of response is illustrated by *A. hortensis* (Fig. 10.5, left). In this species there was a simultaneous fall in the fraction allocated to the roots and an increase in the shoot (and total) yield as the temperature was raised from 15° day/10 °C night to 24° day/10 °C night. Further increases in temperature caused a sharp reduction in shoot (and total) yield and a marked rise in the fraction allocated to the roots. Similar, but less striking, responses were obtained with the coastal species, *A. triangularis* and *A. glabriuscula.* A different response pattern was found with *A. expansa* from the hot San Joaquin Valley of California. This C_4 species requires

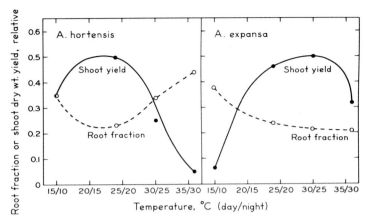

Fig. 10.5. Effect of growth temperature regime (shoot and root) on shoot yield and dry matter partitioning to roots in two contrasting *Atriplex* species. (Based on unpublished experiments of O. Björkman conducted in the Canberra phytotron during December 1971–January 1972)

much higher temperatures for optimum dry matter production than *A. hortensis* and the two coastal *Atriplex* species. As was also the case with the Australian inland species studied by Nicholls (1972) and with *A. rosea* and *A. hymenelytra* (Björkman, unpublished), the fraction allocated to the roots continued to fall as the temperature was increased to 36° day/31 °C night even though this temperature somewhat exceeded the optimum for shoot (and total) yield.

Although the fact that root and shoot temperatures covaried in the above experiments complicates the interpretation of the observed responses, the results do not appear inconsistent with the general hypothesis proposed by Davidson (1969a) that photosynthate is partitioned into growth of roots and shoots in a manner which permits a balanced functional economy throughout the plant (see below). Davidson obtained experimental evidence in support of this hypothesis from studies of the influence of root temperature on photosynthate partitioning and foliage yield in a number of pasture species, grown in the same aerial environments. Some of his results are summarized in Fig. 10.6. The response of the subtropical plants to root temperature is distinct from that of the winter annuals, in a manner resembling the difference in temperature response between *A. hortensis* and *A. expansa* (Fig. 10.5). In the subtropical plants the highest foliage yield and the lowest allocation to roots are obtained at the highest experimental soil temperature used. In the winter annuals, on the other hand, the foliage yield reaches an optimum at a soil temperature of about 20 °C, and allocation to the roots reaches a minimum at about 25 °C soil temperature. The sharp fall in foliage yield at high soil temperatures was accompanied by an increased allocation to the roots. As shown in Fig. 10.7, in each group of species the fraction of dry matter allocated to the roots tended to be proportional to the displacement of the soil temperature from the optimum in a curvilinear fashion. Davidson (1969a) interpreted these data to mean that photosynthate is partitioned relatively more into root growth as the soil temperature departs from the optimum – either above or below. The plant thus appears to compensate for temperature-

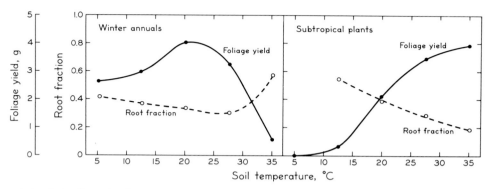

Fig. 10.6. Effect of soil temperature on foliage yield and dry matter partitioning to roots in a uniform aerial environment for certain winter annual and subtropical species. (Redrawn in modified form from Davidson, 1969a)

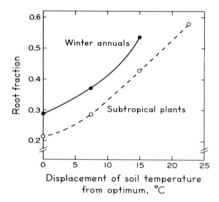

Fig. 10.7. Relationship between root/shoot partitioning and displacement of soil temperature from optimum for certain winter annual and subtropical species. (Redrawn from Davidson, 1969a)

dependent changes in root activity by increasing or decreasing the relative size of the root system. Under conditions where the leaves are in an environment which permits rapid rates of photosynthesis but where soil temperature is either below or above the optimum for root activity, root growth proceeds with little accompanying leaf growth. On the basis of these results, Davidson (1969a) concluded that photosynthate was partitioned between root and shoot growth in inverse proportion to the relative rates of physiological functions in these tissues.

Root mass × rate of absorption α leaf mass × rate of photosynthesis   (10.2)

In this way partitioning of photosynthate (and also of mineral nutrients) into the growth of roots permits the plant to maintain a proper balance between the quantities of water and mineral nutrients absorbed by the roots and the amount of carbohydrate gained and water lost during photosynthesis and accompanying transpiration by the leaves. This hypothetical mechanism provides an explanation not only of the effect of temperature on root growth but also explains the effect of other environmental factors such as soil nutrient level (Chap. 10.4). However, Davidson's hypothesis does not deal with the casual or molecular mechanisms that underlie the observed responses. A treatment of this topic is beyond the scope of this book and the reader is referred to the papers of Cooper and Thornley (1976) and Thornley (1977) which review the approaches that have been made toward developing quantitative treatments of root:shoot interaction on the basis of the component plant processes of transport and chemical conversion. These authors provide a comprehensive discussion of physiological concepts such as sources and sinks, active and passive transport, substrate utilization, and the involvement of hormonal control. They also provide a detailed mathematical treatment of component processes and how these concepts may be incorporated into a mathematical model of photosynthate partitioning.

10.4 Productivity in Relation to Mineral Nutrition and Salinity

Although there is a wealth of information in the agricultural and agronomic literature which describes the response of photosynthetic production and its allocation as a function of soil nutrient and water status, no comparable body of data exists for plants growing in natural habitats. In simple terms, maximum growth and productivity would be expected when the allocation of photosynthate to new photosynthetic surface was a maximum, or in other words, when the diversion of metabolites to roots was the minimum required to support root growth and the nutrient absorption activity commensurate with the requirements of the shoot. Such expectations are appropriate for an agricultural system with unlimited nutrient and water resources. In many natural habitats such conditions are unlikely to be met for many species for more than a short time in their life cycle.

The principal difficulty in the interpretation of these interactions in natural systems is the enormous range of responses among species of diverse growth forms which presumably accounts for the capacities for high productivity in some habitats and the limits for survival in other habitats. Not only are components of performance and survival processes widely different between species in different habitats, but they may vary greatly within a species throughout its life cycle. There is tremendous scope for speculation and modeling of these processes, but there is yet very little experimental evidence against which such exercises can be evaluated. In this section we can do little more than point to ways in which these studies may be developed in the nutrient- and water-limited habitats occupied by *Atriplex* spp.

In general the increased growth which accompanies the fertilization of crop and pasture species is associated with a decrease in the ratio of root/shoot. Davidson (1969b) concluded from his studies with *Lolium* and *Trifolium* that growth of shoots relative to roots was decreased by deficiencies of nutrients and by water stress. This hypothesis that photosynthate is partitioned to roots in inverse proportion to their functional activity can be stated more directly in that, provided roots receive adequate photosynthate to allow them to sustain the nutrient demands of the shoot, photosynthate will be preferentially used to create new photosynthetic surface [Eq. (10.2)].

Cook (1973) grew *Atriplex vesicaria* in its native soil and in the same soil with added nitrogen (100 ppm) and phosphorus (50 ppm). The stimulated growth in nutrient treatments was associated with a reduction in the root/shoot ratio as observed in crop species, although the dry weight of both roots and shoots increased. These experiments were done at a range of temperatures and Fig. 10.8 shows that in the nutrient treatments root/shoot ratio was insensitive to temperature but in the native soil this ratio declined as a result of increased shoot growth at 25° and 30 °C. It is conceivable that stimulation of PO_4^{3-} transport to the shoot at these temperatures permits increased photosynthesis and shoot growth which decreases the root/shoot ratio. We noted earlier (Chap. 7.3.5) that seedlings of these plants were unusually efficient in the translocation of PO_4^{3-} from roots to shoots and that the efficiency of this transport process in *A. vesicaria* increases markedly between 20°–30 °C (Fig. 7.12).

Differences in the efficiency of nutrient transport processes between *Atriplex* and wheat may account for the different partitioning of growth in these species (Cook, 1973). At low temperatures the root/shoot ratio of wheat was four times greater than that of *A. vesicaria* (Fig. 10.8). The processes which permit *A. vesicaria* to maintain much greater photosynthetic surfaces than wheat have not been evaluated, but because wheat is much less efficient in transporting PO_4^{3-} to the shoot than *A. vesicaria,* photosynthesis and productivity of wheat could be limited by demands for the formation of a larger root system. Moreover, C_4 plants such as *A. vesicaria* are able to achieve higher photosynthetic rates at lower leaf protein concentrations than are C_3 plants (Chap. 9.5.2). Thus efficient transport and utilization of nutrients may make a substantial contribution to the different productivity and allocation patterns shown in Fig. 10.8.

One of the most regularly observed growth responses in halophytes is the stimulation of growth by low levels of NaCl in culture solutions or in soil (Chaps. 7.5.6; 8.2.4). The stimulation in growth was not observed when *A. halimus* was maintained in humid environments (Gale et al., 1970) and in dry environments the growth stimulation was associated with an improvement in the water balance of the shoot as a result of osmotic adjustment. Kaplan and Gale (1972) showed that absorption of salt allowed the shoots of *A. halimus* to maintain positive turgor throughout the day whereas leaves of plants not supplied salt spent a large part of the day at zero turgor. The daily course of photosynthesis and of stomatal conductance was not measured in these experiments but it is likely that stomata closed at zero turgor and restricted the total amount of CO_2 fixed in the control compared with the salt-treated plants. Reduced cell turgor and reduced photosynthesis would together result in a reduction in leaf area in control plants relative to salt-treated plants. As discussed previously, photosynthesis in this halophyte is not sensitive to the presence of high salt levels in leaf tissue, i.e., nonstomatal components of photosynthesis are insensitive to salt. Thus growth should be a function of leaf area and photosynthesis rate.

Figure 10.9, which has been reconstructed from two sets of experiments, suggests that this expectation is fulfilled and that the growth response to salinity is of the same form as the total photosynthetic capacity, calculated as the products of

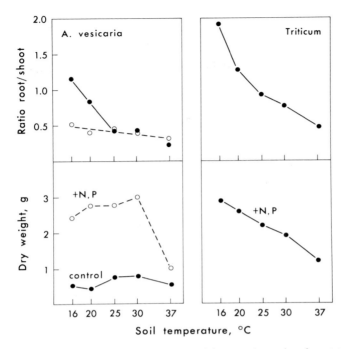

Fig. 10.8. Influence of soil temperature and nutrition on the ratio of root to shoot and on growth of *A. vesicaria* in its native soil (●) and fertilized soil (○), and on *Triticum* ssp. in fertilized soil.(Drawn from data of Cook, 1973)

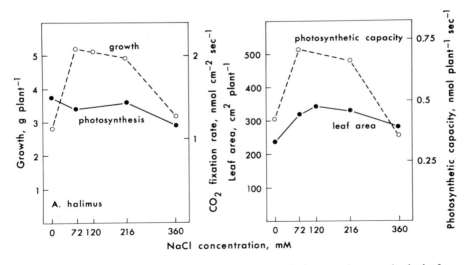

Fig. 10.9. Influence of NaCl concentration in culture solution on photosynthesis, leaf area photosynthetic capacity on a leaf area basis and growth of *A. halimus*. (Drawn from data of Gale et al., 1970, and Kaplan and Gale, 1972)

Table 10.6. Changes in total dry weight (DW) and in the root/shoot ratio (R/S) in *Atriplex* spp. grown in culture solutions with NaCl

NaCl (mM)	(a) A. nummularia		(b) A. vesicaria		(c) A. glabriuscula	
	DW (g)	R/S	DW (g)	R/S	DW (g)	R/S
0	1.04	0.17	10.6	0.22	2.9	0.24
50	—	—	22.8	0.13	2.9	0.18
100	1.40	0.12	—	—	—	—
300	1.14	0.13	11.7	0.17	—	—
700	—	—	—	—	1.71	0.26

(a) Greenway, 1968
(b) Black, 1960
(c) Osmond, unpublished

leaf area and photosynthetic rate. Stomatal closure during part of the day in control plants would simply accentuate the growth stimulation due to NaCl. The decline in growth at higher salinities is the product of reduced leaf area which is presumably a morphological response to high internal salt concentrations. The growth responses shown in Fig. 10.9 are presumably accompanied by changes in the allocation of carbon between root and shoot. That is, the increased leaf area is achieved by the retention of photosynthate in the shoot. In a number of experiments with different *Atriplex* spp. root weights remain more or less unchanged over the range of NaCl concentrations which result in growth stimulations (Table 10.6). Thus the root/shoot ratios decline as growth is stimulated and increase again at higher concentrations of NaCl which lead to a reduction in growth.

The growth of nonhalophytes is inhibited by all concentrations of NaCl and the inhibition of growth is greater in dry than in humid atmospheres. Hoffman et al. (1971) showed that growth of cotton was more sensitive to NaCl in dry atmospheres even though plants maintained positive turgor in all treatments. As noted earlier by Gale et al. (1967), photosynthesis in cotton is reduced by salinity, principally as a result of interference of salt with nonstomatal components of photosynthetic processes. In this nonhalophyte an increase in external NaCl at constant atmospheric humidity deficit (transpiration rate) or in humidity deficit at constant NaCl concentration would increase the NaCl transported to the shoot, reducing photosynthesis and growth. The root/shoot ratio in cotton increased as growth was depressed, both when the growth depression was due to increasing humidity deficit or to increasing salinity (Hoffman et al., 1971).

Although it is possible to put together the above observations in a plausible account of interactions between ion and nutrient absorption, water relations, and photosynthesis which account for the observed growth response in culture solutions, it is another matter to demonstrate that such interactions are significant under natural conditions. Experiments by Williams (1972) suggest that the growth of halophytes in soil at low water potentials is enhanced by salinity, that the enhanced growth is due to changes in the partitioning of photosynthate between

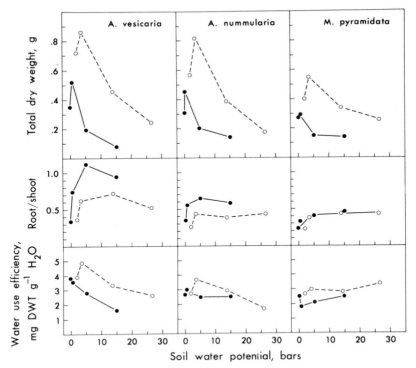

Fig. 10.10. Effect of water potential in soils with 3 mM NaCl (●) and the same soils with 300 mM NaCl (○) on growth, root/shoot ratios, and water use efficiency of arid shrubland species grown in their natural soils under controlled conditions. (Drawn from data of Williams, 1972)

shoot and root, rather than to changes in photosynthetic rate, and that salinity increases the water use efficiency of these plants under arid conditions.

Williams measured the interaction between soil water potential and salinity during the growth of two species of *Atriplex* and of *Maireana* in a nonsaline sandy loam soil native to these plants. The moisture characteristic of the soil was established and an injection system devised to add water throughout the profile. Before sowing, pots were either treated with a nutrient solution containing 3 mM NaCl or with nutrients +300 mM NaCl. Germinated seed was allowed to establish at field capacity for seven days and the soils were then allowed to dry to predetermined water contents corresponding to saturation, field capacity −5 and −15 bar. In the saline series the NaCl concentrations correspond to 46, 80, 210, and 270 mM at each water content which decreased soil water potential by a further 2, 3.3, 8.8, and 11.3 bar, respectively. The growth and water use of these plants in a greenhouse (25° day/20 °C night; 40%–70% R.H.; full sunlight) was followed between days 28 and 60 after planting, soil water being replenishing by syringe at regular intervals.

Some of Williams' observations are shown in Fig. 10.10. Growth by *Atriplex* and *Maireana* was depressed by decreasing soil water potential. In the presence of salt, however, growth was substantially stimulated and although growth declined with increasing total soil water potential it did not decline below growth in the absence of salt. Williams noted that relative growth rate increased and decreased with total growth, as expected, but that the difference in growth between the nonsaline and saline treatments was not due to changes in relative growth rate. The reason for this is evident in the change in the root/shoot ratio. In *A. vesicaria* and *A. nummularia*, salt treatment results in a substantial decrease in the root/shoot ratio, presumably associated with increased leaf area and greater overall photosynthesis in the plant as a whole. At the same time water stress increased the ratio of root to shoot in the presence and absence of salt, consistent with the decreased net growth with increasing stress. The greater growth at field capacity compared with the soil saturation treatments presumably reflects complications due to poor soil aeration in the saturated soil. Soil aeration seems to have been more important for *Atriplex* ssp. than for *Maireana* spp.

The increase in water use efficiency following salt treatment noted in these and other experiments is presumably due to a reduction in transpiration per unit leaf area as discussed previously (Chap. 9.5.1). The absorption of salt and the reduced transpiration evidently permit cells in the shoots to maintain positive turgor at low water potential.

This interlocking control of ion absorption with the water relations in halophytes maintains normal photosynthesis and, by changes in deployment of photosynthates, results in increased growth under saline conditions. Williams' experiments suggest that the productivity of halophytes may, under some conditions, be stimulated by the presence of salt in the soil. There is no evidence, to our knowledge, that the biomass of species such as *A. halimus* or *A. vesicaria* is higher in more saline natural habitats. The possibility remains, however, that interactions of physiological processes in halophytes may be such that the performance of these plants in saline habitats is better than in nonsaline habitats.

The more commonly accepted view is that few plants are obligate halophytes (Barbour, 1970) and that many halophytes are excluded from nonsaline habitats because they are poor competitors with nonhalophytes. That is, the absorption of salt extends the tolerance limits of halophytes, enabling them to occupy saline habitats in which their performance is suboptimal, but nevertheless rather better than that of nonhalophytes. Barbour (1978b) demonstrated this point quite clearly with the saltmarsh halophyte *Jumea carnosa* from the California coast, which was grown in competition with *Lolium perenne*. Under nonsaline conditions, *Lolium* depressed the growth of *Jumea* by 52% when grown in competition. In saline treatments, growth of *Lolium* was depressed three times as much as growth of *Jumea* and no significant competitive effect of *Lolium* or *Jumea* was detected.

Similar relationships are evident in the seasonal changes in photosynthetic capacity of *Atriplex confertifolia* and *Ceratoides lanata* in the Great Basin desert. These two species are both xerophytes but *A. confertifolia* has the capacity for substantial NaCl uptake, as described earlier (Chap. 7.4) whereas *C. lanata* does not (Moore et al., 1972a). The absorption of salt in *A. confertifolia* ensures that this species maintains turgor at leaf water potentials of -70 bars during the hot summer months (Fig. 8.22). The osmotic potential of *C. lanata* remains in the vicinity of -40 bar (Moore et al., 1972a) and plant water potential commonly falls below

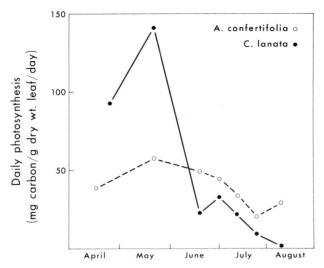

Fig. 10.11. Seasonal progression of daily photosynthetic carbon assimilation measured for *Atriplex* and *Ceratoides* growing in their natural habitat in the Curlew Valley, Utah (M.M. Caldwell, personal communication)

− 50 bar during the day. Under these conditions photosynthesis declined to negligible levels in *C. lanata*, whereas in *A. confertifolia* photosynthesis continued at about half the maximum rate (Fig. 10.11). It is obvious that the halophytic properties of *A. confertifolia* do not permit superior performance of this species in this habitat under favorable conditions. Instead, they extend the tolerance limits, permitting continued low-level productivity into the dry summer months at a time when the nonhalophyte *C. lanata* has ceased production altogether. The extension of tolerance limits of *A. confertifolia* depends on more than the halophytic properties of this species. It depends on root growth and activity patterns in the soil (Chap. 8.2.3) as well as on properties of photosynthesis itself. Some of these interactions are explained in more detail later (Chap. 10.6).

10.5 Productivity in Relation to Water

Because the principal component of plant water relations, in quantitative terms, is the water lost by transpiration and because this water loss is inevitably associated with the opportunity for net CO_2 uptake by terrestrial vegetation, the relationship between productivity and water use should be quite direct. That this is so can be appreciated by the successful development of large-scale correlation and simulation models which show an acceptable level of predictability at appropriate scales (Leith, 1976; Van Keulen et al., 1976). As recently emphasized (Fischer and Turner, 1978) the relevant water parameters to be considered in the relationship between productivity and water are transpiration and water use efficiency:

$$\text{productivity} = \text{transpiration} \times \text{water use efficiency} . \qquad (10.3)$$

Depending on the scale of the inquiry and its objectives, however, it may be more appropriate to relate productivity to gross parameters such as water supply or to small-scale parameters such as the water status of individual plants. The usefulness of the simple relationship shown above depends on the ease of measurement and extrapolation of the parameters, transpiration and water use efficiency. In this section we shall explore these properties of plants in scales of space and time ranging about 10^4 to 10^8, that is from the individual plant through to plant communities. The particular emphasis will be on arid shrubland vegetation, for obvious reasons. Comprehensive treatments of plant productivity in arid and semi-arid environments are found in the recent reviews by Evenari et al. (1976) and Fischer and Turner (1978).

In preceding chapters (Chaps. 8.2 and 9.4.1) we discussed the efficiency of water utilization during photosynthesis of C_3 and C_4 plants and its dependence of environmental factors. We concluded that at least in the short term, and in the absence of severe stress, C_4 plants are two to three times as efficient as C_3 plants in terms of the amount of CO_2 fixed per amount of water transpired. We also concluded that this superiority of C_4 plants occurs over a wide range of environmental conditions and that the underlying cause of this superiority is that the C_4 pathway increases the efficiency by which low intercellular CO_2 pressures can be utilized in photosynthesis. This increased photosynthetic water efficiency may be achieved either by an increased rate of photosynthesis with no difference in transpiration rate or by a decreased transpiration rate with no difference in the rate of photosynthesis, or a combination of both. Thus, depending on species and environmental conditions, the superior photosynthetic water efficiency of C_4 plants may or may not be accompanied by a greater daily rate of dry matter production.

Numerous comparative growth experiments show that the higher photosynthetic water use efficiency of C_4 plants is matched by a similarly increased water use efficiency in terms of dry matter production. Water use efficiencies based on dry matter accumulation are necessarily lower than those based on daytime photosynthesis of single leaves, primarily because of respiratory losses by the whole plant during the night, and by nonphotosynthetic tissue during the day. Some water is also likely to be lost as a result of nighttime transpiration, and evaporation from the soil surface, even in experiments where plants are grown in containers. Another factor contributing to lower apparent water use efficiencies in experiments based on determinations of dry matter accumulation is that the harvest of below ground parts often is incomplete. Such underestimates of water use efficiency tend to be even greater in open-field experiments, especially in situations where the evaporative demand is high and a substantial fraction of the water is lost before it is absorbed by the plant.

On these grounds one would expect the measured water use efficiency to decline as one progresses from the single leaf to the whole plant to a community of plants and this is indeed the case.

For example, Ludlow and Wilson (1972) compared the water use efficiencies for leaves, whole plants, and swards of C_4 pasture grasses and C_3 legumes grown under identical environmental conditions. In the C_4 grasses water use efficiency determined for leaf, plant, and sward were 20, 5, and 3.1 mg dry matter g^{-1} water. Corresponding values for the C_3 plants were 8.7, 2.7, and 1.4 mg dry matter g^{-1}

water. Because of the considerations mentioned above, one would expect the relative difference in water use efficiency between the C_3 and the C_4 species to diminish as one progresses from the leaf to the plant and sward. However, this effect is not pronounced in these experiments. The C_4 species maintained about twice as high water use efficiencies as the C_3 species at all three levels. Similar results were obtained by Slatyer (1970) for *Atriplex* species (see below).

Extensive early comparative determinations by Shantz and Piemeisel (1927) of the water efficiency of plants grown in containers in the field at Akron, Colorado included 10 C_4 species (6 grasses and 4 dicotyledonous species). Both groups were represented by cultivated as well as by wild species of considerable ecological diversity. With no exception the C_4 species exhibited superior water use efficiencies. The mean water use efficiency of the C_4 species was 2.1 times that of the C_3 species.

More recent comparative measurements of water use efficiency of dry matter production have included several C_3 and C_4 desert grasses and shrubs, among them *A. canescens* (Dwyer and DeGarmo, 1970). When these plants were grown in containers in a greenhouse, the difference between the C_4 and C_3 species was similar to that observed with species of more mesic origin, with *A. canescens* having about twice the water use efficiency of C_3 desert shrubs such as *Prosopis juliflora* and *Larrea divaricata*.

In the previous chapter (Chap. 9.4.1) we concluded that C_3 and C_4 species of *Atriplex* exhibit similar differences in intrinsic photosynthetic water use efficiency as those found between C_3 and C_4 plants in general. Among the examples used to illustrate this was Slatyer's (1970) comparison between the inland C_4 annual, *A. spongiosa,* and the C_3 coastal annual, *A. triangularis* (Fig. 9.31). In addition to his determinations of photosynthesis and transpiration on single leaves, Slatyer also measured the dry matter production and total water use of the greenhouse-grown seedlings at the same intervals throughout the 23-day period. The results of these growth experiments are shown in Fig. 10.12.

During the first part of the experiment C_4 *A. spongiosa* had a considerably higher net assimilation rate but a lower transpiration rate than C_3 *A. triangularis,* and water use efficiency of dry matter production was about twice as high in the C_4 species. At the end of the experiments net assimilation rate was about equal in the two species, but because of its lower transpiration rate the C_4 species maintained a superior water use efficiency. In a comparison between these results and the instantaneous water use efficiencies determined in gas exchange measurements (converted to a dry matter basis) we find, as in Ludlow's comparison between C_4 grasses and C_3 legumes, the water use efficiencies are reduced as we move from instantaneous measurements on single leaves to longer-term measurements on whole plants, but the superior water use efficiency of the C_4 species compared to the C_3 species is evident at each scale.

It is also of interest to note that in Slatyer's experiments, water use efficiency of C_4 *A. spongiosa* remained about twice that of C_3 *A. triangularis* when the water contents of the leaves was varied over a wide range. The water status of the plants thus had little impact on the relative water use efficiency between the two species even though it strongly affected stomatal conductance and hence the rates of photosynthesis. As mentioned in Chaps. 8.2.1 and 9.4.1 the "actual" water use efficiency of any given plant is the product of its "intrinsic" water use efficiency and

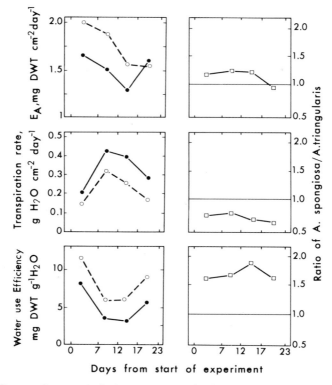

Fig. 10.12. Course of net assimilation rate, transpiration rate, and water use efficiency for *A. spongiosa* (C$_4$) (○) and *A. triangularis* (C$_3$) (●) during the experimental period. The ratios of each of these parameters are given on the *right-hand side* of the figure. (Redrawn in rearranged form from Slatyer, 1970)

the water vapor pressure difference between the leaves and the surrounding air. The actual water use efficiency (expressed as the short-term photosynthetic water use efficiency or as longer-term water use efficiency of dry matter production) thus depends on the environment in which the plants are growing or the conditions under which experiments were conducted. This is undoubtedly the principal reason for the large variation in "actual" water use efficiency in the literature. We can thus expect that for any given plant the water use efficiency will decline with increasing water vapor deficit and this has been demonstrated experimentally in studies such as those of deWit (1958) and Van Keulen (1975). In these studies the considerable seasonal variation in actual water use efficiency could be fully accounted for by the seasonal changes in water vapor deficit.

In addition to water vapor pressure deficit, temperature may also be expected to influence water use efficiency (Chap. 9.4.1). Using growth analysis methods, Downes (1969) determined the effect of temperature on net assimilation rate, transpiration rate, and water use efficiency of dry matter production for several C$_3$ and C$_4$ grass species. In order to minimize the effect of an increasing temperature on the water vapor pressure difference between the leaves and the air, the water vapor

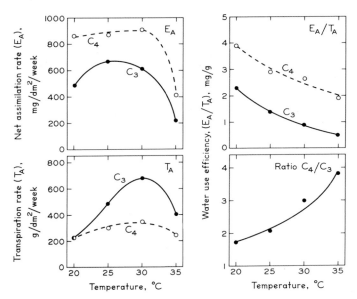

Fig. 10.13. Effect of temperature on net assimilation rate E_A, transpiration rate T_A, water use efficiency E_A/T_A, for *Lolium multiflorum* (C_3) and *Pennisetum clandestinum* (C_4), grown in the same environments. The ratio of water use efficiences between the C_4 and C_3 species is also shown. (Drawn from data of Downes, 1969)

deficit of the air was kept constant over the temperature range used. Figure 10.13 compares the response of the temperate C_3 species *Lolium multiflorum* with that of the tropical C_4 species *Pennisetum clandestinum*. Similar differences were obtained between the other C_3 and C_4 grasses investigated. As shown in the right blocks of Fig. 10.13 the water use efficiency of the C_4 species exceeded that of the C_3 species at all temperatures. The superiority of the C_4 species increased dramatically with increased growth temperature, ranging from a ratio of about 1.8 at 20 °C to almost 4.0 at 35 °C. As shown in the left blocks of Fig. 10.13 this difference in temperature response of water use efficiency is attributable to both E_A and T_A. The C_4 species had higher E_A values than the C_3 species at all temperatures while conversely, the T_A values of the C_3 species exceeded those of the C_4 species, except at the lowest measurement temperature, and the dependence of T_A on temperature was more pronounced in the C_3 species.

As Fisher and Turner (1978) have pointed out, there are many reasons for expecting that the water use efficiency may differ between species or genotypes but surprisingly, there is as yet little or no experimental evidence that this is so, other than those existing between C_3 and C_4 (and CAM) plants. Container experiments as well as open-field determinations show remarkably similar water use efficiencies among plants of different life forms and ecological origins when the values are adjusted for differences in water vapor pressure deficits. It could be that unknown compensating factors act to maintain a constant water use efficiency. However, more experimental data are needed before it can be definitely concluded that water use efficiency is constant among plants within each type of photosynthetic CO_2

fixation pathway. It is particularly important that such experiments include conditions under which the plants are subject to environmental stress.

The fact that C_4 plants have a higher water use efficiency than C_3 plants implies that, other things being equal, starting from a young seedling in a water-limited environment, the C_4 species would make more growth. Equation 10.3 predicts that in the absence of direct composition for water the higher water use of the C_4 plant would enable it to produce twice as much as the C_3 plant before the water supply is depleted. There is little direct experimental work pertaining to the question of what impact the superior water use efficiency of C_4 plants has on the productivity and reproductive ability of plants in water-limited environments. The comparative growth study with *Atriplex* species (Nobs et al., 1972) is one of the few attempts that have been made to elucidate this question.

One of the species used in this study was the C_4 annual *A. rosea*. Although the species is widely distributed in the warm interior valleys of California, it is also very common around San Francisco Bay and in this case the experimental plants were grown from seed collected from a population naturally growing along the shores of this bay where it grows in sites adjacent to populations of the C_3 annual *A. triangularis*. The *A. triangularis* material used in the studies was grown from seed collected from a population growing on the Pacific coast about 30 km south of San Francisco. The photosynthetic water use efficiency of *A. rosea* is about 2.3 times that of *A. triangularis* (Chap. 9.4.1).

The performance of these two species was compared in transplant gardens at Stanford (close to San Francisco Bay) and at Half Moon Bay on the Pacific coast (close to the habitat of the *A. triangularis* population). The climate at the coastal garden is characterized by mild temperatures and summer fog. The water vapor pressure deficits are much lower than in the Stanford garden where summer days are typically clear and warm. Although winter rainfall is greater in the coastal site, neither site receives significant summer rain.

Seeds of the two species were sown in the Stanford garden in February and seedlings were transplanted into the coastal garden in early April. All plantings were irrigated until April 30 to ensure uniform seedling establishment. Thereafter one-half of the rows of each species were left unwatered, while the other half received thorough and frequent irrigations. A sample of nine plants from each species, garden, and treatment was harvested and dried and the phenological development was followed at two-week intervals throughout the growing season.

The seasonal course of dry matter production of the two species (on a logarithmic scale) is shown in Fig. 10.14. Both species grew rapidly in the coastal garden and the final dry matter yields were comparable. The relative growth rates and dry matter yields in the coastal garden were totally unaffected by irrigation in either species. As shown in Table 10.8 irrigation also had no effect on the time of flowering or the production of ripe fruit, and both species produced prodigious crops of ripe seed even in the unirrigated treatment. Obviously, the soil water content remained sufficiently high to prevent any drought stress in both the C_3 and C_4 species in this low evaporation climate and the higher water use efficiency of the C_4 species was of no consequence to either dry matter production or reproductive output.

In the Stanford garden relative growth rate and final dry matter yields also were similar in the two species and both produced massive seed crops as long as the plants

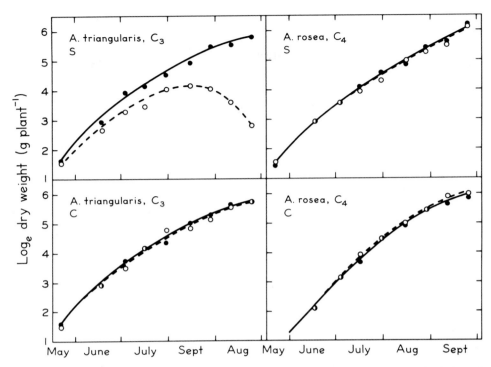

Fig. 10.14. Progression of dry matter production of *A. triangularis* and *A. rosea* under irrigated and unirrigated conditions at Stanford (S) and at a coastal (C) garden. (Based on data from Nobst et al., 1972)

were irrigated (Fig. 10.14, Tables 10.7 and 10.8). However, the species showed dramatic differences in their responses to irrigation. During the first month following the termination of irrigation one-half of the rows of the watered and unwatered plant of the C_3 species had similar growth rates but subsequently the relative growth rate of the unwatered plants showed a progressive decline and growth ceased in mid-August, after which the plants gradually died. By late September the final average dry weight of the irrigated plants was 20 times the average for unirrigated plants. Although the irrigated and unirrigated plants flowered at the same time the unirrigated plants died before they were able to produce ripe seed. This failure was obviously due to severe water stress. As in the coastal garden, irrigation had no effect on the C_4 species in the Stanford garden, either in terms of dry matter production, growth habit, flowering, or seed production. The C_4 species successfully completed its life cycle with massive seed production even in the absence of irrigation.

The reasons for the much greater ability of the C_4 species to cope with the limited water supply and relatively high water vapor pressure deficits in the Stanford garden could of course be several. Possibly, the C_4 species may be capable of deeper penetration into the soil thereby increasing the water supply, or it may have a greater tolerance to low tissue water potential. However, there is no evidence

Table 10.7. Dry weights (g plant^{-1}) of unirrigated (U) and irrigated (I) *A. triangularis* and
A. rosea in the Stanford garden. (Data from Nobs et al., 1972)

Date	*A. triangularis*			*A. rosea*		
	I	U	U/I	I	U	U/I
June 17	19	15	0.78	19	18	0.95
July 16	63	32	0.51	60	52	0.86
Aug. 15	148	63	0.43	128	141	1.11
Sept. 12	270	37	0.14	270	245	0.90

Table 10.8. Date of onset of flowering and production of ripe fruit of irrigated and unirrigated
A. triangularis and *A. rosea* plants grown at a coastal site at Half Moon Bay and an inland
site at Stanford, California. (Data from Nobs et al., 1972)

Species	Treatment	Coastal site		Stanford site	
		Flowers	Ripe fruit	Flowers	Ripe fruit
A. triangularis	Irrigated	June 4	Aug. 13	July 12	Sept. 4
C$_3$	Unirrigated	June 6	Aug. 13	July 9	None
A. rosea	Irrigated	June 4	Sept. 20	July 1	Sept. 7
C$_4$	Unirrigated	June 11	Sept. 25	July 3	Sept. 2

that *A. rosea* and *A. triangularis* are markedly different in either of these characteristics and it seems logical to interpret the observed responses to limited water
supply in terms of the superior water use efficiency of the C$_4$ species.

Since the two species exhibited similar rates of dry matter production under
irrigated conditions it is evident that *A.rosea* in these experiments used less water
per unit time than *A.triangularis*, thereby postponing the onset of low soil water
potentials until it had completed its life cycle. It should, however, be pointed out
that this strategy is not necessarily advantageous under all conditions. First, since at
least a portion of the water that is not used by the plant may not be available for
later use because it is lost by direct evaporation from the soil. Second, in a situation
of a direct competition for the same water supply, water conserved by a frugal plant
(such as *A.rosea*) would presumably be expended by a greedier competitor (such as
A.triangularis).

These speculations have yet to be evaluated in appropriately designed
competition experiments or in natural situations where annual C$_3$ and C$_4$ species
compete for limited water resources. There are many examples of C$_3$ and C$_4$ species
which are in apparent competition in ruderal communities (Fig. 5.12) yet the only
study to date has been based on perennial species of *Atriplex* (C$_4$) and *Ceratoides*
(C$_3$) in the Great Basin Desert. We shall develop this comparison in some detail in
the following section as we proceed to expand the above discussion into the
productivity of plant communities.

10.6 Productivity in Plant Communities

In the previous sections we have been primarily concerned with the productivity of individual plants in response to different environmental factors. For these purposes, measures such as relative growth rate (R_w) are appropriate, but at larger scales it is usual to express productivity in terms of land surface and to expand the time base of measurement appropriately. Thus instead of R_w for individual plants expressed in $g\,g^{-1}$ dry wt. day^{-1} productivity is expressed in g dry wt. $m^{-2}\,year^{-1}$. An average value for the carbon content of plant dry matter is $0.42\,g\,C\,g^{-1}$ dry wt. and to convert from CO_2 fixation rates, the factors $0.65\,g$ dry wt. g^{-1} CO_2 and $35.2\,mmol\,CO_2\,g^{-1}$ dry wt. are useful. The estimation of community productivity is a very approximate and difficult exercise because it is seldom practicable to obtain accurate estimates of biomass or production by harvest methods. For many purposes these approximations are adequate, but just as small differences in relative growth rate of an individual plant, or in the ratio of allocation to root and shoot can have enormous impact on its productivity and form, so the differences in productivity of individual elements of a plant community in response to environment may have a large impact on the state of the community and its productivity.

10.6.1 Annual Patterns of Productivity in Cool-Desert Shrub Communities

The *Atriplex-Ceratoides* community of the Great Basin desert contains both mixed and almost pure stands of *A.confertifolia* and *C.lanata* growing on halomorphic soils in almost identical physical environments (Chap. 5.1.1). It should be immediately obvious that two species which have so effectively occupied this cool desert habitat, and which are evidently in equilibrium with this environment, may have evolved subtly different adaptations to this environment. The fact that *A.confertifolia* is a C_4 plant, and *C.lanata* is a C_3 plant, represents one such difference which could have significant impact on the performance of these shrubs. The shrubs are of similar growth form and the cool desert habitat is relatively simple in that plants respond to a combination of winter precipitation and seasonal temperature change in a single large pulse of primary production which begins in spring. Responses to summer precipitation are slight (Caldwell et al., 1977b; Hodgkinson et al., 1978). This community is well suited for an analysis of the ways in which C_3 and C_4 pathways of photosynthesis contribute to the productivity of these perennial shrubs.

The different pattern of biological activity in these shrubs is due to the ontogenetic differences between species. In *A.confertifolia* the period of regreening of leaves and twig elongation starts earlier and lasts longer than in *C.lanata*. In *C.lanata* flowering and fruit development was essentially complete in July and shrubs entered immediately into summer dormancy (M.M. Caldwell, personal communication). Flowering and fruit development of *A.confertifolia* continues in a large part of the population over a longer period, and overall *A.confertifolia* maintains a longer period of biological activity than *C.lanata*.

Photosynthesis and transpiration were recorded at frequent intervals throughout 1970 for plants growing in the field. The highest photosynthetic rates measured in the field were $1.4\,nmol\,cm^{-2}\,s^{-1}$ for *A. confertifolia* and $1.8\,nmol\,cm^{-2}\,s^{-1}$ for

C.lanata. Figure 10.11 showed that during 1970 the daily carbon gain was often greater in the C_3 species, particularly when plant water potential was high, but that the C_4 species maintained photosynthesis for a longer period through the dry summer months. The longer period of photosynthetic activity in *A.confertifolia* was presumably associated with the differences in phenology described above and is probably a consequence of the superior salt absorption and water relations properties of this species described previously (Chaps. 7.4.6; 8.2.4).

In 1970 precipitation was only 9% greater than the long-term average and Caldwell et al. (1977b) devised an interpolative model based on the 1970 data to determine the seasonal progression of photosynthesis and water use. This model showed good agreement against the 1970 data set and was then used to compute the seasonal progression of photosynthesis and water use from meteorological data for a good year (1974; precipitation 32% above average and a bad year (1973; 20% below average). These calculations (Fig. 10.15) confirmed the principal feature of seasonal CO_2 assimilation for these species; namely, earlier and more rapid CO_2 assimilation in the C_3 plant and prolonged, less rapid CO_2 assimilation in the C_4 species. These differences were most pronounced in the dry year (1974) in which it was estimated that *A.confertifolia* fixed 61% of its total annual carbon after June 1 and expended about half of its water in doing so. The C_3 plant *C.lanata* gained 33% of its carbon after June 1, but 54% of its water was expended after this time. Water use efficiency of the C_4 plant was twice that of the C_3 plant during summer. Caldwell et al. (1977b) show, rather surprisingly, that the average daily water use efficiency of *Atriplex* (C_4) and *Ceratoides* (C_3) was not substantially different during April and May, the season of most active growth. Whether this result stems from measurements taken on different days, under different conditions, or whether it reflects the difficulties of controlling evaporative demand and measuring transpiration from leaves at low temperature and high vapor pressure deficits, is not known. In spite of this, the superior water use of the C_4 plant during summer was obvious and because this plant remained photosynthetically active at this time, the overall seasonal water use efficiencies were also much higher for the C_4 plant (4.3 vs. 2.9 mg dry wt. g^{-1} H_2O; Chap. 9.4.1).

Although *Ceratoides* was rather more profligate with water early in the season when it was available, *Atriplex* continued to expend water extracted from deep in the profile when it remained active into the hot summer months. As a consequence, the two communities showed nearly identical water expense accounts (approximately 50% of the precipitation was transpired) when these were calculated from soil water content data or from the modeled transpiration behavior of the plants. We have no evidence, but might reasonably predict, that in a competitive situation the free expenditure of water by *C.lanata* might reduce the water available later to *A.confertifolia* and hence restrict its marginal advantage in terms of production (Fig. 10.16). The extent to which the slightly greater diversion of carbon to roots in *Atriplex,* and the longer and deeper penetration of soil water by this species (Chap. 8.2.3) could compensate in a competitive situation is unknown. The contribution of these processes to the performance of these species may well vary in wet and dry years.

Using the interpolative model, the average annual carbon fixation (1973/1974) was only marginally higher in the *A.confertifolia* community (240 g cm^{-2} yr^{-1}) than

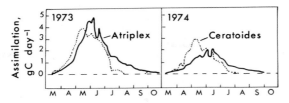

Fig. 10.15. Simulated seasonal progression of daily carbon assimilation of *Atriplex* and *Ceratoides* in their natural habitat. (Redrawn from Caldwell et al., 1977 b)

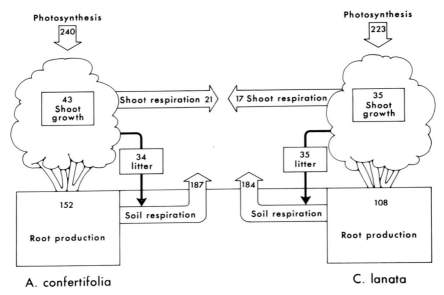

Fig. 10.16. Annual carbon budget for *Atriplex* and *Ceratiodes* communities in the Great Basin Desert, Utah. (Redrawn from Caldwell et al., 1977 b)

in the *C. lanata* community (223 g cm^{-2} yr^{-1}). This was due partly to the 40% higher leaf area displayed per unit ground area in the *Atriplex* community which compensated for the often lower rates of CO_2 assimilation on a leaf area basis in the C_4 plant.

Caldwell et al. (1977b) estimated the partitioning of the carbon assimilated by *Atriplex* and *Ceratoides* between shoots, roots, and litter from measurements of biomass distribution and their turnover. The annual carbon flux budget for the two communities, based on independent techniques and data sets, is shown in Fig. 10.16. None of the carbon flux estimates was derived by subtraction and consequently, a small imbalance was found. The carbon flux budgets for the two communities are remarkably similar overall, with the bulk of the assimilated carbon being diverted belowground as noted previously (Table 5.10). The slightly higher CO_2 assimilation capacity of the C_4 community was translated into a 37% higher incorporation of carbon into aboveground and belowground pools. The reasons for this are not clear. This could be accounted for by the differences in photosynthesis alone, but Caldwell et al. suggest that the higher maintenance respiration costs or greater carbon loss belowground in *Ceratoides* may also contribute.

The studies of Caldwell et al. (1977b) describe a heroic effort to evaluate small-scale physiological and biochemical processes (C_3 and C_4 photosynthesis) in the context of larger-scale ecological processes. They show that one intrinsic property of the C_4 pathway (higher water use efficiency), when coupled with osmotic adjustment in the shoot and greater allocation of photosynthate to deeper root systems providing access to soil water, permits *A. confertifolia* to remain active in carbon assimilation when its C_3 counterpart is inactive. Whether the higher intrinsic water use efficiency of the C_4 pathway actually means that *A. confertifolia* can afford both to maintain a larger assimilating and transpiring surface and to build a larger, costly water acquisition system, is a matter of conjecture.

10.6.2 Long-Term Productivity in Arid Ecosystems

The significance of the different patterns of physiological activity described above for the long-term productivity of arid ecosystems has yet to be evaluated. There are few long-standing field trials in arid regions that permit conclusions as to which of the different patterns of resource utilization found in *Atriplex* and *Ceratoides* is better suited in terms of production or persistence. Some observations (Hutchings and Stewart, 1953) indicate that *Atriplex* populations in the Great Basin desert suffer a higher mortality in severe droughts. The higher mortality may arise because these plants remain active in the summer, whereas *Ceratoides* ceases activity in the late spring and is summer-dormant. A 30-year comparison of changes in plant cover in the Desert Experimental Range, Utah, showed that *Ceratoides* displaced *Atriplex* to some extent in ungrazed plots (Holmgren and Hutchings, 1972). With moderate grazing, however, both species increased their contribution to total plant cover.

Few studies of growth in any natural vegetation have extended more than one or two decades, but in arid regions of the United States and Australia there have been notable attempts to follow long-term changes. At the Desert Experimental Range in Utah, biomass and productivity have been followed for 12 years. Hutchings and Stewart (1953) described estimates of productivity based on harvest and nondestructive methods and some of their data for an *A. confertifolia* shrubland are shown in Fig. 10.17. Total biomass in this community is closely tied to annual precipitation and the biomass of *A. confertifolia* comprises 30%–50% of the total in all years. These data reflect the seasonality of precipitation in this cool desert environment. Thus, for the period 1934–1947, an average of 40% of precipitation (not including snow) was received between October and March, a period of low temperature, low evaporation and low biological activity. Effectively, the soil profile is recharged with water during the cool winter months. Most of the plant production occurs in the spring during which a further 28% of the average annual precipitation was recorded. We have discussed the seasonality and patterns of ecological processes in these communities previously (Chap. 5.1.1), noting that because of

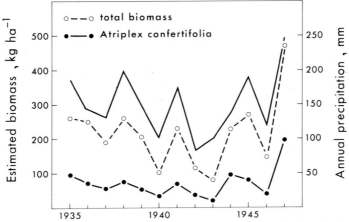

Fig. 10.17. Annual rainfall (———) and biomass of *A. confertifolia* shrublands in the Desert Experimental Range, Utah. (Redrawn from Hutchings and Stewart, 1953)

this interaction of low temperature and precipitation patterns, the productivity of this ecosystem may be thought of as responding to a single large pulse of precipitation.

The average aboveground productivity of the *A. confertifolia* community at the Desert Experimental Range 1935–1947 was 195 g dry wt. $m^{-2} yr^{-1}$ (Table 10.9). If corrected for a root/shoot ratio of 7 for *A. confertifolia* alone (Caldwell and Camp, 1974), this corresponds to 579 g dry wt. $m^{-2} yr^{-1}$, of which 131 g dry wt. $m^{-2} yr^{-1}$ is

Table 10.9. Average long-term productivity of *Atriplex*-dominated arid shrublands. (Data of Noble, 1977 and Hutchings and Stewart, 1953)

Site and Component	Productivity (g $m^{-2} yr^{-1}$)	
	Aboveground	Total
Koonamore (average 1925–75)		
A. vesicaria	1	
M. sedifolia	3	
Annual	22	
	26	52[a]
Koonamore (best year)		
A. vesicaria	6.5	
M. sedifolia	24	
Annuals	160	
	190.5	381[a]
Desert Experimental Range (average 1935–47)		
A. confertifolia	64	
Other	131	
	195	579[b]

[a]   Assuming root/shoot 0.5
[b]   Assuming root/shoot 7.0 for *A. confertifolia*

due to annual species. This estimate of 448 g dry wt. m^{-2} yr^{-1} is in close agreement
with the estimate of Caldwell et al. for *A. confertifolia* in Curlew Valley (200 g C m^{-2}
yr^{-1} = 470 g dry wt. m^{-2} yr^{-1}). Whittaker (1975) estimated that the productivity of
arid ecosystems range from 10–250 g dry wt. m^{-2} yr^{-1} and obviously, the cool
desert is one of the most productive.

Longer-term measurements are available for the Jornada Experimental Range
in New Mexico (Buffington and Herbel, 1965) and from Koonamore Vegetation
Reserve in South Australia (Chap. 5.1.1). The latter is a warm desert community
dominated by an *Atriplex vesicaria, Maireana sedifolia* shrubland which is
homologous with that of the cool desert, an important comparison for our
purposes. Noble (1977) has developed a method for estimating biomass changes
from long-term, nonrigorously controlled photo-point records from Koonamore
for the period 1926–1972. Consecutive pairs of photographs taken in May or mid-
August were examined and an assessment of whether biomass of particular taxa
had increased or decreased was made. These estimates were then made on photographs taken at two-, three-, and four-years intervals. Repeated reevaluations of
the assessment were made from time to time and resulted in 85% agreement as to
whether there was an increase, decrease, or no change in biomass. The accumulated
phase shifts constructed in this way from the photographic data were then
combined to indicate the long-term trends 1926–1971, and current photographic
records were calibrated by harvest techniques. This method provides estimates of
aboveground biomass change only, but total production may be estimated from
data on biomass distribution. The method was calibrated against detailed records
of annual plant numbers in some quadrats over a five-year period and against
harvest weight data for a large number of *A. vesicaria* and *M. sedifolia* shrubs.

Figure 10.18 shows the relative biomass of *A. vesicaria* (two sets of photo-point
data), *M. sedifolia,* and annual plants in quadrats 10A and 100 for the years
1926–1972, together with the three-monthly seasonal rainfall recorded at the
Koonamore. The most pronounced fluctuations are found, of course, in the annual
vegetation. In the period 1926–1954 there were six periods of high annual plant
biomass. These correspond to the times at which 85 mm or more precipitation was
recorded in a 30-day period, and although there were three additional records of
75–85 mm rainfall in summer, this rainfall was insufficient to cause a major
response in the annuals. The response of perennial vegetation was similar to that for
the annuals, except that *A. vesicaria* showed little response to the 1946 summer
rainfall following the prolonged droughts of the 1940's. Presumably this rainfall
was insufficient to wet the soil profile to the extent required for the growth of
perennial shrubs.

The biomass data analyzed by Noble represent the integration of productivity
and the loss of aboveground nonwoody materials. The productivity of the shrubs
can be deduced from the response of biomass to prior rainfall. Noble calculated
Spearman rank correlation coefficients for the lagged rainfall (1 to 60 months prior
to record) with the biomass of *A. vesicaria, M. sedifolia* and annuals in quadrat 100.
As expected, annual plant biomass was best correlated with rainfall in the previous
12 months (Fig. 10.19). For *A. vesicaria* the best correlation was observed with the
previous 42 months and for *M. sedifolia,* with the previous 24 months. The carry-
over effects illustrated in this way may be interpreted in terms of the life span of

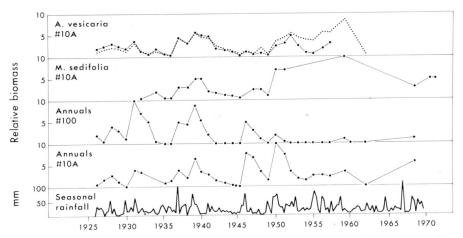

Fig. 10.18. Changes in relative biomass of perennial shrubs (two sites for *A. vesicaria*) and annuals at Koonamore Vegetation Reserve, South Australia, in response to seasonal rainfall. Data based on analysis of photopoint records. (Redrawn from Noble, 1977)

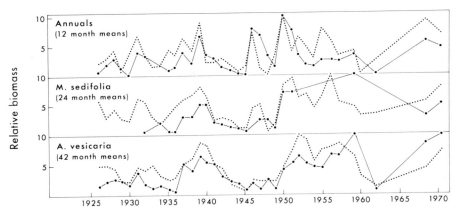

Fig. 10.19. Changes in relative biomass (• — •) of perennial shrub species and annual vegetation at the Koonamore Vegetation Reserve, South Australia, in response to lagged rainfall data (·····). (Redrawn from Noble, 1977)

shoots of the perennial species. Leaves remain on the woody shoots of *A. vesicaria* for several years and presumably retain activity. On the other hand, the majority of the leaves of *M. sedifolia* are formed on nonwoody shoots and are dropped during periods of low rainfall.

We noted earlier that *A. vesicaria* growing on the Riverine Plain was able to respond to precipitation at any time of the year. Noble's analysis of the Koonamore data shows that although biomass is best correlated with winter rainfall, biomass increment (productivity) is best correlated with summer rainfall. The productivity of *A. vesicaria*, like the annuals, responds most to substantial summer rains.

Although rain in the cooler winter and spring contributes to growth, the productivity pattern is that of a pulse, most frequently triggered by summer rains.

This pattern is distinctly different from the spring flush observed in the Great Basin communities, which respond only marginally to later summer rainfall (Hodgkinson et al., 1978). The temperature response functions of photosynthesis in *A. vesicaria* and *A. confertifolia* are not remarkably different (Caldwell et al., 1977a), so that thermal limitation of photosynthesis in *A. confertifolia* is unlikely to account for the absence of a summer growth flush in response to precipitation. It is likely that the pulses of productivity in summer, which are a feature of the Koonamore data, simply reflect the fact that summer rainfall events of such magnitude (75–100 mm) are rare in the Great Basin. On the other hand, recharge of the soil water reserves during winter is by no means assured in the *A. vesicaria* shrublands and there is no regular spring flush against which the summer pulse can be measured. This irregular response means that the long-term average (1925–1972) productivity of the Koonamore shrublands is very much lower than that estimated in the Great Basin (Table 10.9). In the best year, productivity at Koonamore approached that estimated in the *A. confertifolia, C. lanata* community, but the greater part of this was due to annuals, not to *A. vesicaria*.

These comparisons depend very much on the root/shoot ratio applied to the calculations and it is possible that the low ratios of root/shoot biomass in Australian arid shrublands underestimate the proportion of production diverted belowground. The data for both communities may be compared with the productivity in young plantations of irrigated *A. vesicaria* and *A. nummularia* studied by Jones and Hodgkinson (1970). Under these ideal conditions above- and belowground productivity was approximately 500 g dry wt. m^{-2} yr^{-1} (7200 kg ha^{-1} in 1.3 yr); not vastly different from the estimates for productivity of *A. confertifolia* and the Great Basin, but substantially greater than average productivity of *A. vesicaria* in natural habitats (Table 10.9).

Thus the *Atriplex* communities of the warm deserts in Australia are among the least productive and those of the cool deserts of the Great Basin in North America are among the most productive of arid systems. When we take into account the tremendous diversion of photosynthates to the root systems in the Great Basin communities, these are about as productive as the grazing lands of North America (Caldwell, 1975). More meaningful comparisons of the productivity in semi-arid systems have been made by Fischer and Turner (1978). These authors gathered data for four different semi-arid plant communities receiving comparable precipitations which confirm the relationship given in Eq. (10.3). That is, productivity of different growth forms in dry regions, ranging from native shrubland or pasture to cultivated C_3 or C_4 crop species, does not differ much if allowance is made for transpiration, root/shoot allocation ratios and the higher water use efficiency of C_4 photosynthesis. The product of water use efficiency and transpiration [i.e., productivity, Eq. (10.3)] ranges from 11.4–13.9 g m^{-2} day^{-1} in the wild and cultivated C_3 plants, and is approximately twice this value for the C_4 crop plant maize (23.1 g m^{-2} day^{-1}).

These data confirm the point made earlier that for species with the same carbon pathway of photosynthesis, water use efficiency is not much altered by growth form. Fischer and Turner (1978) note that these data do not support the commonly held view that plants of semi-arid habitats "balance their water economy at the expense of reduced productivity", a view which itself "overlooks the fact that water not transpired is water lost in terms of productivity ... and the fact that aridoactive species do not exercise their greater control over water loss until there is very little water left." We have dwelt at length on the performance of C$_3$ and C$_4$ species such as *Cammisonia* and *Tidestromia* which attain the highest recorded photosynthetic rates in semi-arid habitats when water is readily available. Free spending of water resources when they are available reduces the prospect of water loss to competitors and by soil evaporation. Furthermore, it is likely to be an effective strategy because water use efficiency is likely to be highest during the favorable ambient conditions which prevail when the soil is wet.

There is unlimited scope for informed modeling of the plant productivity in different environments and for speculation as to the goals of the games plants play in natural habitats. If we assume with Fischer and Turner (1978) that as a result of natural selection, features of plant life in undisturbed ecosystems are purposeful, studies such as those of Caldwell et al. (1977b) remind us that maximization of productivity is not itself a necessary prerequisite for survival. On the other hand, in water-limited habitats the efficiency of water use associated with the C$_4$ pathway of photosynthesis enables species such as *A. confertifolia* to exploit a niche in space and time not available to C$_3$ plants of similar life form. By showing that many notions of the adaptive value of C$_4$ photosynthesis in other habitats were applicable in the arid shrublands of the Great Basin, Caldwell and his colleagues remind us that it is almost impossible to assert if, or to what degree, a given trait contributes to the productivity or survival of a species without supporting experimental evidence (Eckardt, 1975). The generation of many of these notions may be attributed more to Bacon's demons, identified by Kamen (1963) as the intellectual frailties of premature generalization, specious rationalization and sematic confusion, than to experimental evidence. Caldwells' experiments confirm Eckardt's view that to understand adaptation is to account for the ways components of biological systems are fitted together and are expressed in both performance and survival attributes which maintain plants in communities over long periods of time.

10.7 Effect of Future Increase in CO$_2$ Level on Water Use and Productivity

In concluding the last chapter of this book it seems appropriate to turn our attention to the future and briefly consider the possible responses of plants to predicted changes in a very important environmental factor, namely the atmospheric level of CO$_2$. In the foregoing parts of this book we have repeatedly emphasized that the present CO$_2$ level is suboptimal (and the O$_2$ level supraoptimal) for photosynthesis and primary productivity of the great majority of plants. We have also discussed at some length the paramount consequences of this situation to the water requirement for growth of land plants, and to the evolution of C$_4$ photosynthesis, and other mechanisms which enable plants to alleviate some of the consequences of a low atmospheric CO$_2$ level.

It is a widely held view that reduction of atmospheric CO_2 to the present low level and the high O_2 level are primarily the result of past photosynthetic activity. There is now much evidence that the CO_2 level is increasing as a result of the massive release of CO_2 as man burns the vast store of past photosynthate (fossil carbon). On a geological scale this reversal may be only of short duration (10^8–10^9 s) but its consequences for mankind in the next century may be dramatic.

It is well established that the release rate of fossil carbon into the atmosphere as CO_2 has been increasing exponentially since the start of the industrial revolution. The current world production of CO_2 from burning fossil fuel amounts to about 6×10^9 t year^{-1} compared with about 0.2×10^9 t year^{-1} a century ago and the annual growth rate is about 4.3% (Keeling, 1973; Rotty, 1973). The cumulative amount of CO_2 produced from fossil fuel burning in this period of time has been about twice as great as the increase in the amount of CO_2 in the atmosphere, suggesting that about one-half of the CO_2 released has been taken up and stored in the oceans and the terrestial vegetation (Baes et al., 1977). It is possible that the rapid rate of clearing of tropical forests is another important source of CO_2 and this clearing also causes reduction in the strength of the biota as a sink for CO_2 (Woodwell et al., 1978). In any case, the increase in the amount of atmospheric CO_2 during the past 100 years has been about one-half of that produced from fossil fuel and has shown a similar growth rate. As a result, the mean global atmospheric concentration, or partial pressure of CO_2, has increased from about 290 µbar in pre-industrial times to the present 336 µbar and estimates predict that it will continue to rise for another century (Baes et al., 1977). How much it will increase obviously depends on the future rate of fossil fuel burning and deforestation, which are determined by man, as well as on the manner in which the terrestial vegetation and the oceans respond to resulting flux of CO_2. Conservative estimates, assuming a dramatic decrease in the annual growth rate of CO_2 production caused by man, predict that the atmospheric CO_2 pressure will nearly double in the next 100 years. Estimates assuming a continuation of the present annual growth rate of 4.3% in CO_2 production from fossil fuels, reduced in proportion to the total fossil fuel supply that has been used, predict that more than 50% of all fossil fuel will be released in less than a century. Should the latter projection be realized atmospheric CO_2 pressure would rise to about four times the present level in less than 100 years (Baes et al., 1977).

It is widely thought that the principal effect of an increased CO_2 would be a rising mean global temperature, perhaps by $1°$ to $5°C$ per doubling of the atmospheric CO_2 level (Schneider, 1975). If the higher estimate is correct, a doubling of the CO_2 level could result in a climatic change of unprecedented rapidity which thus would be viewed as catastrophic (at least to man if not to the global vegetation). It should be pointed out however, that current projections of the possible impact of an increased CO_2 level on climate are very uncertain.

Here we shall ignore the indirect effects on plants that may be caused by possible climatic changes such as increased temperature and altered precipitation patterns and focus on an attempt to assess the direct impact of an increased CO_2 level on plant productivity.

Photosynthesis and Productivity. We have repeatedly stated that a wealth of experimental evidence demonstrates that the rate by which photosynthesis can proceed in general is strongly limited by the low CO_2 pressure in the present atmosphere even under conditions that permit the stomata to stay fully open. C_4 plants are an exception since their metabolic CO_2 concentration mechanism allows photosynthesis to proceed near to CO_2 saturation even in normal air. However, under conditions where the stomatal conductance is low (such as is the case when the plant is under water stress) the intercellular CO_2 pressure falls to a low level, and photosynthesis is also markedly CO_2-limited in C_4 plants.

For reasons discussed in Chap. 9, the extent of photosynthetic enhancement by CO_2 enrichment strongly depends on environmental conditions. For any given C_3 plant the enhancement is smallest at low temperature and continuously increases with increased temperature (Fig. 9.18). As a result, the photosynthetic temperature optimum shifts upward when the CO_2 pressure is increased. One would thus predict the beneficial effect of a future increase in CO_2 to be least in cool and greatest in warm climates. It also follows that a warming effect of increased CO_2 on the climate would be complemented by an upward shift of the temperature dependence curve of photosynthesis.

Photosynthetic enhancement by CO_2 enrichment is also dependent on light so that the enhancement is maximal at saturating light levels. However, as shown in Fig. 9.16b, in C_3 plants CO_2 enhancement of photosynthesis in C_3 plants is substantial even under completely rate-limiting light.

Present information does not permit quantitative predictions of the potential enhancement of the integrated daily photosynthetic activity that would result from increased CO_2 pressure. Rough estimates indicate that under conditions of unlimited water supply a doubling in CO_2 would increase photosynthesis in C_3 plants by at least 15% and at most 80% in the warm-temperature and tropical zones (Björkman, 1978). These limits may be transgressed in certain biomes. The photosynthetic activity of C_4 species would be only marginally affected under conditions where water use efficiency is inconsequential (see below).

Numerous studies have been made on the effect of CO_2 enrichment on productivity. Most of these studies have been primarily done for practical purposes with greenhouse flower crops and vegetables. Indeed, "CO_2 fertilization" is commonly used to enhance yield in commercial greenhouse crop production. The response of growth to CO_2 enrichment has also been studied for several important agricultural crops but only a few limited studies have included wild plants, and our understanding of the relationship between the effects of CO_2 on photosynthesis and productivity is fragmentary even for the most extensively studied species. Although impressive, the increase in relative growth rate generally is smaller than might be predicted on the basis of short-term photosynthesis measurements, at least under conditions where the rates of photosynthesis and growth are high even without added CO_2. This difference may be attributable in part to a change in allocation patterns, and an increase in the starch content of the leaves, perhaps caused by an insufficient capacity to translocate the increased production of photosynthate. However, partial stomatal closure in response to a sustained increase in CO_2 may prove a very important factor in explaining differences between observed increases in relative growth and those predicted on the basis of short-term photosynthesis measurements. The consequences of such stomatal responses to CO_2 pressure are discussed below.

Water Use Efficiency and Productivity. Perhaps the most important aspect of an increased CO_2 level is its effect on water use efficiency. As was mentioned in a previous section, under conditions where water is limiting:

productivity $=$ water use efficiency $\times$ water consumed.

The increased photosynthetic potential conferred by an increased atmospheric CO_2 level may be used by the plant to increase the rate of photosynthesis while maintaining the same rate of water loss. In this case the water use efficiency will increase and more dry matter will be produced over a given period of time. Alternatively, the plant may respond to an increased CO_2 level by maintaining a constant photosynthetic rate while decreasing the rate of water loss. Also in this case the water use efficiency will increase and, in a water limited environment, productivity will increase as a result of an extension of the growing period.

The two contrasting modes of response to an increased CO_2 level are illustrated in Figs. 10.20 and 10.21. In Fig. 10.20 it is assumed that stomatal conductance and hence transpiration rate is unaffected by an increased atmospheric CO_2 pressure. Photosynthesis and hence the water use efficiency increase in response to the increased intercellular CO_2 pressure in a curvilinear fashion. Doubling of the present ambient CO_2 pressure approximately doubles the water use efficiency of C_3 plants while an increase in CO_2 to three times the present level causes a 2.4-fold increase in water use efficiency. As expected, CO_2 enrichment has only a small effect on the water use efficiency of the C_4 plant and at a CO_2 pressure three times the present, the C_3 plant has at least as high water use efficiency as the C_4 plant. This is in contrast to the response shown in Fig. 10.21, where it is assumed that stomatal conductance responds so that the intercellular CO_2 pressure, and hence photosynthesis, remain constant when the ambient CO_2 pressure is increased. In this case, the water use efficiencies of both the C_3 and the C_4 plants increase linearly with increased CO_2 pressure. A doubling of the CO_2 pressure causes an almost 4-fold increase in water efficiency of the C_3 plant and a 2.4-fold increase in that of the C_4 plant. Although the relative advantage in water use efficiency of the C_4 in comparison with the C_3 plant declines as the CO_2 is increased, the C_4 plant still remains superior at all CO_2 pressures.

There is considerable evidence from short-term experiments that stomatal conductance tends to decrease when the CO_2 pressure is increased. An example of the actual response to CO_2 of photosynthesis and stomatal conductance of a C_3 species *(Larrea divaricata)* is shown in Fig. 10.22. The response is inter-mediate between the extremes shown in Figs. 10.20 and 10.21. While the stomatal conductance declined as the CO_2 pressure was increased, the stomatal adjustment was not sufficient to maintain a constant intercellular CO_2 pressure. Consequently, increased CO_2 caused a simultaneous increase in photosynthetic rate and a reduction in transpiration rate. As a result the water use efficiency (P/T) increased linearly with CO_2. A doubling in CO_2 pressure increased photosynthesis by a factor of 1.8 and water use efficiency by a factor of 2.5. A 3-fold increase in CO_2 resulted in a 2.2-fold increase in photosynthesis and a 4-fold increase in water use efficiency.

In similar experiments with the C_3 species *Nerium oleander* grown at a 30° day/20 °C night regime, doubling the CO_2 level during growth resulted in a 50% higher photosynthetic rate, 25% lower transpiration and a doubling of the water use

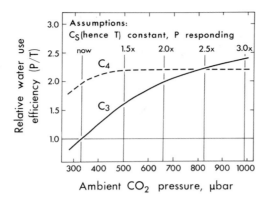

Fig. 10.20. Predicted relationship between water use efficiency and ambient CO_2 pressure for a C_3 plant (*Larrea divaricata*) and a C_4 plant (*Tidestromia oblongiolia*) at 40 °C, assuming that stomatal conductance is independent of CO_2 pressure. Calculations are based on experimentally determined relationships between photosynthetic rate and intercellular CO_2 pressures. (O. Björkman, in preparation)

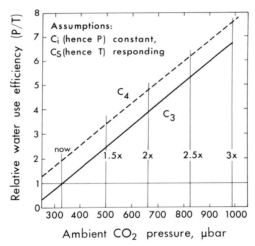

Fig. 10.21. Predicted relationship between water use efficiency and ambient CO_2 pressure for the same plants as in Fig. 10.20, but assuming that stomatal conductance adjusts to an increased CO_2 pressure so that the intercellular CO_2 pressure remains constant. (O. Björkman, in preparation)

efficiency (O. Björkman, in preparation). These differences were maintained over a period of at least two months at the two CO_2 levels and there was no evidence that the differences were diminishing with time. Wong (1979) found that in cotton plants grown with ample nitrogen nutrient, increasing the CO_2 from 330 to 640 μbar caused a 50% increase in photosynthetic rate, a 25% decrease in transpiration rate and a doubling in water use efficiency, just as in *N. oleander*. The increase in photosynthetic rate due to CO_2 enrichment was smaller when the cotton plants were

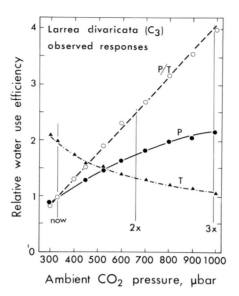

Fig. 10.22. Experimentally determined responses of photosynthesis, transpiration and water use efficiency to short-term increases in ambient CO_2 pressure for the C_3 species *Larrea divaricata*. (O. Björkman, in preparation)

grown at low concentrations of nitrogen nutrient but a doubling of the CO_2 level nevertheless doubled the water use efficiency also in these plants. As expected, increasing the CO_2 pressure during the growth of maize (a C_4 plant) caused only a small increase in photosynthetic rate but water use efficiency doubled when the CO_2 pressure was raised from 330 to 640 µbar. The water use efficiency of maize (C_4) was twice that of the cotton plants (C_3) at any given CO_2 pressure (Wong, 1979). Similar results were obtained with *Tidestromia oblongifolia* (C_4) grown at a 45°/32 °C regime (O. Björkman, in preparation). A threefold enrichment in CO_2 had no significant effect on either the photosynthetic rate or the relative growth rate but caused a strong reduction in transpiration rate and hence greatly improved the water use efficiency.

These results are in agreement with the general observation that stomata of C_4 plants may be especially responsible to increases in CO_2 pressure. Moreover, there is evidence that the CO_2 sensitivity of the stomata of a given plant is influenced by other environmental factors, especially water relations. Short-term experiments indicate that stomatal sensitivity to CO_2 increases with decreasing water potential (e.g., Raschke, 1976). This and other evidence obtained in experiments conducted at the scale 10^3–10^6 s suggest that plants may be able to take full advantage of an increase in CO_2 to maximize water use efficiency and productivity in water-limited environments. Longer-term experiments are needed to evaluate these plant responses and in particular to determine how plants manage their potential for increased productivity.

Gifford (1977) studied the growth of wheat to maturity in air enriched to 600 µbar CO_2 under conditions of adequate nutrition and water supply. He found that primary productivity was greatly enhanced. However, the developing grain was the tissue most able to respond to improved supply of assimilate throughout the growth period. Grain production was enhanced by 44% in these experiments. Subsequently it was shown that when water supply was limiting, the effect of increasing the CO$_2$ concentration to 600 μbar on the grain yield of wheat was much greater than when water was fully available (Gifford, 1979). In the minimal water treatment no grain was produced at 330 μbar but yet at elevated CO$_2$, grain was produced and yield was about 20% of that in well-watered treatments. Water use efficiency for overall growth was greatest in the low water treatments and was increased almost 50% by elevated CO$_2$ concentration. In well-irrigated treatments CO$_2$ enrichment increased water use efficiency by only about 25% (Gifford, 1979).

If extrapolations of these results obtained from a limited number of species are valid on the scale 10^7–10^{10} s, then the future rise in atmospheric CO$_2$ may well enable plants to occupy habitats which in the absence of CO$_2$ enrichment would be too arid to support plant life, to permit a much denser vegetation in semi-arid regions that would otherwise be only sparsely vegetated, or to greatly extend the duration of the growing season in regions characterized by seasonal drought. The impact of elevated CO$_2$ concentration on vegetative production and on reproductive capacity in wild plants may well alter the relationships between species in different habitats. We seem to be confronted with an inevitable increase in atmospheric CO$_2$ concentration and with an experiment of grand proportions. This experiment is likely to tax our predictive ingenuity, and to teach us a good deal about integrated physiological processes as the basis of plant ecology.

Epilog

"As the hypotheses have to depend on assumptions, the more implausible these appear to be, dare I say, the more stimulated we are to contrive further experiments, for there is still a deal to worry about. In the end, when everything is settled, few of us perhaps will really desire to look back on it all."

(Hill, 1965)

As we said in the Introduction, a principal aim of this book is to illustrate how far physiology and ecology, including their sometimes apparently disparate methodologies and underlying philosophies, might be integrated to provide a coherent appraisal of the biology of a group of organisms. A critical reader may question how much of an integration we have provided in attempting to present our *Atriplex* scenario, and thus in this brief epilog we comment on some aspects of our original objectives.

One of our original objectives was to illustrate the importance of recognizing the relationships between state and process (the latter defined as a sequence of causal transitions between states) and scale-dependence, (in both space and time) in achieving an integrated approach to biological and particularly ecophysiological research. While we have demonstrated to our satisfaction that explanations of biological phenomena are essentially scale-dependent, we have not always found ourselves comfortable when analyzing processes in the sense we have defined them above. This difficulty arises from the fact that, although we can sometimes recognize sequential changes from states (as in, for example, plant growth or ecological successions), our recognition of these sequences does not of itself say anything about causation.

We may most easily categorize two concepts of "process" by saying that the term has been used to define both a sequence of states (as in the example of ecological succession) and a sequence of interactions that determine a corresponding sequence of states (as in the sequential biochemical steps of the photosynthetic "process"). Thus while it is abundantly clear from several examples presented in earlier chapters that ecological "processes" can be described from time-based observations of state or inferred from age-stratified spatial data, the analysis of "process" in the stricter sense of our definition depends on appropriate experimental manipulation. That this is the essence of physiological ecology was stated emphatically by Schimper (1903): "The ecology of plant distribution will succeed in opening out new paths only on condition that it leans closely on experimental physiology".

It is evident that most of larger-scale ecological endeavor cannot lend itself to experimental analysis or verification in a traditional laboratory-based context. It is on these larger scales that other methodological strategies, such as serial correlation (e.g., Usher, 1973; Pielou, 1974), transition matrices (e.g., Horn, 1976) or pattern analysis in the wider sense (e.g., Williams, 1976) come into their own, providing techniques for the analysis of "natural experiments" designed on the basis of field-gathered data. Although these numerical approaches may provide directly valuable ecological information for particular purposes (e.g., in population management, land-use planning, etc.), their special utility to a physiological ecologist is the hypotheses they generate, linking states by processes, and the evidence they provide for different behavior and alternative processes in species populations within a defined ecological system. The central purpose of studying complex systems is not to seek generalities which are true of all systems or all species, but rather to search for the causes of different behavior between different species or systems (Levins, 1968).

It is this last point which brings us to the crux of this book and indeed what we believe to be the hub of physiological ecology; namely that a prime objective of this discipline is to seek a better understanding of individual organisms in relation to their environment – in short, to better understand adaptation.

As Harper (1977) has recently emphasized, the demonstration of adaptation requires that an organism be shown to be relatively more fit than its potential competitors, and that this relative fitness – defined as the numbers of descendants left by an individual relative to its fellows – can be determined by any activity that confers a competitive advantage in a particular environmental context. This insistence on seeking evidence of adaptation through criteria of individual rather than population fitness underlies the importance we attach to the study of different behavior between individual organisms at an appropriate environmental scale in ecophysiological research programs.

Harper's definition of adaptation is a good deal more demanding than that which we adopted in the Introduction. The emphasis there was on the ways components of biological systems are fitted together in particular habitats, and the ways they respond to particular environmental perturbations. This emphasis reflects our preoccupation as physiologists with performance response functions, rather than our concern as population biologists with survival response functions. In this book we may fairly claim to have explored many examples of adaptation in the former sense, but can we really point to many proven examples of the latter?

We would have to concede that evidence of adaptation is acceptable only when an organism can be shown to be better fitted, by virtue of its physiological processes, to survive in its environment than either its progenitors or its present competitors. The adaptation could be expressed in the physiological processes which enable it to perform and grow so as to exclude competitors, or to reproduce more successfully. The adaptation may be expressed in the physiological processes which enable the organism to persist and reproduce in extreme environments. Strictly speaking, to demonstrate adaptation we must provide evidence of superior fitness of one genotype compared with some alternative genotype or model. Rarely in physiological plant ecology are comparisons made with the unsuccessful genotypes under the same environmental conditions. Although the novel physiological processes found in organisms from different habitats have been widely explored, rarely have consequences of these attributes been assessed in comparative experiments of performance and survival. Clearly, there is still a deal to worry about.

On this basis then we argue that physiological ecologists should concentrate their attention on those processes that determine, and provide evidence of fitness in the strict sense. This amounts to stressing again the importance that we attach to the experimental approaches adopted by Turesson, Hall and Clements and Clausen and Hiesey, which were described in Chap. 3. For all that has been said here and elsewhere about the adaptive significance of the C_4 pathway of photosynthesis, for example, the reciprocal transplant experiments with C_3 and C_4 *Atriplex* spp. described in Chap. 10 are the only tests of adaptation of C_3 and C_4 plants, in the strict sense, so far conducted.

If atmospheric carbon dioxide levels continue to increase as predicted, and if we prepare the correct state descriptions, we have an unparalleled opportunity to assess many wider questions of photosynthetic processes as a basis for adaptation. This is but one of many grand experiments that man is arranging for his future. We may, in our lifetime, have the opportunity to more fully appreciate Harper's comment that "adaptation is always retrospective". Our insistence throughout this book on experimental evidence for adaptation was motivated by a suspicion that among practicing physiological ecologists, adaptation is frequently extrapolative. We may have protested too much, but if we are not stimulated to contrive further experiments, this suspicion may well be confirmed to our discredit.

In concluding this book it is appropriate to pay some compliment to the genus *Atriplex,* which has enigmatically allowed us to paint many of its leaves but has yet to give us more than a glimpse of the whole conceptual tree. According to Levins (1968) there are at least seven ways in which a taxon can maximize fitness in an uncertain environment; through ecotypic differentiation, genetic polymorphism, maintaining a high degree of genetic variance, formation of distinct breeding groups, development of somatic polymorphism, evolution of phenotypic plasticity, and heterozygotic superiority. It is clear that the genus *Atriplex* has employed most of these devices in achieving its evident ecological success in many parts of the world.

References

Abel GH, MacKenzie AJ (1964) Salt tolerance of soybean varieties during germination and later growth. Crop Sci 4:157–160

Adams FW (1974) Soil solution. In: Carson EW (ed) The plant root and its environment. University Press of Virginia, Charlottesville, pp 441–481

Aellen P (1940a) Die *Atriplex*-Arten des Orients. Bot Jahrb 70:1–66

Aellen P (1940b) *Atriplex* und *Blackiella* in Süd Afrika. Bot Jahrb 70:383–401

Aellen P (1960) Chenopodiaceae. In: Heigi M (ed) Illustrierte Flora von Mitteleuropa, B 3(2). Carl Hauser Verlag, München, pp 664–693

Aellen P (1968) Drei neue *Atriplex*-Arten aus Argentinien. Dansk Bot Ark 22:179–185

Aellen P, Hulme BA (1964) *Atriplex*. In: Tutin TG, Heywood VH, Burges NA, Valentine DH, Walters SM, Webb DA (eds) Flora Europea, vol I. Cambridge University Press, Cambridge, pp 95–97

Alberte RS, Thornber JP (1977) Water stress effects on the content and organization of chlorophyll in mesophyll and bundle-sheath chloroplasts of maize. Plant Physiol 59:351–353

Alieva SA, Tageeva SV, Tairbekov MG, Kasatkina VS, Vagabova ME (1971) Structural and functional conditions of the chloroplasts as a function of water regime. Sov Plant Physiol 18:416–422

Allan HH (1961) Flora of New Zealand. RE Owen, Wellington

Anderson DJ (1967) Studies on structure in plant communities. V. Pattern in *Atriplex vesicaria* communities in south-eastern Australia. Aust J Bot 15:451–458

Anderson DJ (1970) Analysis of pattern in *Atriplex vesicaria* communities from the Riverine Plain of New South Wales. In: Jones R (ed) Biology of *Atriplex*. CSIRO, Canberra, pp 63–68

Anderson DJ (1971a) Pattern in desert perennials. J Ecol 59:555–560

Anderson DJ (1971b) Spatial patterns in some Australian dryland plant communities. Int Symp Stat Ecol 1:271–286

Anderson DJ (1971c) Hierarchies and integration in ecology. Proc Ecol Soc Aust 6:1–6

Anderson DJ, Jacobs SWL, Malik AR (1969) Studies on structure in plant communities. VI. The significance of pattern evaluation in some Australian dry-land vegetation types. Aust J Bot 17:315–322

Anderson DJ, Perry RA, Leigh JH (1972) Some perspectives on shrub/environment interactions. In: McKell CM, Blaisdell JP, Goodin JR (eds) Wildland shrubs – their biology and utilization. USDA Technical Report INT-1, Ogden, pp 172–181

Anderson RH (1930) Notes on the Australian species of the genus *Atriplex*. Proc Linn Soc NSW 55:493–505

Anderson WP, Willcocks DS, Wright BJ (1977) Electrophysical measurements on the root of *Atriplex hastata*. J Exp Bot 28:894–901

Antonovics J, Bradshaw AD, Turner RG (1971) Heavy metal tolerance in plants. Adv Ecol Res 7:1–85

Armond PA, Staehelin LA, Arntzen CJ (1977) Spatial relationship of photosystem I, photosystem II, and the light harvesting complex in chloroplast membranes. J Cell Biol 73:400–418

Armond PA, Schreiber U, Björkman O (1978) Photosynthetic acclimation to temperature in the desert shrub, *Larrea divaricata*. II. Light-harvesting effeciency and electron transport. Plant Physiol 61:411–415

Ashby WC, Beadle NCW (1957) Studies on halophytes III. Salinity factors in the growth of Australian saltbushes. Ecology 38:344–352

Atkinson MR, Findlay GP, Hope AB, Pitman MG, Saddler HWD, West KR (1967) Salt regulation in the mangroves *Rhizophora mucronata* Lam. and *Aegialitis annulata* R.Br Aust J Biol Sci 20:589–599

Austenfeld F-A (1974) Der Einfluß des NaCl und anderer Alkalisalze auf die Nitratredukta-seaktivität von *Salicornia europea* L. Z Pflanzenphysiol 71:288–296

Baes DF, Goeller HE, Olson JS, Rotty RM (1977) Carbon dioxide and climate: the uncontrolled experiment. Am Sci 65:310–320

Baker DA, Hall JL (eds) (1975) Ion transport in plant cells and tissues. North Holland, Amsterdam

Baker HG (1965) Characteristics and modes of origin of weeds. In: Baker HG, Stebbins GL (eds) The genetics of colonizing species. Academic Press, London New York, pp 147–168

Baker HG (1974) The evolution of weeds. Ann Rev Ecol Syst 5:1–24

Ballantine JEM, Forde BJ (1970) The effect of light intensity and temperature on plant growth and chloroplast ultrastructure in soybean. Am J Bot 57:1150–1159

Barber DA (1968) Micro-organisms and the inorganic nutrition of higher plants. Ann Rev Plant Physiol 19:71–88

Barber SA (1974) Influence of the plant root on ion movement in the soil. In: Carson EW (ed) The plant root and its environment. University Press of Virginia, Charlottesville, pp 525–564

Barbour MG (1970) Is any angiosperm an obligate halophyte. Am Midl Nat 84:105–120

Barbour MG (1973) Desert dogma re-examined: root/shoot productivity and plant spacing. Am Midl Nat 89:41–57

Barbour MG (1978a) Salt spray as a microenvironmental factor in the distribution of beach plants at Point Reyes, California. Oecologia 32:213–224

Barbour MG (1978b) The effect of competition and salinity on the growth of a saltmarsh plant species. Oecologia 37:93–99

Barbour MG, De Jong TM (1977) Response of West Coast beach taxa to saltspray, seawater inundation and soil salinity. Bull Torrey Bot Club 104:29–34

Barbour MG, De Jong, TM, Johnson AF (1976) Synecology of beach vegetation along the Pacific coast of the United States of America: a first approximation. J Biogeogr 3:55–69

Barker S, Lange RT (1970) Population ecology of *Atriplex* under sheep stocking. In: Jones R (ed) The biology of *Atriplex*. CSIRO, Canberra, pp 105–120

Barley KP (1970) The configuration of the root system in relation to nutrient uptake. Adv Agron 22:159–201

Barrs HD (1968) Determination of water deficits in plant tissues. In: Kozlowski TT (ed) Water deficits and plant growth I. Academic Press, London New York, pp 236–368

Barrs HD (1973) Controlled environment studies of the effects of variable atmospheric water stress on photosynthesis, transpiration, and water status of *Z. mays* L. and other species. In: Slatyer RO (ed) Plant response to climatic factors (Proc Uppsala Symp 1970). UNESCO, Paris, pp 249–258

Bassett TJ (1969) IOPB chromosome number report. Taxon 18:310

Bassett TJ, Crompton CW (1973) The genus *Atriplex* (Chenopodiaceae) in Canada and Alaska. III. Three hexaploid annuals: *A.subspicata*, *A.gmelinii*, and *A.alaskensis*. Can J Bot 51:1715–1723

Bassham JA, Calvin M (1962) The photosynthesis of carbon compounds. WA Benjamin, New York

Beadle NCW (1948) The vegetation and pastures of New South Wales. Government Printer, Sydney

Beadle NCW (1952) Studies on halophytes. I. The germination of the seeds and establishment of the seedlings of five species of *Atriplex* in Australia. Ecology 33:49–62

Beadle NCW, Tchan YT (1955) Nitrogen economy in arid and semi-arid plant communities. I. The environment and general considerations. Linn Soc NSW Proc 80:97–104

Beadle NCW, Whalley RDB, Gibson JB (1957) Studies on halophytes II. Analytic data on the mineral constituents of three species of *Atriplex* and their accompanying soils in Australia. Ecology 38:340–344

Beer S, Shomer-Ilan A, Waisel Y (1975) Salt-stimulated phosphoenolpyruvate carboxylase in *Cakile maritima*. Physiol Plant 34:293–295

Begg JE, Torssell BWR (1974) Diaphotomastic and parahelionastic leaf movements in *Stylosanthes humilis* H.B.K. (Townsville stylo) In: Bieleski RL, Ferguson AR, Cresswell MM (eds) Mechanisms of regulation of plant growth. Royal Society of New Zealand, Wellington, pp 277–283

Begg JE, Turner NC (1976) Crop water deficits. Adv Agron 28:161–217

Belford HS, Thompson WF (1976) DNA sequence comparisons in *Atriplex*. Carnegie Inst Wash Yearb 75:362–367

Belford HS, Thompson WF (1977) Single-copy DNA sequence comparisons in *Atriplex*. Carnegie Inst Wash Yearb 76:246–252

Belford HS, Thompson WF (1979) Single copy DNA homologies and the phylogeny of *Atriplex*. Carnegie Inst Wash Yearb 78:217–223

Bender MM (1968) Mass spectrometric studies of carbon-13 variations in corn and other grasses. Radiocarbon 10:468–472

Benedict CR (1978) Nature of obligate photoautotrophy. Ann Rev Plant Physiol 29:67–93

Bennert WH, Mooney HA (1979) The water relations of some desert plants in Death Valley, California. Flora 168:405–427

Bentham G, Hooker JD (1924) Handbook of the British flora. 7th edn. revised by Rendle AR. Reeve, Ashford

Bernstein L (1961) Osmotic adjustment of plants to saline media. I. Steady state. Am J Bot 48:909–918

Berry J, Farquhar GD (1978) The CO_2 concentrating function of C_4 photosynthesis. A biochemical model. In: Hall DO, Coombs J, Goodwin TW (eds) Photosynthesis 77. The Biochemical Society, London, pp 119–131

Berry JA, Fork DC, Garrison S (1975) Mechanistic studies of thermal damage to leaves. Carnegie Inst Wash Yearb 74:751–760

Berry JA, Osmond CB, Lorimer GH (1978) Fixation of $^{18}O_2$ during photorespiration: Kinetic and steady state studies of the photorespiratory carbon oxidation cycle with intact leaves and isolated chloroplasts of C_3 plants. Plant Physiol 62:954–967

Billings WD (1949) The shadscale vegetation of Nevada and eastern California in relation to climate and soils. Am Midl Nat 42:87–109

Billings WD, Morris RJ (1951) Reflection of visible and infrared rediation from leaves of different ecological groups. Am J Bot 38:327–331

Binet P (1965) Etude des quelques aspects physiologiques de la germination chez *Atriplex tornabeni* Tin. Bull Soc Bot Nord Fr 18:40–55

Binet P (1966) Proprietes physiologiques fondamentales des semences d'*Atriplex babingtonii* Woods. Bull Soc Bot Nord Fr 19:121–137

Binet P (1969) Action des complexes formées par les "hautes de plage" et les "dunes embryonnaires" sur *Atriplex babingtonii* Woods. Oecol Plant 2:63–78

Bisalputra T (1960) Anatomical and morphological studies in the Chenopodiaceae. I. Inflorescence of *Atriplex* and *Bassia*. Aust J Bot 8:226–243

Bjerregaard RS (1971) The nitrogen budget of two salt desert shrub plant communities of western Utah. PhD Thesis. Utah State University, Logan

Björkman O (1966a) Carboxydismutase activity in relation to light-saturated rate of photosynthesis in plants from exposed and shaded habitats. Carnegie Inst Wash Yearb 65:454–459

Björkman O (1966b) The effect of oxygen concentration on photosynthesis in higher plants. Physiol Plant 19:618–633

Björkman O (1968a) Carboxydismutase activity in shade-adapted and sun-adapted species of higher plants. Carnegie Inst Wash Yearb 67:487–488

Björkman O (1968b) Further studies on differentiation of photosynthetic properties in sun and shade ecotypes of *Solidago virgaurea* L. Physiol Plant 21:84–89

432   References

Björkman O (1971a) Interaction between the effects of oxygen and CO_2 concentration on quantum yield and light-saturated rate of photosynthesis in leaves of *Atriplex patula* ssp. *spicata*. Carnegie Inst Wash Yearb 70:520–626

Björkman O (1971b) Comparative photosynthetic CO_2 exchange in higher plants. In: Hatch MD, Osmond CB, Slatyer RO (eds) Photosynthesis and photorespiration. Wiley Interscience, New York, pp 18–32

Björkman O (1973) Comparative studies on photosynthesis in higher plants. In: Giese A (ed) Current topics in photobiology, photochemistry, and photophysiology, vol 8. Academic Press, London New York, pp 1–63

Björkman O (1975) Environmental and biological control of photosynthesis: inaugural address. In: Marcelle R (ed) Environmental and biological control of photosynthesis. Junk, The Hague, pp 1–16

Björkman O (1976) Adaptive and genetic aspects of C_4 photosynthesis. In: Burris RH, Black CC (eds) CO_2 metabolism and plant productivity. University Park Press, Baltimore, pp 287–309

Björkman O (1978) Response of the biota to increased carbon dioxide. Research Development Paper for the US Department of Energy, Washington DC

Björkman O, Badger M (1977) Thermal stability of photosynthetic enzymes in heat- and cool-adapted C_4 species. Carnegie Inst Wash Yearb 76:346–354

Björkman O, Gauhl E (1969) Carboxydismutase activity in plants with and without β-carboxylation photosynthesis. Planta 88:197–203

Björkman O, Holmgren P (1963) Adaptability of the photosynthetic apparatus to light intensity in ecotypes from exposed and shaded habitats. Physiol Plant 16:889–914

Björkman O, Holmgren P (1966) Photosynthetic adaptation to light intensity in plants native to shaded and exposed habitats. Physiol Plant 19:854–859

Björkman O, Ludlow MM (1972) Characterization of the light climate on the floor of a Queensland rainforest. Carnegie Inst Wash Yearb 71:85–94

Björkman O, Pearcy RW (1971) Effect of growth temperature on the temperature dependence of photosynthesis *in vivo* and on CO_2 fixation by carboxydismutase *in vitro* in C_3 and C_4 species. Carnegie Inst Wash Yearb 70:511–520

Björkman O, Gauhl E, Nobs MA (1970) Comparative studies of *Atriplex* species with and without β-carboxylation photosynthesis. Carnegie Inst Wash Yearb 68:620–633

Björkman O, Nobs MA, Berry JA (1971) Further studies on hybrids between C_3 and C_4 species of *Atriplex*. Carnegie Inst Wash Yearb 70:507–511

Björkman O, Boardman NK, Anderson JM, Thorne SW, Goodchild DJ, Pyliotis NA (1972a) Effect of light intensity during growth of *Atriplex patula* on the capacity of photosynthetic reactions, chloroplast components, and structure. Carnegie Inst Wash Yearb 71:115–135

Björkman O, Pearcy RW, Harrison AT, Mooney HA (1972b) Photosynthetic adaptation to high temperatures: a field study in Death Valley, California. Science 172:786–789

Björkman O, Mahall B, Nobs MA, Ward W, Nicholson F, Mooney HA (1974a) Growth responses of plants from contrasting thermal environments. An analysis of the temperature dependence of growth under controlled conditions. Carnegie Inst Wash Yearb 73:757–767

Björkman O, Nobs M, Mooney HA, Troughton J, Berry J, Nicholson F, Ward W (1974b) Growth responses of plants from habitats with contrasting thermal environments. Transplant studies in the Death Valley and the Bodega Head experimental gardens. Carnegie Inst Wash Yearb 73:748–757

Björkman O, Troughton J, Nobs MA (1974c) Photosynthesis in relation to leaf structure. In: Basic mechanisms in plant morphogenesis. Brookhaven Symp Biol 25:206–226

Björkman O, Mooney HA, Ehleringer J (1975) Photosynthetic responses of plants from habitats with contrasting thermal environments: Comparison of photosynthetic characteristics of intact plants. Carnegie Inst Wash Yearb 74:743–748

Björkman O, Boynton J, Berry J (1976) Comparison of the heat stability of photosynthesis, chloroplast membrane reactions, photosynthetic enzymes, and soluble protein in leaves of heat-adapted and cool-adapted C_4 species. Carnegie Inst Wash Yearb 75:400–407

Björkman O, Badger M, Armond PA (1978) Thermal acclimation of photosynthesis: Effect of growth temperature on photosynthetic characteristics and components of the photosynthetic apparatus in *Nerium oleander*. Carnegie Inst Wash Yearb 77:262–282

Björkman O, Badger M, Armond O (1979) Response and adaptation of photosynthesis to high temperatures. In: Turner NC, Kramer PJ (eds) (1979) Stress physiology. Academic Press, London New York (in press)

Black CA (1968) Soil-plant relationships. Wiley, New York

Black JM (1948) Flora of South Australia, Part II. Casuarinaceae-Euphorbiaceae, 2nd edn. Government Printer, Adelaide, pp 255–521

Black RF (1954) The leaf anatomy of Australian members of the genus *Atriplex*. I. *Atriplex vesicaria* Heward and *A.nummularia* Lindl. Aust J Bot 2:269–286

Black RF (1956) Effect of NaCl in water culture on the ion uptake and growth of *Atriplex hastata*, L. Aust J Biol Sci 9:67–80

Black RF (1958) The effect of sodium chloride on leaf succulence and area of *Atriplex hastata* L. Aust J Bot 6:306–321

Black RF (1960) Effects of NaCl on the ion uptake and growth of *Atriplex vesicaria* Heward. Aust J Biol Sci 13:249–266

Blackman GE (1956) Influence of light and temperature on leaf growth. In: Milthorpe FL (ed) The growth of leaves. Butterworths, London, pp 151–169

Blackman GE (1961) Responses to environmental factors by plants in the vegetative phase. In: Zarrow MX (ed) Growth in living systems. Basic Books, New York, pp 525–526

Blackman GE (1968) The application of the concepts of growth analysis to the assessment of productivity. In: Eckardt F (ed) Functioning of terrestrial ecosystems at the primary productivity level. UNESCO, Paris, pp 243–260

Blackman GE, Black JN (1959) Physiological and ecological studies in the analysis of plant environment. XII. The role of the light factor in limiting growth. Ann Bot 23:131–145

Blackman GE, Wilson GL (1951a) Physiological and ecological studies in the analysis of plant environment. VI. The constancy for different species of a logarithmic relationship between net assimilation rate and light intensity and its ecological significance. Ann Bot NS 15:63–94

Blackman GE, Wilson GL (1951b) Physiological and ecological studies of plant environment. VII. An analysis of the differential effects of light intensity on the net assimilation rate, leaf-area ratio, and relative growth rate of different species. Ann Bot NS 15:373–408

Blauer AC, Plummer AP, McArthur ED, Stevens R, Giunta BC (1976) Characteristics and hybridization of important intermountain shrubs. II. Chenopod family. US Dept. Agric. Forest Service Research Paper INT-177, 42p. Intermountain Forest and Range Experiment Stations, Ogden, Utah

Blevins DG, Hiatt AJ, Lowe RH (1974) The influence of nitrate and chloride uptake on expressed sap pH, organic acid synthesis, and potassium accumulation in higher plants. Plant Physiol 54:82–87

Bloom AJ, Troughton JH (1979) High productivity and photosynthetic flexibility in a CAM plant. Oecologia 38:35–43

Blum A, Sullivan CY (1972) A laboratory method for monitoring net photosynthesis in leaf segments under controlled water stress experiments with *Sorghum*. Photosynthetica 6:18–23

Blumenthal-Goldsmith S, Poljakoff-Mayber J (1968) Effect of substrate salinity on growth and on sub-microscopic structure of leaf cells of *Atriplex halimus* L. Aust J Bot 16:469–478

Boag S, Brownell PF (1979) Properties of photosynthesis in Na deficient C_4 plants. Aust J Plant Physiol 6:431–434

Boardman NK (1977) Comparative photosynthesis of sun and shade plants. Ann Rev Plant Physiol 28:355–377

Bottrill DE, Possingham JV, Kriedemann PE (1970) The effect of nutrient deficiencies on photosynthesis and respiration in spinach. Plant Soil 32:424–438

Boucard J, Ungar IA (1976) Hormonal control of germination under saline conditions of three halophytic taxa in the genus *Suaeda*. Physiol Plant 37:143–148

Bowen GD, Cartwright B (1977) Mechanisms and models of plant nutrition. In: Russell JS, Greacen GL (eds) Soil factors in crop production in a semi arid environment. University of Queensland Press, St Lucia, pp 197–223

Boyer JS (1970) Differing sensitivity of photosynthesis to low leaf water potentials in corn and soybean. Plant Physiol 46:236–239

Boyer JS (1971) Nonstomatal inhibition of photosynthesis in sunflower at low leaf water potentials and high light intensities. Plant Physiol 48:532–536

Boyer JS, Bowen BL (1970) Inhibition of oxygen evolution in chloroplasts isolated from leaves with low water potentials. Plant Physiol 45:612–615

Boyer JS, Potter JR (1973) Chloroplast response to low leaf water potentials. I. Effect of turgor. Plant Physiol 51:989–992

Boynton J, Nobs MA, Björkman O, Pearcy RW (1970) Hybrids between *Atriplex* species with and without β-carboxylation photosynthesis. Leaf anatomy and ultrastructure. Carnegie Inst Wash Yearb 69:629–632

Brian RC, Cattlin ND (1968) The surface structure of leaves of *Chenopodium album* L. Ann Bot 32:609–612

Briese DT (1974) Ecological studies on an ant community in a semiarid habitat (with emphasis on seed harvesting species). PhD Thesis. Australian National University, Canberra

Briggs GE (1967) Movement of water in plants. Blackwells Scientific Publications, Oxford

Briggs GE, Hope AB, Robertson RN (1961) Electrolytes and plant cells. Blackwell Scientific Publications, Oxford

Brooking JR, Taylor AO (1973) Plants under climatic stress. V. Chilling and light effects on radiocarbon exchange between photosynthetic intermediates of *Sorghum*. Plant Physiol 52:180–182

Brown RH (1976) Characteristics related to photosynthesis and photorespiration of *Panicum milioides*. In: Burris RH, Black CC (eds) CO_2 metabolism and plant productivity. University Park Press, Baltimore, pp 311–325

Brown RH (1978) A difference in the N use efficiency in C_3 and C_4 plants and its implications in adaptation and evolution. Crop Sci 18:93–98

Brown RH, Brown WV (1975) Photosynthetic characteristics of *Panicum milioides* a species with reduced photorespiration. Crop Sci 15:681–685

Brownell PF (1965) Sodium as an essential micronutrient element for a higher plant *(Atriplex vesicaria)*. Plant Physiol 40:460–468

Brownell PF (1968) Sodium as an essential micronutrient element for some higher plants. Plant Soil 28:161–164

Brownell PF, Crossland CJ (1972) The requirement for sodium as a micronutrient by species having the C_4 dicarboxylic photosynthetic pathway. Plant Physiol 49:794–497

Brownell PF, Crossland CJ (1974) Growth responses to sodium by *Bryophyllum tubiflorum* under conditions inducing crassulacean acid metabolism. Plant Physiol 54:416–417

Brownell PF, Wood JG (1975) Sodium as an essential micronutrient element for *Atriplex vesicaria* Heward. Nature (London) 179:635–636

Buffington LC, Herbel CH (1965) Vegetational changes on a semi-desert grassland range from 1858 to 1963. Ecol Monogr 35:139–164

Bunce JA (1977) Nonstomatal inhibition of photosynthesis at low water potentials of intact leaves from a variety of habitats. Plant Physiol 59:348–350

Burbidge NT (1945) Germination studies of Australian chenopodiaceae with special reference to the conditions necessary for regeneration I. *Atriplex vesicaria* Heward. Trans R Soc South Aust 69:73–84

Burbidge NT (1960) The phytogeography of the Australian region. Aust J Bot 8:72–211

Burrows WH (1972) Productivity of an arid zone shrub *(Eremophila gilesii)* community in south-western Queensland. Aust J Bot 20:317–329

Buxton GL (1967) The Riverina 1961–1891: an Australian regional study. Melbourne University Press, Melbourne

Cabrera AL (1957) La vegetacion de la Puna Argentina. In: Revisita de Investigaciones Agricolas, vol XI. Buenos Aires, p 317

Caldwell MM (1974) Physiology of desert halophytes. In: Reimold RJ, Queen WJ (eds) Ecology of halophytes. Academic Press, London New York, pp 355–378

Caldwell MM (1975) Primary production of grazing lands. In: Cooper JP (ed) Photosynthesis and productivity in different environments, IBP 3. Cambridge University Press, Cambridge, pp 41–73

Caldwell MM (1976) Root extension and water absorption. In: Lange OL, Kappen L, Schulze E-D (eds) Water and plant life: problems and modern approaches. Springer, Berlin Heidelberg New York, pp 63–85

Caldwell MM, Camp LB (1974) Below ground productivity of two cool desert communities. Oecologia 17:123–130

Caldwell MM, Osmond CB, Nott DL (1977a) C_4 pathway photosynthesis at low temperature in cold-tolerant *Atriplex* species. Plant Physiol 60:157–164

Caldwell MM, White RS, Moore RT, Camp LB (1977b) Carbon balance, productivity, and water use of cold-winter desert shrub communities dominated by C_3 and C_4 species. Oecologia 29:275–300

Camacho-B, Hall SE, Kaufmann MR (1974) Efficiency and regulation of water transport in some woody and herbaceous species. Plant Physiol 54:169–172

Carey R, Berry JA (1976) Effects of temperature on respiration and uptake of Rb^+ ion by roots of barley and corn. Carnegie Inst Wash Yearb 75:433–438

Carolin RC, Jacobs S, Vesk M (1975) Leaf structure in the Chenopodiaceae. Bot Jahrb Syst 95:226–255

Carrodus BB, Specht RL (1965) Factors affecting the relative distribution of *Atriplex vesicaria* and *Kochia sedifolia* in the arid zone of South Australia. Aust J Bot 13:419–433

Carter OG, Lathwell DJ (1967) Effect of chloride on phosphorus uptake by corn roots. Agron J 59:250–253

Cavaco A (1954) Chénopodiacées. In: Humbert H (ed) Flore de Madagascar et des Camores. Firmin-Didot et Cie, Paris

Chapman VJ (1960) Salt marshes and salt deserts of the world. Hill, London

Chapman VJ (1964) Coastal vegetation. Pergamon, Oxford

Charles-Edwards DA, Ludwig LJ (1974) A model for leaf photosynthesis by C_3 plant species. Ann Bot 38:921–930

Charles-Edwards DA, Ludwig JL (1975) The basis of expression of leaf photosynthetic activities. In: Marcelle R (ed). Environmental and biological control of photosynthesis. Junk, The Hague, pp 37–44

Charley JL (1959) Soil salinity – vegetation patterns in western New South Wales and their modification by overgrazing. PhD Thesis. University of New England, Armidale

Charley JL (1972) The role of shrubs in nutrient cycling. In: McKell CM, Blaisdell JP, Goodin JR (eds) Wildland shrubs – their biology and utilization. USDA Forest Service INT-1, Ogden, pp 182–203

Charley JL (1978) Mineral cycling in rangeland ecosystems. In: Sosebee RE (ed) Rangeland plant physiology. Range Sci Ser 4. Society for Range Management, Denver, pp 215–256

Charley JL, Cowling SW (1968) Changes in soil nutrient status resulting from overgrazing and their consequences in plant communities of semi-arid areas. Proc Ecol Soc Aust 3:28–38

Charley JL, McGarity JW (1964) High soil nitrate-levels in patterned saltbush communities. Nature (London) 201:1351–1352

Charley JL, West NE (1975) Plant induced chemical patterns in some shrub-dominated semi-desert ecosystems of Utah. J Ecol 63:945–964

Chatterton NJ, McKell CM (1969) *Atriplex polycarpa*. I. Germination and growth as affected by sodium chloride in water cultures. Agron J 61:448–450

Chatterton NJ, McKell CM, Goodin JR, Bingham FT (1969) *Atriplex polycarpa*: II. Germination and growth in water cultures containing high levels of boron. Agron J 61:451–453

Chatterton NJ, McKell CM, Bingham FT, Clawson WJ (1970) Absorption of Na, Cl, and B by desert saltbush in relation to composition of nutrient solution culture. Agron J 62:351–353

Chatterton NJ, Goodin JP, Duncan D (1971) Nitrogen metabolism in *Atriplex polycarpa* as affected by substrate nitrogen and NaCl salinity. Agron J 63:271–274

Ching TM (1972) Metabolism of germinating seeds. In: Kozlowski TT (ed) Seed biology, vol II. Academic Press, London New York, pp 103–208

Clapham AR, Tutin TG, Warburg EF (1962) Flora of the British Isles. University Press, Cambridge

Clarkson DT (1974) Ion transport and cell structure in plants. McGraw Hill, London

Clausen J (1922) Studies on the collective species *Viola tricolor* L. Bot Tidsskr 37:363–416

Clausen J, Hiesey WM (1958) Experimental studies on the nature of species. IV. Genetic structure of ecological races. Publ Carnegie Inst No 615, Washington DC

Clausen J, Keck DD, Hiesey WM (1940) Experimental studies on the nature of species. I. Effect of varied environments on western North American plants. Publ Carnegie Inst No 520, Washington DC

Clausen J, Keck DD, Hiesey WM (1945) Experimental studies on the nature of species. II. Plant evolution through amphiploidy and autoploidy, with examples from the Madiinae. Publ Carnegie Inst No 564, Washington DC

Clausen J, Keck DD, Hiesey WM (1948) Experimental studies on the nature of species. III. Environmental responses of climatic races of *Achillea*. Publ Carnegie Inst No 581, Washington DC

Clements RJ, Ludlow MM (1977) Frost avoidance and frost resistance in *Centrosema virginianum*. J Appl Ecol 14:551–566

Collatz GJ (1977) Influence of certain environmental factors on photosynthesis and photorespiration in *Simmondsia chinensis*. Planta 134:127–132

Collatz J, Ferrar P, Slatyer RO (1976) Effects of water stress and differential hardening treatments on photosynthetic characteristics of a xeromorphic shrub, *Eucalyptus socialis*, F. Muell. Oecologia 23:95–105

Collins GN (1901) Seeds of commercial saltbushes. USDA Div of Bot Bull, vol 27. US Gov Printing Office, Washington

Connor DJ, Tunstall BR (1968) Tissue water relations for brigalow and mulga. Aust J Bot 16:487–490

Cook MG (1973) Effects of root temperature on nutrient uptake in *Atriplex* spp. BSc Hons Thesis. University of New England, Armidale

Cooper AJ, Thornley JHM (1976) Response of dry matter partitioning, growth, and carbon and nitrogen levels in the tomato plant to changes in root temperature: Experiment and theory. Ann Bot 40:1139–1152

Cornelius DR, Hylton LO (1969) Influence of temperature and leachate on germination of *Atriplex polycarpa*. Agron J 61:209–211

Courtice A (1971) Fluctuating plant numbers in three species of *Atriplex* L. PhD Thesis. University of Sydney, Sydney

Cowan IR (1965) Transport of water in the soil-plant-atmosphere system. J Appl Ecol 2:221–239

Cowan IR (1972) An electrical analogue of evaporation from, and flow of water in plants. Planta 106:221–226

Cowan IR (1977) Stomatal behaviour and Environment. In: Preston RD, Woolhouse HW (eds) Advances in botanical research, vol IV. Academic Press, London New York, pp 117–228

Cowan IR, Farquhar GD (1977) Stomatal function in relation to leaf metabolism and environment. In: Jennings DH (ed) Integration of activity in the higher plant. SEB Symp 31. Cambridge University Press, Cambridge, pp 471–505

Cowan IR, Milthorpe FL (1968) Plant factors influencing the water status of plant tissues. In: Kozlowski TT (ed) Water deficits and plant growth, vol I. Academic Press, London New York, pp 137–193

Cowling SW (1969) A study of vegetation activity patterns in a semi arid environment. PhD Thesis. University of New England, Armidale

Cozic-Trichet J, Goas M (1969) Etude taxonomique de quelques *Atriplex*: Les acides amines et amides libres. CR Acad Sci Ser D (Paris) 268:1034–1037

Cram WJ (1973a) Chloride fluxes in cells of the isolated root cortex of *Zea mays*. Aust J Biol Sci 26:757–779

Cram WJ (1973b) Internal factors regulating the nitrate and chloride influx in plant cells. J Exp Bot 24:328–341

Cram WJ (1976) Negative feedback regulation of transport in cells. The maintenance of turgor, volume, and nutrient supply. In: Lüttge U, Pitman MG (eds) Transport in plants II. Encyclopedia of plant physiology (New Series), vol II A. Springer, Berlin Heidelberg New York, pp 284–316

Crisp MD (1975) Changes in arid zone vegetation at Koonamore, South Australia. PhD Thesis. University of Adelaide, Adelaide

Currie JA (1973) The seed-soil system. In: Heydecker W (ed) Seed ecology. Butterworths, London, pp 463–478

Dainty J (1976) Water relations of plant cells. In: Lüttge U, Pitman MG (eds) Transport in plants II. Encyclopedia of plant physiology. New Series, vol II A. Springer, Berlin Heidelberg New York, pp 12–35

Darlington CD, Wylie AP (1956) Chromosome atlas of flowering plants. MacMillan, New York

Da Silva JV, Naylor AW, Kramer PJ (1974) Some ultrastructural and enzymatic effects of water stress in cotton *(Gossypium hirsutum, L.)* leaves. Proc Natl Acad Sci USA 71:3243–3247

Davidson RL (1969a) Effect of root/leaf temperature differentials on root/shoot ratios in some pasture grasses and clover. Ann Bot 33:561–569

Davidson RL (1969b) Effects of soil nutrients and moisture on root/shoot ratio in *Lolium perenne* L. and *Trifolium repens* L. Ann Bot 33:571–577

Davis PH (1965) Flora of Turkey and the East Aegean Islands. Edinburgh University Press, Edinburgh

Davis SD, McCree KJ (1978) Photosynthetic rate and diffusion conductance as a function of age in leaves of bean plants. Crop Sci 18:280–282

De Castro D, Fontes FC (1946) Primeiro contacto citólogico com a flora halófila dos salgados de Sacavém. Broteria 15:38–46

De Jong TM (1977) Water and salinity relations of California beach species. PhD Thesis. University of California, Davis

Dewet JMJ (1971) Reversible tetraploidy as an evolutionary mechanism. Evolution 25:545–548

Dewey DR (1960) Salt tolerance of 25 strains of *Agropyron*. Agron J 52:631–635

Dewey DR (1962) Breeding crested wheatgrass for salt tolerance. Crop Sci 2:403–407

De Wit CT (1958) Transpiration and crop yields. Versl Landbouwk Onderz 646:1–88

De Wit CT (1970) Dynamic concepts in biology. In: Prediction and measurement of photosynthetic productivity. Center for Agricultural Publishing and Documentation, Wageningen, pp 17–23

De Wit CT, Dijkshoorn W, Noggle JC (1963) Ionic balance and the growth of plants. Versl Landbouwk Onderz 69:1–68

Dijkshoorn W (1962) Metabolic regulation of the alkaline effect of nitrate utilization in plants. Nature (London) 194:165–167

Dobzhansky TW (1970) Genetics of the evolutionary process. Columbia Univ Press, New York

Doley D, Trivett NB (1974) Effects of low water potential on transpiration and photosynthesis in Mitchell grass *(Astrebla lappacea)*. Aust J Plant Physiol 1:539–550

Downes RW (1969) Differences in transpiration rates between tropical and temperate grasses under controlled conditions. Planta 88:261–273

Downes RW (1970) Effect of light intensity and leaf temperature on photosynthesis and transpiration in wheat and sorghum. Aust J Biol Sci 23:775–782

Downton WJS (1971) The chloroplasts and mitochondria of bundle sheath cells in relation to C_4 photosynthesis. In: Hatch MD, Osmond CB, Slatyer RO (eds) Photosynthesis and photorespiration. Wiley-Interscience, New York, pp 419–425

Downton WJS (1977) Photosynthesis in salt-stressed grapevines. Aust J Plant Physiol 4:183–192

Downton WJS, Bisalputra T, Tregunna EB (1969) The distribution and ultrastructure of chloroplasts in leaves differing in photosynthetic carbon metabolism. II. *Atriplex rosea* and *Atriplex hastata* (Chenopodiaceae). Can J Bot 47:915–919

Drobnick R, Plummer AP (1966) Progress in browse hybridization in Utah. Proc Conf West
    State Game and Fish Comm 46:203–211
Duncan WH (1974) Vascular halophytes of the Atlantic and Gulf coasts of North America,
    north of Mexico. In: Reimold RJ, Queen WH (eds) Ecology of halophytes. Academic
    Press, London New York, pp 23–50
Duniway JM (1971) Water relations of *Fusarium* wilt in tomato. Physiol Plant Pathol
    1:537–546
Dwyer DD, De Garmo HC (1970) Greenhouse productivity and water-use efficiency of
    selected desert shrubs and grasses under four moisture levels. New Mexico State
    University. Agric Exp Stn Bull 570:15
Eagles CF, Treharne KJ (1969) Photosynthetic activity of *Dactylis glomerata* L. in different
    light regimes. Photosynthetica 3:29–38
Eaton FM (1927) The water requirement and cell-sap concentration of Australian saltbush
    and wheat as related to the salinity of the soil. Am J Bot 14:212–226
Eckardt FE (1975) Functioning of the biosphere at the primary production level –
    objectives and achievements. In: Cooper JP (ed) Photosynthesis and productivity in
    different environments, IBP 3. Cambridge University Press, Cambridge, pp 173–185
Edwards MM (1973) Seed dormancy and seed environment – internal oxygen relationship.
    In: Heydecker W (ed) Seed ecology. Butterworths, London, pp 169–188
Edye LA, Kiers HJ (1966) Variation in maturity, stolon development and frost resistance of
    Glycine javanica. Aust J Exp Agric Anim Husb 6:380–387
Ehleringer J (1977) The adaptive significance of leaf hairs in a desert shrub. PhD Thesis.
    Stanford University, Stanford
Ehleringer J, Björkman O (1977) Quantum yields for CO_2 uptake in C_3 and C_4 plants. Plant
    Physiol 59:86–90
Eichler HJ (1965) Supplement to JM Black's Flora of South Australia. Government Printer,
    Adelaide
Ellern SJ, Samish YB, Lachover D (1974) Salt and oxalic acid content of leaves of the
    saltbush *Atriplex halimus* in the Northern Negev. J Range Manage 27:267–271
Ellis RP (1974) The significance of the occurrence of both Kranz and non-Kranz leaf
    anatomy in the grass species *Alloteropsis semialata*. S Afr J Sci 70:169–173
Emmett JM, Walker DA (1973) Thermal uncoupling in chloroplasts. Inhibition of
    photophosphorylation without depression of light-induced pH change. Arch Biochem
    Biophys 157:106–113
Epstein E (1972) Mineral nutrition of plants – principles and perspectives. Wiley, New York
Evenari M, Schulze, E-D, Kappen L, Buschbom U, Lange OL (1976) Plant production in
    arid and semi-arid areas. In: Lange OL, Kappen L, Schulze E-D (eds) Water and plant
    life: problems and approaches. Springer, Berlin Heidelberg New York, pp 439–451
Everson RG (1970) Carbonic anhydrase and CO_2 fixation in isolated chloroplasts.
    Phytochemistry 9:25–32
Fanous MA (1967) Test for drought resistance in Pearl Millet *(Pennisetum typoideum)*.
    Agron J 59:337–340
Farquhar GD (1978) Feedforward responses of stomata to humidity. Aust J Plant Physiol
    5:787–800
Farquhar GD (1979a) Models describing the kinetics of ribulose biphosphate carboxylase
    oxygenase. Arch Biochem 193:456–468
Farquhar GD (1979b) Carbon assimilation in relation to transpiration and fluxes of
    ammonia. In: Marcelle R, Clijsters H, Van Poncke (eds) Photosynthesis and plant
    development, pp 321–328. Junk, The Hague
Farquhar GD, von Caemmerer S, Berry JA (1979) A biochemical model of photosynthesis
    Planta (in press)
Feierabend J, Schrader-Reichardt, U (1976) Biochemical differentiation of plastids and
    other organelles in rye leaves with a high-temperature-induced deficiency of plastid
    ribosomes. Planta 129:133–145
Fellows RJ, Boyer JS (1976) Structure and activity of chloroplasts of sunflower leaves having
    various water potentials. Planta 132:229–239
Fernandez OA, Caldwell MM (1975) Phenology and dynamics of root growth of three cool
    semi-desert shrubs under field conditions. J Ecol 63:703–714

Findlay GP, Hope AB (1976) Electrical properties of plant cells: Methods and findings. In: Lüttge U, Pitman MG (eds) Transport in plants II. Encyclopedia of plant physiology (New Series), vol II A. Springer, Berlin Heidelberg New York, pp 53–92

Fireman M, Haywood HE (1952) Indicator significance of some shrubs in the Escalante Desert, Utah. Bot Gaz 114:143–155

Fischer RA, Turner NC (1978) Plant productivity in the arid and semiarid zones. Ann Rev Plant Physiol 29:277–317

Flavell RB, Bennett MD, Smith JB, Smith DB (1974) Genome size and the proportion of repeated nucleotide sequence DNA in plants. Biochem Genet 12:257–269

Flowers TJ (1972a) Salt tolerance in *Suaeda maritima* L. (Dum). The effect of sodium chloride on growth, respiration, and soluble enzymes in a comparative study with *Pisum*. J Exp Bot 23:310–321

Flowers TJ (1972b) The effect of sodium chloride on enzyme activities from four halophyte species of Chenopodiaceae. Phytochemistry 11:1881–1886

Flowers TJ (1975) Halophytes. In: Baker DA, Hall JL (eds) Ion transport in plant cells and tissues. North Holland, Amsterdam, pp 309–334

Flowers TJ, Ward ME, Hall JL (1976) Salt tolerance in the halophyte *Suaeda maritima:* some properties of malate dehydrogenase. Philos Trans R Soc London Ser B 273:523–540

Flowers TJ, Troke PF, Yeo AR (1977) The mechanism of salt tolerance in halophytes. Ann Rev Plant Physiol 28:89–121

Fork DC, Murata N, Sato N (1979) Effect of growth temperature on the lipid and fatty acid composition, and the dependence on temperature of light-induced redox reactions of cytochrome f of light energy redistribution in the thermophilic blue-green alga *Synechococcus lividus*. Plant Physiol 63:524–530

Fowler FG, Fowler HW (1949) The pocket Oxford dictionary of current English. 4th edn revised. The University Press, Oxford

Foy CD, Flemming AL, Burns GF, Armiger WH (1967) Characterization of differential aluminum tolerance among varieties of wheat and barley. Soil Sci Soc Am Proc 31:513–521

Foy CD, Orellana RG, Scheartz JW, Fleming AL (1974) Responses of sunflower genotypes to aluminum in acid soil and nutrient solution. Agron J 66:293–296

Franclet A, Le Houérou H (1971) Le *Atriplex* en Tunisie et en Afrique du Nord, F.A.O. Technical Report 7. FAO, Rome

Frankton C, Bassett TJ (1968) The genus *Atriplex* (Chenopodiaceae) in Canada. I. Three introduced species: *A. heterosperma, A. oblongifolia* and *A. hortensis*. Can J Bot 46:1309–1313

Frankton C, Bassett TJ (1970) The genus *Atriplex* (Chenopodiaceae) in Canada. II. Four native western annuals; *A. argentea, A. truncata, A. powellii*, and *A. dioica*. Can J Bot 48:981–989

French CS (1967) Photosynthesis. In: Haskins CP (ed) The search for understanding. Carnegie Institution, Washington DC, pp 155–178

Friedman J, Orshan G (1975) The distribution, emergence, and survival of seedlings of *Artemisia herba-alba* Asso in the Negev Desert of Israel in relation to distance from adult plants. J Ecol 63:627–632

Froggatt WW (1910) Insects which damage saltbush. Agric Gaz NSW 21:465–471

Fry KE (1970) Some factors affecting the Hill reaction activity in cotton chloroplasts. Plant Physiol 45:465–469

Gadella TWJ, Kliphuis E (1966) Chromosome numbers of flowering plants in the Netherlands. II. Proc K Ned Akad Wet Ser C 69:541–556

Gaff DF (1977) Desiccation tolerant plants of Southern Africa. Oecologia 31:95–109

Gaff DF, Churchill DM (1976) *Borya nitida* Labill. – an Australian species in the Liliaceae with desiccation tolerant leaves. Aust J Bot 24:209–224

Gale J, Poljakoff-Mayber A (1970) Interrelationships between growth and photosynthesis of saltbush (*Atriplex halimus* L.) grown in saline media. Aust J Biol Sci 23:937–945

Gale J, Kohl HC, Hagan RM (1967) Changes in the water balance and photosynthesis of onion, bean, and cotton plants under saline conditions. Physiol Plant 20:408–420

Gale J, Naaman R, Poljakoff-Mayber A (1970) Growth of *Atriplex halimus* in sodium chloride salinated culture solutions as affected by the relative humidity of the air. Aust J Biol Sci 23:947–952

Gates CT, Muirhead W (1967) Studies of the tolerance of *Atriplex* species 1. Environmental characteristics and plant responses of *A. vesicaria, A. nummularia,* and *A. semibaccata.* Aust J Exp Agric Anim Husb 7:39–49

Gates DH, Stoddart LA, Cook CW (1956) Soil as a factor influencing plant distribution on salt dunes of Utah. Ecol Monogr 26:155–175

Gates DM (1968) Transpiration and leaf temperature. Ann Rev Plant Physiol 19:211–238

Gauhl E (1969) Differential photosynthetic performance of *Solanum dulcamara* ecotypes from shaded and exposed habitats. Carnegie Inst Wash Yearb 67:482–487

Gauhl E (1970) Leaf factors affecting the rate of light-saturated photosynthesis in ecotypes of *Solanum dulcamara.* Carnegie Inst Wash Yearb 68:633–636

Gauhl E (1976) Photosynthetic response to varying light intensity in ecotypes of *Solanum dulcamara* L. from shaded and exposed environments. Oecologia 22:275–286

Gifford RM (1974) A comparison of potential photosynthesis, productivity, and yield of plant species with differing photosynthetic metabolism. Aust J Plant Physiol 1:107–117

Gifford RM (1977) Growth pattern, carbon dioxide exchange and dry weight distribution in wheat growing under different photosynthetic environments. Aust J Plant Physiol 4:99–110

Gifford RM (1979) Growth and yield of CO_2 enriched wheat under water limited conditions. Aust J Plant Physiol 6:367–378

Giles KL, Beardsell MF, Cohen D (1974) Cellular and ultrastructural changes in mesophyll and bundle-sheath cells of maize in response to water stress. Plant Physiol 54:208–212

Giles KL, Cohen D, Beardsell MF (1976) Effects of water stress on the ultrastructure of leaf cells of *Sorghum bicolor.* Plant Physiol 57:11–14

Godwin H (1956) The history of the British flora: a factual basis for phytogeography. Cambridge University Press, Cambridge

Good R (1974) The geography of the flowering plants. Longmans, London

Goodall DW (1954) Objective methods for the classification of vegetation. III. An essay in the use of factor analysis. Aust J Bot 1:304–324

Goodman PJ (1973) Physiological and ecotypic adaptations of plants to salt desert conditions in Utah. J Ecol 61:473–494

Goodman PJ, Caldwell MM (1971) Shrub ecotypes in a salt desert. Nature (London) 232:571–572

Gorham E (1958) Soluble salts in dune sands from Blakeney Point in Norfolk. J Ecol 46:373–379

Grahl H, Wild A (1972) Die Variabilität der Größe der Photosyntheseeinheit bei Licht- und Schattenpflanzen. Z Pflanzenphysiol 67:443–453

Grant V (1964) The architecture of the germ plasm. Wiley, New York

Grant V (1966) The selective origin of incompatibility barriers in the plant genus *Gilia.* Am Nat 100:99–118

Greenway H (1962) Plant response to saline substrates. I. Growth and ion uptake of several varieties of *Hordeum vulgare* during and after NaCl treatment. Aust J Biol Sci 15:16–38

Greenway H (1965) Plant responses to saline substrates IV. Chloride uptake by *Hordeum vulgare* as affected by inhibitors, transpiration, and nutrients in the medium. Aust J Biol Sci 18:249–268

Greenway H (1968) Growth stimulation by high chloride concentrations in halophytes. Isr J Bot 17:169–177

Greenway H (1973) Salinity, plant growth and metabolism. J Aust Inst Agric Sci 39:24–34

Greenway H, Osmond CB (1970) Ion relations, growth and metabolism of *Atriplex* at high external electrolyte concentrations. In: Jones R (ed) The biology of *Atriplex.* CSIRO, Canberra, pp 49–56

Greenway H, Osmond CB (1972) Salt responses of enzymes from species differing in salt tolerance. Plant physiol 49:256–259

Greenway H, Gunn A, Thomas DA (1966) Plant response to saline substrates VIII. Regulation of ion concentrations in salt-sensitive and halophytic species. Aust J Biol Sci 19:741–756

Gregor JW (1944) The ecotype. Biol Rev 19:29–30

Gregor JW (1946) Ecotypic differentiation. New Phytol 45:254–270

Gregor JW (1956) Adaptation and ecotypic components. Proc R Soc London Ser B 145:333–337

Greig-Smith P (1952) The use of random and contiguous quadrats in the study of structure of plant communities. Ann Bot (London) 16:293–316

Greig-Smith P (1957) Quantitative plant ecology. Butterworths, London

Greig-Smith P (1961) Data on pattern within plant communities. I. The analysis of pattern. J Ecol 49:695–702

Gulmon SL, Mooney HA (1977) Spatial and temporal relationships between two desert shrubs, *Atriplex hymenelytra* and *Tidestromia oblongifolia* in Death Valley, California. J Ecol 65:831–838

Gunning BES (1977) Transfer cells and their roles in transport of solutes in plants. Sci Prog (Oxford) 64:539–568

Gustafsson M (1970) Studies in the Aegean flora XVII. Variation and distribution of *Atriplex* plasmodesmata. Springer, Berlin Heidelberg New York

Gustafsson M (1970) Studies in the Aegean flora XVII. Variation and distribution of *Atriplex recurva* D'Urv. Bot Not 123:371–383

Gustafsson M (1972) Distribution and effect of paracentric inversions in populations of *Atriplex longipes*. Hereditas 71:173–194

Gustafsson M (1973a) Evolutionary trends in the *Atriplex triangularis* group in Scandinavia. I. Hybrid sterility and chromosomal differentiation. Bot Not 126:345–392

Gustafsson M (1973b) Evolutionary trends in the *Atriplex triangularis* group in Scandinavia. II. Spontaneous hybridization in relation to reproductive isolation. Bot Not 126:398–416

Gustafsson M (1974) Evolutionary trends in the *Atriplex triangularis* group of Scandinavia. III. The effects of population size and introgression on chromosomal differentiation. Bot Not 127:125–148

Gustafsson M (1976) Evolutionary trends in the *Atriplex prostrata* group of Scandinavia. IV. Taxonomy and morphological variation. Opera Bot 39:1–63

Haberlandt G (1884) Physiological plant anatomy (Translated from 4th German edn by M Drummond, 1914). MacMillan, London

Hacker JB, Forde BJ, Gow JM (1974) Simulated frosting of tropical grasses. Aust J Agric Res 25:45–57

Hall AE, Björkman O (1975) Model of leaf photosynthesis and respiration. In: Gates DM, Schmerl RB (eds) Perspectives in biophysical ecology. Springer, Berlin Heidelberg New York, pp 55–72

Hall AE, Schulze E-D, Lange OL (1976) Current perspectives of steady-state stomatal responses to environment. In: Lange OL, Kappen L, Schulze E-D (eds) Water and plant life, problems and modern approaches. Springer, Berlin Heidelberg New York, pp 168–188

Hall DO, Coombs J, Goodwin TW (eds) (1978) Photosynthesis 77: Proceedings of the fourth international congress of photosynthesis. The Biochemical Society, London

Hall EA, Specht RL, Eardley CM (1964) Regeneration of the vegetation on Koonamore Vegetation Reserve, 1926–1962. Aust J Bot 12:205–264

Hall HM, Clements FE (1923) The phylogenetic method of taxonomy: The North American species of *Artemisia, Chrysothamnus,* and *Atriplex*. Publ Carnegie Inst No 326

Hall JL, Flowers TJ (1973) The effect of salt on protein synthesis in the halophyte *Suaeda maritima*. Planta 110:361–368

Hanson CA (1962) Perennial *Atriplex* of Utah and the northern deserts. MS Thesis, Brigham Young Univ., Provo, Utah. 133 pp

Hanson-Porath E, Poljakoff-Mayber A (1969) The effect of salinity on the malic dehydrogenase of pea roots. Plant Physiol 44:1031–1034

Harper JL (1967) A Darwinian approach to plant ecology. J Ecol 55:247–270

Harper JL (1977) Population biology of plants. Academic Press, London New York

Harper JL, White J (1974) The demography of plants. Ann Rev Ecol Syst 5:419–463

Harper JL, Lovell PH, Moore KG (1970) The shapes and sizes of seeds. Ann Rev Ecol Syst 1:327–356

Harrington JF (1972) Seed storage and longevity. In: Kozlowski TT (ed) Seed biology. Academic Press, London New York, pp 145–245

Hasick DJ (1979) Heat and water vapour fluxes in an arid zone community. In: Graetz D, Howes KMW (eds) Studies of the Australian Arid Zone, Part 4. Chenopod Shrublands. CSIRO, Melbourne, pp 54–60

Hatch MD (1971) The C_4-pathway of photosynthesis. Evidence for an intermediate pool of carbon dioxide and the identity of the donor C_4-dicarboxylic acid. Biochem J 125:425–432

Hatch MD, Osmond CB (1976) Compartmentation and transport in C_4 photosynthesis. In: Heber U, Stocking CR (eds) Transport in plants III. Encyclopedia of plant physiology. (New Series), vol III. Springer, Berlin Heidelberg New York, pp 144–184

Hatch MD, Slack CR (1970) Photosynthetic CO_2 fixation pathways. Ann Rev Plant Physiol 21:141–162

Hatch MD, Osmond CB, Troughton JH, Björkman O (1972) Physiological and biochemical characteristics of C_3 and C_4 Atriplex species and hybrids in relation to the evolution of the C_4 pathway. Carnegie Inst Wash Yearb 71:135–141

Hattersley PW, Watson L, Osmond CB (1977) In-situ immunofluorescent labelling of ribulose-1,5-bisphosphate carboxylase in leaves of C_3 and C_4 plants. Aust J Plant Physiol 4:523–539

Heichel GH, Musgrave RB (1970) Photosynthetic response to drought in maize. Philipp Agric 54:102–114

Heiser CB, Whittaker TW (1948) Chromosome numbers, polyploidy, and growth habit in California weeds. Am J Bot 35:179–187

Heslop-Harrison J (1964) Forty years of genecology. Adv Ecol Res 2:159–247

Hiesey WM, Nobs MA, Björkman O (1971) Experimental studies on the nature of species. V. Biosystematics, genetics, and physiological ecology of the erythranthe section of Mimulus. Publ Carnegie Inst No 628, Washington DC

Hill AE, Hill BS (1976) Elimination processes by glands: mineral ions. In: Lüttge U, Pitman MG (eds) Transport in plants II. Encyclopedia of plant physiology (New Series), vol II B. Springer, Berlin Heidelberg New York, pp 225–243

Hill R (1965) The biochemist's green mansions: the photosynthetic-electron transport chain in plants. Essays Biochem 1:120–151

Hillel D (1972) Soil moisture and seed germination. In: Kozlowski TT (ed) Water deficits and plant growth III. Academic Press, London New York, pp 65–89

Hiroi T, Monsi M (1963) Physiological and ecological analyses of shade tolerance of plants. 3. Effect of shading on growth attributes of Helianthus annuus. Bot Mag Tokyo 76:121–129

Hochachka PW, Somero G (1973) Strategies of biochemical adaptation. WB Saunders, Philadelphia

Hodges TK (1973) Ion absorption by plant roots. Adv Agron 25:163–207

Hodges TK (1976) ATPases associated with membranes of plant cells. In: Lüttge U, Pitman MG (eds) Transport in plants II. Encyclopedia of plant physiology (New Series), vol II A. Springer, Berlin Heidelberg New York, pp 260–283

Hodgkinson KC, Johnson PS, Norton BE (1978) Influence of summer rainfall on root and shoot growth of a cold-winter desert shrub, Atriplex confertifolia. Oecologia 34:353–362

Hoffman GJ, Rawlins ST, Garber MJ, Cullen EM (1971) Water relations and growth of cotton as influenced by salinity and relative humidity. Agron J 63:822–826

Holmgren P (1968) Leaf factors affecting light-saturated photosynthesis in ecotypes from exposed and shaded habitats cultivated under two light regimes. Physiol Plant 21:676–698

Holmgren RC, Hutchings SS (1972) Salt desert shrub response to grazing use. In: McKell CM, Blaisdell JP, Goodin JR (eds) Wildland Shrubs – their biology and utilization. US Forest Service, Ogden, pp 153–164

Holton RW, Blecker HH, Onore M (1964) Effect of growth temperature on the fatty acid composition of a blue-green alga. Phytochemistry 3:595–602

Horn HS (1976) Succession. In: May RM (ed) Theoretical ecology. Blackwell Publications, Oxford, pp 187–204

Hsiao TC (1973) Plant responses to water stress. Ann Rev Plant Physiol 24:519–570

Hsiao TC, Acevedo E, Fereres E, Henderson DW (1976) Water stress, growth and osmotic adjustment. Philos Trans R Soc London Ser B 273:479–500

Huber W, Sankhla N (1976) C_4 pathway and regulation of the balance between C_4 and C_3 metabolism. In: Lange OL, Kappen L, Schulze E-D (eds) Water and plant life: problems and modern approaches. Springer, Berlin Heidelberg New York, pp 335–363

Hughes AP, Evans GC (1962) Plant growth and the aerial environment. II. Effect of light intensity on *Impatiens parviflora*. New Phytol 61:154–174

Hulme BA (1957) Studies on some British species of *Atriplex* L. PhD Thesis. University of Edinburgh, Edinburgh

Hulten E (1958) The amphi-atlantic plants and their phytographical connections. Almqvist and Wiskell, Stockholm

Hunt CB (1966) Plant ecology of Death Valley, California. Geological Survey Professional Paper 509. US Gov Printing Office, Washington DC

Hutchings SS, Stewart G (1953) Increasing forage yields and sheep production on Intermountain winter ranges. Circ 925. US Dep Agric, 63 pp

Hutchinson GE (1965) The ecological theater and the evolutionary play. Yale University Press, New Haven

Jackson WA, Flesher D, Hageman RH (1973) Nitrate uptake by dark grown corn seedlings: some characteristics of apparent induction. Plant Physiol 51:120–127

Jann RC, Amen RD (1977) What is germination? In: Khan AA (ed) The physiology and biochemistry of seed dormancy and germination. North Holland, Amsterdam, pp 7–28

Jefferies RL (1972) Aspects of saltmarsh ecology with particular reference to inorganic plant nutrition. In: Barnes RSK, Green J (eds) The estuarine environment. Applied Science Publishers, London, pp 61–85

Jefferies RL (1973) The ionic relations of seedlings of the halophyte *Triglochin maritima* L. In: Anderson WP (ed) Ion transport in plants. Academic Press, London New York, pp 297–231

Jefferies RL (1977) Growth responses of coastal halophytes to inorganic nitrogen. J Ecol 65:847–865

Jefferies RL, Perkins N (1977) The effects on the vegetation of the additions of inorganic nutrients to salt marsh soils at Stiffkey, Norfolk. J Ecol 65:867–882

Jennings DH (1968) Halophytes, succulence and sodium in plants – a unified theory. New Phytol 67:899–911

Jenny H (1966) Pathways of ions into roots according to diffusion models. Plant Soil 25:265–289

Jeschke WD (1977) K^+-Na^+ selectivity in roots, localization of selective fluxes and their regulation. In: Marre E, Ciferri O (eds) Regulation of cell membrane activities in plants. Elsevier, Amsterdam, pp 63–78

Jeschke WD, Stelter W (1976) Measurement of longitudinal ion profiles in single roots of *Hordeum* and *Atriplex* by use of flameless atomic absorption spectroscopy. Planta 128:107–112

Jessup RW (1951) The soils, geology, and vegetation of northwestern South Australia. Trans R Soc South Aust 74:189–273

Jessup RW (1969) Soil salinity in saltbush country of northeastern South Australia. Trans R Soc South Aust 93:69–78

Jones EM (1975) Taxonomic studies of the genus *Atriplex* (Chenopodiaceae) in Britain. Watsonia 10:233–251

Jones LW, Kok B (1966a) Photoinhibition of chloroplast reactions. I. Kinetics and action spectra. Plant Physiol 41:1037–1043

Jones LW, Kok B (1966b) Photoinhibition of chloroplast reactions. II. Multiple effects. Plant Physiol 41:1044–1049

Jones MM, Turner NC (1978) Osmotic adjustment in leaves of *Sorghum* in response to water deficits. Plant Physiol 61:122–126

Jones MM, Turner NC, Osmond CB (1979) Mechanisms of drought resistance. In: Paleg LG, Aspinall DA (eds) The physiology and biochemistry of drought resistance. Academic Press, London New York (in press)

Jones RM (1969a) Soil moisture and salinity under bladder saltbush *(Atriplex vesicaria)* pastures in the New South Wales Riverine plain. Aust J Exp Agric Anim Husb 9:603–609

Jones RM (1969b) Mortality of some tropical grasses and legumes following frosting in the first winter after sowing. Trop Grassl 3:57–63

Jones R (ed) (1970) The biology of *Atriplex*. CSIRO, Canberra

Jones R, Hodgkinson KC (1970) Root growth of rangeland chenopods: morphology and production of *Atriplex nummularia* and *Atriplex vesicaria*. In: Jones R (ed) The biology of *Atriplex*. CSIRO, Canberra, pp 77–85

Jones R, Hodgkinson KC, Rixon AJ (1970) Growth and productivity in rangeland species of *Atriplex*. In: Jones R (ed) The biology of *Atriplex*. CSIRO, Canberra, pp 31–42

Jones RJ, Seawright AA, Little DA (1970) Oxalate poisoning in animals grazing the tropical grass *Setaria sphacelata*. J Aust Inst Agric Sci 36:41–43

Jordan WR, Ritchie JT (1971) Influence of soil water stress on evaporation, root absorption, and internal water status of cotton. Plant Physiol 48:783–788

Kadman-Zahavy A (1955) Notes on the germination of *Atriplex rosea*. Bull Res Counc Isr 4:375–378

Kamen MD (1963) Primary processes in photosynthesis. Academic Press, London New York

Kaplan A, Gale J (1972) Effect of sodium chloride salinity on the water balance of *Atriplex halimus*. Aust J Biol Sci 25:895–903

Kassas M (1966) Plant life in deserts. In: Hills ES (ed) Arid lands – a geographical appraisal. Methuen, London, pp 145–180

Kaufmann MR (1976a) Stomatal response of Engelmann spruce to humidity, light, and water stress. Plant Physiol 57:898–901

Kaufmann MR (1976a) Water transport through plants: current perspectives. In: Wardlaw IF, Passioura JB (eds) Transport and transfer processes in plants. Academic Press, London New York, pp 313–327

Keck RW, Boyer JS (1974) Chloroplast response to low leaf water potentials. III. Differing inhibition of electron transport and photophosphorylation. Plant Physiol 53:474–479

Keeling CD (1973) Industrial production of carbon dioxide from fossil fuel and limestone. Tellus 25:174

Kenagy GJ (1972) Saltbush leaves: excision of hypersaline tissue by a Kangaroo rat. Science 178:1094–1096

Kennedy RA, Laetsch WM (1974) Plant species intermediate for C_3, C_4 photosynthesis. Science 184:1087–1089

Khan AA (ed) (1977) The physiology and biochemistry of seed dormancy and germination. North Holland, Amsterdam

Kirkby EA (1969) Ion uptake and ionic balance in plants in relation to the form of nitrogen nutrition. In: Rorison IH (ed) Ecological aspects of the mineral nutrition of plants. Blackwell, Oxford, pp 215–235

Kluge M, Ting IP (1978) Crassulacean acid metabolism: Analysis of an ecological adaptation. Ecological Studies 30. Springer, Berlin Heidelberg New York

Koestler A, Smythies JR (eds) (1972) Beyond reductionism. Hutchinson, London

Kohne DE (1970) Evolution of higher organism DNA. Q Rev Biophys 3:327–375

Koller D (1957) Germination-regulating mechanisms in some desert seeds. IV. *Atriplex dimorphostegia*, Kar. et Kir. Ecology 38:1–13

Koller D (1970) Analysis of the dual action of white light on germination of *Atriplex dimorphostegia* (Chenopodiaceae). Isr J Bot 19:499–516

Koller D (1972) Environmental control of seed germination. In: Kozlowski TT (ed) Seed biology vol II. Academic Press, London New York, pp 1–101

Koller D, Tadmor NH, Hillel D (1958) Experiments in the propagation of *Atriplex halimus* for desert pasture and soil conservation. Isr J Agric Res 9:83–106

Komarov VL (ed) (1970) Flora of the USSR, vol VI, Centrospermae. Israel Program for Scientific Translations. Jerusalem

Kramer D, Anderson WP, Preston JL (1978) Transfer cells in the root epidermis of *Atriplex* in a response to salinity: a comparative cytological and x-ray microprobe investigation. Aust J Plant Physiol 5:739–747

Kramer PJ (1969) Plant and soil water relationship: a modern synthesis. McGraw-Hill, New York

Krause GH, Santarius KA (1975) Relative thermostability of the chloroplast envelope. Planta 127:285–299

Kruckeberg AR (1969) Soil diversity and the distribution of plants with examples from western North America. Madrõno 20:129–154

Ku SB, Edwards GE (1978) Oxygen inhibition of photosynthesis III. Temperature dependence of quantum yield and its relation to O_2/CO_2 solubility ratio. Planta 140:1–6

Kurkova EB, Motorina MV (1974) Chloroplast ultrastructure and photosynthesis at different rates of dehydration. Sov Plant Physiol 21:28–31

Kuroiwa S, Hiroi T, Takada K, Monsi M (1964) Distribution ratio of net photosynthate to photosynthetic and nonphotosynthetic systems in shaded plants. Bot Mag Tokyo 77:37–42

Kylin A, Quatrano RS (1975) Metabolic and biochemical aspects of salt tolerance. In: Poljakoff Mayber A, Gale J (eds) Plants in saline habitats. Springer, Berlin Heidelberg New York, pp 147–167

La Cour LF (1931) Improvements in every day technique in plant cytology. J R Microsc Soc 51:49–126

Lafever HN, Campbell LG, Foy CD (1977) Differential response of wheat cultivars to Al. Agron J 69:563–568

Laing WA, Ogren WL, Hageman RH (1974) Regulation of soybean net photosynthetic CO_2 fixation by the interaction of CO_2, O_2, and ribulose-1,5-diphosphate carboxylase. Plant Physiol 54:678–685

Lanyi JK (1974) Salt dependant properties of proteins from extremely halophilic bacteria. Bacteriol Rev 38:272–290

Larsen H (1962) Halophilism. In: Gunsalus IC, Stanier RY (eds) The bacteria, a treatise on structure and function, vol IV. Academic Press, London New York, pp 297–342

Läuchli A (1976a) Apoplasmic transport in tissues. In: Lüttge U, Pitman MG (eds) Transport in plants II. Encyclopedia of plant physiology (New Series), vol II B. Springer, Berlin Heidelberg New York, pp 3–34

Läuchli A (1976b) Symplastic transport and ion release to the xylem. In: Wardlaw IF, Passioura JB (eds) Transport and transfer processes in plants. Academic Press, London New York, pp 101–112

Lawlor DW (1976) Water stress induced changes in photosynthesis, photorespiration, respiration, and CO_2 compensation concentration of wheat. Photosynthetica 10:378–387

Leigh JH, Mulham WE (1966) Selection of diet by sheep grazing semi-arid pastures on the Riverine Plain. Aust J Exp Agric Anim Husb 6:460–467

Leon H, Alain H (1951) Flore de Cuba, vol II. Dicotiledoneas: Casuarinaceas a Meliaceas. P Fernadez, La Habara

Lerman JC (1975) How to interpret variations in the carbon isotope ratio of plants: biologic and environmental effects. In: Marcelle R (ed) Environmental and biological control of photosynthesis. W Junk, The Hague, pp 323–335

Levins R (1968) Evolution in changing environments. Princeton University Press, Princeton

Levitt J (1972) Responses of plants to environmental stresses. Academic Press, London New York

Lexander K, Carlsson R, Schalin V, Simonsson Å, Lundberg T (1970) Quantities and qualities of leaf protein concentration from wild species and crop species grown under controlled conditions. Ann Appl Biol 66:193–216

Lieth H (1976) The use of correlation models to predict primary productivity from precipitation or evapotranspiration. In: Lange OL, Kappen L, Schulze E-D (eds) Water and plant life: problems and modern approaches. Springer, Berlin Heidelberg New York, pp 392–407

Lockhart JA (1963) Photomorphogenesis in plants. Adv Front Plant Sci 1:1–44

Longstreth DJ, Nobel PS (1978) Salinity induced changes in leaf anatomy and photosynthesis. Plant Physiol 61:594

Longstreth DJ, Strain BR (1977) Effects of salinity and illumination on photosynthesis and water balance of *Spartina alterniflora* Loisel. Oecologia 31:191–200

Lorimer GH, Woo KC, Berry JA, Osmond CB (1978) The C_2 photorespiratory carbon oxidation cycle in leaves of higher plants, pathway and consequences. In: Hall DO, Coombs J, Goodwin TW (eds). Photosynthesis 77. The Biochemical Society, London, pp 311–322

Louwerse W, Van der Zweerde W (1977) Photosynthesis, transpiration and leaf morphology of *Phaseolus vulgaris* and *Zea mays* grown at different irradiances in artificial and sunlight. Photosynthetica 11:11–21

Löve A, Löve D (1956) Cytotaxonomical conspectus of the Islandic flora. Acta Horti Gotob 20:65–291

Ludlow MM (1976) Ecophysiology of C_4 grasses. In: Lange OL, Kappen L, Schulze E-D (eds) Water and plant life: problems and modern approaches. Springer, Berlin Heidelberg New York, pp 364–386

Ludlow MM, Ng TT (1976) Effect of water deficit on carbon dioxide exchange and leaf elongation rate of *Panicum maximum* var. *trichoglume*. Aust J Plant Physiol 3:401–413

Ludlow MM, Wilson GL (1971) Photosynthesis of tropical pasture plants. I. Illuminance, carbon dioxide concentration, leaf temperature and leaf-air vapour pressure difference. Aust J Biol Sci 24:449–470

Ludlow MM, Wilson GL (1978) Photosynthesis of tropical pasture plants. IV. Basis and consequences of differences between grasses and legumes. Aust J Biol Sci 25:1133–1145

Lundell CL (1969) Flora of Texas, vol II. Texas Research Foundation, Renner

Lüttge U (1971) Structure and function of plant glands. Ann Rev Plant Physiol 22:23–44

Lüttge U, Osmond CB (1970) Ion absorption in *Atriplex* leaf tissue III. Site and metabolic control of light-dependant chloride secretion to epidermal bladders. Aust J Biol Sci 23:17–25

Lüttge U, Pitman MG (ed) (1976) Transport in plants II: Encyclopedia of plant physiology (New Series), vol II A. Springer, Berlin Heidelberg New York

Lycklama JC (1963) The absorption of ammonium and nitrate by perennial rye-grass. Acta Bot Neerl 12:361–424

Lyons JM (1973) Chilling injury in plants. Ann Rev Plant Physiol 24:445–456

MacArthur R, Connell J (1966) The biology of populations. John Wiley and Sons, New York

MacDonald KB, Barbour MG (1974) Beach and saltmarsh vegetation of the North American Pacific coast. In: Reimold RJ, Queen WH (eds) Ecology of halophytes. Academic Press, London New York, pp 175–233

MacFarlane WV (1971) Salinity and the whole animal. In: Talsma T, Philip JR (eds) Salinity and water use. MacMillan, London, pp 161–178

Mahall BE, Park RB (1976a) The ecotone between *Spartina foliosa* Trin. and *Salicornia virginica* L. in salt marshes of northern San Francisco Bay II. Soil water and salinity. J Ecol 64:793–809

Mahall BE, Park RB (1976b) The ecotone between *Spartina foliosa* Trin and *Salicornia virginica* L. in salt marshes of northern San Francisco Bay III. Soil aeration and tidal immersion. J Ecol 64:811–819

Malik AR (1970) Studies on structure in saltbush and bluebush communities in south-eastern Australia. PhD Thesis. University of Sydney, Sydney

Malik AR, Anderson DJ (1971) An analysis of *Atriplex inflata* populations from western New South Wales. Aust J Bot 19:191–202

Malik AR, Anderson DJ, Myerscough PJ (1976) Studies on structure in plant communities VII. Field and experimental analyses of *Atriplex vesicaria* populations from the Riverine Plain of New South Wales. Aust J Bot 24:265–280

Mallott PG, Davy AJ, Jeffries RL, Hutton MJ (1975) Carbon dioxide exchange in leaves of *Spartina anglica* Hubbard. Oecologia 20:351–358

Martin EV, Clements FE (1939) Adaptation and origin in the plant world. I. Factors and functions in coastal dunes. Publ Carnegie Inst, No 521, Washington DC

Martin HA (1973) Palynology and historical ecology of some cave excavations in the Australian Nullarbor. Aust J Bot 21:283–316

Mayer AM (1977) Metabolic control of germination. In: Khan AA (ed) The physiology and biochemistry of seed dormancy and germination. North Holland, Amsterdam, pp 357–384

McArthur ED (1977) Environmentally induced changes of sex expression in *Atriplex canescens*. Heredity 38:97–103

McCree KJ (1976) The role of dark respiration in the carbon economy of a plant. In: Black CC, Burris RH (eds) CO₂ metabolism and plant productivity. University Park Press, Baltimore, pp 177–184

McKell CM, Perrier ER, Stebbins GL (1960) Responses of two subspecies of orchardgrass (*Dactylis glomerata* subsp. *lusitanica* and *judaica*) to increasing soil moisture stress. Ecology 41:772–778

McMillan C (1959) Salt tolerance within a *Typha* population. Am J Bot 46:521–526

McNaughton SJ (1966) Ecotype function in the *Typha* community-type. Ecol Monogr 36:297–325

McWilliam JR, Naylor AW (1967) Interaction of temperature and light in the synthesis of chlorophyll in corn. Plant Physiol 42:1711–1715

Mederski HJ, Chen LH, Curry RB (1975) Effect of leaf water deficit on stomatal and nonstomatal regulation of net carbon dioxide assimilation. Plant Physiol 55:589–593

Medina E (1970) Relationships between nitrogen level, photosynthetic capacity and carboxydismutase activity in *Atriplex patula* leaves. Carnegie Inst Wash Yearb 69:655–662

Medina E (1971) Effect of nitrogen supply and light intensity during growth on the photosynthetic capacity and carboxydismutase activity of leaves of *Atriplex patula ssp. hastata*. Carnegie Inst Wash Yearb 70:551–559

Meidner H, Sheriff D (1976) Water and plants. Blackie, Glasgow

Meusel H, Jäeger E, Weinhert E (1965) Vergleichende Chronologie der Zentraleuropäischen Flora. Gustav Fischer Verlag, Jena

Miller RS (1957) Observations on the status of ecology. Ecology 38:353–354

Mohanty P, Boyer JS (1976) Chloroplast response to low leaf water potentials. IV. Quantum yield is reduced. Plant Physiol 57:704–709

Moldau H (1973) Effect of various water regimes on stomatal and mesophyll conductance of bean leaves. Photosynthetica 7:1–7

Monsi N (1968) Mathematical models of plant communities. In: Eckhardt F (ed) Functioning of terrestrial ecosystems at the primary production level. UNESCO, Paris, pp 131–149

Monteith JL (1975) Vegetation and the atmosphere, vol I. Academic Press, London New York

Mooney HA (1974) Plant forms in relation to environment. In: Billings WD, Strain BR (eds) Vegetation and environment. Handbook in vegetation science, Part 6. Junk, The Hague, pp 113–122

Mooney HA, Björkman O, Troughton JH (1974) Seasonal changes in the leaf characteristics of the desert shrub *Atriplex hymenelytra*. Carnegie Inst Wash Yearb 73:846–852

Mooney HA, Björkman O, Ehleringer J, Berry JA (1976) Photosynthetic capacity of in situ Death Valley plants. Carnegie Inst Wash Yearb 75:410–413

Mooney HA, Björkman O, Collatz GJ (1977a) Photosynthetic acclimation to temperature and water stress in the desert shrub, *Larrea divaricata*. Carnegie Inst Wash Yearb 76:328–335

Mooney HA, Ehleringer J, Björkman O (1977b) The energy balance of leaves of the evergreen desert shrub *Atriplex hymenelytra*. Oecologia 29:301–310

Mooney HA, Björkman O, Collatz GJ (1978) Photosynthetic acclimation to temperature in the desert shrub, *Larrea divaricata*. I. Carbon exchange characteristics of intact leaves. Plant Physiol 61:406–410

Moore CWE (1953) The vegetation of the south-eastern Riverina, New South Wales. I. The climax communities. Aust J Bot 1:485–547

Moore RT, Caldwell MM (1972) Field use of thermocouple psychrometers in desert soils. In: Brown RW, Van Haveren BP (eds) Psychrometry in water relations research. Utah Agric Exp Station, Logan, pp 115–169

Moore RT, Breckle SW, Caldwell MM (1972a) Mineral ion composition and osmotic relations of *Atriplex confertifolia* and *Eurotia lanata*. Oecologia 11:67–78

Moore RT, White RS, Caldwell MM (1972b) Transpiration of *Atriplex confertifolia* and *Eurotia lanata* in relation to soil, plant, and atmospheric moisture stresses. Can J Bot 50:2411–2418

Moran R (1975) *Atriplex frankenoides* (Chenopodiaceae) a new species from Baja California, Mexico. Phytologia 30:1–4

Moser H (1934) Untersuchungen über die Blattstruktur von *Atriplex*-Arten und ihre Beziehungen zur Systematik. Beih Bot Zentralbl 52:378–388

Mozafar A, Goodin JR (1970) Vesiculated hairs: a mechanism for salt tolerance in *Atriplex halimus* L. Plant Physiol 45:62–65

Mozafar A, Goodin JR, Oertli JJ (1970) Sodium and potassium interactions in increasing the salt tolerance of *Atriplex halimus* L.: II. Na$^+$ and K$^+$ uptake characteristics. Agron J 62:481–484

Mukohata Y (1973) Thermal denaturation of thylakoids and inactivation of photophosphorylation in isolated spinach chloroplasts. In: Nakao N, Packer L (eds) Organization of energy tranducing membranes. University Park Press, Baltimore, pp 219–237

Mulligan GA (1957) Chromosome numbers of Canadian weeds. I. Can J Bot 35:779–789

Mulligan GA (1965) IOPB Chromosome number Report V. Taxon 14:191–196

Murata N, Fork DC (1975) Temperature dependence of chlorophyll a fluorescence in relation to the physical phase of membrane lipids in algae and higher plants. Plant Physiol 56:719–796

Murata N, Fork DC (1977) Temperature dependence of the light-induced spectral shift of carotenoids in *Cyanidium caldarum* and higher plant leaves. Biochim Biophys Acta 461:365–378

Murata N, Troughton JH, Fork DC (1975) Relationship between the transition of the physical phase of membrane lipids and photosynthetic parameters in *Anacystis nidulans* and lettuce and spinach chloroplasts. Plant Physiol 56:508–517

Nátr L (1975) Influence of mineral nutrition on photosynthesis and the use of assimilates. In: Cooper JP (ed) Photosynthesis and productivity in different environments. University Press, Cambridge, pp 537–555

Neales TF, Treharne KJ, Wareing PF (1971) A relationship between net photosynthesis, diffusive resistance and carboxylating enzyme activity in bean leaves. In: Hatch MD, Osmond CB, Slatyer RO (eds) Photosynthesis and photorespiration. Wiley-Interscience, New York, pp 89–96

Nevins DJ, Loomis RS (1970) Nitrogen nutrition and photosynthesis in sugar beet. Crop Sci 10:21–25

Newman EI (1974) Root and soil water relations. In: Carson EW (ed) The plant root and its environment. The University Press of Virginia, Charlottesville, pp 363–440

Newman EI (1976) Water movement through root systems. Philos Trans R Soc London Ser B 273:463–478

Newsome AE (1971) The ecology of red Kangaroos. Aust J Zool 16:32–50

Nicholls AO (1972) An analysis of the growth of seedlings of four *Atriplex* L. species in controlled environments. PhD Thesis. University of Melbourne, Melbourne

Nicholls AO, Calder DM (1973) Comments on the use of regression analysis for the study of plant growth. New Phytol 72:571–581

Nieman RH (1962) Some effects of sodium chloride on growth photosynthesis and respiration of twelve crop plants. Bot Gaz 123:279–285

Nir I, Poljakoff-Mayber A (1967) Effect of water stress on the photochemical activity of chloroplasts. Nature (London) 213:418–419

Nobel PS (1970) Plant cell physiology: a physicochemical approach. Freeman, San Francisco

Nobel PS (1974) Boundary layers of air adjacent to cylinders. Estimation of effective thickness and measurement on plant material. Plant Physiol 54:177–181

Nobel PS, Zaragosa LJ, Smith WK (1975) Relationship between mesophyll surface area, photosynthetic rate, and illumination level during development for leaves of *Plectranthus parviflorus* Henekel. Plant Physiol 55:1067–1070

Noble IR (1977) Long term biomass dynamics in an arid chenopod shrub community at Koonamore, South Australia. Aust J Bot 25:639–653

Nobs MA (1975) Chromosome numbers in *Atriplex*. Carnegie Inst Yearb 74:762–765

Nobs MA (1976) Hybridization in *Atriplex*. Carnegie Inst Wash Yearb 75:421–423

Nobs MA (1978) Chromosome numbers in *Atriplex*. Carnegie Inst Wash Yearb 77:240–241

Nobs MA, Hagar WA (1974) Analysis of germination and flowering rates of dimorphic seeds from *Atriplex hortensis*. Carnegie Inst Wash Yearb 73:859–864

Nobs MA, Björkman O, Pearcy RW (1971) Hybrids between *Atriplex* species with and without β-carboxylation photosynthesis. Cytogenetic and morphological characteristics. Carnegie Inst Wash Yearb 69:625–629

Nobs MA, Pearcy RW, Berry JA, Nicholson F (1972) Reciprocal transplant responses of C_3 and C_4 Atriplexes. Carnegie Inst Wash Yearb 71:164–169

Noy-Meir I (1970) Component analysis of semi-arid vegetation in south-eastern Australia. PhD Thesis. Australian National University, Canberra

Noy-Meir I (1973) Desert ecosystems: environment and producers. Ann Rev Ecol Syst 4:25–51

Noy-Meir I, Anderson DJ (1970) Multiple pattern analysis, or multiscale ordination: pathway to a vegetation hologram? Proc Int Symp Stat Ecol III:207–225

Noy-Meir I, Ginzburg BZ (1969) An analysis of the water potential isotherm in plant tissues. II. Comparative studies on leaves of different types. Aust J Biol Sci 22:35–52

Nye PH, Tinker PB (1977) Solute movement in the soil-root system. Blackwell Scientific Publications, Oxford

Oertli JJ (1968) Extracellular salt accumulation, a possible mechanism of salt injury in plants. Agrochimica 12:461–469

Oku T, Tomita G (1971) Effect of polyethylene glycol on heat inactivation of the Hill reaction. Biochem Biophys Res Commun 44:948–962

Osmond CB (1963) Oxalates and ionic equilibria in Australian saltbushes. Nature (London) 198:503–504

Osmond CB (1965) Related aspects of ion absorption and acid metabolism in *Atriplex spongiosa* F.v.M. PhD Thesis. University of Adelaide, Adelaide

Osmond CB (1966) Divalent cation absorption and interaction in *Atriplex*. Aust J Biol Sci 19:37–48

Osmond CB (1967a) Acid metabolism in *Atriplex* I. Regulation of oxalate synthesis by the apparent excess cation absorption in leaf tissue. Aust J Biol Sci 20:575–587

Osmond CB (1967b) β-carboxylation photosynthesis in *Atriplex*. Biochim Biophys Acta 141:197–199

Osmond CB (1968) Ion absorption in *Atriplex* leaf tissue I. Absorption by leaf mesophyll cells. Aust J Biol Sci 21:1119–1130

Osmond CB (1974) Leaf anatomy of Australian saltbushes in relation to photosynthetic pathways. Aust J Bot 22:39–44

Osmond CB (1976) Ion absorption and carbon metabolism in cells of higher plants. In: Lüttge U, Pitman MG (eds) Transport in plants II. Encyclopedia of plant physiology (New Series), vol II A. Springer, Berlin Heidelberg New York, pp 345–372

Osmond CB (1978) Crassulacean acid metabolism – a curiosity in context. Ann Rev Plant Physiol 29:379–414

Osmond CB (1979) Ion uptake, transport and excretion. In: Perry RA, Goodall DW (eds) Arid-land ecosystems: Structure, functioning and management. IBP 16, vol I. Cambridge University Press, Cambridge, pp 607–625

Osmond CB, Björkman O (1972) Simultaneous measurements of oxygen effects on net photosynthesis and glycolate metabolism in C_3 and C_4 species of *Atriplex*. Carnegie Inst Wash Yearb 71:141–148

Osmond CB, Greenway H (1973) Salt responses of carboxylation enzymes from species differing in salt tolerance. Plant Physiol 49:260–263

Osmond CB, Smith FA (1976) Symplastic transport of metabolites during C_4 photosynthesis. In: Gunning BES, Robards AW (eds) Intercellular communication in plants: studies on plasmodesmata. Springer, Berlin Heidelberg New York, pp 229–241

Osmond CB, Ziegler H (1975) Schwere Pflanzen und leichte Pflanzen: Stabile Isotope im Photosynthesestoffwechsel und in der biochemischen Ökologie. Naturwiss Rundsch 28:323–328

Osmond CB, Lüttge U, West KR, Pallaghy CK, Shacher-Hill B (1969a) Ion absorption in *Atriplex* leaf tissue. II. Secretion of ions to epidermal bladders. Aust J Biol Sci 22:797–814

Osmond CB, Troughton JH, Goodchild DJ (1969b) Physiological, biochemical, and structural studies of photosynthesis and photorespiration in two species of *Atriplex*. Z Pflanzenphysiol 61:218–237

O'Toole JC, Ozbun JL, Wallace DH (1977) Photosynthetic response to water stress in *Phaseolus vulgaris*. Physiol Plant 40:111–114

Pallaghy CK (1973) Electron probe microanalysis of potassium and chloride in freeze-substituted leaf sections of *Zea mays*. Aust J Biol Sci 26:1015–1034

Palmer JH, Trickett ES, Linacre ET (1964) Transpiration response of *A. nummularia* Lindl. and upland cotton vegetation to soil moisture stress. Agric Meteorol 1:282–293

Parr-Smith GA (1977) Taxonomy of *Atriplex vesicaria* (Heward ex Benth) (Chenopodiaceae) and related species. PhD Thesis. University of Melbourne, Melbourne, 221 pp

Parr-Smith GA, Calder DM (1979) *Atriplex vesicaria*: Taxonomy and variation. In: Graetz D, Howes KMW (eds) Studies on the Australian arid zone. Part 4, Chenopod shrublands. CSIRO, Melbourne

Passioura JB (1976) The control of water movement through plants. In: Wardlaw IF, Passioura JB (eds) Transport and transfer processes in plants. Academic Press, London New York, pp 373–380

Patterson BD, Murata T, Graham D (1976) Electrolyte leakage induced by chilling in *Passiflora* species tolerant to different climates. Aust J Plant Physiol 3:435–442

Patterson BD, Kenrick JR, Raison JK (1978) Lipids of chill-sensitive and chill-resistant *Passiflora* species: Fatty acid composition and temperature dependence of spin label motion. Phytochemistry 17:1089–1092

Pearcy RW (1976) Temperature effects on growth and CO_2 exchange rates in coastal and desert races of *Atriplex lentiformis*. Oecologia 26:245–255

Pearcy RW (1977) Acclimation of photosynthetic and respiratory CO_2 exchange to growth temperatures in *Atriplex lentiformis* (Torr.) Wats. Plant Physiol 59:795–799

Pearcy RW (1978) Effect of growth temperature on the fatty acid composition of the leaf lipids in *Atriplex lentiformis* (Torr.) Wats. Plant Physiol 61:484–486

Pearcy RW, Björkman O (1971) Hybrids between *Atriplex* species with and without β-carboxylation photosynthesis. Biochemical characteristics. Carnegie Inst Wash Yearb 69:632–640

Pearcy RW, Harrison AT (1974) Comparative photosynthetic and respiratory gas exchange characteristics of *Atriplex lentiformis* (Torr.) Wats. in coastal and desert habitats. Ecology 55:1104–1111

Pearcy RW, Harrison AT, Mooney HA, Björkman O (1974) Seasonal changes in net photosynthesis of *Atriplex hymenelytra* shrubs growing in Death Valley, California. Oecologia 17:111–121

Pearcy RW, Berry JA, Fork DC (1977) Effects of growth temperature on the thermal stability of the photosynthetic apparatus of *Atriplex lentiformis* (Torr.) Wats. Plant Physiol 59:873–878

Peaslee DW, Moss DN (1968) Stomatal conductivities in K-deficient leaves of maize (*Zea mays* L.). Crop Sci 8:427–430

Peisker M (1974) A model describing the influence of oxygen on photosynthetic carboxylation. Photosynthetica 8:47–50

Penning de Vries FWT (1975) The cost of maintenance processes in plant cells. Ann Bot 39:77–92

Perring FH, Walters SM (eds) (1962) Atlas of the British flora. Nelson, London

Philpott J, Troughton JH (1974) Photosynthetic mechanisms and leaf anatomy of hot desert plants. Carnegie Inst Wash Yearb 73:790–793

Pielou EC (1974) Population and community ecology. Gordon and Breach, New York

Pitman MG (1965) Transpiration and the selective uptake of potassium by barley seedlings (*Hordeum vulgare* cv. "Bolivia"). Aust J Biol Sci 18:987–998

Pitman MG (1977) Ion transport into the xylem. Ann Rev Plant Physiol 28:71–88

Pitman MG, Cram WJ (1977) Regulation of ion content in whole plants. In: Jennings DH (ed) Integration of activity in the higher plant, SEB Symposium 31. Cambridge University Press, Cambridge, pp 391–424

Pizarro CM (1959) Sinopsis de la flora Chilena. Ediciones de la Universidad de Chile, Chile

Plaut Z (1971) Inhibition of photosynthetic carbon dioxide fixation in isolated chloroplasts exposed to reduced osmotic potentials. Plant Physiol 48:591–595

Plaut Z, Bravdo B (1973) Response of carbon dioxide fixation to water stress. Plant Physiol 52:28–32

Plummer AP, Jensen RL, Stapley HD (1957) Job completion report for game forage revegetation project W-82-R2. Utah State Dept. Fish and Game Inf. Bull 1956–1957. 128 pp

Poljakoff-Mayber A (1975) Morphological and anatomical changes in plants as a response to salinity stress. In: Poljakoff-Mayber A, Gale J (eds) Plants in saline environments. Springer, Berlin Heidelberg New York, pp 97–117

Poljakoff-Mayber A, Gale J (1975) Plants in saline environments. Springer, Berlin Heidelberg New York

Poljakoff-Mayber A, Greenway H (1974) Effect of high sodium chloride concentrations in the growth medium on the activity of glucose-6-phosphate dehydrogenase from pea roots. Aust J Plant Physiol 1:483–489

Pólya L (1948) Chromosome numbers of certain alkali plants. Arch Biol Hung 18:145–148

Pólya L (1949) Chromosome numbers of some Hungarian plants. Acta Geobot Hung 6:124–137

Poore MED (1962) The method of successive approximation in descriptive ecology. Adv Ecol Res 1:35–68

Potter JR, Boyer JS (1973) Chloroplast response to low leaf water potentials. II. Role of osmotic potentials Plant Physiol 51:993–997

Powles SB, Osmond CB (1978) Inhibition of the capacity and efficiency of photosynthesis in bean leaflets illuminated in the absence of CO_2 at low O_2 concentrations – a protective role for photorespiration. Aust J Plant Physiol 5:619–629

Powles SB, Osmond CB, Thorne SW (1979) Photoinhibition in leaves of C_3 plants illuminated under conditions which prevent photorespiration. Plant Physiol, 64:982–988

Probst R (1949) Wolladventivflora Mitteleuropas. Vogt Schild, Solothurn

Proctor J (1975) The ecology of serpentine soils. Adv Ecol Res 9:255–366

Purer EA (1936) Studies of certain sand dune plants of Southern California. Ecol Monogr 6:1–87

Pyykko M (1966) The leaf anatomy of East Patagonian xeromorphic plants. Ann Bot Fenn 3:453–622

Quirk JP (1971) Chemistry of saline soils and their physical properties. In: Talsma T, Philip JR (eds) Salinity and water use. MacMillan, London, pp 79–91

Raison JK (1973) Temperature-induced phase changes in membrane lipids and their influence on metabolic regulation. In: Rate control of biological processes. Soc Exp Biol Symp 27. Cambridge University Press, Cambridge, pp 485–512

Raison JK (1974) A biochemical explanation of low-temperature stress in tropical and sub-tropical plants. In: Bieleski RL, Ferguson AR, Cresswell MM (eds) Mechanisms of regulation of plant growth, Bull 12. Royal Soc New Zealand: Wellington, pp 487–497

Raison JK, Lyons JM, Mehlhorn RJ, Keith AD (1971) Temperature-induced phase changes in mitochondrial membranes detected by spin labeling. J Biol Chem 246:4036–4040

Randall PJ, Bouma D (1973) Zinc deficiency, carbonic anhydrase, and photosynthesis in leaves of spinach. Plant Physiol 52:229–232

Ranwell DS (1972) Ecology of salt marshes and sand dunes. Chapman and Hall, London

Rao KP, Rains DW (1976) Nitrate absorption by barley. I. Kinetics and energetics. Plant Physiol 57:55–58

Raschke K (1960) Heat transfer between the plant and the environment. Ann Rev Plant Physiol 11:111–126

Raschke K (1975) Stomatal action. Ann Rev Plant Physiol 26:309–340

Raschke K (1976) How stomata resolve the dilemma of opposing priorities. Philos Trans R Soc London Ser B 273:551–560

Raven JA, Smith FA (1976) Nitrogen assimilation and transport in vascular land plants in relation to intracellular pH regulation. New Phytol 76:415–431

Rawson HM, Begg JE, Woodward RG (1977) The effect of atmospheric humidity on photosynthesis, transpiration, and water use efficiency of leaves of several plant species. Planta 134:5–10

Redshaw AJ, Meidner H (1972) Effects of water stress on the resistance to uptake of CO_2 in tobacco. J Exp Bot 23:229–240

Rees WJ, Sidrak CH (1956) Plant nutrition on fly ash. Plant Soil 8:141–159

Reisenauer HM (1966) Mineral nutrients in soil solution. In: Altman PL, Dittmer DS (eds) Environmental biology. Federation of American Societies for Experimental Biology, Bethesda, pp 507–508

Richards FJ (1969) The quantitative analysis of growth. In: Stewart FC (ed) Plant physiology, vol V A. Academic Press, London New York

Richards LA (ed) (1954) Saline and alkaline soils. Wash USDA Agric Handb 60:1954

Ridley HN (1930) The dispersal of plants throughout the world. Reeve, Ashford

Rixon AJ (1971) Oxygen uptake and nitrification by soil within a grazed *Atriplex vesicaria* community in semi-arid rangeland. J Range Manage 24:435–439

Robards AW, Clarkson DT (1976) The role of plasmodesmata in the transport of water and nutrients across roots. In: Gunning BES, Robards AW (eds) Intercellular communication in plants. Springer, Berlin Heidelberg New York, pp 181–202

Roberts RC (1950) Chemical effects of salt-tolerant shrubs on soils. 4th Int Congr Soil Sci 1:404–406

Robinson JB, Smith FA (1970) Chloride influx into citrus leaf slices. Aust J Biol Sci 23:953–960

Rodin LE, Bazilevich NI (1965) Production and mineral cycling in terrestrial vegetation. (Transl ed Fogg GE). Oliver and Boyd, Edinburgh

Rogers RW, Lange RT, Nicholas DJD (1966) Nitrogen fixation by lichens in arid soil crusts. Nature (London) 209:96–97

Rorison IH (ed) (1969) Ecological aspects of the mineral nutrition of plants. Blackwell Scientific Publications, Oxford

Rosenberg NJ (1974) Microclimate: the biological environment. Wiley Interscience, New York

Rotty RM (1973) Global production of CO_2 from fossil fuels and possible changes in the world's climate. Am Soc Mech Eng Pap 73- Power-11

Rush DW, Epstein W (1976) Genotypic responses to salinity. Differences between salt-sensitive and salt-tolerant genotypes of the tomato. Plant Physiol 57:162–166

Ryle GJA, Hesketh JD (1969) Carbon dioxide uptake in nitrogen deficient plants. Crop Sci 9:451–454

Sachs J von (1887) Lectures on the physiology of plants. (English Transl Marshall Ward H). Clarendon Press, Oxford

Sankary MN, Barbour MG (1972) Autecology of *Atriplex polycarpa* from California. Ecology 53:1155–1162

Santarius KA (1967) Das Verhalten von CO_2 Assimilation, NADP- und ATP-Synthese intakter Blattzellen in Abhängigkeit vom Wassergehalt. Planta 73:228–242

Santarius KA (1973) The protective effect of sugars on chloroplast membranes during temperature and water stress and its relationship to frost desiccation and heat resistance. Planta 113:105–114

Sayre RJ, Kennedy RA (1977) Ecotypic differences in the C_3 and C_4 photosynthetic activity in *Mollugo verticillata*, a $C_3 - C_4$ intermediate. Planta 134:257–262

Schimper AFW (1903) Plant geography upon a physiological basis, (English Transl. Fischer WR), revised and edited Groom P, Balfour IB. Clarendon Press, Oxford

Schnarrenberger C, Fock H (1976) The glycollate pathway. In: Heber U, Stocking CR (eds) Transport in plants III. Encyclopedia of plant physiology. New Series, vol III. Springer, Berlin Heidelberg New York, pp 185–234

Schneider SH (1975) On the carbon dioxide-climate confusion. J Atmos Sci 32:2060

Scholander PF, Hammel HT, Bradstreet ED, Hemmingsen EA (1965) Sap pressure in vascular plants. Science 148:339–346

Schönherr J (1976) Water permeability of cuticular membranes. In: Lange OL, Kappen L, Schulze E-D (eds) Water and plant life, problems and modern approaches. Springer, Berlin Heidelberg New York, pp 148–159

Schreiber U, Berry JA (1977) Heat-induced changes of chlorophyll fluorescence in intact leaves correlated with damage of the photosynthetic apparatus. Planta 136:233–238

Schulze E-D, Lange OL, Buschbom U, Kappen L, Evenari M (1972) Stomatal responses to changes in humidity in plants growing in the desert. Planta 108:259–270

Schulze E-D, Lange OL, Evenari M, Kappen L, Buschbom U (1974) The role of air humidity and leaf temperature in controlling stomatal resistance of *Prunus armeniaca* L. under desert conditions. I. A simulation of the daily course of stomatal resistance. Oecologia 17:159–170

Schulze E-D, Lange OL, Kappen L, Evenari M, Buschbom U (1975a) The role of air humidity and leaf temperature in regulating stomatal resistance of *Prunus armeniaca* L. under desert conditions. II. The significance of leaf water status and internal carbon dioxide concentration. Oecologia 18:219–233

Schulze E-D, Lange OL, Evenari M, Kappen L, Buschbom U (1975b) The role of air humidity and temperature in controlling stomatal resistance of *Prunus armeniaca* L. under desert conditions. III. The effect on water use efficiency. Oecologia 19:303–314

Schulze E-D, Lange OL, Kappen L, Buschbom U, Evenari M (1977) Stomatal responses to changes in temperature at increasing water stress. Planta 110:29–42

Scott D (1974) Description of relationships between plants and environment. In: Billings WD, Strain BR (eds) Vegetation and environment. Handbook of vegetation science, Part 6. Junk, The Hague, pp 49–69

Scott-Russell R (1977) Plant root systems: their function and interaction with the soil. McGraw Hill, London

Shainberg I (1975) Salinity of soils – effects of salinity on the physics and chemistry of soils. In: Poljakoff-Mayber A, Gale J (eds) Plants in saline environments. Springer, Berlin Heidelberg New York, pp 39–55

Shantz HL, Piemeisel LM (1927) The water requirements of plants at Akron, Colorado. J Agric Res 34:1093–1190

Sharma ML (1973a) Stimulation of drought and its effect on germination of five pasture species. Agron J 65:982–987

Sharma ML (1973b) Soil physical and physicochemical variability induced by *Atriplex nummularia*. J Range Manage 26:426–430

Sharma ML (1976) Soil water regimes and water extraction patterns under two semi-arid (*Atriplex* spp.) communities. Aust J Ecol 1:249–258

Sharma ML (1978) Water use by chenopod shrublands. In: Howes KMW (ed) Studies of the Australian arid zone III. Water in rangelands. CSIRO, Melbourne, pp 139–149

Sharma ML, Tongway DJ (1973) Plant induced salinity patterns in two saltbush (*Atriplex* spp.) communities. J Range Manage 26:121–125

Sharma ML, Tunny J, Tongway DJ (1972) Seasonal changes in sodium and chloride concentration of saltbush (*Atriplex* spp) leaves as related to soil and plant water potential. Aust J Agric Res 23:1007–1019

Shneyour A, Raison JK, Smillie RM (1973) The effect of temperature on the rate of photosynthetic electron transfer in chloroplasts of chilling-sensitive and chilling resistant plants. Biochim Biophys Acta 292:152–161

Shomer-Ilan A, Waisel Y (1973) The effect of sodium chloride on the balance between C_3 and C_4 carbon fixation pathways. Physiol Plant 29:190–193

Shreve F (1942) The desert vegetation of North America. Bot Rev 8:195–246

Simon EW, Minchin A, McMenamin MM, Smith JM (1976) The low temperature limit for seed germination. New Phytol 77:301–311

Sinclair R, Thomas DA (1970) Optical properties of leaves of some species in arid South Australia. Aust J Bot 18:261–273

Singh JS (1969) *Cassia tora*. Final technical report, PL 480/Ecology research project. Botany Department, Banaras Hindu University, pp 6–69

Singh KD, Gopal B (1973) The effects of photoperiod and light intensity on the growth of some weeds of crop fields. In: Slatyer RO (ed) Plant response to climatic factors. UNESCO, Paris, pp 73–75

Sinha RP (1968) Effect of artificial shading on the growth and morphology of *Stellaria* species. In: Misra A, Gopal B (eds) Proc Symp Recent Adv Trop Ecol. International Society for Tropical Ecology, Varanasi, pp 252–266

Skaykewich CF, Williams J (1971) Resistance to water absorption in germinating rapeseed (*Brassica napus* L.). J Exp Bot 22:19–24

Skene DS (1974) Chloroplast structure in mature apple leaves grown under different levels of illumination and their response to illumination. Proc R Soc London Ser B 186:75–78

Slack CR, Roughan RG, Bassett HCM (1974) Selective inhibition of mesophyll chloroplast development in some C_4 pathway species by low night temperature. Planta 118:67–73

Slatyer RO (1961) Effects of several osmotic substances on the water relation of tomato. Aust J Biol Sci 14:519–540

Slatyer RO (1967) Plant-water relationships. Academic Press, London New York

Slatyer RO (1970) Comparative photosynthesis, growth, and transpiration of two species of Atriplex. Planta 93:175–189

Slatyer RO (1973) Effects of short periods of water stress on leaf photosynthesis. In: Slatyer RO (ed) Plant response to climatic factors. UNESCO, Paris, pp 271–276

Smaoui MA (1971) Differentiation des trichomes chez Atriplex halimus L. CR Acad Sci Paris Ser D 273:1268–1271

Smith BN, Robbins MJ (1974) Evolution of C_4 photosynthesis: An assessment based on $^{13}C/^{12}C$ ratios and Kranz anatomy. In: Avron M (ed) 3rd Int Congr Photosynth. Elsevier, Amsterdam, pp 1579–1587

Smith FA (1973) The internal control of nitrate uptake into excised barley roots with differing salt content. New Phytol 72:769–782

Smith WK (1978) Temperatures of desert plants: another perspective on the adaptability of leaf size. Science 201:614–616

Spanswick RM (1976) Symplastic transport in tissues. In: Lüttge U, Pitman MG (eds) Transport in plants II, Encyclopedia of plant physiology (New Series), vol II B. Springer, Berlin Heidelberg New York, pp 35–53

Specht RL (1972) The vegetation of South Australia. Second edn. Government Printer, Adelaide

Stebbins GL (1971) Adaptive radiation of reproductive characteristics in angiosperms, II: seeds and seedlings. Ann Rev Ecol Syst 2:237–260

Stebbins L (1950) Variation and evolution in plants. Columbia University Press, New York

Stein DB, Thompson WF (1975) DNA hybridization and evolutionary relationships in three Osmunda species. Science 189:888–890

Stewart GR, Lee JA, Orebamjo TO (1972) Nitrogen metabolism of halophytes I. Nitrate reductase activity in Suaeda maritima. New Phytol 71:263–267

Stewart GR, Lee JA, Orebamjo TO (1973) Nitrogen metabolism of halophytes II. Nitrate availability and utilization. New Phytologist 12:539–546

Stirling EC, Zietz AHC (1900) Description of the manus and pes of Diprotodon australis Owen. Mem R Soc South Aust 1:1–40

Storey R, Wyn Jones RG (1979) Responses of Atriplex spongiosa and Suaeda monoica to salinity. Plant Physiol 63:156–162

Stowe LG, Teeri JA (1978) The geographic distribution of C_4 species of the dicotyledonae in relation to climate. Am Nat 112:609–623

Stutz HC, Melby JM, Livingston GK (1975) Evolutionary studies of Atriplex: A relic gigas diploid population of Atriplex canescens. Am J Bot 62:236–245

Sullivan CY (1972) Mechanisms of heat and drought resistance in grain sorghum and methods of measurement. In: Rao NGP, House LR (eds) Sorghum in the Seventies. IBH Publishing CO, New Delhi, pp 247–264

Sutcliffe JF (1976) Regulation in the whole plant. In: Lüttge U, Pitman MG (eds) Transport in plants II, Encyclopedia of plant physiology (New Series), vol II B. Springer, Berlin Heidelberg New York, pp 394–417

Täckholm V (1956) Students flora of Egypt. Anglo-Egyptian Bookshop, Cairo

Tarimoto TT (1969) Differential response of sugarcane varieties to osmotic pressures of saline media. Crop Sci 9:683–688

Tarnavschi IT (1948) Die Chromosomenzahlen der Anthophyten-Flora von Rumänien mit einem Ausblick auf das Polyploidie-Problem. Bull Jard Mus Bot Univ Cluj 28:1–130

Taschereau PM (1972) Taxonomy and distribution of Atriplex sp. in Nova Scotia. Can J Bot 50:1571–1594

Taylor AO, Craig S (1971) Plants under climatic stress. II. Low temperature, high light effects on chloroplast ultra-structure. Plant Physiol 47:719–725

Taylor AO, Rowley JA (1971) Plants under climatic stress. I. Low temperature, high light effects on photosynthesis. Plant Physiol 47:713–718

Teeri JA, Stowe LG (1976) Climatic patterns and the distribution of C_4 grasses in North America. Oecologia 23:1–12

Terry N, Ulrich A (1973) Effects of potassium deficiency on the photosynthesis and respiration of leaves of sugar beet. Plant Physiol 51:43–47

Thomas GW (1974) Chemical reactions controlling soil solution electrolyte concentration. In: Carson EW (ed) The plant root and its environment. University Press of Virginia, Charlottesville, pp 485–506

Thompson WW (1975) The structure and function of salt glands. In: Poljakoff-Mayber A, Gale J (eds) Plants in saline environments. Springer, Berlin Heidelberg New York, pp 118–146

Thornley JHM (1977) Root: Shoot interactions. In: Jennings DH (ed) Integration of activity in the higher plant SEB Symposium 31. Cambridge University Press, Cambridge, pp 367–387

Ting IP, Osmond CB (1973a) Photosynthetic P-enolpyruvate carboxylases. Characteristics of alloenzymes from leaves of C_3 and C_4 plants. Plant Physiol 51:439–447

Ting IP, Osmond CB (1973b) Multiple forms of plant phosphoenolpyruvate carboxylase associated with different metabolic pathways. Plant Physiol 51:448–453

Tolkien JRR (1964) Tree and leaf. George Allen and Unwin, London

Trebst A, Avron M (1977) Photosynthesis I. Photosynthetic electron transport and photophosphorylation. Encyclopedia of plant physiology (New Series), vol V. Springer, Berlin Heidelberg New York

Troughton JH (1969) Plant water stress and carbon dioxide exchange of cotton leaves. Aust J Biol Sci 22:289–302

Troughton JH, Card KA (1974) Leaf anatomy of *Atriplex buchananii*. NZ J Bot 12:167–177

Troughton JH, Slatyer RO (1969) Plant water status, leaf temperature and calculated mesophyll resistance to carbon dioxide of cotton leaves. Aust J Biol Sci 22:815–837

Troughton JH, Hendy CH, Card KA (1971) Carbon isotope fractionation in *Atriplex* spp. Z Pflanzenphysiol 65:461–464

Troughton JH, Card KA, Hendy CH (1974a) Photosynthetic pathways and carbon isotope discrimination by plants. Carnegie Inst Wash Yearb 73:768–780

Troughton JH, Wells PV, Mooney HA (1974b) Photosynthetic mechanisms in ancient C_3 and CAM species. Carnegie Inst Wash Yearb 73:812–816

Trumble HC (1932) Preliminary investigations on the cultivation of indigenous saltbushes (*Atriplex* spp.) in an area of winter rainfall and summer drought. J Counc Sci Ind Res 5:152–161

Turesson G (1919) The cause of plagiotropy in maritime shore plants. Contributions from the plant ecology station, Hallands Väderö 1. Lunds Univ Årsskr Adv. 2 16:2

Turesson G (1922a) The species and the variety as ecological units. Hereditas 3:100–113

Turesson G (1922b) The genotypical response of the plant species to the habitat. Hereditas 3:211–350

Turesson G (1923) The scope and import of genecology. Hereditas 4:171–176

Turesson G (1925) The plant species in relation to habitat and climate. Hereditas 6:147–236

Turesson G (1930) The selective effect of climate upon the plant species. Hereditas 14:99–152

Turner NC (1974) Stomatal behaviour and water status of maize, sorghum, and tobacco under field conditions. II. At low soil water potential. Plant Physiol 53:360–365

Turrell FM (1936) The area of the internal exposed surfaces of dicotyledon leaves. Am J Bot 23:255–264

Ulbrich E (1934) Chenopodiaceae. In: Engler A, Prantl K (eds) Die natürlichen Pflanzenfamilien, B 16c. Wilhelm Engelmann, Leipzig, pp 379–584

Ungar IA (1971) *Atriplex patula* var *hastata* seed dimorphism. Rhodora 73:548–551

Ungar IA (1974) Inland halophytes of the United States. In: Reimold RJ, Queen WH (eds) Ecology of halophytes. Academic Press, London New York, pp 235–305

Uotila P (1975) Sokerijuurikas ja sen rikkakasuit. Luonnon Tutkija 79:81–84

Uotila P (1977) *Atriplex prostrata* susp. *polonica*, and *Chenopodium serotinum*. Ann Bot Fenn 14:197–198

Usher MB (1973) Biological management and conservation. Chapman and Hall, London

Vaadia Y (1976) Plant hormones and water stress. Philos Trans Roy Soc London Ser B 273:513–522

Valentine I, Nagorcka BN (1979) Contour patterning in *Atriplex vesicaria* communities. In: Graetz D, Howes KMW (eds) Studies of the Australian Arid Zone, Part 4. Chenopod Shrublands. CSIRO, Melbourne, pp 61–74

Van den Honert TM (1948) Water transport in plants as a catenary process. Discuss Faraday Soc 3:146–153

Van der Meijden R (1970) Biosystematic notes on *Atriplex patula* L., *A.hastata* L., and *A.littoralis* L. (Chenopodiaceae). Blumea 18:53–63

Van der Valk AG (1974) Environmental factors controlling the distribution of forbs on coastal dunes in Cape Hatteras National Seashore. Can J Bot 52:1057–1073

Van Keulen H (1975) Simulation of Water use and Herbage Growth in Arid Regions. Wageningen Centre for Agricultural Publishing Documentation, p 176

Van Keulen H, de Wit, CT, Lof H (1976) The use of simulation models for productivity studies in arid regions. In: Lange OL, Kappen L, Schulze E-D (eds) Water and plant life: problems and approaches. Springer, Berlin Heidelberg New York, pp 408–420

Waddington CH (1972) The theory of evolution today. In: Koestler A, Smythies JR (eds) Beyond reductionism. Hutchinson, London, pp 357–395

Waisel Y (1972) Biology of halophytes. Academic Press, London New York

Walker NA, Pitman MG (1976) Measurement of fluxes across membranes. In: Lüttge U, Pitman MG (eds) Transport in plants II. Encyclopedia of plant physiology, vol II A. Springer, Berlin Heidelberg New York, pp 93–126

Walter H (1971) Ecology of tropical and subtropical vegetation. Oliver and Boyd, Edinburgh

Walter H, Lieth H (1961) Klimadiagramm Weltatlas. Gustav Fischer Verlag, Jena

Walter H, Stadelmann E (1974) A new approach to the water relations of desert plants. In: Brown GW Jr (ed) Desert biology, vol II. Academic Press, London New York, pp 213–310

Ward JM (1967a) Studies in ecology on a shell barrier beach. I. Physiology and vegetation of shell barrier beaches. Vegetatio 14:240–297

Ward JM (1967b) Studies in ecology on a shell barrier beach. II. Physical factors of shell barrier environment. Vegetatio 14:298–342

Ward JM (1967c) Studies in ecology on a shell barrier beach. III. Chemical factors of the environment. Vegetatio 15:77–112

Warren-Wilson J (1967) Effect of seasonal variation in radiation and temperature on net assimilation and growth rates in an arid climate. Ann Bot NS 31:41–57

Wassink EC (1969) Effect of light intensity on dry matter production and morphogenesis of *Iris* "Wedgwood" as compared with *gladiolus* and Tulip. Meded Landbhoogesch Wageningen 69:1–17

Weatherley PE, Slatyer RO (1957) Relationship between relative turgidity and diffusion pressure deficit in leaves. Nature (London) 179:1085–1086

Webb KL, Burley JWA (1965) Dark fixation of $^{14}CO_2$ by obligate and facultative salt marsh halophytes. Can J Bot 43:281–285

Weber WA (1950) A new species and subgenus of *Atriplex* from south-western Colorado. Madrono 10:187–191

Webley DM, Eastwood DJ, Gimingham CH (1952) Development of soil microflora in relation to plant succession on sand-dunes including the "rhizosphere" flora associated with colonizing species. J Ecol 40:168–178

Weimberg R (1970) Enzyme levels in pea seedlings grown on highly saline media. Plant Physiol 46:466–470

Weiss PA (1972) The living system: determinism stratified. In: Koestler A, Symthies JR (eds) Beyond reductionism. Hutchinson, London, pp 3–42

Welkie GW, Caldwell M (1970) Leaf anatomy of species in some dicotyledon families as related to the C_3 and C_4 pathways of carbon fixation. Can J Bot 48:2135–2146

Whitehead FH (1973) The relationship between light intensity and reproductive capacity. In: Slatyer RO (ed) Plant response to climatic factors. UNESCO, Paris, pp 73–75

Whitehead FH, Myerscough PJ (1962) Growth analysis of plants. The ratio of mean relative growth rate to mean relative rate of leaf area increase. New Phytol 61:314

Whiteman PC, Koller D (1967) Species characteristics in whole plant resistances to water vapor and CO_2 diffusion. J Appl Ecol 4:363–377

Whiteman PL, Koller D (1967) Interactions of carbon dioxide concentration, light intensity and temperature on plant resistances to water vapor and carbon dioxide diffusion. New Phytol 66:463–473

Whittaker RH (ed) (1973) Handbook of Vegetation Science. V. Ordination and classification of communities. Junk, The Hague

Whittaker RH (1975) Communities and Ecosystems. MacMillan, New York

Wiggins IL, Porter DM (1971) Flora of the Galapagos Islands. Stanford University Press, Stanford

Wight JR, Nichols JT (1966) Effects of harvester ants on production of a saltbush community. J Range Manage 19:68–71

Wild A, Skarlou V, Clement CR, Snaydon RW (1974) Comparison of potassium uptake by four plant species grown in sand and flowing solution culture. J Appl Ecol 11:801–812

Williams DG (1972) Ecological studies on shrub-steppe of the western Riverina, New South Wales. PhD Thesis. Australian National University, Canberra

Williams DG, Anderson DJ, Slater KR (1978) The influence of sheep on pattern and process in *Atriplex vesicaria* populations from the Riverine Plain of New South Wales. Aust J Bot 26:381–392

Williams OB, Oxley RE (1979) Historical aspects of the use of chenopod shrublands. In: Graetz D, Howes KMW (eds) Studies of the Australian Arid Zone, Part 4. Chenopod Shrublands. CSIRO, Melbourne, pp 5–16

Williams WT (1968) Numbers, taxonomy, and judgement. Bot Rev 33:379–386

Williams WT (ed) (1976) Pattern analysis in agricultural science. CSIRO and Elsevier, Melbourne Amsterdam

Willis AJ, Folkes BF, Hope-Simpson JF, Yemm EW (1959) Braunton Burrows: the dune system and its vegetation. I. J Ecol 47:1–24

Winge Ö (1917) The chromosomes. Their numbers and general importance. CR Trav Lab Carlsberg 13:131–275

Winter K (1978) Photosynthetic and water relationships of higher plants in a saline environment. In: Jefferies RL, Davy AJ (eds) Ecological processes in coastal environments. Blackwell, Oxford, pp 257–320

Winter K, Lüttge U (1976) Balance between C_3 and CAM pathway of photosynthesis. In: Lange OL, Kappen L, Schulze E-D (eds) Water and plant life: problems and modern approaches. Springer, Berlin Heidelberg New York, pp 323–334

Winter K, Lüttge U, Winter E, Troughton JH (1978) Seasonal shift from C_3 photosynthesis to crassulacean acid metabolism in *Mesembryanthemum crystallinum* in its native environment. Oecologia 34:225–237

Witter MB (1947) A comparative cytological study of three species of Chenopodiaceae. Bull Torrey Bot Club 74:443–452

Wong SC (1979) Plant responses to increasing atmospheric CO_2 concentration. Oecologia 44:68–74

Wong SC, Cowan IR, Farquhar GD (1978) Leaf conductance in relation to assimilation in *Eucalyptus pauciflora* Sieb. ex Spreng: the influence of irradiance and partial pressure of CO_2. Plant Physiol 62:670–674

Woo KC, Osmond CB (1976) Glycine decarboxylase in mitochondria isolated from spinach leaves. Aust J Plant Physiol 3:771–785

Woo KC, Berry JA, Turner GL (1978) Release and refixation of ammonia during photorespiration. Carnegie Inst Wash Yearb 77:241–245

Wood JG (1925) The selective absorption of chlorine ions; and the absorption of water by the leaves in the genus *Atriplex*. Aust J Exp Biol Med Sci 2:45–56

Wood JG (1936) Regeneration of the vegetation on the Koonamore Vegetation Reserve, 1926 to 1936. Trans R Soc South Aust 60:96–111

Wood JG, Sibly PM (1952) Carbonic anhydrase activity in plants in relation to zinc content. Aust J Sci Res B 5:244–255

Woodell SRJ, Mooney HA, Hill AJ (1969) The behaviour of *Larrea divaricata* (creosote bush) in response to rainfall in California. J Ecol 57:37–44

Woodhouse LR (1964) Regeneration of perennial saltbush (*Atriplex vesicaria* Heward) in western New South Wales, Australia. Trop Ecol 5:1–9

Woodwell GM, Whittaker RH, Reiners WA, Likens GE, Delwiche CC, Botkin DB (1978) The biota and the world carbon budget. Science 199:141–146

Woolley JT, Stoller EW (1978) Light penetration and light-induced seed germination in soil. Plant Physiol 61:597–600

World Food and Nutrition Study (1977) Supporting papers, vol II. National Academy of Sciences, Washington DC, pp 74–84

Wright CH (1912) Chenopodiaceae. In: Thiselton-Dyer WT (ed) Flora capensis, vol V. Lovell Reeve, London

Wuenscher JE (1974) The ecological niche and vegetation dynamics. In: Billings WD, Strain BR (eds) Vegetation and environment: Handbook of vegetation science, Part 6. Junk, The Hague, pp 39–45

Wulff HD (1936) Die Polysomatie der Chenopodiaceen. Planta 26:275–290

Wyn Jones RG, Storey R, Leigh RA, Ahmad N, Pollard A (1977) A hypothesis on cytoplasmic osmoregulation. In: Marrè E, Ciferri O (eds) Regulation of cell membrane activities in plants. North Holland, Amsterdam, pp 121–136

Yoshida S, Coronel V (1976) Nitrogen nutrition, leaf resistance, and leaf photosynthetic rate of the rice plant. Soil Sci Plant Nutr (Tokyo) 22:207–211

Zeiger E, Hepler PK (1977) Light and stomatal function: Blue light stimulates swelling of guard cell chloroplasts. Science 196:887–889

Zimmerman U (1978) Physics of turgor – and osmoregulation. Ann Rev Plant Physiol 29:121–148

Additional References Added in Proof

Albert R, Kinzel H (1973) Unterscheidung von Physiotypen bei Halophyten des Neusiedler-seegebietes (Österreich). Z Pflanzenphysiol 70:138–157

Albert R, Popp M (1980) Zur Rolle der löslichen Kohlenhydrate in Halophyten des Neusiedlersee-Gebietes (Österreich). Oecol Plant 13:27–42

Bewley JD, Black M (1978) Physiology and biochemistry of seeds in relation to germination. Vol. Springer, Berlin, Heidelberg, New York

Cavalieri AJ, Huang AHC (1977) Effect of NaCl on the in vitro activity of malate dehydrogenase in salt marsh halophytes of the U.S. Physiol Plant 41:79–84

Evans GC (1972) The quantitative analysis of plant growth. Blackwells, Oxford

Glass ADM, Perley JE (1980) Varietal differences in potassium uptake by barley. Plant Physiol 65:160–164

Grime JP (1979) Plant strategies and vegetation processes. Wiley, New York, London

Wong SC, Cowan IR, Farquhar GD (1979) Stomatal conductance correlates with photo-synthetic capacity. Nature (London) 282:424–426

Taxonomic Index

(For the most part, only species used in experimental studies have been cited, and then only at the first mention in relation to specific experiments)

Acacia spp.   112
A. harpophylla   272
Achillea   50
Aegialitis annulata   223
Agrostis tenuis   60
Allenrolfea spp.   64, 119
A. occidentalis   121
Allium cepa   368
Alloteropsis semialata   96
Alnus rugosa   275, 356
Alocasia macrorrhiza   302, 313
Amaranthus spinosus   382
A. tricolor   376
Anabasis spp.   119
Anagalis arvensis   382
Anthoxanthum odoratum   204
Artemisia spp.   112, 119
A. herb-alba   177
Arthrocnemum spp.   126
Astrebla lappacea   356
Atriplex acuminata   33, 36, 128
A. alaskensis   31
A. angulata   129
A. argentea   15, 317
A. australasica   38
A. californica   88, 90, 369
A. calotheca   75, 127
A. canescens   16, 72, 91, 120, 405
A. cinerea   124, 126, 199
A. conduplicata   129, 219
A. confertifolia   23, 92, 112, 118, 144, 148, 222, 244, 249, 257, 261, 264, 275, 283, 286, 334, 354, 403, 411, 415
A. dimorphostegia   167
A. dioeca   88
A. expansa   71, 88, 318, 352, 392, 395
A. fruticulosa   90, 93
A. glabriuscula   13, 17, 75, 125, 127, 131, 161, 167, 184, 226, 235, 317, 324, 342, 349, 352, 375, 376, 386, 392
A. gmelinii   71, 208, 210
A. halimus   114, 119, 126, 201, 230, 241, 271, 276, 278, 286, 288, 352, 361, 368, 398

A. hastata   13, 15
A. heterosperma   17, 71, 90, 130, 177, 317, 352
A. holocarpa   177
A. hortensis   17, 19, 36, 88, 93, 128, 159, 177, 260, 317, 352, 376, 392, 395
A. hymenelytra   74, 120, 222, 233, 258, 266, 276, 278, 281, 321, 329, 349, 352, 358, 364, 366, 387, 392
A. inflata   131, 143, 161, 200, 376
A. laciniata   13
A. latifolia   128, 262
A. lentiformis   56, 114, 119, 266, 278, 322, 325, 329, 342, 387
A. leptocarpa   199, 376
A. leucophylla   125, 126, 196, 257, 369
A. lindleyi   159, 180, 376, 392
A. littoralis   13, 33, 42, 88, 262
A. longipes   75, 127
A. monolifera   88
A. novae-zealandiae   123, 161
A. nummularia   19, 47, 112, 144, 148, 159, 196, 210, 225, 230, 237, 256, 286, 376, 385
A. nuttallii   63, 119, 180, 222
A. paludosa   37, 47, 71, 126, 376
A. patula   13, 15, 19, 75, 88, 128
A. pentandra   36, 89, 125
A. phyllostegia   21, 89, 93, 119
A. pleiantha   37, 88
A. polycarpa   15, 63, 120, 166, 210, 221, 256, 272, 322
A. prostrata   24
A. recurva   88
A. rosea   17, 33, 71, 90, 93, 99, 129, 167, 226, 235, 298, 317, 352, 387, 392, 410
A. sabulosa   7, 33, 86, 90, 93, 125, 317, 324, 338, 342, 344, 349, 352, 375, 386, 392
A. semibaccata   71, 89, 127, 161, 164, 224, 285, 375, 376
A. serenana   71, 130, 317, 352
A. spicata   88

A. spongiosa   5, 16, 19, 161, 166, 172, 200, 212, 215, 234, 238, 240, 298, 352, 376, 392, 405
A. stipitata   5, 148, 187, 282
A. subcordata   36
A. subspicata   31, 71, 128
A. tartarica   88, 161, 168, 178
A. triangularis   13, 17, 36, 62, 75, 88, 90, 93, 99, 123, 127, 131, 161, 198, 201, 208, 212, 216, 222, 225, 232, 235, 275, 280, 302, 307, 313, 317, 322, 352, 371, 374, 386, 392, 405, 410
A. vesicaria   3, 13, 17, 37, 47, 63, 89, 112, 118, 136, 139, 144, 148, 159, 166, 173, 182, 187, 199, 210, 219, 221, 225, 234, 244, 256, 268, 282, 334, 339, 376, 385, 398, 415

Bassia spp.   135
Beta vulgaris   358, 371, 376
Blackiella   13
Borya nitida   272

Camissonia spp.   122, 419
Centrosema virginianum   54
Ceratoides lanata   64, 92, 114, 149, 244, 354, 403, 411
Ceratonia siliqua   271
Chamaenerion angustifolium   382
Chenopodium spp.   14
C. album   20
Chloris barbata   376
Citrus spp.   205

Dactylis glomerata   54, 58
Danthonia caespitosa   159, 385
Datura stamonium   382
Distichilis spicata   121

Echinochloa crus-galli   278, 352
E. utilis   376
Encelia california   44, 317
E. farinosa   44, 121, 261, 317, 320
Endolopsis   13
Enneapogon spp.   248
Eremophila gilesii   149, 243
Euatriplex   88
Eucalyptus spp.   112
E. pauciflora   274
E. socialis   356, 363

Filipendula ulmaria   382

Gossypium spp.   355, 368, 400, 424
Grayia brandegei   92
G. spinosa   92

Halimione   14, 119
Halocnemum spp.   119
Hammada scorparia   277
Helianthus annuus   277, 352, 359, 362, 384
Hordeum spp.   204

Impatiens parviflora   382
Iris pseudocarpus   383

Jumea carnosa   402

Kochia spp.   119
K. childsii   376
K. indica   278

Larrea divaricata   121, 145, 276, 278, 324, 329, 336, 342, 349, 356, 363, 366, 423
Limonium spp.   210
Lolium multiflorum   407
L. perenne   402
Lycopersicon esculentum   62, 224, 335, 355
L. cheesmanii   62, 223

Maireana spp.   112
M. aphylla   248
M. pyramidata   136, 182, 378, 401
M. sedifolia   137, 415
Melandrium spp.   72
Mesembryanthemum crystallinum   9, 240
Mimulus cardinalis   48
M. lewisii   48
Mollugo verticillata   97, 108
Morrisiella   13

Nerium oleander   293, 330, 337, 422

Obione   13, 88
Oxyria spp.   51

Panicum hians   96
P. laxum   96
P. maximum   356, 363
P. milioides   96, 108
P. typhoideum   58
Passiflora caerulea   55
P. edulis   55
Pennisetum clandestinum   407
P. typhoides   174
Phaseolus vulgaris   239, 316, 335, 368, 374
Phragmites communis   63, 302
Pisum spp.   212, 359
Plantago lanceolata   317
P. maritima   226
Potentilla spp.   50
Prosopis juliflora   405
Prunus armeniaca   267, 277

Pterochiton   13
Pulvinaria spp.   180

Rhagodia spp.   14
Rhizophora mucronata   223
Rumex spp.   72
R. acetosa   54

Salicornia spp.   64, 119
Salix spp.   72
Salsola kali   22
Sarcobatus spp.   119
S. vermiculatus   92
Scirpus maritimus   82, 302
Sclerocalymma   88
Senniella   13
Sesame indicum   277
Simmondsia chinensis   275, 218, 352, 356, 363
Solanum dulcamera   52, 301
Solidago virgaurea   52, 301
Sorghum spp.   56, 59, 272, 352, 361
Spinacea oleracea   359

Stellaria media   382
Suaeda spp.   64, 119
S. maritima   216, 227, 229

Tamarix spp.   119
Teutliopsis   88
Theleophyton   13
Themeda australis   56, 335
Tidestromia oblongifolia   121, 278, 290, 317, 324, 329, 338, 342, 344, 349, 358, 375, 387, 392, 423
Triglochin maritima   208, 227
Triticum aestivum   352, 355, 424
Typha angustifolia   62
T. latifolia   62

Vauquelinia californica   356
Vigna luteola   277
Vitis vinifera   368

Xanthium spp.   212

Zea mays   335, 355, 360, 371, 374, 384, 424
Zygophyllum dumosum   277

Subject Index

absorptance   44, 281, 320
acclimation   291
– potential   292
– to light   302
– to temperature   328
– to water stress   362
active uptake   207, 213
adaptation   9, 38, 427
albedo   268
allocation of photosynthate   379
altitudinal transects   48
aluminum tolerance   61, 192
ammonium uptake   211
ancestral stock   36
animal vectors   178
apoplastic pathway   197, 229, 234, 264, 272
arid, halophytic communities   118
– shrublands   112
atmospheric CO_2 concentration, changes in   419
ATP ases   210
Australian coastal habitats   37, 124, 129

Barrier Ranges   113, 149, 194, 245, 248
biomass distribution   147
– estimation   416
Bodega Head   56, 74, 121, 257, 279, 386
boron tolerance   222
boundary layer   260
bracteoles   4, 158, 167
bundle sheath cells (see also PCR cells)   5, 22, 86, 100, 300, 360
burning of fossil fuel   420

$\delta^{13}C$ value   23, 97, 108, 300
C_3 photosynthesis (see also PCR cycle and Calvin-Benson cycle),   9, 20, 296
– – intrinsic water use efficiency of   352
– – O_2 effects on   308
– – quantum yield of   317
C_4 photosynthesis,   8, 21, 39, 84, 298
– – and CO_2 concentration   300, 351
– – and light intensity   320
– – and temperature   326
– – and water use   351
– – geographic distribution of   33, 86
– – inheritance of   97
– – intrinsic water use efficiency of   352
– – mechanism of   86, 108, 300
– – Na^+ requirement for   220, 376
– – taxonomic distribution of   85
C_3/C_4 hybrids   23, 96, 98
C_3/C_4 "intermediate forms"   96, 174, 240
calcicoles/calcifuges   191
Calvin-Benson cycle   86, 296
carbon dioxide, compensation point   297, 315
– –, concentrating mechanism   86, 101, 300, 320, 327
– –, fertilization   421
– –, global concentration of   419
– isotope discrimination   23, 96, 108, 300
cation excess   200, 213
cell membranes and electrolyte leakage   55, 341
– – – ion uptake   202
cell wall area in leaves   312
– – chelation of heavy metals   60
– – elasticity   271
chilling injury   55, 333
chloride transport   227
– uptake   207, 214, 226
chlorophyll bleaching   312
chloroplast fluorescence   6, 335, 342
– membranes, components of   294
– –, lipids in   336
– –, responses to temperature   341
– –, – to water stress   360
– ultrastructure   293, 310, 361, 373
chromosomal differentiation   83
chromosome numbers   67, 104
clay colloids   194
climate diagrams   116, 121
climatic change   420
– races   48
coastal communities   122
cold adaptation   323
communities   111

community productivity 411
compartmentation of ions 201
compatible solutes 240, 288
competition during ion uptake 214
– for water 145, 266
– in saline habitats 402
– with parent shrub 178
conductance to CO_2 diffusion 273, 311. 347
– to water flow in mesophyll cells 289
– – – – in roots 289
– – – – in seeds 158, 162
– – – – in soil 158, 265
– – – – in stomata 273, 289
crassulacean acid metabolism 9, 296
cultivation 128
Curlew Valley 113, 245, 354, 403
cuticular conductance 263
cytogenetics 81, 98, 103
cytoplasmic ionic concentration 236
– K^+/Na^+ ratio 231
– water relations 240, 288
cytotaxonomy 31, 48, 56

daily course of CO_2 fixation 304, 350
– – – energy balance 269
– – – stomatal conductance 267
– – – transpiration 267
– – – water potential 256, 267, 289, 365
dark CO_2 fixation 239
– respiration 302, 341, 379
Death Valley 56, 74, 121, 258, 266, 278,
  321, 328, 363, 366, 386
debris, and seedling establishment 178
demography 175
density dependent thinning 189
depth of seed burial 170
Desert Experimental Range 113, 415
desiccation tolerance 58, 272
diffusion of CO_2 311, 347
– – ions 194, 206
– – water vapor 261
dimorphic seed, germination of 162
– –, longevity of 164
– –, production of 18, 177
dioecy-monoecy 74
DNA sequences 92
driftline 123, 152
drought resistance 58, 272
drying profile in soil 171, 258
dune communities 123

ecological races 48
ecotypes 42
ecotypic differentiation and drought 58
– – – heavy metals 59, 222
– – – light 52, 301
– – – morphological features 43

– – – salinity 62, 216
– – – temperature 54, 345
edaphic factors 59, 119, 141, 158, 170, 192,
  258
– forms 17, 47, 62
effective precipitation 117, 186
endodermis 197, 229, 264
energy balance of leaves 45, 269, 281
– – of plants 269
enzyme distribution, intercellular 22, 102,
  108, 300
– –, intracellular 297, 300
– responses to electrolytes 237
– – – temperature 337, 345
epidermal trichomes 18, 44, 199, 222, 233,
  261, 282, 321
establishment of seedlings and metabolic
  changes 173
– – – – nutrition 171
– – – – population dynamics 184
– – – – seed size 170
– – – – water relations 170
evaporation sites 263
evolutionary divergence 93
– relationships 37, 66, 84

fertility of hybrids 80
flavanols 13
food plants 128, 248
foredunes 126, 152
fraction-1 protein 309, 372, 375
frost responses 49, 55
fructose bisphosphatase 338
fruit morphology 14, 158

genecological differentiation 41, 48, 59
genetic coherence 47, 51, 55
– relationships 12, 66
geographical distribution, analysis of 25,
  31, 33
– – in relation to temperature and aridity
  35
germination and aeration 167
– – bracteoles 159
– – herbivores 179
– – light 166
– – salinity 160
– – temperature 164
– – water potential 160
– inhibitors 168
gigas forms 72
gilgais 120, 142, 151
glucose-6-P dehydrogenase 238
grana 292
grazing 5, 32, 73, 179, 248
Great Basin Desert 63, 72, 112, 115, 149,
  245, 283, 354, 411

growth analysis   380
– responses to atmospheric CO₂
   concentration   421
– – – light   380
– – – nutrition   397
– – – salinity   240, 400
– – – temperature   386, 399, 407
– – – water relations   402, 408
guard cells   261, 274

halophytes   118, 126, 222
harvester ants   180
heat adapted plants   323
– stability of enzymes   346
– – – membranes   335, 342
– tolerance   56, 340
heavy metal tolerance   59, 62, 222
herbivores   142, 179, 248
hierarchies   10
hormonal control of ion transport   206,
   228
– – – seed germination   168
humidity and equivalent water potential
   254
– – stomatal conductance   277
hybridization barriers, prezygotic   77, 92
– – zygotic   79, 92
– experiments   44, 48, 76, 79, 89, 97
– in nature   73, 76, 91

imbibition   156, 164
immersion, in seawater   125
intercellular CO₂ concentration   273, 348,
   374, 422
"intermediate" C₃/C₄ photosynthesis   96,
   174, 240
intraspecific differentiation   23, 42, 59, 96
intrinsic water use efficiency   348, 365, 374
introgression of genes   82
ion absorption in cell metabolism   210
– – – natural habitats   199, 216
– – – whole plants   225
– balance   200, 213
– excretion   201, 209, 234
– fluxes   204, 231
– leakage   341
– transport   206, 226
– uptake by leaves   205, 212, 232
– – by roots   205, 208, 228
– –, feedback control of   204
– – isotherms   204
isocitrate dehydrogenase   238

kangaroo rats   249
Koonamore Vegetation Reserve   5, 113,
   178, 184, 416
"Kranz" anatomy   7, 17, 22, 33, 86, 91, 96,
   100, 107, 234, 299

latent heat exchange   268
leaching of salt from bracteoles   160
– – – – litter   248
leaf anatomy   21, 100, 107, 234, 298
– area ratio   380
– morphology   20, 54, 77, 100, 106, 260,
   306
– protein content   216, 307, 309, 345, 372
– temperature and inclination   280
– – – light   46, 323
– – – pubescence   45, 281, 321
– – – stomatal conductance   46, 281
– – – transpiration   281
– – – wind speed   261, 323
– thickness and light   306
– – – salinity   235
– weight ratio   380
levels of organization   3
light acclimation   301, 309, 314
– and growth   380
– – ion uptake   209
– – photosynthesis   301
– – root/shoot ratio   382
– – seed germination   166
– -harvesting chlorophyll   294
– penetration of soil   166
– reactions of photosynthesis   292
lipids in cuticles   263
– –membranes   335
litter   149, 244
longevity   186

maintenance respiration   305, 379
malate dehydrogenases   237, 298, 337
mangroves   223
matric potential of cytoplasm   270
– – – seeds   157
– – – soils   159, 165, 259
membrane potentials   208
– stability   341, 361
– structure   293
– synthesis   165
mesophyll cells (see also PCA cells)   22, 86,
   99, 106, 300, 360
– cell chloroplasts   300
metabolic development   156, 174
microbiological activity   152, 246
microclimate   39, 44, 269
micronutrients   220, 376
microsporogenesis   103
microtopography   119, 141
migration   36, 178
mineral cycling   242
mineralization   246
mortality during establishment   186
– – maturity   186, 414
mychorrhiza   202

net assimilation rate   380, 406
niche specification   8
Niggle's leaves   1, 11, 154, 428
nitrate reductase   211, 216, 240, 284
– uptake   211
nitrogen cycles   244
– fixation   246
– transport   212
– utilization efficiency   217, 375
nutrient distribution in soil   151, 172, 244
– – – vegetation   244
– withdrawl   243

optimal stomatal behavior   273, 350, 374
ordination of soil properties   137
– – taxonomic characters   13
– – vegetation   132
origins of Atriplex   36
osmotic adjustment   224, 240, 285, 364, 369, 398
– pressure   170, 214, 224, 240, 270, 285
oxalate synthesis   200, 213
– toxicity   249
oxygen effects on CO_2 fixation   298, 310
– – – photosynthesis   297
– – – quantum yield   317
– – – seed germination   168

parent shrub competition   178
pattern analysis of nutrients   150
– – – soil properties   141
– – – vegetation   138
– intensity   146
PCA and PCR cells   86, 100, 102, 300
PCO cycle   296, 315
PCR cycle enzymes   337, 345
PEP carboxylase   86, 97, 100, 107, 237, 300, 339, 345
performance   9, 39, 49, 386
phase transition temperature   55, 166, 335
phenology   118, 125
phosphate cycles   244
– in soils   151, 193, 195
– transport   220
– uptake   173, 218
photochemical reactions   292
photoinhibition   304, 312
photophosphorylation   295
photorespiration   297, 315
photosynthate partitioning and light   382
– – – nutrition   398
– – – salinity   400
– – – temperature   394
photosynthesis and CO_2 concentration   324, 326, 349, 357, 423

– – light   46, 52, 301, 312, 316, 380
– – micronutrients   376
– – nonstomatal factors   312, 362, 368
– – nutrition   371
– – O_2 concentration   298
– – salinity   368
– – stomatal conductance   46, 273, 308, 348
– – temperature   56, 321
– – water stress   347
photosynthetic acclimation   57, 301, 328
– apparatus   294, 307
– electron transport   294, 307, 360
– pathways   9, 33, 86, 96, 174, 240, 296
– reaction centers   294, 313
– units   294, 307
photosystems   294, 307, 343, 360
phylogeny   87, 93
phytochrome   167
plasmodesmata   197, 264, 300
poikilohydric plants   157, 255, 272
pollen analysis   38, 128
polymorphic characters   17, 20, 75, 82
polyploidy   56
population dynamics   5, 142, 175, 186
potassium deficiency   371
– uptake   207
potassium/sodium ratio   202, 209, 230
precipitation patterns   117, 183, 415, 417
principal components analysis   133
process descriptions   2
productivity and CO_2 concentration   419
– – light   380
– – temperature   386
– – water   306
protoplasmic tolerance to salt   236, 285
– – – low water potential   285
pubescence   18, 44, 281

quantum yield in C_3/C_4 plants   317
– – in shade plants   302
– – O_2 effects on   317
– –, temperature effects on   319, 342
– –, water stress effects on   358, 368

rabbits   142, 179
rainforest   302
regularity in arid vegetation   144
relative growth rate   307, 380, 408
reproductive behavior/strategy   175, 389, 410
– vigor   180
rhizosphere flora   152
Riverine Plain   113, 125, 148, 181, 194, 199, 384

root biomass   149, 413
– extension growth   183, 282
– hairs   264
root/shoot ratio and light   382
– – – nutrition   399
– – – salinity   401
– – – temperature   395
– – – waterstress   400
root/soil interface and ion absorption   202
– – – water uptake   264, 283
ruderals   128
RuP$_2$ carboxylase/oxygenase   97, 102, 108, 296, 307, 337, 345, 372

salinity   62, 222
– effects on development   174, 235
– – – enzyme activity   239
– – – germination   159
– – – growth   399
– – – photosynthesis   368
– – – soil properties   194
– – – water relations   287, 401
salt cycles   147, 247
– glands   19, 198, 201, 209, 234
– load (for animals)   248
– marshes   126, 196, 216
– spray   125
– tolerance   62, 223, 236
sand dunes, coastal   123, 152, 196, 257
– – interior   72
saponins   168
Scandinavian coastal habitats   42, 77, 121, 124, 322
seasonal patterns of ion content   287
– – – photosynthesis   122, 354, 366, 403, 413
– – – salinity   125, 287
– – – water potential   257, 283, 366
seawater, composition of   193
seed coat   158, 163
– dimorphism   18, 161
– dispersal   177
– floatation   178
– herbivory   180
– longevity   164
– morphology   17, 158
– production   389
– weight   176
seedling establishment   169
selectivity for K$^+$/Na$^+$ 209, 230
serpentine soils   61
sex expression   74
– ratio   73, 179
shade clones and species   52, 301, 312
shading effects   381
shell beaches   125

sodium absorption ratio   142, 195
– as an essential element   220, 376
– uptake   207, 221
soil, nitrogen and phosphorus   116, 151, 193, 195, 244
– salinity   116, 118, 120, 125, 151, 193
– solution   192
– water, content and potential   259, 283
– – penetration   195
– – profile   258, 264
– – and vegetation patterns   137, 144
Sonoran Desert   39, 44
space-time scales   3, 8, 39, 111, 132, 138, 153, 156, 249, 252, 292
species diversification   32
– list   24
specific leaf area   380
– – weight   305
stability of ecosystems   248, 414, 425
Stanford garden   48, 408
state descriptions   2
stochastic environmental processes   8, 117, 248, 417
stomata   262
stomatal conductance and CO$_2$ 272, 311, 347
– – – light   275
– – – temperature   275
– – – vapor pressure difference   277
– – – water potential   267, 276, 357
– – – – use efficiency   349
– frequency   312
– regulation   347, 350
– vs. nonstomatal control of photosynthesis   355, 368
stratification of seed   164
stroma   292
succulence   235
sun clones and species   52, 301, 312
surface albedo   268
symplastic pathway   197, 228, 234, 264, 300

temperature acclimation   328, 343
temperature and cellular integrity   340
– – enzymes   337
– – germination   165
– – ion absorption   219
– – mebranes   335
– – productivity   386
– – water vapor pressure   255
thermal stability   57, 327, 332, 340
threshold stomatal conductance   276
thylakoid membranes   292, 361
tolerance limits   8, 39, 49, 55, 335
transfer cells   198

transpiration and assimilation   273, 289,
   347
– – boundary layer   260
– – leaf temperature   350
– – vapor pressure difference   278
–, daily course of   267, 350
–, stomatal control of   261
transplant experiments   49, 74, 387
tropical barriers   36
turgor pressure   170, 204, 236, 270, 289
– regulation   240, 270, 285

vapor pressure difference   255, 278
vegetation analysis   4, 132
– patterns   119, 139, 144, 147

water availability   146, 257, 283
– flow   260
– penetration in soil   196
– potential   157, 254
– – and germination   160

– – – osmotic adjustment   284
– – – photosynthesis   365
– – – seedling establishment   170
– – – water use efficiency   405
– release curves   271
– stress and dimorphic seed   177
– – – photosynthesis   355
– – – productivity   403
– – – root activity   283
– – – stomatal conductance   276
– uptake by seed   157
water use efficiency   347, 364, 402, 405, 412
weeds   128
wind speed   260, 268
"Wolladventiv" flora   32, 36

x-ray microanalysis   201, 233
xylem pathway for ion transport   198
– – – water flow   263

zinc deficiency   376

Ecological Studies

Analysis and Synthesis

Editors:
W. D. Billings, F. Golley,
O. L. Lange, J. S. Olson

Volume 30

M. Kluge, I. P. Ting

Crassulacean Acid Metabolism

Analysis of an Ecological Adaptation

1978. 112 figures, 21 tables. XII, 210 pages
ISBN 3-540-08979-9

The acid metabolism of Crassulaceae and other succulents, known today as Crassulacean Acid Metabolism (CAM), has developed as an adaptation of plants to arid zones. CAM, which has fascinated plant physiologists for more than 150 years, is understood to be a modification of photosynthetic carbon assimilation, similar to C_4 synthesis. This book represents the first comprehensive treatment of this phenomenon and covers taxonomic, morphological-anatomic, biochemical, physiological, and ecological aspects. The work discusses also the significance of CAM plants in agriculture and technology.

Springer-Verlag
Berlin
Heidelberg
New York

Contents: Taxonomy and Geographical Distribution of CAM Plants. — Morphology, Anatomy, and Ultrastructure of CAM Plants. — The Metabolic Pathway of CAM. — Control and Modification of CAM. — Gas Exchange of CAM Plants. – Ecology, Productivity and Economic Use of CAM Plants.

Ecological Studies

Analysis and Synthesis

Editors:
W. D. Billings, F. Golley,
O. L. Lange, J. S. Olson

A Selection

Springer-Verlag
Berlin
Heidelberg
New York

Volume 19

Water and Plant Life

Problems and Modern Approaches

Editors: O. L. Lange, L. Kappen, E.-D. Schulze
1976. 178 figures, 66 tables. XX, 536 pages.
ISBN 3-540-07838-X

Contents: Fundamentals of Plant Water Relations. —
Water Uptake and Soil Water Relations. — Transpiration and
Its Regulation. — Direct and Indirect Water Stress. — Water
Relations and CO_2 Fixation Types. — Water Relations and
Productivity. — Water and Vegetation Patterns.

Volume 21
V. Y. Alexandrov

Cells, Molecules, and Temperature

Conformational Flexibility of Macromolecules and Ecological
Adaptation

Translated from the Russian by V. A. Bernstam

1977. 74 figures, 30 tables. XI, 330 pages
ISBN 3-540-08026-0

Contents: Modificational Changes of the Primary
Thermoresistance of Cells. — Genotypic Changes of the
Primary Thermoresistance of Cells. — Variations in Thermo-
stability of Protoplasmic Proteins as a Basis for Changes in the
Level of the Primary Cellular Thermoresistance. — Adaptive
Modifications of Conformational Flexibility of Macromole-
cules as a Basis for Changes of the Protein Thermostability. —
The Plausible Points of Application of the Natural Selection
During Alteration of a Correspondence Between the Level of
Conformational Flexibility of Protein Molecules and the
Temperature Ecology of a Species. — Plausible Mechanisms of
Regulation of the Level of Conformational Flexibility of
Proteins. — Thermostability of Nucleic Acids and the
Temperature Environments of Species' Life. — Saturation of
Fatty Acids and the Temperature Conditions of Life.

Volume 24
J. N. Kremer, S. W. Nixon

A Coastal Marine Ecosystem

Simulation and Analysis

1978. 80 figures, 13 tables. XI, 217 pages
ISBN 3-540-08365-0

Contents: Perspectives. — The Narragansett Bay Model. —
Theoretical Formulations: Physical Forcing Functions. Phytoplankton. Zooplankton. Additional Compartments. — Simulation and Analysis: Mathematical Considerations and the Computer Program. The Tidal Mixing Model.
The Standard Run. The Role of Biological Detail. Sensitivity
and Stability. Applications and Limitations.